Ethnomedicinal
Plants
Revitalization of Traditional
Knowledge of Herbs

Ethnomedicinal Plants

Revitalization of Traditional Knowledge of Herbs

Editors
Mahendra Rai
Deepak Acharya
José Luis Ríos

CRC Press
Taylor & Francis Group
Boca Raton London New York

CRC Press is an imprint of the
Taylor & Francis Group, an **informa** business

CRC Press
Taylor & Francis Group
6000 Broken Sound Parkway NW, Suite 300
Boca Raton, FL 33487-2742

First issued in paperback 2019

ISBN-13: 978-1-57808-696-2 (hbk)
ISBN-13: 978-0-367-38307-7 (pbk)

Library of Congress Cataloging-in-Publication Data

Ethnomedicinal plants : revitalization of traditional knowledge of herbs
/ editors: M. Rai, D. Acharya, Jose Luis Rios. -- 1st ed.
    p. cm.
  Includes index.
  ISBN 978-1-57808-696-2 (hardcover)
1.   Herbs--Therapeutic use. 2.   Medicinal plants. 3. Ethnobotany. 4.
Traditional medicine. I. Rai, M. K. II. Acharya, Deepak. III. Rios,
Jose Luis. IV. Title: Revitalization of traditional knowledge of herbs.
  RM666.H33E88 2010
  615'.321--dc22

                                                          2010035107

Visit the Taylor & Francis Web site at
http://www.taylorandfrancis.com

and the CRC Press Web site at
http://www.crcpress.com

Preface

Since the beginning of civilization, people have been using plants for medicines. A discussion of human life on this planet would not be complete without a look at the role of plants. Ethnobotany is the study of how people of a particular culture and region make use of indigenous plants. In fact, medicine and botany have always had a close relationship. Many of the drugs today, have been derived from plant sources. However, as modern medicine advances, chemically-synthesized drugs have replaced plants as the source of most medicinal agents. Research on plant sources is still receiving attention these days and they are used as the main basis of drug development.

There is a pressing need for revitalization of traditional knowledge of the plants used by rural and tribal people. Biotechnological approaches will prove to be boon, which can help validation and value addition of prominent herbal practices. Many fatal diseases like AIDS, cancer and swine flue have emerged and need proper treatment. Ethnomedicinal plants can be screened and modern approaches can be used for analysis of herbs. This book consists of 17 chapters, covering ethnomedicinal plants from Mexico, Brazil, West Indies, Lebanon, Nepal, India, Costa Rica, Tunisia, Cameroon, Norway and Spain. It includes ethnomedicinal uses of medicinal plants, their bioactivity and the role of bioinformatics and molecular biology in ethnomedicinal plants research.

The book could be an essential reading for botanists, medicos, Ayurvedic experts, traditional healers, pharmacologists and common people who are interested in curative properties of healing herbs.

Finally, we are thankful to Mr. Raju Primlani for his help and suggestions for the book. I wish to thank my research students—Alka Karwa, Aniket Gade, Ravindra Ade, Avinash Ingle, Dyaneshwar Rathod, Alka Yadav, Vaibhav Tiwari, Jayendra Kesharwani, and Swapnil Gaikwad for their help and support during the preparation of the book. MKR wishes to thank his daughters—Shivangi, Shivani and son Aditya for moral support during the editing of the book.

Contents

List of Contributors

M.E. Carretero Accame
Department of Pharmacology, School of Pharmacy,
Universidad Complutense de Madrid. Pza Ramón y Cajal s/n, 28040
Madrid, Spain.
Tel: +34 91 394 18 71
Fax: +33 91 394 17 26
E-mail: *meca@farm.ucm.es*

Deepak Acharya
Abhumka Herbal Pvt. Ltd.,
5th Floor, Shreeji Chambers, Behind Cargo Motors CG Road,
Ahmedabad 380 006, Gujarat, India.
Tel: 91 79 30077811 to 19
E-mail: *deep_acharya@rediffmail.com*

Mahmoud Amor
Laboratory of Pharmacology, Faculty of Pharmacy-Monastir 5000,
Tunisia.
Tel: +216 73 461 000
Fax: +216 73 461 830
E-mail: *phytopathos@yahoo.fr*

Raquel de Luna Antonio
Department of Psychobiology, Universidade Federal de São Paulo,
Rua Botucatu, 862-1° andar, Edificio de Ciências Biomédicas,
CEP 04023-062-São Paulo, SP, Brazil.
Tel: +55 11 2149-0155
E-mail: *luna.raquel@gmail.com*

Shandesh Bhattarai
Nepal Academy of Science and Technology, GPO Box 3323, Kathmandu,
Nepal.
Tel: I 977 1 9841408803 (cell)
E-mail: *bhattaraishandesh@ yahoo.com*

Dinesh Raj Bhuju
Nepal Academy of Science and Technology, GPO Box 3323, Kathmandu, Nepal.
Tel: +977 1 5547716
Fax: +977 1 5547713
E-mail: *dineshbhuju@gmail.com*

M.P. Gómez-Serranillos Cuadrado
Department of Pharmacology, School of Pharmacy,
Universidad Complutense de Madrid.
Pza Ramón y Cajal s/n, 28040 Madrid, Spain.
Tel: +34 91 394 18 71
Fax: +33 91 394 17 26

Thomas Efferth
Department of Pharmaceutical Biology, Institute of Pharmacy,
University of Mainz,
Staudinger Weg 5; 55128 Mainz, Germany.
Tel: 49-6131-3925751
Fax: 49-6131-3923752
E-mail: *efferth@uni-mainz.de*

Sabreen F. Fostok
Department of Biology, Faculty of Arts and Sciences and Nature Conservation Center for Sustainable Futures (IBSAR),
American University of Beirut, Beirut, Lebanon.
Tel: +961 70 969126
Fax: +961 1 374374, ext 3888
E-mail: *sff07@aub.edu.lb*

Jitendra Gaikwad
Department of Chemistry and Biomolecular Sciences,
Macquarie University, Sydney, NSW 2109, Australia.
Tel: +61 2 9850 8276
Fax: +61 2 9850 8313
E-mail: *jitendra.gaikwad@mq.edu.au*

Karla Georges
Veterinary Public Health, The University of West Indies,
Faculty of Medical Sciences, School of Veterinary Medicine, EWMSC
Mount Hope, Trinidad and Tobago, West Indies. 1(868)6452640.
Fax: 1 (868) 645 7428
E-mail: *karla.georges@sta.uwi.edu*

Rosa Martha Perez Gutierrez
Escuela Superior de Ingeníeria Química e Industrias extractivas IPN. Av,
Instituto Politecnico Nacional S/N, Unidad Profesional Adolfo López
Mateos, Mexico D.F.
Tel: 05557529349
Fax: 0557529349
E-mail: *rmpg@prodigy.net.mx*

Jörg D. Hoheisel
Functional Genome Analysis, German Cancer Research Center (DKFZ),
Im Neuenheimer Feld 580, 69120 Heidelberg, Germany.
Tel: [49] (6221)42-4680
Fax: [49] (6221)42-4687

M.T. Ortega Hernandez-Agero
Department of Pharmacology, School of Pharmacy,
Universidad Complutense de Madrid. Pza Ramón y Cajal s/n, 28040
Madrid, Spain.
Tel: +34 91 394 18 71
Fax: +33 91 394 17 26

Joanne Jamie
Department of Chemistry and Biomolecular Sciences,
Macquarie University, Sydney, NSW 2109, Australia.
Tel: +61 2 9850 8276
Fax: +61 2 9850 8313
E-mail: *joanne.jamie@mq.edu.au*

Mahmud Tareq Hassan Khan
GenØk—Center for Biosafety, FellesLab, MHB, University of Tromsø,
9037 Tromsø, Norway.
E-mail: *mahmud.khan@uit.no, mthkhan2002@yahoo.com*

Jim Kohen
Department of Biological Sciences,
Macquarie University, Sydney, NSW 2109, Australia.
Tel: +61 2 9850 8138
Fax: +61 2 9850 8245
E-mail: *jim.kohen@mq.edu.au*

Cheryl Lans
PO Box 72045 Sasamat, Vancouver, V6R4P2, Canada.
E-mail: *cher?lans@netscape.net*

Tamiris Andrade Mediros
Department of Psychobiology, Universidade Federal de São Paulo, Rua
Botucatu, 862-1° andar, Edificio de Ciências Biomédicas,
CEP 04023-062-São Paulo-SP, Brazil.
Tel: +55 11 2149-0155

Gerardo Mora
Centro de Investigaciones en Productos Naturales (CIPRONA);
Universidad de Costa Rica, 2060 San José, Costa Rica.
Tel:+50625113001
Fax: +50622259866
E-mail: *gamora@racsa.co.cr*

Chouchane Nabil
Laboratory of Pharmacology, Faculty of Pharmacy-Monastir 5000,
Tunisia.
Tel: +216 73 461 000
Fax: +216 73 461 830
E-mail: *Nabil.Chouchane@fphm.rnu.tn*

Julino A.R. Soares Neto
Department of Preventive Medicine, Universidade Federal de São Paulo,
Rua Borges Lagoa, 1341-1° andar, CEP 04038-034-São Paulo, SP, Brazil.
Tel: +55 11 2149-0155

Dieudonné Njamen
Department of Animal Biology and Physiology,
Faculty of Science, University of Yaounde 1,
PO Box 812 Yaounde, Cameroon.
Tel: +237 79 42 47 10
E-mail: *dnjamen@Gmail.com*

Rafael Ocampo
Bougainvillea Extractos Naturales,
S.A. Apartado Postal 764-3100 Santo Domingo de Heredia, Costa Rica.
E-mail: *quassia@racsa.co.cr*

O.M. Palomino Ruiz-Poveda
Department of Pharmacology, School of Pharmacy,
Universidad Complutense de Madrid. Pza Ramón y Cajal s/n, 28040
Madrid, Spain.
Tel: +34 91 394 18 71
Fax: +33 91 394 17 26

Mahendra Rai
Department of Biotechnology, SGB Amravati University,
Amravati 444 602, Maharashtra, India.
Tel: 91-721-2667380
Fax: 91-721-2660949
E-mail: *mkrai123@rediffmail.com*

Shoba Ranganathan
Department of Chemistry and Biomolecular Sciences and ARC Centre of
Excellence in Bioinformatics, Macquarie University, Sydney. NSW 2109,
Australia
and
Department of Biochemistry, Yong Loo Lin School of Medicine,
National University of Singapore, Singapore.
Tel: +61 2 9850 6262
Fax: +61 2 9850 8313
E-mail: *shoba.ranganathan@mq.edu.au*

Arun Rijal
Director, Plant People Protection,
P.O. Box 4326, Bauddha Mahankal-6, Kathmandu, Nepal.
Tel: +977 1 4470779
E-mail: *arunrijal@yahoo.com*

José Luis Ríos
Department de Farmacologia, Facultat de Farmacia,
Universitat de Valencia. Av. Vicent Andrés Estellés s/n 46100 Burjassot,
Valencia, Spain.
Tel: +34 963544973
Fax: +34 963544498
E-mail: *riosjl@uv.es*

Eliana Rodrigues
Department of Preventive Medicine, Universidade Federal de São Paulo,
Rua Borges Lagoa, 1341-1° andar, CEP 04038-034-São Paulo, SP, Brazil.
Tel: +55 11 2149-0155
E-mail: *68.eliana@gmail.com*

Nayara Scalco
Department of Psychobiology, Universidade Federal de São Paulo,
Rua Botucatu, 862-1° andar Edificio de Ciências Biomédicas, CEP
04023-062-São Paulo, SP, Brazil.
Tel: +55 11 2149-0155

Rabih S. Talhouk
Department of Biology, Faculty of Arts and Sciences and Nature
Conservation Center for Sustainable Futures (IBSAR),
American University of Beirut, Beirut, Lebanon.
Tel: +961 1 374374, ext 3895
Fax: +961 1 374374, ext 3888
E-mail: *rtalhouk@aub.edu.lb*

José Carlos Tavares Carvalho
Laboratório de Pesquisa em Fármacos, Universidade Federal do Amapá,
Rod. JK km 02 CEP 68902-280, Macapá-AP, Brazil.
Tel: +55(96)33121742
Fax: +55(96)33121740
E-mail: *farmacos@unifap.br*

Subramanyam Vemulpad
Department of Chiropractic, Macquarie University, Sydney, NSW 2109,
Australia.
Tel: +61 2 9850 9385
Fax: +61 2 9850 9389
E-mail: *subramanyam.vemulpad@mq.edu.au*

Borgi Wahida
Laboratory of Pharmacology, Faculty of Pharmacy-Monastir 5000, Tunisia.
Tel: +216 73 461 000
Fax: +216 73 461 830
E-mail: *wahida_b2002@yahoo.fr*

Antonios N. Wehbe
Department of Biology, Faculty of Arts and Sciences and Nature
Conservation Center for Sustainable Futures (IBSAR), American
University of Beirut, Beirut, Lebanon.
Tel: +961 3 144327
E-mail: *anw02@aub.edu.lb*

Karen Wilson
National Herbarium of NSW, Royal Botanic Gardens, Sydney NSW 2000,
Australia.
Tel: +61 2 9231 8137
Fax: +61 2 9251 7231
E-mail: *karen.wilson@rbgsyd.nsw.gov.au*

Mahmoud Youns
Department of Pharmaceutical Biology, Institute of Pharmacy,
University of Mainz, Staudinger Weg 5;55128 Mainz, Germany.
Tel: 49-6131-3924237
Fax: 49-6131-3923752

Chapter 1

Ethnomedicinal Plants: Progress and the Future of Drug Development

José Luis Ríos

Introduction

Ethnobotany studies the relationships between people and plants and includes various aspects of how plants are used as food, cosmetics, textiles, in gardening, and as medicine. In contrast, ethnopharmacology studies the pharmacological aspects of a given culture's medical treatments and their social appeal, concentrating especially on the bio-evaluation of the effectiveness of traditional medicines. Ethnopharmacology is related to ethnobotany in part because many pharmaceuticals come from plants; however, ethnopharmacological research also includes drugs and medicines from animal, fungal, microbial, and mineral sources. Ethnomedicine is, of course, closely related to both fields, but is based on ancient written sources along with knowledge and practices that have been handed down orally over the centuries. It is also akin to traditional medicine or medical anthropology in that it studies the perception of and context in which traditional medicines are used. For its part, ethnopharmacy is a broad interdisciplinary science focusing on the perception, use, and management of pharmaceuticals in a specific society. Finally, pharmacognosy is the study of medicines derived from natural sources or "the study of the physical, chemical, biochemical and biological properties of drugs, drug substances

Departament de Farmacologia, Facultat de Farmacia, Universitat de Valencia. Av. Vicent Andrés Estellés s/n. 46100 Burjassot, Valencia, Spain.
E-mail: *riosjl@uv.es*

or potential drugs or drug substances of natural origin as well as the search for new drugs from natural sources" (web 1) while phytotherapy is the study of the use of extracts of natural origin as medicines or health-promoting agents, known as phytomedicines, and their clinical use in phytotherapy or herbal medicine (Heinrich et al. 2004).

In former times, natural products were the origin of all medicinal drugs; however, in the last century, synthetic chemistry and biotechnology techniques have offered alternatives to natural sources (Harvey 2009). At first, the focus on these new methodologies for conducting drug research had a negative impact on the study of natural products from plant origin. Nevertheless, the past few decades have witnessed a renewed interest in the field. This, coupled with the application of new technologies to natural product research, has led to the discovery of relevant new active compounds such as paclitaxel and others.

This chapter will review the progress of ethnomedicinal plant research over the past century and will attempt to predict what the future holds for this ever developing subject area. Special attention will be given to new methodologies, techniques, and advances in the field and their potential application for the future development of medicinal plant research.

Natural products versus medicinal plants

Medicinal plants and their extracts comprise the natural sources of treatments used in ethnomedicine and phytotherapy. They are also the source of natural products which can be utilized as medicinal agents containing series for new compounds and drugs. Medicinal plant research should thus focus on two objectives: phytomedicine and natural products. The latter are of great importance as a major source of therapeutic drugs, including antibiotics and immunosuppressive drugs from microbiological sources, as well as steroids, prostaglandins, and peptide hormones from animal sources. Plants are also a good source for other types of drugs, either directly or through semi synthesis, including the steroidal hormones recently obtained through saponins from species of *Agave* and other sources.

Newman and Cragg (2007) reviewed natural products as a source of new drugs over the past 25 years and found that of 1,010 new chemical entities, 124 were biological (peptides or proteins obtained through biotechnological means), 41 were natural products, and 232 were derived from natural products, generally with the aid of semi-synthesis. Another 310 compounds were obtained through pure synthesis. Other interesting groups of chemicals reviewed were compounds synthesized from a pharmacophore from either a natural product (47) or a natural product mimetic (107). Within the strict confines of the field of natural products (without biological or semisynthetic derivatives), the most relevant pharmacological group comprised antibacterials (10) followed by anticancer (9), immunosuppressant (5), immunostimulant (3), hypocholesterolemic (3), and antiparasitic (2) agents. The remaining groups contained one compound each: analgesic, anti-Alzheimer's disease, antiallergic, antiarrhythmic, antidiabetic, antithrombotic, anti-ulcer, hypolipidemic, immunomodulatory, and two drugs against benign prostatic hypertrophy and respiratory distress syndrome (Cragg et al. 1997, Newman et al. 2003, Newman and Cragg 2007). In a complementary review, Butler (2008) examined natural compounds and drugs derived from them in late-stage clinical development and found that out of 37 compounds, 3 were products obtained directly from natural sources, 18 were natural product-derived, and 16 were semi-synthetic natural products.

Past

Ethnobotany and ethnopharmacology have given rise to various ethnomedicines over the years and, taken together, they have contributed to the discovery of many important plant-derived drugs. Before the advent of high-throughput screening and the post-genomic era, more than 80% of drug substances were either natural products or inspired by a natural compound. Although to a lesser extent, the field still continues to produce new drugs; indeed, information presented on sources of new drugs from 1981 to 2007 indicates that almost half of the drugs approved since 1994 have been based on natural products (Harvey 2008). Moreover, traditional medicines that make use of herbal drugs are used throughout the world in accordance with practices that have been developed following specific rules over the centuries. In many cases, modern Western science has corroborated the proper use of diverse traditional ethnomedicinal plants including ginkgo, ginseng, and centella, which have become a part of many modern therapies after thorough investigations establishing their quality, security, and safety.

Eastern ethnomedicinal plants introduced in western medicine

Traditional medicine with herbal drugs exists in every part of the world, but is especially popular in China, India, and Europe. The philosophies of these traditional medicines bear some resemblance to each other, but all differ widely from modern Western medicine (Vogel 1991). Many medicinal plants and crude drugs used in traditional medicine have gradually been included into Western medicine, but legislation in the developed countries of Europe and America requires that medicinal herbal drugs fulfill international requirements of quality, safety, and efficacy to be included in their *vade mecum*. Although the procedure for developing herbal drugs for worldwide use is necessarily different from that of synthetic drugs, the former have the tremendous advantage of being readily available, usually at a reasonable cost, to patients living in the geographical areas where such drugs have been traditionally used.

Ginkgo seeds for example, have been listed as a source of medicine since the earliest records of Chinese medicine, while the leaves which were recommended for medicinal use as early as 1509, are still used in the form of teas. Nowadays, extracts of ginkgo leaves in the form of film-coated tablets, oral liquids, or injectable solutions are widely used in official medicines of Europe and America (Chan et al. 2007). *Ginkgo biloba*, for example, is one of the most widely sold products in health food stores in both the United States and Europe, with sales in the United States exceeding $100 million in 1996 (Chan et al. 2007) and $249 million in 2006 (web 2). Indeed, a diverse array of enriched-extracts (water-acetone or water-ethanol) of ginkgo leaves is now commercially available, all standardized based on their flavonoid or terpene trilactone content. The standardized commercial products, including EGb 761 and LI 1370, contain neither biflavones nor alkylphenol or alkylbenzoic acid derivatives, which have allergic, immunotoxic, and other undesirable properties.

Ginseng root has been used for over 2000 years as a general cure-all that promotes longevity. The efficacy of ginseng was well-known in Asia by the nineteenth century, but more recently there has been a renewed interest in research on ginseng for its medicinal properties (Attele et al. 1999). Thus, the beneficial pharmacological effects of this root on the central nervous system as well as on the cardiovascular, endocrine, and immune systems have all been demonstrated. There are different species of *Panax* used with the common name of ginseng. Of these, *Panax ginseng* (Asian ginseng), *Panax quinquefolius* (American ginseng), and *Panax japonicus* (Japanese ginseng) are the most common and possess similar chemical and pharmacological properties. Another species, *Eleutherococcus senticosus* (Siberian ginseng), is also fairly common, but its chemistry is quite different (Davydov and Krikorian 2000). All have been introduced into Western medicine after the requisite studies demonstrating their safety and efficacy.

Gotu kola (*Centella asiatica*) is another plant that has long been used in Asian medicine to treat various diseases, including skin disorders. It was introduced into European medicine as an effective wound healing agent as well as for treating skin lesions, venous insufficiency, and varicose veins. In contrast to many other medicinal plants, gotu kola has been subjected to extensive experimental and clinical investigations. Studies done in accordance with standardized scientific criteria have shown it to have a positive effect in the treatment of venous insufficiency and striae gravidarum, as well as to be effective in the treatment of wound-healing disturbances (Brinkhaus et al. 2000). Some of the plant's principles (asiaticoside, asiatic acid, madecassic acid, and madecassoside) are generally used in dermatological preparations in Western countries. In the case of madecassoside, it was recently demonstrated that an oral administration of this compound at different doses facilitated wound closure in a time-dependent manner, reaching its peak effect on day 20 at the highest dose of 24 mg/kg. Several mechanisms could be involved in this effect including antioxidative activity, collagen synthesis, and angiogenesis (Liu et al. 2008).

Natural products from ethnomedicinal sources in western medicine

While Eastern medicine is based on ethnopharmacological knowledge, the progress of Western medicine is in large part supported by new synthetic drugs. However, many relevant therapeutic groups are based on natural products which were first isolated and then used as pharmacological agents. For example, both relevant groups of analgesics—opioids and non-steroidal anti-inflammatory drugs—are directly derived from natural products, namely morphine and salicylic acid, respectively. Even before the discovery of opioid receptors and endorphins, the analgesic effects of opium and morphine were well known. The isolation of morphine from poppies (*Papaver somniferum*) by Sertürner and the subsequent studies on potential derivatives with similar properties but without negative side effects led to the semi-synthesis of heroin, which was originally used as a potent antitussive. Later research led to diverse semi-synthetic or synthetic derivatives including pentazocine, pethidine, fentanyl, and methadone. In addition, the study and elucidation of new receptors, mediators, agonists, and antagonists of the opioid system gave rise not only to the development of a broad group of potent active compounds, but also to a vast increase in knowledge of the physiological system of pain and neurotransmission mechanisms. Thus, for example, researchers discovered that animal organisms produce endorphins (**endo**genous mo**rphin**e-like substances), which have a relevant role in both pain and neurotransmission. On the other hand, the isolation of salicin and salicylic acid from willows (*Salix* species) or *Spiraea ulmaria* (syn. *Filipendula ulmaria*) produced one of the kings of modern medicine and pharmacy, aspirin (A*cetyl*-**spir**a*ea*-**in**). This analgesic is not only the most relevant non-steroidal anti-inflammatory drug available to date, but is has also served as the first head of various series for many new groups of peripheral analgesic drugs. Indeed, numerous different series of derivatives have been obtained from research on aspirin, and many enzymes and mediators implicated in pain, fever, and inflammation have been synthesized as a result (Heinrich et al. 2004, Rishton 2008).

Other relevant pharmacological groups introduced into modern therapies after studies with natural products are calcium antagonists and drugs against malaria. In the case of the former, verapamil was developed from studies of papaverine, a non-analgesic alkaloid obtained from opium. This natural alkaloid has spasmolitic properties and acts as a Ca^{2+} channel blocker; its different modifications allowed researchers to obtain verapamil, the first specific calcium antagonist. In the case of anti-malarial drugs, isolation of quinine from *Cinchona* species in the eighteenth and nineteenth centuries led to the successful treatment of malaria around the world. Modifications to quinine's chemical structure were used, in turn, as a template for the synthesis of new drugs, such as chloroquine and mefloquine, which are the bases for the treatment and prevention of malaria today (Heinrich et al. 2004).

Statins, which constitute a relevant group of hypocholesterolemic drugs, were also obtained from natural sources, including species of *Penicillium*, *Monascus*, and *Aspergillus*. Another ethnopharmacological phenomenon, namely red yeast rice or *pin yin* (*Monascus purpureus*), has been used in traditional Chinese medicine since the Tang Dynasty (880 AD) for different digestive and circulatory problems. The drug is used in various forms in traditional Chinese medicine, for example as a powder (*zhi tai*) or an alcoholic extract (*xue zhi kang*) (web 3). A meta-analysis from different clinical trials demonstrated that red yeast rice reduces levels of low density lipoprotein (LDL)-cholesterol and triglycerides while increasing high density lipoprotein (HDL) levels (Liu et al. 2006). Other authors have reported on the beneficial effects of this crude drug against other chronic pathologies, such as cardiovascular disease. Of these studies, the most interesting is a prospective clinical trial with a purified extract of red yeast rice (*xue zhi kang*) on patients with previous myocardial infarction. The researchers concluded that long-term therapy with this extract significantly decreased both the recurrence of coronary events and the occurrence of new cardiovascular events and death while simultaneously improving lipoprotein regulation. Moreover, the extract was safe and well tolerated (Lu et al. 2008). The key principles of the extract are monacolins, principally monacolin K, which was isolated and commercialized with the name lovastatin and which has served as the head of a series of hypocholesterolemic drugs, the statins. Other similar derivatives have also been isolated from natural sources. These include mevastatin from *Penicillium citrinum* and *P. brevicompactum* and simvastatin from *Monascus ruber* and *Aspergillus terreus*. Others have been obtained through microbiological transformations of mevastatin, as is the case with pravastatin from *Streptomyces carbophilus* (Heinrich et al. 2004). These types of compounds are relevant examples for establishing future guidelines for research into new ethnomedicinal plant sources which can serve as potential heads of series for new active drugs.

Other series of natural products obtained from natural sources but identified through various means include phytoestrogens, resveratrol, silybin, and capsaicin, which are all currently used in Western medicine for treating specific pathologies and physiological states. Thus, the effects of phytoestrogens from different sources were examined and found to have different pathways. For example, the effect of *Trifolium subterraneum* on the sterility of ewes in Australia has been reported, as has the effect of soya on the incidence of sexual cancer of women in Asia with respect to Western women. The latter effect was examined and its relationship with food demonstrated, especially with the consumption of soya and other species rich in isoflavones (Adlercreutz et al. 1987, Adlercreutz 1990, Adlercreutz et al. 1992). Silymarin (a mixture of flavanolignans from *Silybum marianum*) and its principal compound silybin are widely used in Western medicine as liver protectors due to their antihepatotoxic properties (Stickel and Schuppan 2007), while resveratrol from *Vitis vinifera* leaves was recently introduced as both a pure compound and an enriched extract (Leifert and Abeywardena 2008, Fan et al. 2008). Capsaicin from *Capsicum frutescens* and other species of *Capsicum* is used as an analgesic (Altman and Barkin 2009), as is curcumin from *Curcuma longa*, which has analgesic and anti-inflammatory properties and is used as either a chemical or a phytotherapeutic agent (Anand et al. 2008). In addition to the aforementioned group of relevant drugs obtained from natural sources, there are many examples of natural products that have been instrumental in advances in the fields of physiology, biochemistry, and pharmacology. This is the case of the alkaloids muscarine, nicotine, and tubocuranine and their role in the development of cholinergic transmission and acetylcholine receptors, and of morphine from poppies and its part in furthering our knowledge of opioid receptors. Digitalis glycosides from foxglove were the most relevant tool in the study of Na^+, K^+-ATPase while aspirin, semi-synthesized from salicylic acid, proved to be an excellent tool for discovering and developing cyclooxygenase enzyme subtypes, which, in turn, led to the development of the next generation of selective cyclooxygenase-2 inhibitors (Harvey 2008, Rishton 2008).

Present

The evolution of organic chemistry over the past two centuries has produced an enormous number of semi-synthetic and synthetic compounds with increasingly better and more potent activity against a wide range of diseases. This increased focus on synthesis led to a decline in the relevance of natural products as medicinal agents in the West, with the historic use of phytomedicines in European folk medicine helping to retain only a limited amount of traditional medicinal plant use. In contrast, ethnopharmacology continued to be the principal source of therapeutic drugs in Asian countries such as China, India, Korea, Japan, Malaysia, Indonesia, and others, where traditions were maintained and traditional medicines continued to be used in a high percentage of treatments. It was mainly from these countries, along with a few others, that many plants and crude drugs were incorporated into Western therapeutic medicine in the latter part of the twentieth century. Many of these drugs now have a great degree of acceptance in the West, including Ginseng, Echinacea, Gugul, Garcinia, Barberry, Saw palmetto, African prune, Devil's claw, Cat's claw, Calaguala, and Tee tree oil. However, because the legal restrictions of many Western countries do not allow the direct introduction of most medicinal plants as a medicine or drug, many of them were introduced as herbal or alimentary supplements. Thus, while many European countries now classify these products as drugs, in the USA they are still sold as dietary supplements (Gilani and Atta-ur-Rahman 2005). Moreover, although the European Union established basic quality requirements for what were deemed official medicines from medicinal plants through its Pharmacopoeia, new ethnomedicines are not included in this official codex.

European Union requirements for medicinal plants

On March 31, 2004, the European Union established Directive 2004/24/EC, which requires: applications for authorization to place a medicinal product to be accompanied by a dossier containing particulars and documents relating in particular to the results of physico-chemical, biological or microbiological tests as well as pharmacological and toxicological tests and clinical trials carried out on the product and thus proving its quality, safety and efficacy. However, when: the applicant can demonstrate by detailed references to published scientific literature that the constituent or the constituents of the medicinal product has or have a well-established medicinal use with recognized efficacy and an acceptable level of safety within the meaning of Directive 2001/83/EC, he/she should not be required to provide the results of pre-clinical tests or the results of clinical trials due to the fact that: the long tradition of the medicinal product makes it possible to reduce the need for clinical trials, in so far as the efficacy of the medicinal product is plausible on the basis of long-standing use and experience. Pre-clinical tests do not seem necessary, where the medicinal product on the basis of the information on its traditional use proves not to be harmful in specified conditions of use.

Thus, "the simplified registration should be acceptable only where the herbal medicinal product may rely on a sufficiently long medicinal use in the Community". As a consequence of these regulations, the European

Community defined different groups of medicinal products (Directive 2004/24/EC): Traditional herbal medicines, Herbal medicines, Herbal substances, and Herbal preparations. These directives assume that the future of medicinal plants and their derivatives will be linked with the international requirements for their quality, safety, and efficacy, but the same document acknowledges that there are difficulties in establishing many of these studies and allows for the introduction of different traditional ethnomedicines in Western medicine. One problem is that while modern Pharmacopoeias establish basic assays for quality and safety, studies on efficacy are more difficult to carry out. It must be taken into account that the procedure for developing drugs derived from plants is necessarily different from that for synthetic drugs. In the case of natural medicines, their age-old use could be a first guarantee of their safety and efficacy. The development of research on ethnopharmacology and ethnomedicine must therefore form the basis for the future of new medicinal agents so that the perspectives of traditional medical knowledge and safety can support the present practice of pharmaceutical development.

Studies on new ethnomedicinal plants

One way of introducing new ethnomedicinal plants into modern therapies is through studies of known species used around the world. Many of these plants are widely used in both developed and developing countries, but many are located in remote areas and known only by the local people. The review of local flora and its ethnobotanical use could thus lead to characterizations and localizations of numerous new, potential medicinal plants. Of these, those with a therapeutic use in folk medicine against specific diseases could be of great interest for researchers in the field of natural products and medicinal plants. However, for the research to be valid and coherent, the selection of material and activities should be well established. Previous and thorough prospection of material would be the best starting point for every study; in such cases personal interviews with native people and especially with the medicine man of each area are desirable. Another positive point would be if there were references to the same plant for the same disease by different people from different areas or countries. After this step, positive identification, first by an ethnobotanist and then by field and taxonomical botanists, would allow for the proper identification of the material. Subsequent incorporation of specialists in ethnopharmacology, phytochemistry, and pharmacognosy is also of interest for performing selective phytochemical and pharmacological screening to demonstrate or corroborate the possible activity cited for each plant. In this case, the extractive protocol of its use in folk medicine and its form of administration need to be considered. An isolation-guided screening directed to the main activity would be of interest for future studies, and a complete screening of a given material's pharmacological and biological activities would be welcome for a more thorough insight into potential future uses and possible toxicological events (Malone 1983). In summary, good research in this field calls for the integration of four basic areas: ethnobotany, for delving into and answering all the questions concerning the plant and its medicinal tradition; ethnopharmacy, for exploring all aspects related to the preparation of medicines from crude plant extracts; ethnopharmacology, for resolving questions concerning biological, pharmacological, and therapeutic events; and finally, ethnomedicine, for evaluating all the data with respect to disease and treatment (Labadie 1986). Other welcome members in the research team would include phytochemists, specialists in analytical and organic chemistry, and statisticians. Still other specialists would be necessary depending on the activity to be studied, for example a specialist in microbiology, virology, cancer, or experimental pharmacology could provide much needed expertise.

The study and characterization of the chemical composition of each medicinal plant have two relevant justifications. First, precise knowledge of the plant material or extract is essential for establishing its chemical composition and possible active principles, as well as its other properties, including chemical stability, molecular weight, polarity, and solubility. The second objective is to assess a given compound as a possible marker for future analysis and evaluation of quality.

At present, different biotechnology industries are focusing their efforts on exploring new natural sources for drug discovery. Some of them, for example PharmaMar (Spain), use organisms from the marine environment, while others, such as Indena (Italy), apply this approach to prepare standardized extracts as well as highly purified molecules from medicinal plants. Still others, such as Phytopharm (United Kingdom) directly evaluate the clinical efficacy of standardized plant-derived extracts in diseases where the underlying biochemical causes are not well understood, whereas Galileo Pharmaceuticals (USA) is currently identifying inhibitory principles of inflammatory pathways (Gullo and Hughes 2005).

Future

The future will require novel strategies for developing new compounds from natural origins. The search for new drugs to treat prevalent diseases (AIDS, infectious diseases, cancer, and cardiovascular pathologies) and the use of different strategies for modifying active structures (combinatorial chemistry and biochemistry, microbial transformation, or engineered hairy root cultures for valuable natural products) must progress in tandem to ensure a more complete and precise knowledge of the chemistry of plant extracts and their pharmacology as well as that of the resulting isolated natural products. Such research will greatly enhance the utility of plant extracts and their compounds in modern medicine. There are several challenges involved in leading ethnomedicine into the twenty-first century. These include the preservation of the biodiversity of the rainforest and the ocean, the development of new integrated global information systems on the use of medicinal plants, the enhancement of drug discovery technology, the broadening of our understanding of the chemical diversity of natural products, the assurance of their safety and efficacy, and the development of new facilities for this kind of research (Cordell and Colvard 2005).

Sustainability and mass bioprospecting

Increasing interest in phytomedicine has paved the way for the introduction of several drugs from higher plants into conventional medical therapy during the past two decades. Such is the case with paclitaxel, a diterpenoid originally obtained from the bark of *Taxus brevifolia* but now produced semi-synthetically from a precursor found in the needles of *Taxus baccata*, a rapidly renewable source (Gilani and Atta-ur-Rahman 2005). This breakthrough was a clear focus for subsequent studies and exploitation insofar as sustainability in the cultivation of plant material is a highly relevant criterion for the future success of ethnopharmacology. Once a traditional medicine or its products becomes a major commercial entity, wildcrafting could eliminate it if agricultural techniques for crop development are not introduced. To avoid this possibility, the World Health Organization (WHO) has issued a set of guidelines for good agricultural and collection practices (GACP) for medicinal plants. These are specifically focused on the protection of medicinal plants and the promotion of their cultivation, collection, and use in a sustainable manner which conserves both the medicinal plant and the environment (Cordell and Colvard 2005). With regard to sustainability, the use of solvents for primary extraction should also be considered.

Another relevant landmark in this field concerns the recognition of ownership of indigenous and traditional knowledge. The Declaration of Belém (International Congress of Ethnobiology, Brazil, 1988) and the Code of Ethics of the International Society of Ethnobiology (2005) states that "indigenous peoples, traditional societies and local communities have a right to self determination (or local determination for traditional and local communities) and that researchers and associated organizations will acknowledge and respect such rights" and that they "must be fairly and adequately compensated for their contribution to ethnobiological research activities and outcomes involving their knowledge". Thus, traditional and indigenous knowledge are to be considered both inventive and intellectual and therefore worthy of protection in all legal, ethical, and professional frameworks, which shall represent all such knowledge as property of its holders, who will be duly compensated for the utilization and conversion of such knowledge into a tangible product (Soejarto et al. 2005).

Future trends in ethnopharmacology

Etkin and Elisabetsky (2005) identified a number of important issues for future research in the field of ethnopharmacology. First, advances in laboratory and clinical sciences should continue in order to better characterize the constituents and activities of medicinal plants and other substances; this includes altering pharmacological profiles depending not only on variations in the collection and storage of plants, but also on the preparation and administration of a given medicine. It also implies the study of interactions among the various constituents of medicinal preparations. The second point calls for a more comprehensive analysis after the ingestion of phytochemicals to determine their effect in the maintenance and improvement of body functions and/or disease prevention. The third point specifies that the dosage schedules for indigenous medicines should be clarified since traditional therapies involve the regular ingestion of low doses of an active substance over a significant period of time. Significantly, traditional dosages are not usually evaluated in studies of medicinal plant extracts or substances as a part of new drug screening/development programs. In the fourth point, the researchers suggest that the application of rigorous ethnographic field methodologies refined over the past several decades should improve the comprehension of how healing is approached in diverse cultures. The fifth point underscores the importance of ecological factors for a better understanding of resource management strategies, which are based in part on topography, soil composition, canopy cover, UV radiation, rainfall, and the proximity of other plant and animal species. The sixth point emphasizes that drug discovery from natural products will improve health in all world cultures while the seventh points out that the interactions of researchers with national and local governments should be participatory collaborations involving local people in all phases of research. The eighth point explains that the existing and future iterations of the United Nations Convention on Biological Diversity and other ethical issues will be of interest for the design and application of the ethnopharmacological research guide. The ninth point outlines the problems in developing Western-style plant medicines based on local medicinal flora while the tenth point comments on the importance of connecting this research to social, phytochemical, and clinical issues to assure that the results will be translated, integrated, and applied to the indigenous contexts in which people use these plants. Finally, an eleventh point establishes a series of issues for future research (Etkin and Elisabetsky 2005).

Apart from these considerations for future ethnopharmacological and ethnomedicinal studies, several practices need to be established to ensure the proper development of research in the field and to avoid the undesirable effects that improper development could have on both the scientific community and the indigenous people who use the medicinal plants under study. The use of traditional medicines should be rationalized, the natural resources used as traditional medicines should be preserved and developed, the knowledge of hidden and lost traditional medicines should be researched, and the scope of ethnopharmacology should be expanded (Kim 2005). One could argue, however, that the 'new' field for expanding the scope of ethnopharmacology, dubbed by Kim as 'ethnoergogenics' and defined as "performance enhancing products and substances originating from traditional medicines," falls clearly within the framework of ethnopharmacology, at least as it has previously been defined by various authors: "Ethnopharmacology is a multidisciplinary area of research, concerned with the observation, description, and experimental investigation of indigenous drugs and their biological activities," or the "interdisciplinary study of the physiological actions of plant, animal, and other substances used in indigenous medicines of past and present cultures," or the "use of plants, fungi, animals, microorganisms and minerals and their biological and pharmacological effects based on the principles established through international convention," or "the observation and experimental investigation of the biological activities of plant and animal substances" (Soejarto 2005).

About one hundred compounds derived from natural products are currently undergoing clinical trials and at least another hundred similar projects are in preclinical development. They include compounds from plants, microbes, and animals, as well as synthetic or semi-synthetic compounds based on natural products. Despite the advantages of natural products versus synthetic compounds (e.g. molecular weight, oral bioavailability, past research successes), many large pharmaceutical companies remain unenthusiastic about using natural products for drug discovery screening, probably due to difficulties in access and supply, the complexities of natural product chemistry, the inherent slowness of working with natural products, concerns about intellectual property rights, and the expectations associated with the use of collections of compounds prepared with the aid of combinatorial chemistry methods (Harvey 2008).

Is there a future in ethnopharmacological research?

Plants are a good source of chemical compounds and ethnopharmacology is an excellent way to obtain them. As explained above, many relevant drugs originate from a natural compound isolated from a medicinal plant.

Taking into consideration that most estimates calculate that only 20% of all plant species have been chemically or biologically evaluated (Cordell 2003) and that only 25% of all isolated alkaloids, for example, have been evaluated in a single bioassay, the evaluation of a natural product library represents a substantial opportunity in the search for new drugs or new mechanisms of action. Moreover, there are approximately 5,750 different chemical skeletal for natural products; these constitute an excellent perspective for studying the interactions of various natural products with enzymes and receptors (Cordell and Colvard 2005). To date, however, few natural products have been evaluated and the possibility to do so extensively is quite limited. In order to add to and improve these data it is necessary to enhance the structural diversification of these products through a number of techniques, including, for example, both the theoretical and practical methods involved in combinatorial chemistry on strategic natural products, combinatorial biosynthesis (in which the gene sequence is shifted), moving the gene sequence to faster growing organisms, the chemistry of plant extracts, microbial transformation of single or multiple natural products, potentiating the genes of biosynthesis to realize full metabolic capability, solid phase stable enzymes for structure modification and synthesis, the use of plant-associated and difficult to grow fungi and bacteria, the utilization of engineered hairy root cultures for valuable natural products, and the study of the other 80% of medicinal plants which have not been tested chemically or biologically (Cordell and Colvard 2005).

The genetically controlled biosynthesis of natural products is one of the greatest challenges for enhancing the consistent availability of biologically significant natural products (Cordell 2004). By using isolated enzyme systems it is possible to increase the structural diversity of a given natural product and to conduct reactions which have no parallels in organic synthesis.

Automation, nanotechnology, and proteomics should all be included in ethnopharmacological studies and researchers must be prepared to use these techniques appropriately. Microarray assays based on our enhanced knowledge of the human genome should be introduced for evaluating extracts and compounds in order to assess their genetic impact. Moreover, automated real-time polymerase chain reaction (PCR) processes could be used so that large numbers of plant samples can be genetically identified at the same time, which would constitute an essential step in enhancing the quality control of medicinal plants (Cordell and Colvard 2005). To summarize, several challenges for the future of ethnopharmacology should be underscored: the accurate cataloging and preservation of both the bio- and chemo-diversity of rainforests and oceans as well as the cataloging of the eco- and ethno-information on plants and their products, maintaining equitable access to the biome and assuring intellectual property rights, developing both medicinal plant germplasm banks and integrated global information systems on the uses of medicinal plants, and the cultivation of medicinal plants in a sustainable manner. By the same token, it is also necessary to enhance natural product drug discovery technology in the areas of automation, proteomics, and bioassay targets and to evaluate the known natural products with the aid of various bioassays. Proteomics-based, in-field bioassays should also be developed to help establish the chemical diversity of natural products. Of utmost importance is the assurance of the safety and efficacy of traditional medicines. Finally, the scientific community should develop integrated global alliances for medicinal plant product development, establishing the facilities, infrastructure, and personnel to carry out the aforementioned programs (Cordell and Colvard 2005).

Is high-throughput screening of plant extracts and natural products the future of the field?

The advent of high throughput screening and the availability of combinatorial compound libraries were the major impetus for the decline in research on medicinal plants and natural products for industrial purposes. The incompatibility of the natural product drug discovery process with the fast-paced rapid screening identification process currently used by pharmaceutical companies puts the former at a severe disadvantage. However, several smaller biotechnology companies that have focused their efforts on drug discovery from natural products are using new approaches to increase the chemical diversity of their collections (Gullo and Hughes 2005). For many researchers, biochemical assays and high-throughput screening of natural extracts is somehow viewed as less effective or less practical than the high-throughput screening of pure compounds. In fact, massive investment in high-throughput screening campaigns of natural products tends to result in thousands of biochemical situations that require confirmation, evaluation, and optimization to determine whether they are meaningful leads or simply artifacts (Rishton 2008). Because natural extracts contain chemically reactive compounds which are inappropriate for biochemical screening and which tend to modify target proteins covalently, inducing false-positive results, these should be eliminated by means of chemical conditioning treatments or physicochemical methods. For example, the tannins present in many natural extracts often react with proteins, such as enzymes. They can, however, be easily removed using insoluble (Toth and Pavia 2001) or soluble (Daohong et al. 2004) polyvinylpolypyrrolidone (PVPP). Taking all this into consideration, one could assert that the effectiveness of biochemical assays and high-throughput screening technologies in drug discovery since 1990 is somewhat debatable. Indeed, because biochemical assays are too sensitive and too prone to artifacts due to issues involving solubility, aggregation, chemical reactivity, and quenching effects, the pharmaceutical industry's screening methods have gradually changed to include the screening of natural extract mixtures, with functional biological assays being increasingly preferred over biochemical assays (Rishton 2008).

Can combinatorial chemistry be of use in natural products research?

A recent review of newly introduced drugs affirms that of all the compounds originating from high-throughput screening of combinatorial chemistry libraries, only one has reached the market. With the growing realization that the chemical diversity found in natural products is more likely to match that of successful drugs than the diversity of synthetic compound collections, the interest in applying that natural diversity to drug discovery appears to be increasing once again.

From this perspective, more extensive collections of plants and further advances in our ability to culture microbes could provide many novel chemicals for use in drug discovery assays. Although the mostly untapped microbial diversity of marine environments has been duly recognized, continued productive use of terrestrial bacteria has been limited by difficulties in culturing the vast majority of species. To this end, several different methods have been used. For example, molecular biology techniques have been applied, using bacteria to produce drug-like isoprenoid compounds originally isolated from plants, as well as to produce novel flavanones and dihydroflavonols. A bioinformatics approach has also been used to predict which microbes will produce novel chemicals on the basis of the gene sequences encoding polyketide synthesis; this has led to the discovery of novel compounds with potential pharmacological activities. More recently, a metagenomics approach has been employed to access a wider range of synthetic capabilities from bacteria, leading to the discovery of novel compounds with antibiotic activity (Harvey 2008).

The structures of natural products can be used for designing and synthesizing analogues with improved chemical properties. This can be accomplished by applying techniques developed for combinatorial chemistry or through a variety of genetic techniques involving non-natural combinations of biosynthetic enzymes. In fact, combinatorial approaches have been applied to a wide variety of naturally derived chemical scaffolds, including steroids, terpenoids, flavonoids, and alkaloids. In this way, researchers have been able to select and use biologically active natural products for creating libraries; for example, there are now libraries based on the anticholinesterase compound galanthamine and on the anti-inflammatory derivative andrographolide (Harvey 2007). Natural products have thus inspired many developments in organic chemistry, leading to advances in synthetic methodologies and to the possibility of making analogues of an original lead compound with improved pharmacological properties.

Could *in silico* screening of natural products be of use in research?

Drug discovery for a single agent drug has come to be a very inefficient and extremely expensive process, costing more than € 600 million (about US$ 800 million) per drug. Of the millions of compounds screened for their properties, only about 5,000 will be considered for advanced pharmacological development. Of these, only one may become an effective drug.

Moreover, at the phase I clinical trial stage, only one of every twelve compounds is actually marketed (Cordell and Colvard 2005). Because of this, facilities for high-throughput screening are now available at both universities and drug-companies; however, the cost of random screening of large collections of compounds is extremely expensive. For this reason, using *in silico* or virtual screening methods to narrow down the number of compounds used in real analysis is of great interest. It would be tremendously advantageous, for example, to mix the advantages of virtual screening of chemically diverse natural products and their synthetic analogues with the rapid availability of physical samples for testing (Harvey 2008). Our own experience has demonstrated that this interaction can lead to better research since the isolation of natural products in small amounts does not allow for a large screening or selective assays. Thus, in an attempt to demonstrate the viability of several products, we first made a theoretical prediction *in silico* (non published data) for several previously isolated acetophenones from *Helichrysum italicum*. We then demonstrated their anti-inflammatory and analgesic activity (Sala et al. 2003) as predicted by the theoretical screening (non published data). This was only a first attempt, but we intend to develop this methodology in the near future as it could be an important tool in conducting a virtual screening of different natural product libraries, such as that presented in the Dictionary of Natural Products, which lists approximately 150,000 different compounds.

Is it possible to use 'Omic' technologies in phytomedicine?

Ulrich-Merzenich et al. (2007) reviewed this possibility and concluded that omic technologies "allow the simultaneous analysis of complex chains of action and have the potential to relate complex mixtures to complex effects on the different levels of metabolism". There are four kinds of 'omic' technologies which can serve for this purpose: genomics, for a comprehensive description of the genetic information; transcriptomics, for gene expression; proteomics, for the description of proteins; and metabolomics, for metabolite networks (the metabolome is the quantitative complement of all the low-molecular-weight molecules of a cell or organism present in a particular physiological or developmental state) (Hollywood et al. 2006). These new technologies allow for the simultaneous detection of tens of thousands of genes and proteins. Indeed, they represent a challenging complexity for scientific analysis and will undoubtedly open up new perspectives for phytomedical research (Ulrich-Merzenich et al. 2007). Although methodologies for the analysis of mRNA (transcriptome) and protein (proteome) levels are well advanced, with high-throughput methods being routinely used in many laboratories, methodologies for metabolome analysis are less developed, with many research groups still using classic protocols such as HPLC, GC, MS, and NMR to analyze metabolites in metabolome studies (Villas-Bôas et al. 2006, Schnackenberg and Beger 2007).

Metabolic analysis comprises of four different areas: the quantification of specific metabolites, the quantitative and qualitative determination of a group of related compounds or of specific metabolic pathways, the qualitative and quantitative analysis of all metabolites or metabolome, and metabolomic 'fingerprinting', which utilizes NMR, direct injection mass spectrometry, or Fourier transform infrared spectroscopy to analyze crude extracts without a separation step (Ulrich-Merzenich et al. 2007).

Whereas research in the fields of phytochemistry and pharmacognosy has generally focused on the search for active principles in plants and crude drugs (the assumption being that a plant has one or several ingredients which determine its therapeutic effect), phytomedicine assumes that synergy is essential for activity. Moreover, some traditional medicines use preparations with mixtures of plants, which could be considered the 'herbal shotgun' or multitarget approach of herbals as opposed to the 'silver bullet' method of conventional medicine (Williamson 2001), which uses a monotarget approach, employing synthetic drugs to address specific enzymes or receptors. However, even though multitarget treatment methods are generally well accepted, phytomedicines still need to prove their efficacy, provide a rationale behind this efficacy, and demonstrate reproducibility through standardization. One stumbling block to this is the fact that as biogenic products they can have fluctuating compositions of ingredients. Another confusing factor is that in some cases European law allows written histories on the uses of these medicines to be used as valid scientific source material in lieu of more conventional proof of efficacy (Ulrich-Merzenich et al. 2007).

Omic technologies could be useful for purposes of quality assurance and standardization of phytomedicines. For example, correct taxonomic identification of the plant and the identification of the major active principles of the plant, its extract, or multiextract mixtures could be aided by omic technologies. Thus, the combined information from genomics, protcomics, and metabolomics could be of use in obtaining an integral understanding of a given plant. On the other hand, these new analytic platforms substantially increase the dynamic range and number of metabolites and genes that can be detected, which, in turn, creates an increasing need for information technology to transform the data obtained into real biological knowledge. This need has already led to advances in systems biology. In this particular field, omic data—especially metabolomic data—are currently being organized with the aim of creating computer models to simulate biological systems (Ulrich-Merzenich et al. 2007).

One of the major trends in new drug research is the focus on the identification and development of biomarkers that can accurately predict toxicity in preclinical and clinical studies of new chemical entities, which is now considered vital. A technique such as metabolic profiling that has the potential to identify toxicity early in the drug discovery process saves both time and money in the development of new drugs. In general, the most widely used biofluid for toxicity studies up to now has been urine, which is easily obtained and provides information about the workings of the entire system after a toxic insult. The main advantage of using urine or plasma is that because the sample is collected noninvasively, it can be applied in clinical studies. In addition, multiple biofluid samples from a single subject can be collected over a specific time span and because the same animal can be used many different times, the number of animals needed for a toxicological study is much smaller. Another advantage of using a temporal study with the same animal is that it is not necessary to know the pharmacodynamics or pharmacokinetics of the drug before the biomarker investigation. This eliminates the need for preparatory pharmacokinetic research, which also saves both time and money. This is especially important in the early absorption, distribution, metabolism, and excretion-toxicology stage of drug discovery (Schnackenberg and Beger 2007).

Metabolic engineering for obtaining natural products from higher plants

Khan and Ather (2009) reviewed the potential use of metabolic engineering for the fabrication of pharmaceutically central metabolites from microorganisms and higher plants. They cited that the most relevant objective of metabolic engineering and metabolomics is the augmentation of a given cellular phenotype, for example, the over-production of certain metabolites, through the introduction of genetic controls. The completion of the sequencing of several genomes has resulted in an increase in the number of metabolically engineered genes available, which, together with the increase in databases and the *in silico* tools to handle these genomes, aids in the more precise development of metabolomics. In summary, various biotechnological issues in the production of secondary metabolites from plants are of relevance, including those concerning plant cells, tissues, and organ cultures; transgenic plants and organisms; micro-propagation of medicinal plants to obtain endangered species, high-yielding varieties, and metabolically engineered plants; newer sources, such as algae and other photosynthetic marine forms; and safety considerations such as propagation in the cell environment and effects on biodiversity and health. All this only serves to confirm the fact that higher plants are still an invaluable source of natural products with interesting new activities. Moreover, there is a great amount of interest in incorporating other groups of organisms, e.g. marine flora, into the research currently being carried out in this field (Khan and Ather 2009).

Application of DNA microarrays to phytomedicine

Microarray technology was introduced only very recently into the field phytomedicines and is still in its first stages of application (Ulrich-Merzenich et al. 2007). Using a cDNA microarray system containing 1,176 known genes, Wang et al. (2004) demonstrated a dose-dependent effect for herbal preparations containing *Scutellaria baicalensis* and *Dioscorea* spp. (with baicalein and dioscin in a ratio of 1:1) and affirmed the usefulness of this methodology for elucidating the mechanism of the pharmacological functions of herbal preparations. Other similar studies have been published about different phytomedicines, including soya, ginkgo, and chamomile. In addition, microarray analysis has been used for studying the synergistic and antagonistic effects of different compounds found either in plants or in mixtures thereof, as well as their toxicity (Ulrich-Merzenich et al. 2007). These studies show great promise for the future application of microarray technology in phytomedicine as well as in phytotherapy research. Because this methodology makes it possible to identify a compound or mixture of compounds that correlates with a specific activity as well as activities due to synergy and the possible activity of pro-drugs, it is especially useful with regard to standardization and determination of the activity of different extracts on living organisms (Verpoorte et al. 2005). Unfortunately, reproducibility of microarrays on plant extracts is often criticized because the scientific community only accepts chemically defined compounds as causal agents for defined effects. However, with mircoarray technology it is now possible to attribute a reproducible gene expression profile to a complex herbal mixture that is not fully defined chemically (Ulrich-Merzenich et al. 2007). This possibility will be discussed in more depth in another chapter of this book.

Conclusions

In the field of natural products of plant origin, two groups can be clearly differentiated: those phytomedicines comprised of either one extract or a complex mixture of several plants, and those principles isolated or derived from active compounds of plants. In the former case, the application of new technologies focuses on in-depth knowledge of the material in order to establish its safety and quality. In this context, the classic techniques in the field of natural product rescarch are as effective as in the past, and their usefulness in searching for new compounds or studying plant extracts in order to establish their identity and quality remains unquestioned. However, novel techniques and protocols have opened up new paths of investigation which have led to a more complete understanding of materials of plant origin, including phytomedicines. These new technical procedures can be used to develop new active compounds from phytochemicals that have already been isolated. Drug discovery and a better understanding of complex disease mechanisms and therapeutic modalities will increasingly require different approaches using methods and tools from varied disciplines. This will have a profound impact on the systematic creation of large collections of reagents, detection methods, laboratory technology, computer science, and organizational design (Cho et al. 2006). The needs of the near future include a widespread collaboration between experimental scientists, physicians, computer scientists, and mathematicians. Thus, all these systems, techniques, and protocols will modify the future of drug discovery and lead to improvements in the efficacy and success of new phytomedicines. Moreover, the production of plant cells and transgenic hairy root cultures are a potential alternative source for the production of high-value secondary metabolites (Khan and Athcr 2009).

One of the most interesting aspects of the field of natural products is perhaps the knowledge that they constitute a highly attractive source of varied structures with different biological properties and pharmacological activities. Because natural products chemistry is such a powerful tool for optimizing natural leads and generating new diversity from natural scaffolds, libraries of natural products, their derivatives, and compounds prepared from them with the aid of total synthesis constitute a new and exciting source from which to search for novel compounds with a specific mechanism or effect. The amalgamation of these features is likely to become an important strategy in future drug design (Abel et al. 2002, Boldi 2004). In addition, modern technologies including database technology, data-mining technology, chemometrics, brief virtual screening technology, experimental design, innovative design, complexity studies, and bioinformatics technology are all currently being used to study traditional Chinese prescriptions. It is our belief that these new technologies will enlighten researchers on the effects of these age-old drugs and help modernize the study of traditional medicines to further develop them in the future (Long et al. 2007).

References

Abel, U., C. Koch, M. Speitling, and F.G. Hansske. 2002. Modern methods to produce natural-product libraries. Curr. Opin. Chem. Biol. 6: 453–458.

Adlercreutz, H. 1990. Western diet and Western diseases: some hormonal and biochemical mechanisms and associations. Scand. J. Clin. Lab. Invest. Suppl. 201: 3–23.

Adlercreutz, H., K. Höckerstedt, C. Bannwart, S. Bloigu, E. Hämäläinen, T. Fotsis, and A. Ollus. 1987. Effect of dietary components, including lignans and phytoestrogens, on enterohepatic circulation and liver metabolism of estrogens and on sex hormone binding globulin (SHBG). J. Steroid Biochem. 27: 1135–1144.

Adlercreutz, H., Y. Mousavi, and K. Höckerstedt. 1992. Diet and breast cancer. Acta Oncol. 31: 175–181.

Altman, R., and R.L. Barkin. 2009. Topical therapy for osteoarthritis: clinical and pharmacologic perspectives. Postgrad. Med. 121: 139–147.

Anand, P., S.G. Thomas, A.B. Kunnumakkara, C. Sundaram, K.B. Harikumar, B. Sung, S.T. Tharakan, K. Misra, I.K. Priyadarsini, K.N. Rajasekharan, and B.B. Aggarwal. 2008. Biological activities of curcumin and its analogues (Congeners) made by man and Mother Nature. Biochem. Pharmacol. 76: 1590–1611.

Attele, A.S., J.A. Wu, and C.S. Yuan. 1999. Ginseng pharmacology: multiple constituents and multiple actions. Biochem. Pharmacol. 58: 1685–1693.

Boldi, A.M. 2004. Libraries from natural product-like scaffolds. Curr. Opin. Chem. Biol. 8: 281–286.

Brinkhaus, B., M. Lindner, D. Schuppan, and E.G. Hahn. 2000. Chemical, pharmacological and clinical profile of the East Asian medical plant *Centella asiatica*. Phytomedicine 7: 427–448.

Butler, M.S. 2008. Natural products to drugs: natural product-derived compounds in clinical trials. Nat. Prod. Rep. 25: 475–516.

Chan, P.C., Q. Xia, and P.P. Fu. 2007. *Ginkgo biloba* leave extract: biological, medicinal, and toxicological effects. J. Environ. Sci. Health C. Environ. Carcino. Ecotoxicol. Rev. 25: 211–244.

Cho, C.R., M. Labow, M. Reinhardt, J. van Oostrum, and M.C. Peitsch. 2006. The application of systems biology to drug discovery. Curr. Opin. Chem. Biol. 10: 294–302.

Cordell, G.A. 2003. Discovering our gifts from nature, now and in the future. Part II. Rev. Quim. 17: 3–15.

Cordell, G.A. 2004. Accessing our gifts from nature, now and in the future. Part III, Rev. Quim. 19: 33–41.

Cordell, G.A., and M.D. Colvard. 2005. Some thoughts on the future of ethnopharmacology. J. Ethnopharmacol. 100: 5–14.

Cragg, G.M., D.J. Newman, and K.M. Snader. 1997. Natural products in drug discovery and development. J. Nat. Prod. 60: 52–60.

Daohong, W., W. Bochu, L. Biao, D. Chuanren, and Z. Jin. 2004. Extraction of total RNA from *Chrysanthemum* containing high levels of phenolic and carbohydrates. Colloids Surf. B Biointerfaces 36: 111–114.

Davydov, M. and A.D. Krikorian. 2000. *Eleutherococcus senticosus* (Rupr. & Maxim.) Maxim. (Araliaceae) as an adaptogen: a closer look. J. Ethnopharmacol. 72: 345–393.

Directive 2004/24/EC of the European Parliament and of the Council of 31 March 2004. http://ec.europa.eu/enterprise/pharmaceuticals/eudralex/vol-1/dir_2004_24/dir_2004_24_en.pdf (23 July 2009).

Etkin, N.L. and E. Elisabetsky. 2005. Seeking a transdisciplinary and culturally germane science: The future of ethnopharmacology J. Ethnopharmacol. 100: 23–26.

Fan, E., L. Zhang, S. Jiang, and Y. Bai. 2008. Beneficial effects of resveratrol on atherosclerosis. J. Med. Food 11: 610–614.

Gilani, A.H., and Atta-ur-Rahman. 2005. Trends in ethnopharmacology. J. Ethnopharmacol. 100: 43–49.

Gullo, V.P. and D.E. Hughes. 2005. Exploiting new approaches for natural product drug discovery in the biotechnology industry. Drug Discov. Today Technol. 2: 281–286.

Harvey, A.L. 2007. Natural products as a screening resource. Curr. Opin. Chem. Biol. 11: 480–484.

Harvey, A.L. 2008. Natural products in drug discovery. Drug Discov. Today 13: 894–901.

Harvey, A.L. 2009. Bridging the Gap: using natural products in drug discovery research in academia and industry. In: M.C. Carpinella and M. Rai (eds.). Novel Therapeutic Agents from Plants. Science Publishers, Enfield, NH, USA. pp. 167–175.

Heinrich, M., J. Barnes, S. Gibbons, and E.M. Williamson. 2004. Fundamentals of Pharmacognosy and Phytotherapy. Churchill Livingstone, Edinburgh, UK.

Hollywood, K., D.R. Brison, and R. Goodacre. 2006. Metabolomics: current technologies and future trends. Proteomics 6: 4716–4723.

Khan, M.T.H. and A. Ather. 2009. Metabolic engineering for the fabrications of pharmaceutically central metabolites from microorganisms and higher plants. In: M.C. Carpinella and M. Rai (eds.). 2009. Novel Therapeutic Agents from Plants. Science Publishers, Enfield, NH, USA. pp. 368–386.

Kim, H.-S. 2005. Do not put too much value on conventional medicines. J. Ethnopharmacol. 100: 37–39.

Labadie, R.P. 1986. Problems and possibilities in the use of traditional drugs. J. Ethnopharmacol. 15: 221–230.

Leifert, W.R., and M.Y. Abeywardena. 2008. Cardioprotective actions of grape polyphenols. Nutr. Res. 28: 729–737.

Liu, J., J. Zhang, Y. Shi, S. Grimsgaard, T. Alraek, and V. Fønnebø. 2006. Chinese red yeast rice (*Monascus purpureus*) for primary hyperlipidemia: a meta-analysis of randomized controlled trials. Chin. Med. 1:4 (doi:10.1186/1749-8546-1-4).

Liu, M., Y. Dai, Y. Li, Y. Luo, F. Huang, Z. Gong, and Q. Meng. 2008. Madecassoside isolated from *Centella asiatica* herbs facilitates burn wound healing in mice. Planta Med. 74: 809–815.

Long, W., P.X. Liu, and J. Gao. 2007. Application of modern information technology in the study of traditional Chinese medicine prescriptions. Zhongguo Zhong Yao Za Zhi 32: 1260–1263 (PMID: 17879720).

Lu, Z., W. Kou, B. Du, Y. Wu, S. Zhao, O.A. Brusco, J.M. Morgan, D.M. Capuzzi, and Chinese Coronary Secondary Prevention Study Group. 2008. Effect of xuezhikang, an extract from red yeast Chinese rice, on coronary events in a Chinese population with previous myocardial infarction. Am. J. Cardiol. 101: 1689–1693.

Malone, M.H. 1983. The pharmacological evaluation of natural products—general and specific approaches to screening ethnopharmaceuticals. J. Ethnopharmacol. 8: 127–147.

Newman, D.J., and G.M. Cragg. 2007. Natural products as sources of new drugs over the last 25 years. J. Nat. Prod. 70: 52–60.

Newman, D.J., G.M. Cragg, and K.M. Snader. 2003. Natural products as sources of new drugs over the period 1981–2002. J. Nat. Prod. 66: 1022–1037.

Rishton, G.M. 2008. Natural products as a robust source of new drugs and drug leads: past successes and present day issues. Am. J. Cardiol. 101: 43D–49D.

Sala, A., M.C. Recio, G. Schinella, S. Máñez, R.M. Giner, and J.L. Ríos. 2003. A new dual inhibitor of arachidonate metabolism isolated from *Helichrysum italicum*. Eur. J. Pharmacol. 460: 219–226.

Schnackenberg, L.K., and R.D. Beger. 2007. Metabolomic biomarkers: their role in the critical path. Drug Discov. Today Technol. 4: 13–16.
Soejarto, D.D. 2005. Ethnographic component and organism documentation in an ethnopharmacology paper: A "minimun" standard. J. Ethnopharmacol. 100: 27–29.
Soejarto, D.D., H.H.S. Fong, G.T. Tan, H.J. Zhang, C.Y. Ma, S.G. Franzblau, C. Gyllenhaal, M.C. Riley, M.R. Kadushin, J.M. Pezzuto, L.T. Xuan, N.T. Hiep, N.V. Hung, B.M. Vu, P.K. Loc, L.X. Dac, L.T. Binh, N.Q. Chien, N.V. Hai, T.Q. Bich, N.M. Cuong, B. Southavong, K. Sydara, S. Bouamanivong, H.M. Ly, T. Van Thuy, W.C. Rose, and G.R. Dietzman. 2005. Ethnobotany/ethnopharmacology and mass bioprospecting: Issues on intellectual property and benefit-sharing. J. Ethnopharmacol. 100: 15–22.
Stickel, F. and D. Schuppan. 2007. Herbal medicine in the treatment of liver diseases. Dig. Liver Dis. 39: 293–304.
Toth, G.B. and H. Pavia. 2001. Removal of dissolved brown algal phlorotannins using insoluble polyvinylpolypyrrolidone (PVPP). J. Chem. Ecol. 27: 1899–1910.
Ulrich-Merzenich, G., H. Zeitler, D. Jobst, D. Panek, H. Vetter, and H. Wagner. 2007. Application of the "-Omic-" technologies in phytomedicine. Phytomedicine 14: 70–82.
Verpoorte, R., Y.H. Choi, and H.K. Kim. 2005. Ethnopharmacology and systems biology: a perfect holistic match. J. Ethnopharmacol. 100: 53–56.
Villas-Bôas, S.G., S. Noel, G.A. Lane, G. Attwood, and A. Cookson. 2006. Extracellular metabolomics: A metabolic footprinting approach to assess fiber degradation in complex media. Anal. Biochem. 349: 297–305.
Vogel, H.G. 1991. Similarities between various systems of traditional medicine. Considerations for the future of ethnopharmacology. J. Ethnopharmacol. 35: 179–190.
Wang, Z., Q. Du, F. Wang, Z. Liu, B. Li, A. Wang, and Y. Wang. 2004. Microarray analysis of gene expression on herbal glycoside recipes improving deficient ability of spatial learning memory in ischemic mice. J. Neurochem. 88: 1406–1415.
Williamson, E.M. 2001. Synergy and other interactions in phytomedicines. Phytomedicine 8: 401–409.
Web 1. http://www.phcog.org/what.html. (May 7, 2009).
Web 2. http://www.cbsnews.com/stories/2008/11/20/health/webmd/ main4620156.shtml. (May 7, 2009).
Web 3. http://www.mayoclinic.com/health/red-yeast-rice/ns_patient-redyeast. (May 13, 2009).

Chapter 2

Spasmolytic Effect of Constituents from Ethnomedicinal Plants on Smooth Muscle

Rosa Martha Perez Gutierrez

Introduction

Diarrhoea is still one of the major health threats to populations in tropical and subtropical poor countries. The World Health Organization (WHO) has estimated that 3–5 billion cases occur each year (1 billion in children, <5 years old) and about 5 million deaths are due to diarrhoea annually (2.5 million in children, <5 years old) (Lanata and Mendoza 2002). In Mexico, intestinal infectious diseases are the 20th most important causes of death. In the year 2004, many deaths (472,273) were registered and 4180 (0.9%) were caused by these diseases. The most affected were children under 4 years (1167; 28%) and people who were >65 years (1587; 38%). The major impact of these illnesses is morbidity, because it demands primary medical services, hospital-care time and labour days lost.

From a mechanistic perspective, diarrhoea can be caused by an increased osmotic load within the intestine; excessive secretion of electrolytes and water into the intestinal lumen; exudation of protein and fluid from the mucosa; and altered intestinal motility resulting in rapid transit. In most instances, multiple processes are affected simultaneously, leading to a net increase in stool volume and weight accompanied by

Escuela Superior de Ingeniería Química e Industrias extractivas IPN. Av, Instituto Politecnico Nacional S/N, Unidad Profesional Adolfo López Mateos, Mexico D.F.

Corresponding address: Rosa Martha Perez Gutierrez, Punto Fijo No. 16, Col, Torres de Lindavista, C.P. 07708 D.F. Mexico.

E-mail: *rmpg@prodigy.net.mx*

increases in fractional water content. There are several types of etiologies: virus, bacteria, parasites, food poisoning, antibiotics, laxative agents, or certain medical conditions that can also lead to diarrhoea (Pasricha 2006). Despite the etiology, chronic diarrhoea is linked with dehydration fairly quickly and electrolyte-containing solutions are the first choice for treatment.

Furthermore, the most highly used drugs for intestinal disease therapies are frequently very expensive, overused and the inadequate use of antibiotics has lead an important increase in the prevalence of multi drugs-resistant pathogens. Important classes of antidiarrhoeal drugs are opioid agonists, colloidal bismuth compounds, bile salt binding resins, and others such as antiviral, antibacterial, antiprotozoal and antihelmintic drugs (Pasricha 2006).

In developing countries, oral re-hydration therapy saves many thousands of lives every year, while medicinal plants represent an important tool of therapy for diarrhoea. Regulation of smooth muscle contractility is essential for many important biological processes such as tissue perfusion, cardiovascular homeostasis and gastrointestinal motility (Gomes et al. 2005). Antispasmodics or spasmolytics are drugs used to relieve or prevent smooth muscle spasms by reducing the intestinal hypercontractility of smooth muscles. In fact, these drugs allow the gastrointestinal muscles to return to their proper tone, thus reducing abdominal pains and associated symptoms (Hasler 2003). Consequently, antispasmodics are frequently prescribed for a number of gastrointestinal diseases, including irritable bowel syndrome, a condition which affects 10–25% of the population (Jailwala et al. 2000, Farhadi et al. 2001).

The main antispasmodic drugs include antimuscarinic compounds (e.g. the alkaloids derived from the plant belladonna and their synthetic derivatives) and calcium channel blockers (e.g. otilonium and pinaverium) (Pasricha 2001). Since the use of these drugs may be associated with the appearance of unwanted side-effects (dry mouth and urinary retention for antimuscarinic drugs and headache, nausea, vomiting and constipation for calcium blockers), the search for safe, plant-derived antispasmodics is an interesting research field.

Treatment of spastic motility disorders and the search for new therapeutic agents are still a challenge (Achem 2004). A large number of plants have been empirically used for the treatment of different diseases, including diarrhoea, thus, several studies have been undertaken to provide scientific proof to justify the medicinal use of various plants in the treatment of this disease.

The aim of this chapter is an up-to-date and comprehensive analysis of traditional and folklore uses, phyto-constituents and pharmacological reports. A wide range of chemical compounds are presented including alkaloids, flavonoids, coumarins, rotenoids terpenoids and triterpenes isolated with antispasmodic effect from several species in recent times.

Alkaloids

An aporphin alkaloid (Fig. 2.1) isolated from *Nandina domestica* (Berberidaceae) showed inhibition on contractions produced by phenylephrine and 5-hydroxytryptamine in rat isolated intact aorta. It is also reported that depolarizing Ca^{2+} free high KCl 50 mM solution inhibited, in a non-competitive way, the increase in tension evoked by Ca^{2+} with depression of the maximum response.

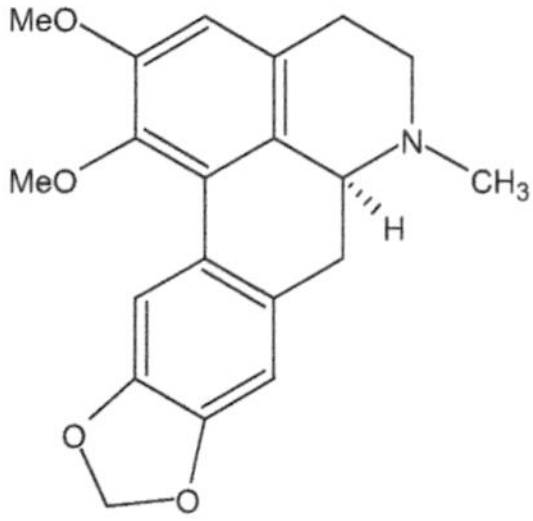

Fig. 2.1. Structure of Nantenine.

On the other hand, (+)-nantenine did not affect the contractile effect caused by okadaic acid, however this alkaloid totally relaxed, in a concentration-dependent fashion, the contractions produced by phorbol 12-myristate 13-acetate (PMA) in endothelium-containing rat aortic rings. These results indicate that the pharmacological effects of (+)-nantenine observed at concentrations lower than 1 µM can be attributed to α_1-adrenergic and 5-HT 2A receptor blocking properties whereas at higher concentrations (>1 µM) the pharmacological activity of this natural compound may be also due to a decrease of Ca^{2+} influx through transmembrane calcium channels (calcium antagonist activity), to an inhibition of protein kinase C (PKC) actions and/or to an α_2-adrenoceptor blockade (Orallo 2003).

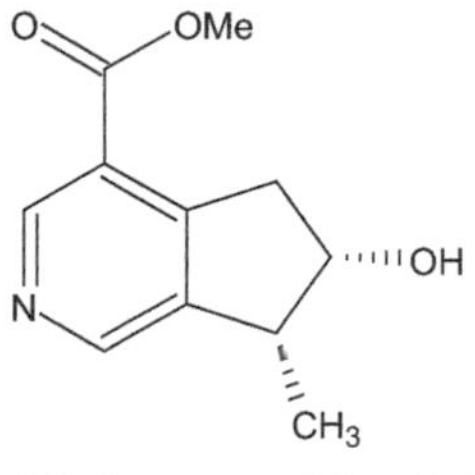

Fig. 2.2. Structure of Cantleyine.

A monoterpene (Fig. 2.2) alkaloid isolated from the root bark of *Strychnos trinervis* (Loganiaceae), relaxed the guinea-pig ileum and guinea-pig trachea, by carbachol and histamine induced contractions; The tonic contractions induced by KCl were also inhibited in a concentration-dependent and reversible manner, suggesting that cantleyine should be acting on voltage-dependent Ca^{2+} channels (da Silva et al. 1999).

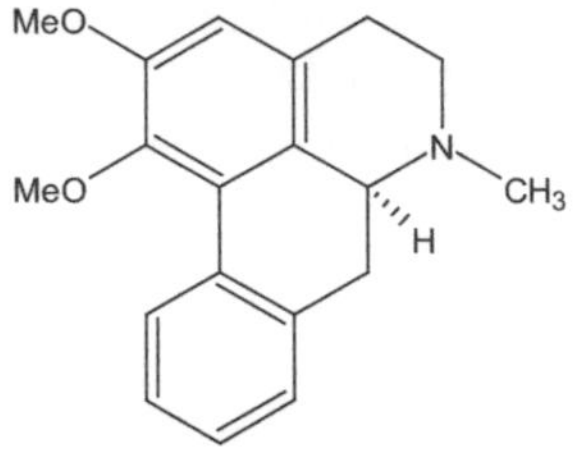

Fig. 2.3. Structure of Atherosperminine.

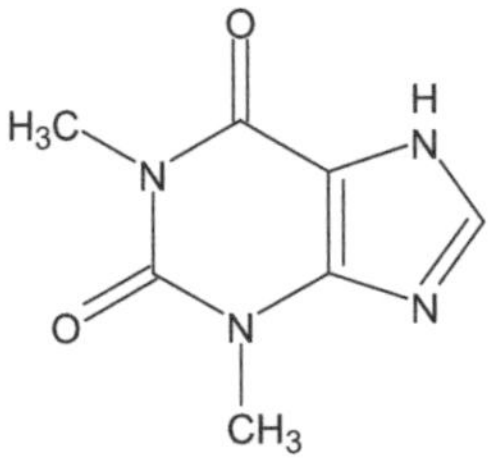

Fig. 2.4. Structure of Theophylline.

Atherosperminine and theophylline were isolated from *Fissistigma glaucescens* (Annonaceae), both inhibited the contractile response caused by carbachol, prostaglandin $F_2\alpha$ ($PGF_2\alpha$), U46619 (thromboxane A_2 analogue), leukotriene C_4 (LTC_4) and Ca^{2+} (in the presence of 120 mM KCl) in a concentration-dependent manner. The inhibition was characterized by a rightwards shift of the concentration-response curves with suppression of the maximal contraction. Moreover, cantleyine potentiated the action of forskolin to increase tissue cAMP content and, in higher concentrations (100 and 250 μM), itself increased tissue cAMP but not cGMP content, inhibited cAMP phosphodiesterase (PDE) but not cGMP in homogenates of guinea-pig trachealis. It is concluded that atherosperminine exerts a non-specific relaxant effect on the trachealis. Its major mechanism of action appears to be inhibition of cAMP perhaps with a minor effect on cGMP at higher concentrations (Lin et al. 1993).

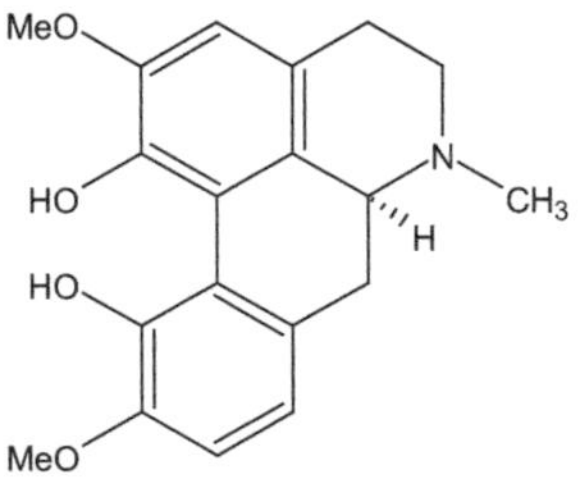

Fig. 2.5. Structure of Corytuberine.

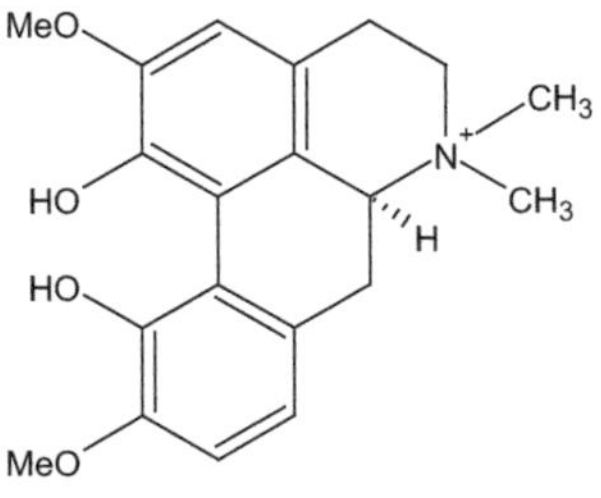

Fig. 2.6. Structure of Magnoflorin.

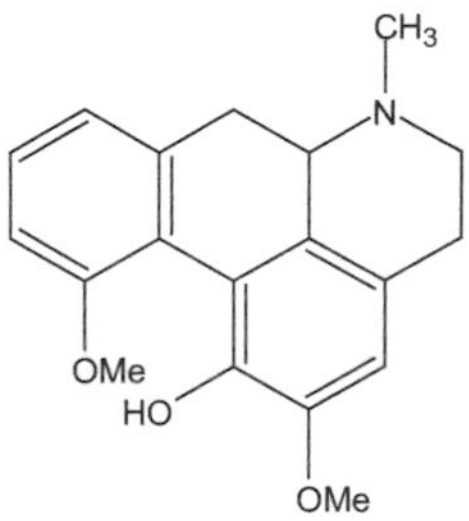

Fig. 2.7. Structure of Isothebaine. 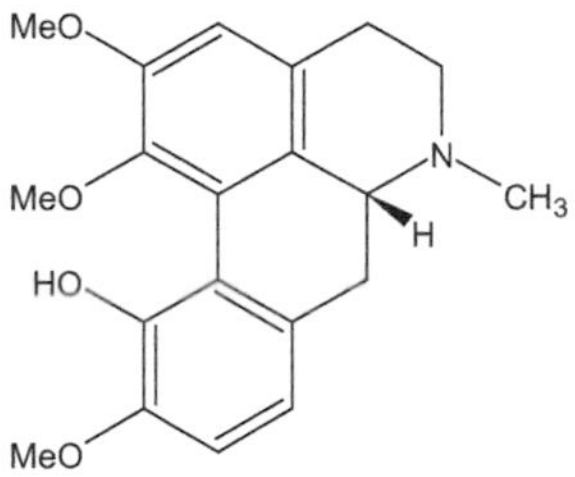   **Fig. 2.8.** Structure of Isocorydine.

The aporphine alkaloids (Fig. 2.5. to 2.8), were isolated from *Mahonia aquifolium* (Berberidaceae). Corytuberine and magnoflorine showed little effect as relaxants in KCl and noradrenaline-induced contractions. They did not inhibit the phenylephrine concentration-response curve (CRC). In addition, both showed relaxant properties in the rat aorta, relaxed the contractions induced by noradrenaline to a greater extent than those induced by KCl and they also inhibited the noradrenaline-induced contraction in calcium-free solution. These alkaloids competitively shifted the phenylephrine CRC to the right and non-competitively the serotonin CRC. They seemed to inhibit both the influx of calcium into the cell, preferentially through receptor-regulated calcium channels, and the release of calcium from intracellular stores. Moreover, they appear to be antagonists of α_1-adrenoceptors. Study of structure-activity relationships of aporphine alkaloids in the inhibition of calcium influx via potential-operated Ca^{2+} channels yielded information indicative of the interaction of these substances with the benzothiazepine receptor (Sotníková et al. 1997).

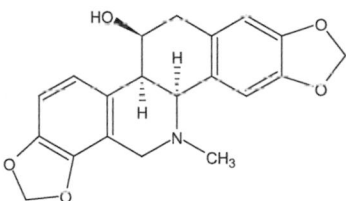

Fig. 2.9. Structure of Chelidonine. 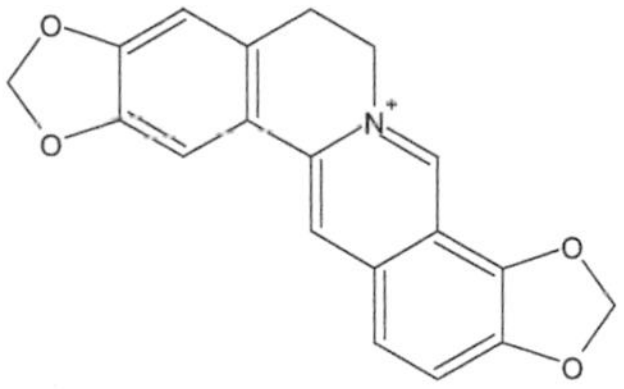   **Fig. 2.10.** Structure of Coptisine.

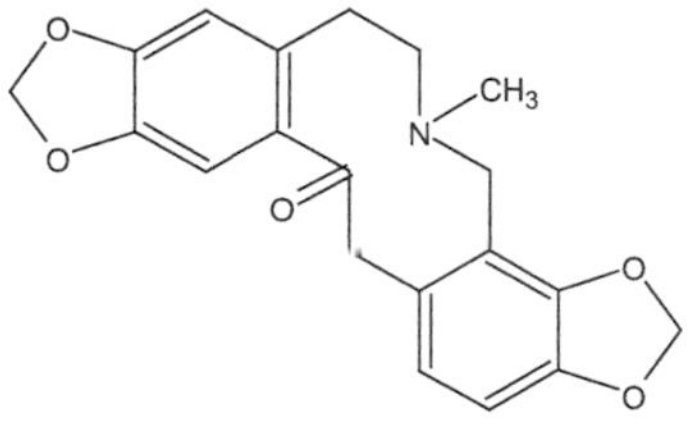

Fig. 2.11. Structure of Protopine.

The alkaloids chelidonine, coptisine and protopine (Fig. 2.9 to 2.11) were isolated from *Chelidonium majus*. (Papaveraceae). In the $BaCl_2$-stimulated ileum of guinea-pigs chelidonine and protopine exhibited the known papaverine-like musculotropic action, whereas coptisine was ineffective in this model. The carbachol and the electric field stimulated contractions were antagonized by all three alkaloids. Coptisine showed competitive antagonist behaviour with a pA_2 value of 5.95. In addition, chelidonine and protopine exhibited a certain degree of non-competitive antagonism. In the electric field the antagonist activities decreased in the order protopine > coptisine > chelidonine (Hiller et al. 1998).

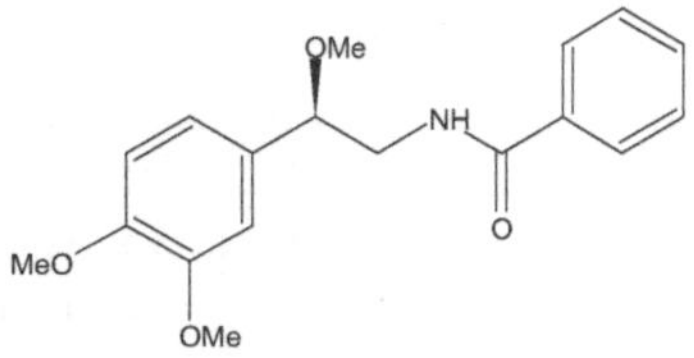

Fig. 2.12. Structure of O-Methylbalsamide. 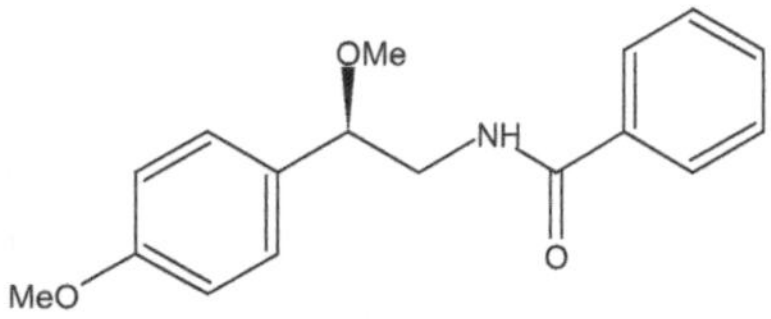   **Fig. 2.13.** Structure of O-Methyltembamide.

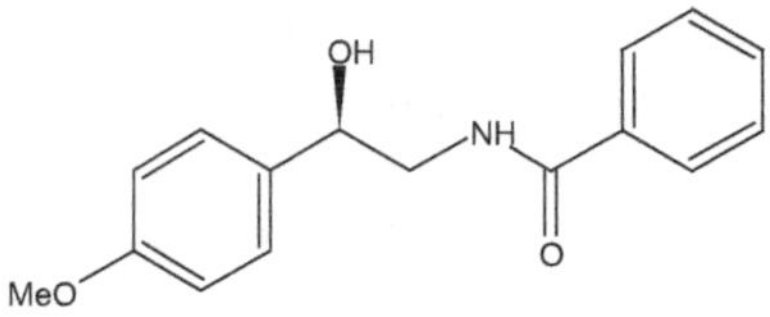

Fig. 2.14. Structure of Tembamide.

O-Methylbalsamide, O-methyltembamide and tembamide were isolated from the stem barks of *Zanthoxylum hyemale* (Rutaceae). The crude extract of *Z. hyemale* acted as a non-competitive antagonist of contractions induced by acetylcholine (ACh) and $BaCl_2$. (R)-Tembamide increased pD_2 of $BaCl_2$-induced contraction. O-methyltembamide did not alter pD_2 of ACh- and $BaCl_2$-induced contractions. In addition, the studied isolated compounds O-methylbalsamide and tembamide did not change ACh- and $BaCl_2$-induced maximal contraction of isolated rat ileum. Therefore, since O-methyltembamide has no effect and the effects of tembamide on isolated rat ileum are different from the crude extract of *Z. hyemale*. These results suggest it is possible that other substances are responsible for the antispasmodic effect of this plant (Moura et al. 2002).

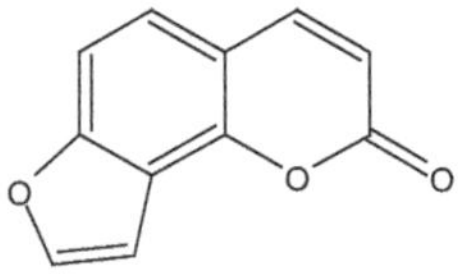

Fig. 2.15. Structure of Angelicin.

Coumarins

Angelicin (Fig. 2.15), which occurs in *Heracleum thomsoni* (Apiaceae) has a relevant effect on a wide variety of smooth muscle preparations from various species (rabbit, rat and cat), and is equally effective *in vitro* and *in vivo*. It has inhibited the contraction induced by a wide range of spasmogenic agents in these tissues. In addition, the non-specific spasmolytic activity of the compound, like papaverine, is evident by its ability to block a variety of spasmogens in the ileum, trachea, bile duct, gall-bladder, and uterus and its being almost equally active against all the spasmogens in any particular preparation. Angelicin further relaxed the rabbit jejunum and cat intestine where both tone and motility were inhibited. The relaxation of the gastrointestinal tract is also evident by the inhibition of the propulsion of charcoal meal in the mouse. The relaxant effect of this coumarin seems to be specific to the smooth muscles since it had no effect on the cardiac and skeletal muscles, thus indicating a selectivity of action (Patnaik et al. 1987).

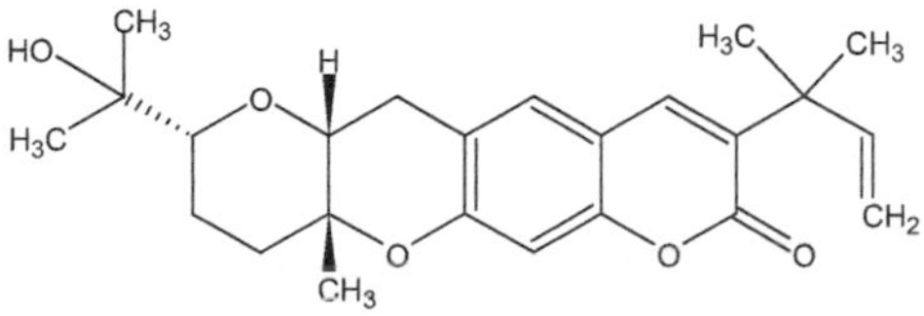

Fig. 2.16. Structure of Clausmarin A.

Clausmarin A, the active principle of *Clausena pentaphylla* (Rutaceae) produced a concentration-dependent nonspecific spasmolytic effect like papaverine. The compound was almost equivalent against the various spasmogens and was about 10 times more potent than papaverine when their IC_{50} values were compared. Both clausmarin A and papaverine produced a concentration-dependent inhibition of the contraction of ileum induced electrically (Patnaik and Dhawan 1982).

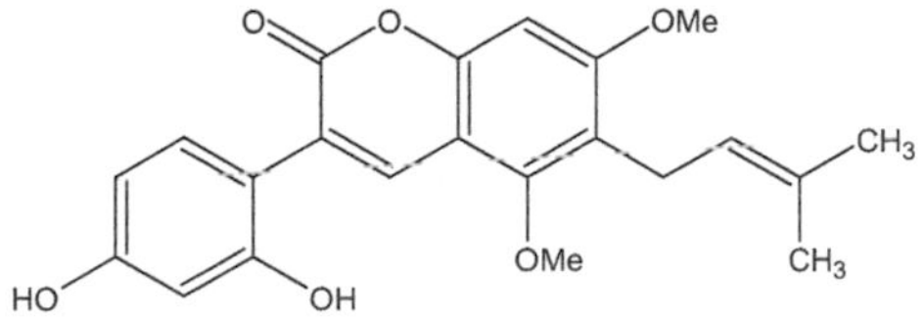

Fig. 2.17. Structure of Glycycoumarin.

Shakuyaku-kanzo-to, a Chinese herbal extract of a mixture plants Paeonia radix and Glycyrrhizae radix is prescribed for abdominal pain in Japan. Shakuyaku-kanzo-to and Kanzo-to (an aqueous extract of Glycyrrhizae radix only), have inhibited the contraction induced by ACh in guinea pig ileum (Maeda et al. 1983). In another report, glycycoumarin was isolated from the aqueous extract of Glycyrrhizae Radix, inhibited the contraction induced by various types of stimulants with almost the same potency as papaverine, a representative antispasmodic for smooth muscle, indicating that the two agents share a similar action (Sato et al. 2006). In another study glycycoumarin is a potent antispasmodic with an IC_{50} value of 2.93 μM for carbamylcholine (CCh)-induced contraction of mouse jejunum, and clarified that its mechanism of action involves inhibition of PDE. Moreover, glycycoumarin increased the content of cAMP, but not cGMP, in mouse jejunum to an extent almost similar to that obtained with papaverine in guinea pig ileum (Kaneda et al. 1998). In conclusion, glycycoumarin, acts as a potent antispasmodic by means of intracellular accumulation of cAMP through the inhibition of PDEs, especially isozyme 3.

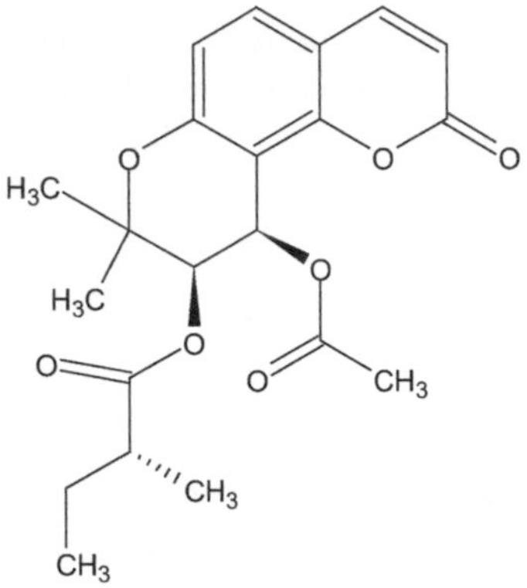

Fig. 2.18. Structure of Visnagan.

Visnagan (Fig. 2.18) was found in *Ammi visnaga* (Apiaceae). Showed antispasmodic activity (Gomes 1953).

Flavonoids

In another study twenty-six natural and synthetic flavonoids, divided into the five classes of flavones, flavonols, flavanones, isoflavones, and chalcones, were tested as to their relaxant effects in guinea pig trachealis. The IC_{50} values of these five classes indicated that flavones were more potent than flavonols. It appears that introduction of a hydroxy group at position C-3 decreases their potency, based on the fact that the IC_{50} values of apigenin and luteolin against histamine, carbachol, or KCl were significantly less than those of kaempferol and quercetin, respectively. Flavones were also more potent than flavanones. It has been suggested that the presence of a double bond between C-2 and C-3 is important for lipid peroxidation-inhibiting (Cos et al. 2001), and for nitric oxide (NO) production-inhibiting activity in LPS-activated RAW 264.7 cells (Kim et al. 1999).

Introduction of a hydroxy group at position C-6 of flavones increases their relaxant activities against these three contractile agents, because the IC_{50} values of 6-hydroxyflavones were significantly less than those of flavones. Adding a hydroxy group at position C-7 of flavones has a similar effect, although the IC_{50} values of 7-hydroxyflavones against carbachol or KCl were similar to those of flavones. The IC_{50} value of 7-hydroxyflavone against histamine was significantly less than that of flavone. It appears that the optimal number of hydroxy groups introduced to the A-ring of flavones is one. If more hydroxy groups are introduced to positions C-5, C-6, and/or C-7 of flavones, then IC_{50} values increase. For example, 5,7-di-hydroxyflavone (chrysin), similar to flavone, had moderate potency in relaxant effects, but 5,6,7,-trihydroxyflavone (baicalein) had almost no activity ($IC_{50} > 300\ \mu M$). This rule was also applicable to flavonols, based on the fact that the IC_{50} values of fisetin against these three contractile agents were significantly less than those of quercetin.

It has been reported that the introduction of a hydroxy group at position C-5 to a 7-hydroxyphenolic flavonoid increased the lipid peroxidation-inhibiting activity of flavones and flavonols (Cos et al. 2001). The IC_{50} values of 7-hydroxyflavone against these three contractile agents did not differ from those of 5,7-dihy-droxyflavone (chrysin). Therefore, mechanisms of the tracheal relaxant effect may differ from those of lipid peroxidation-inhibiting action. It seems that flavones or flavonols with pyrogallol moiety either in the A- or B-ring, respectively, have no activity. This speculation was supported by the results that either baicalein or myricetin had no relaxant effects. It appears that flavonols with ortho-hydroxy groups in the B-ring are more potent than those with meta-hydroxy groups, based on the fact that the IC_{50} values of quercetin against histamine, carbachol, or KCl were significantly less than those of morin, which had almost no relaxant effects.

The activity of 6-hydroxyflavone disappeared if the C-6 hydroxy group of the A-ring was methoxylated, because 6-methoxyflavone had no relaxant effects against these three contractile agents. This finding is consistent with the lipid peroxidation-inhibiting activity of baicalein disappearing if the C-6 and C-7 hydroxy groups are methoxylated (Ko et al. 2003). If the C-4' hydroxy group of the B-ring was methoxylated, the relaxant effect of these flavones was also attenuated or disappeared. This conclusion was supported by the results of IC_{50} values of luteolin or apigenin against these three contractile agents, which were significantly less than those of diosmetin or acacetin, respectively, with the exception that the IC_{50} value of apigenin against carbachol did not differ from that of acacetin (Ko et al. 1999).

The hydroxy group on either the A- or B-ring of flavones and flavonols being methylated will result in lower potency in tracheal relaxant effects.

However, all hydroxy groups on both the A- and B-rings of flavones and flavonols being methoxylated will result in higher potency. This conclusion was supported by the lower IC_{50} values of tangeretin (flavone 5,6,7,2',4'-pentamethyl ether) and quercetin 3,5,7,3',4'-pentamethyl ether. Therefore, the influence of methoxylation in flavones may be similar to that in flavonols. However the C-3 hydroxy group on the C-ring of flavonols, but not flavones which lack this hydroxy group, is methoxylated, the relaxant effects may increase (Ko et al. 1999) and lipid peroxidation-inhibiting potencies of quercetin 3-methyl ether were significantly greater than those of quercetin.

Glycosylation of the hydroxyl group at position C-7 of flavones or flavanones attenuates the relaxant effects. For example, luteolin and diosmetin had lower IC_{50} values against carbachol or KCl than those of luteolin 7-glucoside and diosmin, respectively. Naringenin and hesperetin also had lower IC_{50} values against KCl than naringin and hesperidin, respectively. It is possible that flavonoid gycosides do not penetrate the cell membrane due to their hydrophilicity or bulky glycosyl residues as suggested previously (Namgoong et al. 1994).

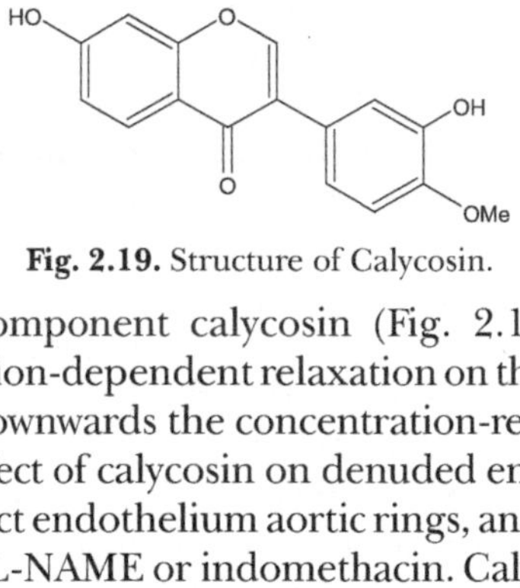

Fig. 2.19. Structure of Calycosin.

A major active component calycosin (Fig. 2.19) of Astragali Radix produced a concentration-dependent relaxation on the tissue pre-contracted using KCl displaced downwards the concentration-response curves of aortic rings. The relaxant effect of calycosin on denuded endothelium aortic rings was the same as on intact endothelium aortic rings, and its vasorelaxant effect was not influenced by L-NAME or indomethacin. Calycosin decreased aortic ring contractions induced by the two agonists (KCl and phenylephrine), this is a vasorelaxant. Its action is endothelium-independent and is unrelated to intracellular Ca^{2+} release. It is a noncompetitive Ca^{2+}channel blocker. The effect of calycosin on Ca^{2+} channel blockade may be different from that of dihydropyridines (Wu et al. 2006).

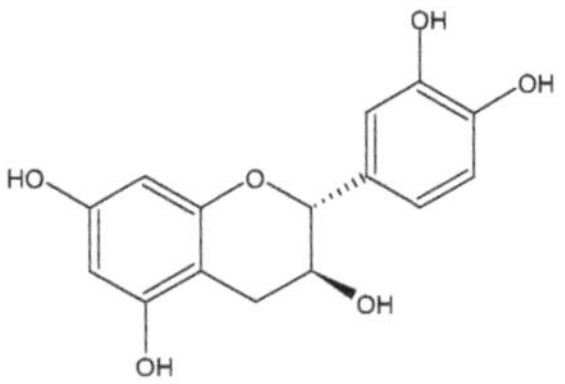

Fig. 2.20. Structure of Catechin.

Structure of Catechin is a well-known flavonoid found in many food plants and often utilized by naturopaths for the symptomatic treatment of several gastrointestinal, respiratory and vascular diseases. Catechin dose-dependently relaxes both spontaneous and high K^+ (80 mM)-induced contraction in rabbit jejunum, showing specificity for the latter by causing a right-ward shift in the Ca^{2+} dose-response curve. Similar results were observed with verapamil, a standard Ca^{2+} channel blocker. Catechin also inhibited high K^+-induced contraction in intact smooth muscle preparations from rat stomach fundus, guinea-pig ileum and guinea-pig trachea. In rat aorta, catechin inhibited phenylephrine and K^+-induced contractions in a similar fashion. In phenylephrine-contracted, endothelium-intact aorta, this vasodilator effect was partially blocked by nomega-nitro-L-arginine methyl ester and atropine, indicating activity at cholinergic receptors and possibly a channel blocker effect at higher doses of catechin. In guinea-pig atria catechin was found inactive. These data suggest that catechin may possess Ca^{2+} antagonist activity in addition to an endothelium-dependent relaxant component in blood vessels (Ghayur et al. 2007).

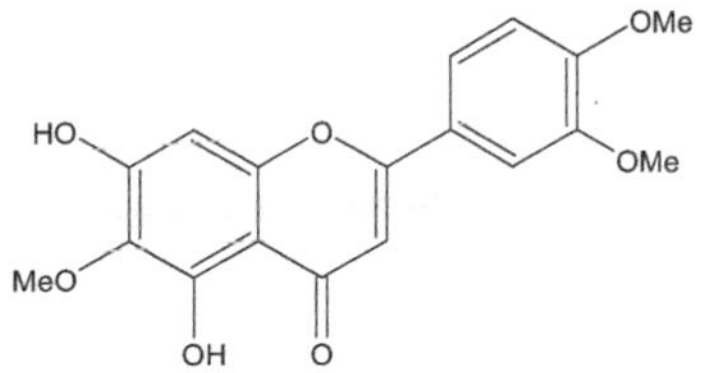

Fig. 2.21. Structure of Eupatilin.

Found in *Artemisia monosperma* (Asteraceae). Eupatilin (Fig. 2.21) inhibited in a reversible manner the phasic contractions and the tone of rat isolated ileum, uterus, and urinary bladder. It relaxed the tonic contractions of the phenylephrine-precontracted pulmonary artery and acetylcholine-precontracted trachea. It also shifted the concentration-effect curves of acetylcholine and calcium chloride on rat isolated ileum, oxytocin concentration-effect curve on uterine smooth muscle and phenylephrine concentration-effect curve on pulmonary artery smooth muscle. These observations indicate that eupatilin shows a nonspecific antispasmodic effect on rat isolated smooth muscle. They also suggest that the inhibitory effect of eupatilin may be mediated by changes in Ca^{2+} metabolism in these preparations (Abu-Niaaj et al. 1996).

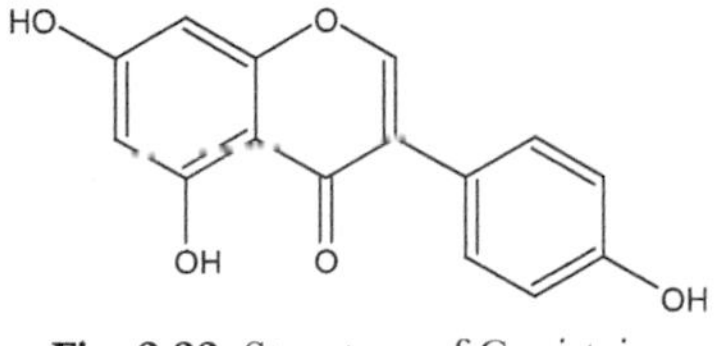

Fig. 2.22. Structure of Genistein.

The major constituent of *Genista tridentata* (Fabaceae), genistein (Fig. 2.22), showed inhibition on contractions induced by agonists and electrical field stimulation of smooth muscle. Genistein inhibited twitches evoked by electrical-stimulation of strips of guinea-pig ileum, inhibited contractions of the guinea-pig ileum by several agonists in a non-selective, antispasmodic action and had no effect on inhibition of 3H-ACh release from ileal myenteric plexus. It also produces an increase in cAMP levels of guinea-pig ileum which resulted in a smooth muscle relaxation which leads us to think that there must be a blockade of its phosphodiesterase (Herrera et al. 1992).

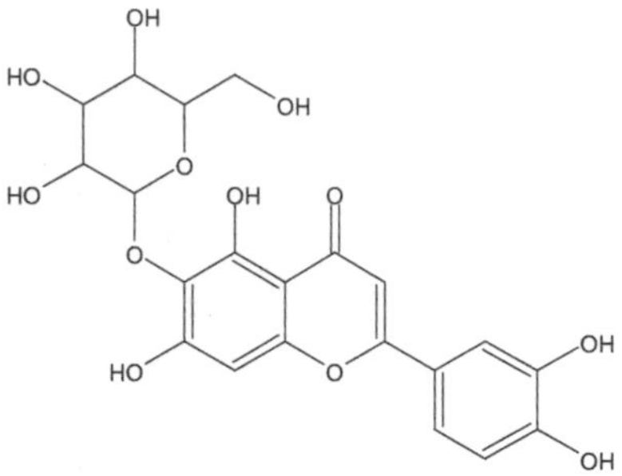

Fig. 2.23. Structure of Isoorientin.

The phytochemical investigation of *Arum palaestinum* (Araceae) resulted in the isolation of two flavone C-glucosides, namely isoorientin (luteolin 6-C-glucoside) and vitexin (apigenin 8-C glucoside). Isoorientin (Fig. 2.23) caused concentration-dependent inhibition of the amplitude and the frequency of the phasic contractions of the rat and guinea-pig uterus but did not affect the isolated aorta, ileum or trachea (Afifi et al. 1999).

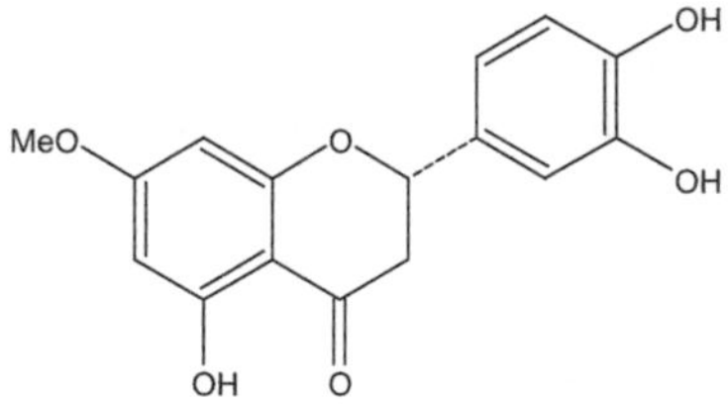

Fig. 2.24. Structure of 7-O-Methyleriodictyol.

7-O-Methyleriodictyol (Fig. 2.24), isolated from *Artemisa monosperma* (Asteraceae) caused concentration-dependent inhibition of the amplitude of the phasic contractions and reduced the tone of the ileum, the uterus, and the urinary bladder. It also caused relaxation of the phenylephrine-precontracted pulmonary artery and the acetylcholine-precontracted trachea. The flavanone also shifted to the right the ACh concentration-effect curve on the ileum and the oxytocin concentration-effect curve on the uterus. The maximum contractions induced by acetylcholine or oxytocin were also inhibited by 7-O-methyleriodictyol (Abu-Niaaj et al. 1993).

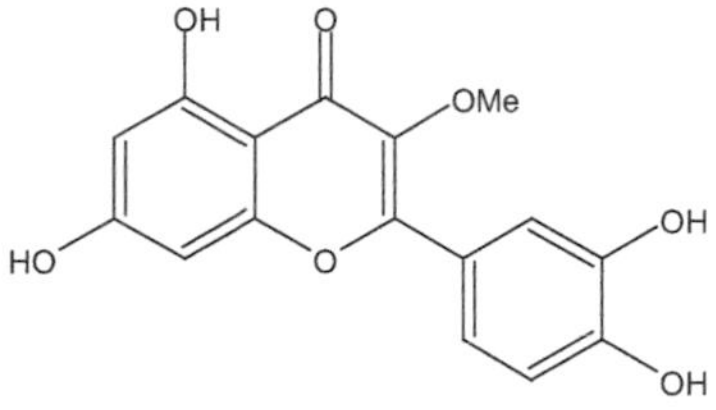

Fig. 2.25. Structure of 3-O-Methoxyquercetin.

The mechanisms of action in 3-O-methoxyquercetin (Fig. 2.25), isolated from *Rhamnus nakaharai* (Rhamnaceae) which is used as a folk medicine for treating constipation, inflammation, tumours and asthma in Taiwan. The relaxant effect of 3-O-methoxyquercetin was unaffected by the removal of epithelium or by the presence of propranolol, 2',5'-dideoxyadenosine, methylene blue, glibenclamide, N(ω)-nitro-L-arginine, or α-chymotrypsin. However, 3-O-methoxyquercetin produced parallel and leftward shifts of the concentration-response curve of forskoline or nitroprusside. 3-O-Methoxyquercetin significantly inhibited cAMP- and cGMP-PDE activities of the trachealis. The results reveal that the mechanisms of relaxant action of 3-O-methoxyquercetin may be due to its inhibitory effects on both PDE activities and its subsequent reducing effect on Ca^{2+} of the trachealis (Ko et al. 2002).

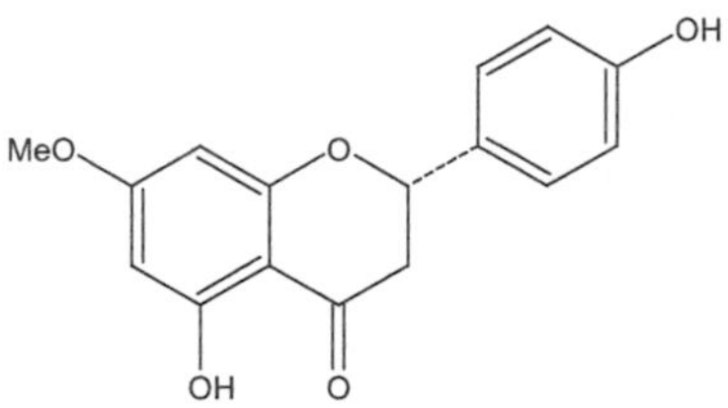

Fig. 2.26. Structure of Sakuranetin.

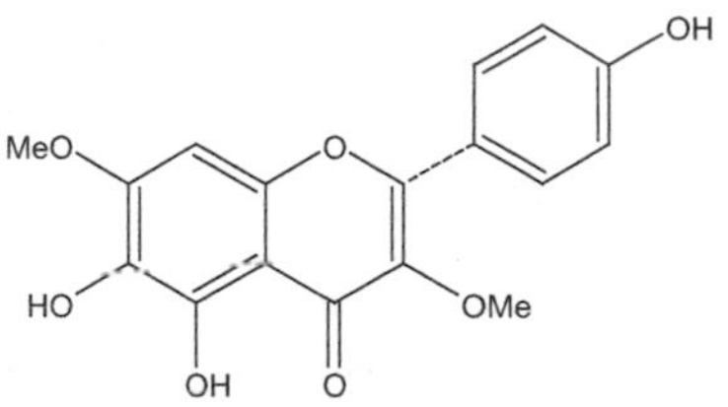

Fig. 2.27. Structure of 6-Hydroxykaempferyl-3,7-dimethyl ether.

Bioassay-directed fractionation of the chloroform-methanol extract of *Dodonaea viscosa* (Sapindaceae) resulted in the isolation of two active spasmolytic principles: Sakuranetin (Fig. 2.26) and 6-hydroxykaempferyl-3,7-dimethyl ether (Fig. 2.27). The isolated compounds elicited a concentration-dependent inhibition of the spontaneous and electrically-induced contractions of guinea-pig ileum. Sakuranetin inhibited the ileum contractions evoked by ACh, histamine, and $BaCl_2$. In addition, both substances were capable of relaxing contractions of rat uterus induced by Ca^{2+} in K^+-depolarizing solution, displacing to the right the concentration-response curves to Ca^{2+}. These results suggest that sakuranetin produces an interference with calcium metabolism in smooth muscle cells (Rojas et al. 1996).

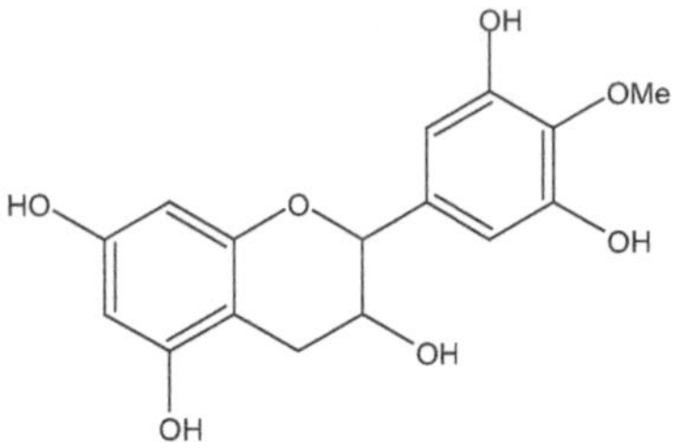

Fig. 2.28. Structure of 3,3´,5,5´,7-Pentahydroxy-4´-methoxy-2,3-*cis*-flavane.

3,3´,5,5´,7-Pentahydroxy-4´-methoxy-2,3-cis-flavane was isolated from the root bark of *Elaeodendron balae* (Celastraceae). This flavane (Fig. 2.28) showed inhibition of the electrically induced twitches of the isolated guinea pig ileum (Go and Leung 1984).

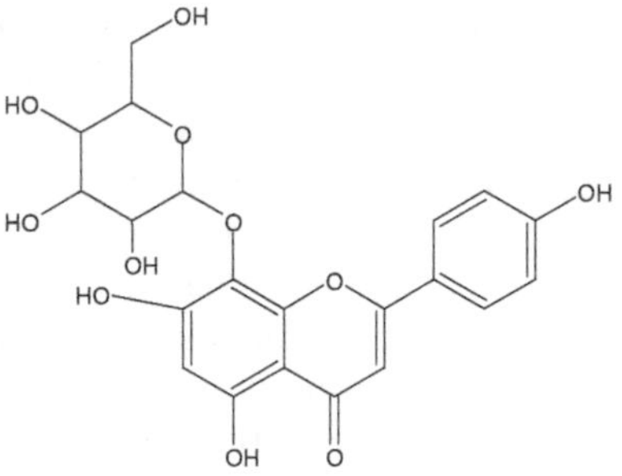

Fig. 2.29. Structure of Vitexin.

The plant *Aloysia citriodora* (Verbenaceae) is widely used for gastrointestinal disorders and as eupeptic in South America. Vitexin (Fig. 2.29) is partially responsible for the effect, since it non-competitively inhibits ACh but not the Ca^{2+} influx. Isovitexin was devoid of activity on rat duodenums (Ragone et al. 2007).

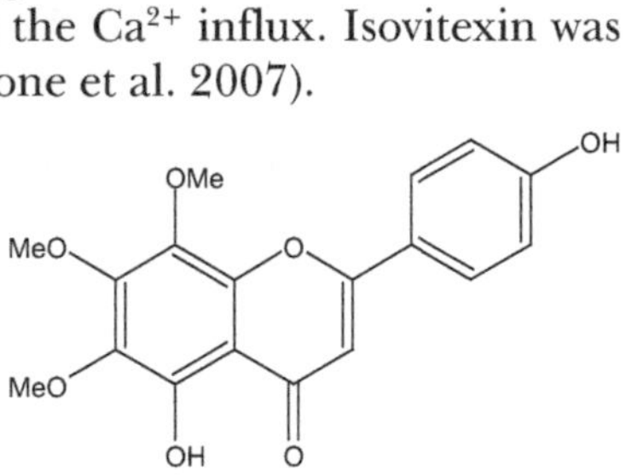

Fig. 2.30. Structure of Xanthomicrol.

KCl-induced uterine smooth muscle contraction results mainly from calcium influx via voltage-sensitive calcium channels. Recently, some other factors have been described as contributing to this contraction such as increase of intracellular concentration of inositol triphosphate (IP_3) or arachidonic acid production. Xanthomicrol (Fig. 2.30) has been suggested to act as calcium antagonists to display spasmolytic action. Other mechanisms such as inducing NO and nitrate increase have also been suggested to explain flavonoid spasmolytic action in different tissues. Xanthomicrol is responsible for the spasmolytic effect of *Brickellia paniculata* (Asteraceae). Xanthomicrol inhibits KCl-induced uterine contractions acting on voltage-sensitive calcium channels, because this contraction is dependent on extracellular calcium. Xanthomicrol effects on uterine smooth muscle were similar to those displayed by verapamil and papaverine, respectively, i.e. xanthomicrol and verapamil inhibited uterine smooth muscle contraction apparently by blocking voltage-sensitive calcium channels (Ponce-Monter et al. 2006).

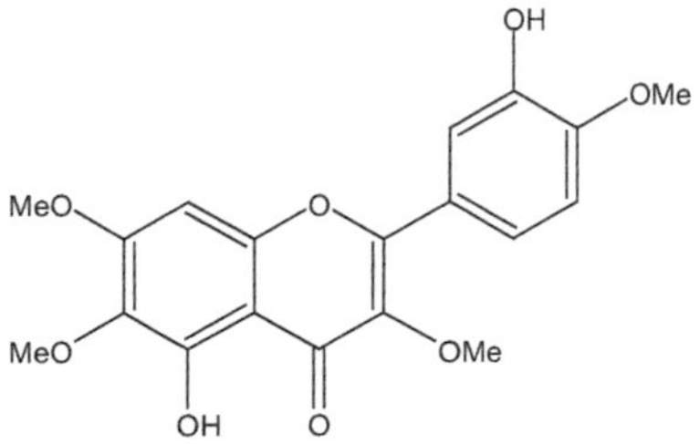

Fig. 2.31. Structure of 3, 6,7,4´,Tetramethoxy,5,3´-dihydroxyflavone.

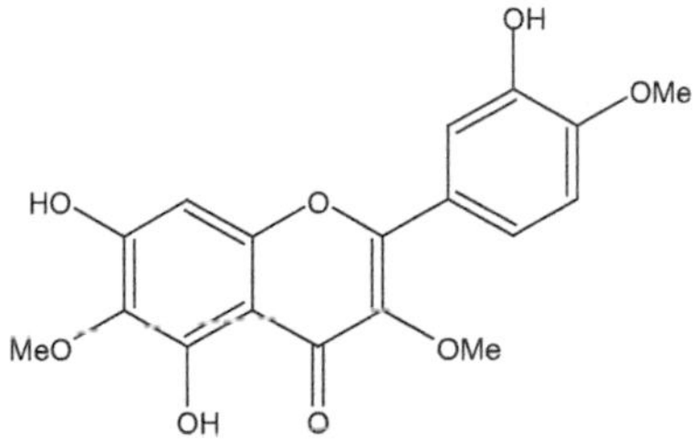

Fig. 2.32. Structure of 3,6,4´-Trimethoxy,5,7,3´-trihidroxyflavone.

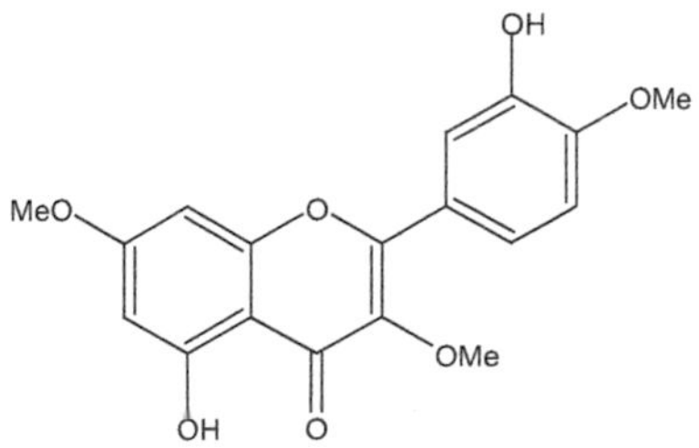

Fig. 2.33. Structure of 3,7,4´-Trimethoxy,5,3´-dihidroxyflavone.

3,6,7,4´,Tetramethoxy,5,3´-dihydroxyflavone   (Fig.   2.31),   3,6,4´-
Trimethoxy,5,7,3´-trihidroxyflavone (Fig. 2.32) and 3,7,4´-Trimethoxy,5,3´-
dihidroxyflavone (Fig. 2.33) with spasmolytic activity have been isolated
from a methanol extract of *Artemisia abrotanum* (Asteraceae), as the principles
primarily responsible for the smooth muscle relaxing activity of this plant. The
flavonols show a dose-dependent relaxing effect on the carbacholine-induced
contraction of guinea-pig trachea, the EC_{50} values for three compounds are
20–30 μmol/l (Bergendorff and Sternerl 1995).

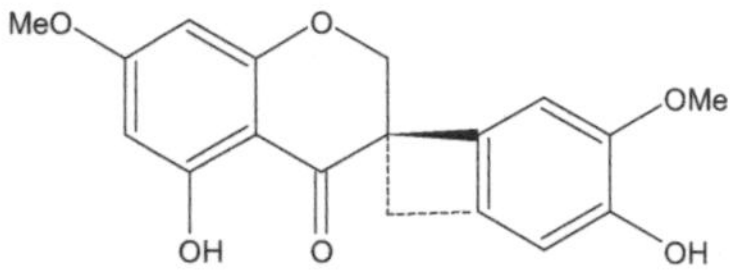

Fig. 2.34. Structure of (R)-3',5-dihydroxy-4',7-dimethoxyspiro [2H-1-benzopyran-3(4H),7'-
bicyclo[4.2.0]-octa[1,3,5]-trien]-4-one.

The vasorelaxing effect of isolate homoisoflavanone (Fig. 2.34) from
Eucomis schiffii (Hyacinthoideae) has been assessed using rat aortic ring
preparations. Compound inhibited the tonic contraction induced by
both 60 mM K^+ and phenylephrine, displayed antispasmodic effects not
reversed by tetraethylammonium. Under precontraction induced with
phenylephrine, the concentration-relaxation curves of either isoprenaline
or sodium nitroprusside shifted to the left. Homoisoflavanone partially
antagonized the contraction induced by PMA. This compound proved
to be an effective vasorelaxing agent, partly acting via the activation of
soluble guanylyl cyclase (Fusi et al. 2008).

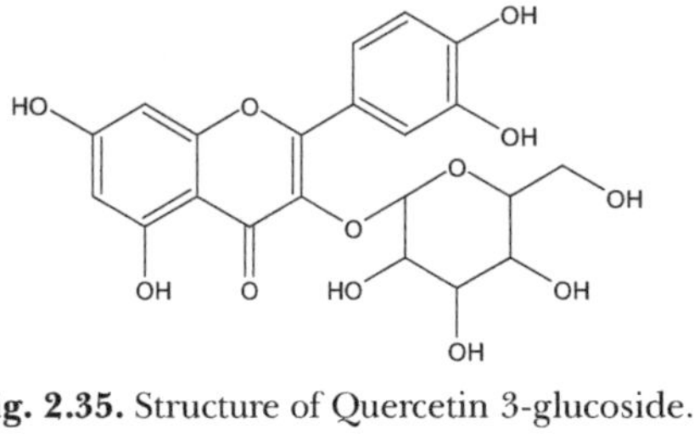

Fig. 2.35. Structure of Quercetin 3-glucoside.

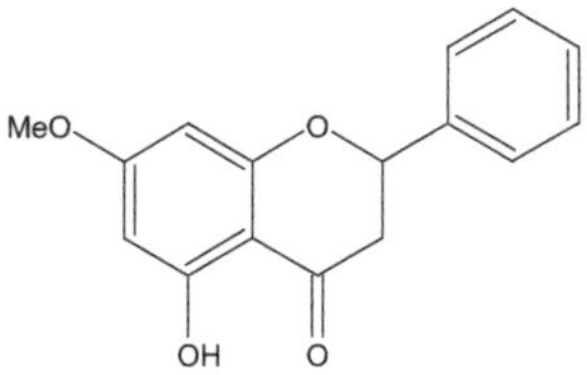

Fig. 2.36. Structure of Pinostrobin.

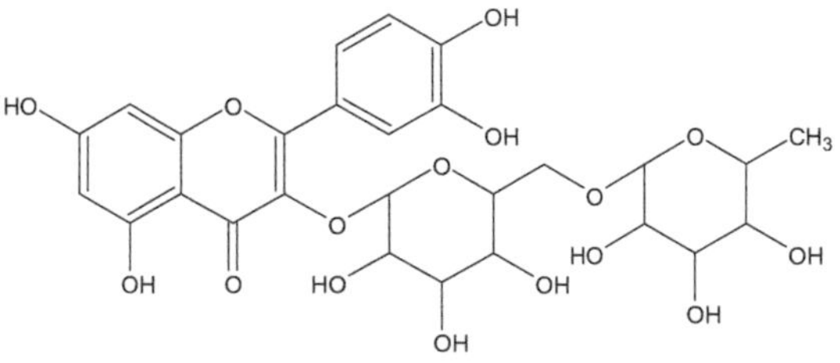

Fig. 2.37. Structure of Rutin.

Activity-guided fractionation of chloroform-methanol extract of *Conyza filaginoides* (Asteraceae) led to the isolation of three flavonoids (Fig. 2.35 to 2.37). Among the tested flavonoids, quercetin 3-glucoside and rutin possessed a significantly high inhibitory activity on the spontaneous contractions of rat ileum, while pinostrobin exerts a rather weak activity (Mata et al. 1997).

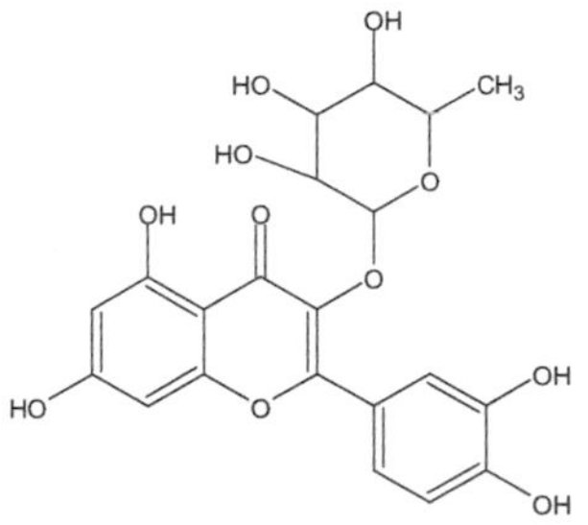

Fig. 2.38. Structure of Quercitrin.

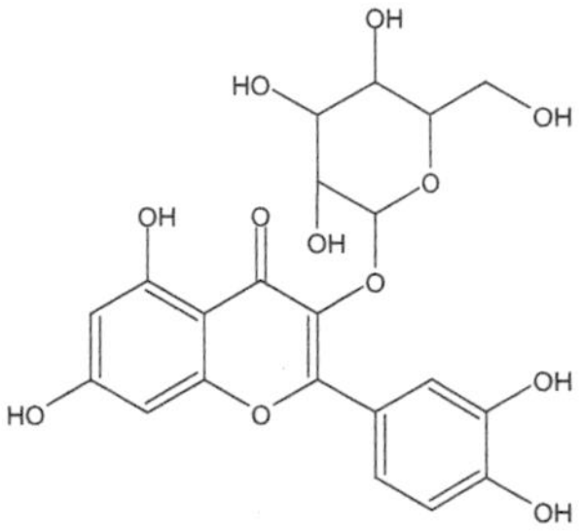

Fig. 2.39. Structure of Isoquercetin.

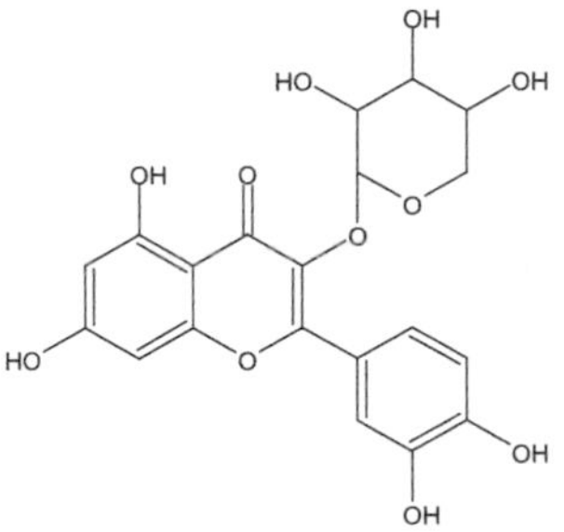

Fig. 2.40. Structure of Guaijaverin.

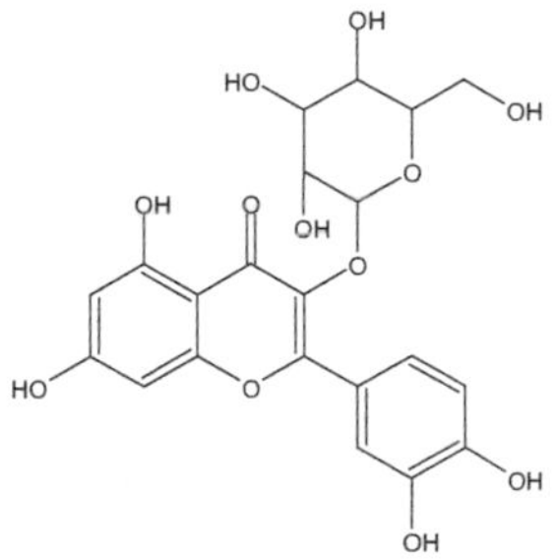

Fig. 2.41. Structure of Hyperin.

The traditional herbal remedy from *Psidium guajava* (Myrtaceae) leaves has been medically proposed in Mexico as an effective treatment for acute diarrhoea. The flavonols isolated (Fig. 2.38 to 2.41) inhibited peristalsis of guinea pig ileum *in vitro* (Lozoya et al. 1994).

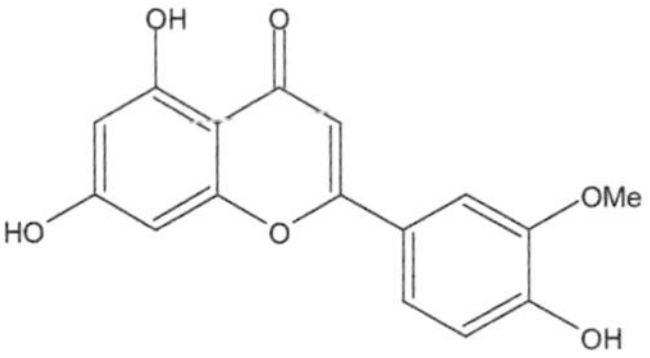

Fig. 2.42. Structure of Chrysoeriol.

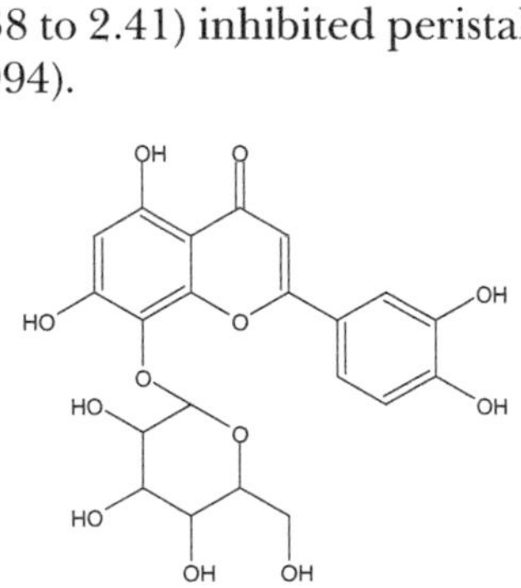

Fig. 2.43. Structure of Orientin.

Rooibos tea (*Aspalathus linearis*, Fabaceae) has been widely used for abdominal spasm and diarrhoea. Its aqueous extract produced relaxation of spontaneous and low K^+ (25 mM)-induced contractions of rabbit jejunum, with weak effect on high K^+ (80 mM)-induced contractions. In the presence of glibenclamide, relaxation of low K^+-induced contractions was prevented. Its constituents, Chrysoeriol (Fig. 2.42), orientin (Fig. 2.43) and vitexin showed a similar pattern of spasmolytic effects to the extract, while rutin was more like verapamil. So Rooibos tea possesses a combination of dominant K(ATP) channel activation and weak Ca^{+2} antagonist mechanisms (Gilani et al. 2006).

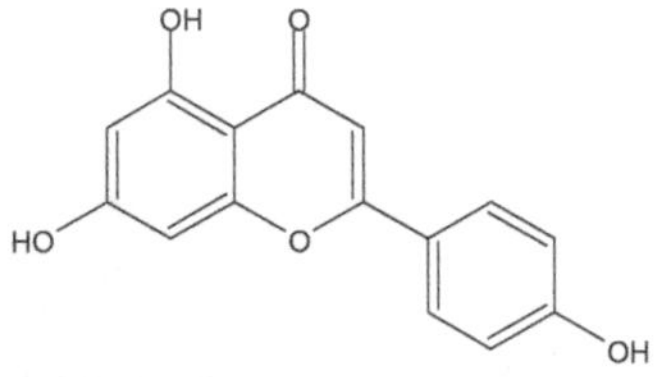

Fig. 2.44. Structure of Apigenin.

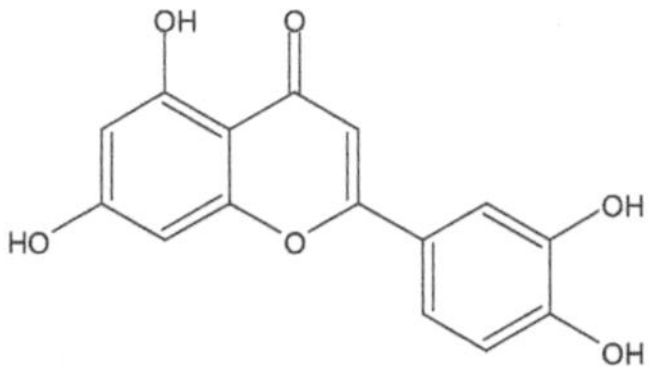

Fig. 2.45. Structure of Luteolin.

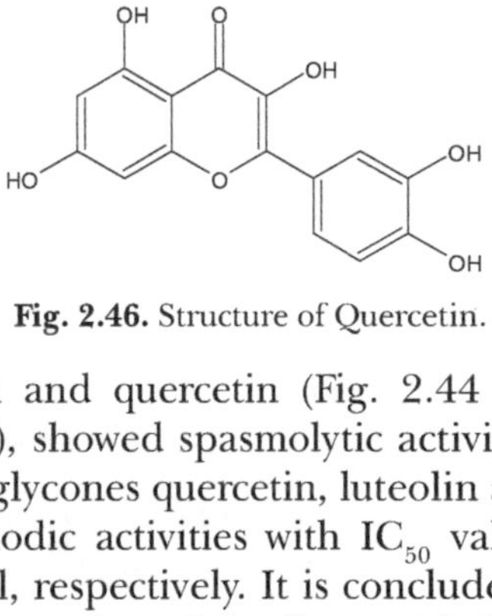

Fig. 2.46. Structure of Quercetin.

Apigenin, luteolin and quercetin (Fig. 2.44 to 2.46) from *Achillea millefolium* (Asteraceae), showed spasmolytic activity on isolated terminal guinea-pig ilea. The aglycones quercetin, luteolin and apigenin exhibited the highest antispasmodic activities with IC_{50} values of 7.8 μmol/l, 9.8 μmol/l and 12.5 μmol/l, respectively. It is concluded that in tea prepared from yarrow the concentration of the flavonoids is high enough to exert a spasmolytic effect in the gut, which is mainly caused by a blockage of the calcium inward current, but additionally also by mediator-antagonistic effects (Lemmens-Gruber et al. 2006).

Rotenoids

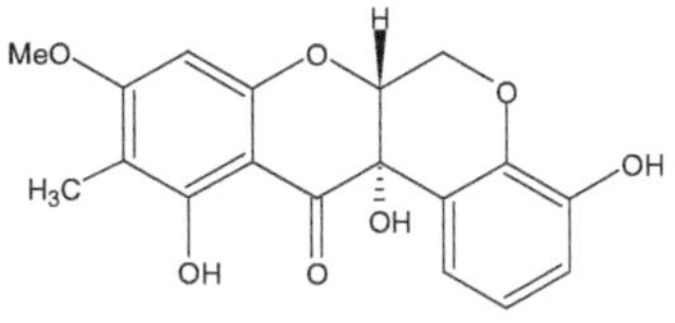

Fig. 2.47. Structure of Boeravinone C.

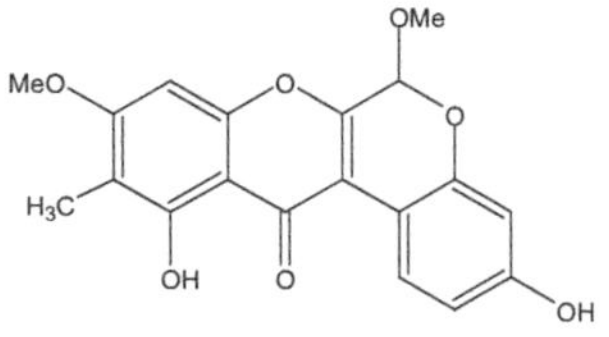

Fig. 2.48. Structure of Boeravinone D.

Fig. 2.49. Structure of Boeravinone E.

Fig. 2.50. Structure of Boeravinone F.

Fig. 2.51. Structure of Boeravinone G.

Fig. 2.52. Structure of Boeravinone H.

Fig. 2.53. Structure of Coccineone B.

Fig. 2.54. Structure of Coccineone E.

Fig. 2.55. Structure of 10-
Demethylboeravinone C.

Fig. 2.56. Structure of 6-*O*-
Demethylberavinone H.

Fig. 2.57. Structure of 9-*O*-Methyl-10-hydroxycoccineone B.

Boerhaavia diffusa (Nyctaginaceae) is an Ayurvedic remedy used traditionally for the treatment of a number of diseases, including those affecting the gastrointestinal tract. Methanol extract obtained from roots of *B. diffusa* exhibited a significant spasmolytic activity in the guinea pig ileum, probably through a direct effect on the smooth muscle. The nonprenylated rotenoids (Figs. 2.47 to 2.57) from *B.diffusa* showed inhibition on intestinal motility. The negative effect of *O*-methylation at C-6 on this activity was observed, only boeravinone E and 6-*O*-demethylberavinone H produced a complete inhibition of ACh-induced contractions. In particular, the activities of compounds boeravinone C-10, 10-demethylboeravinone C, coccineone E, 2'-*O*-methylabronisoflavone and boeravinone F clearly highlight the crucial role of ring B. Indeed, compounds boeravinone C, 10-demethylboeravinone C and coccineone E, showing hydration of the double bond, $\Delta^{6a(12a)}$, regardless of the presence of either a H, a CH_3, or an OCH_3 at C-10, showed reduced or no activity. Finally, the low potencies of coccineone B and 9-O-methyl-10-hydroxycoccineone B indicate that the presence of a hydroxylated pyran ring B is not sufficient for the exhibition of spasmolyitic activity. Most likely, a monohydroxylated ring A and/or a trisubstituted ring D are also needed (Borrelli et al. 2005, 2006).

Terpenoids

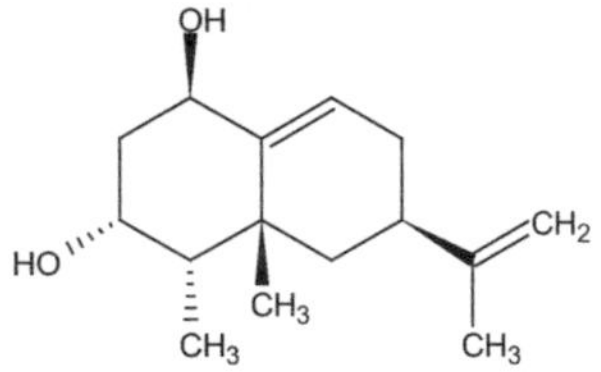

Fig. 2.58. Structure of Capsidiol.

The sesquiterpene capsidiol, isolated from elicited, cultured cells of *Nicotiana silvestris* (Solanaceae), inhibits contraction induced by the agonists, histamine, ACh, bradykinin, and $BaCl_2$ in the isolated guinea-pig ileum. In addition, capsidiol inhibits contraction induced by histamine and carbachol in the isolated guinea-pig trachea in a nonspecific and noncompetitive but concentration-dependent manner. Capsidiol also inhibits prostaglandin synthesis (cyclooxygenase) *in vitro* in a concentration dependent manner (Nasiri et al. 1993).

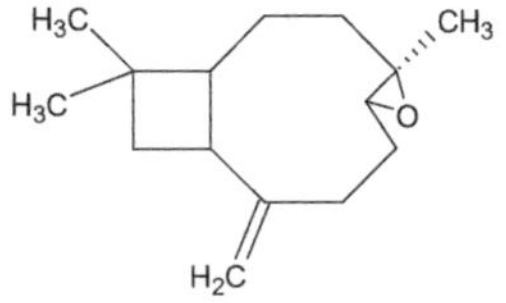

Fig. 2.59. Structure of β-caryophyllene 4,5-α-oxide.

The sesquiterpenoid β-caryophyllene 4,5-α-oxide (Fig. 2.59), isolated from *Conyza filaginoides* (Asteraceae) exhibited a moderate relaxant activity on the spontaneous contractions of rat ileum (Mata et al. 1997).

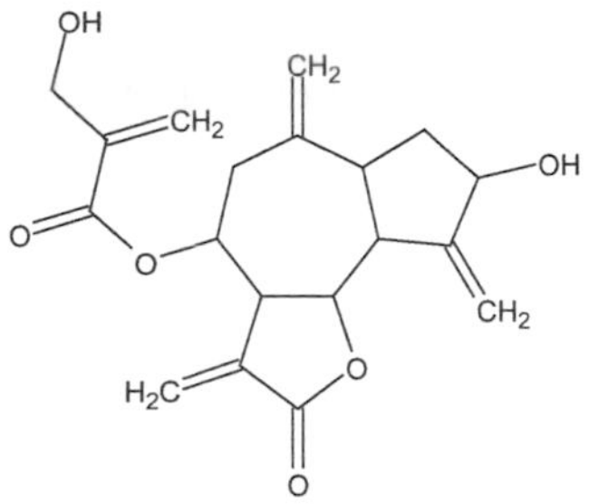

Fig. 2.60. Structure of Cynaropicrin.

Cynara scolymus (Asteraceae), popularly known as 'alcachofa' or 'artichoke', is widely cultivated in Mediterranean and American countries, and its sprout is often eaten as a vegetable. Its leaves are frequently used in folk medicine in many countries, to treat several ailments, including hepatitis, hyperlipidemia, and obesity, dyspeptic disorder, among others. Clinical and pre-clinical trials have confirmed the therapeutic potential of this plant, particularly in the treatment of hepato-biliary dysfunction and digestive complaints, and also its effectiveness in patients with irritable bowel syndrome. Cynropicrin (Fig. 2.60), a sesquiterpene lactone isolated from this species, exhibited potent activity, against guinea-pig ileum contracted by ACh having similar potency of that papaverine, a well-known antispasmodic agent (Emendorfer et al. 2005).

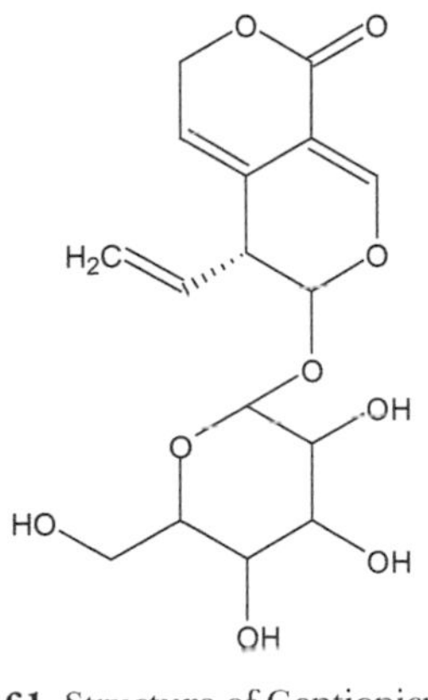

Fig. 2.61. Structure of Gentiopicrin.

Bioassay directed fractionation of chloroform-methanol extract of *Gentiana spathacea* (Gentianaceae) led to the isolation of gentiopicrin (Fig. 2.61), which inhibited, in a concentration-dependent manner, the spontaneous contractions of isolated guinea pig ileum. Contractions induced by histamine, ACh, $BaCl_2$ and KCl on the ileum were also significantly blocked by this monoterpene glucoside, which suggests that this compound might be interfering with calcium influx into the smooth muscle cells (Rojas et al. 2000).

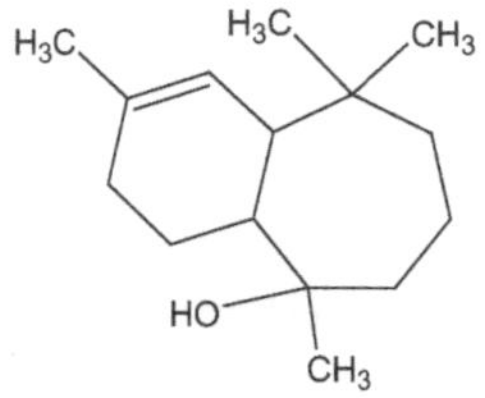

Fig. 2.62. Structure of Himachalol.

Himachalol (Fig. 2.62), has been identified as the major antispasmodic constituent in the wood of *Cedrus deodara* (Pinaceae). It was a more potent antagonist of barium chloride-induced spasm of guinea pig ileum than papaverine but less effective in reverting a similar spasm of rabbit jejunum and had no relaxing effect alone. In the conscious immobilized cat, intragastric administration of himachalol or papaverine produced equal inhibition of carbachol-induced spasm of the intestine. Himachalol was devoid of spasmolytic effect on the bronchial musculature of the guinea pig but was 3.3 times more potent than papaverine in antagonizing epinephrine-induced contraction of the guinea pig seminal vesicle. Intravenous injection in the cat produced a dose-dependent fall in blood pressure and an increased femoral blood flow (Kar et al. 1975).

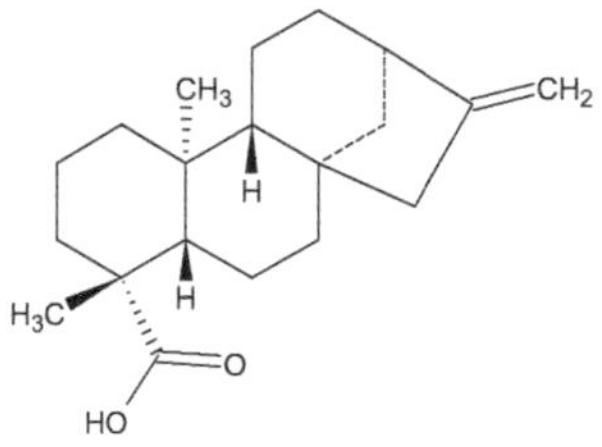

Fig. 2.63. Structure of Kaurenoic acid.

Kaurenoic acid (Fig. 2.63), a diterpene isolated from the oleo-resin of the popular medicinal plant *Copaifera langsdorffii* (Caesalpiniaceae) showed inhibition of the contractile response evoked by agonists $BaCl_2$, ACh and oxytocin on isolated rat uterus, was inhibited in a concentration-related manner. These data indicate that the diterpene, kaurenoic acid exerts a uterine relaxant effect acting principally through calcium blockade and in part, by the opening of ATP-sensitive potassium channels (Cunha et al. 2003).

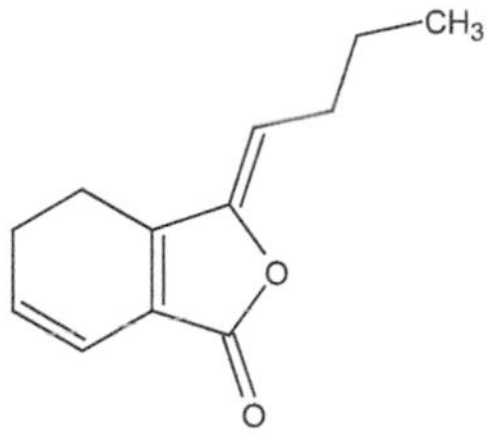

Fig. 2.64. Structure of Ligustilide.

Ligustilide was isolated from neutral oil of *Ligusticum wallichii* Franch. (Apiaceae). Inhibiting rat uterine contractions induced by $PGF_{2\alpha}$, oxytocin and Ach (Ko 1980).

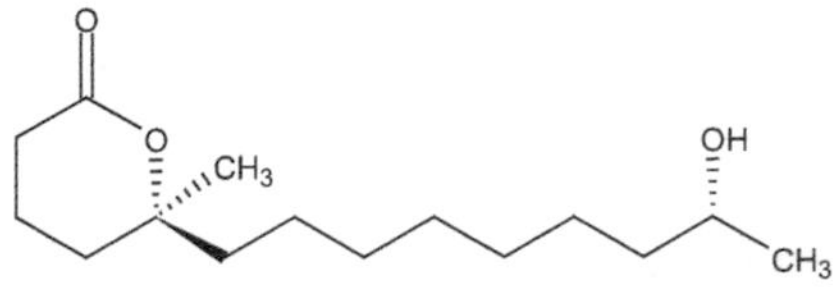

Fig. 2.65. Structure of Oedogonolide.

Found in *Oedogonium capillare* (Oedogoniaceae) was found to antagonize in a concentration-dependent way, the contractions of the rat ileum induced by acetylcholine, histamine and $BaCl_2$. Oedogonolide (Fig. 2.65) possessing both anticholinergic and antistaminic properties (Perez-Gutierrez et al. 2006a).

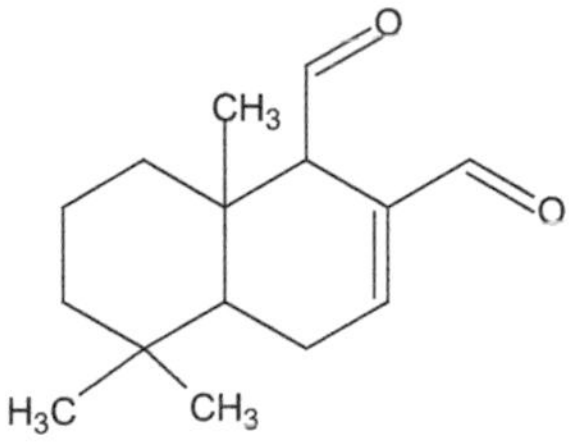

Fig. 2.66. Structure of Polygodial.

The sesquiterpene polygodial, the major constituent isolated from the bark of *Drymis winteri* (Winteraceae) in the guinea pig ileum and trachea *in vitro* caused graded inhibition, associated with rightward displacement of the ACh, histamine and KCl contraction response curves. When assessed in the guinea-pig trachea, polygodial caused significant inhibition of bradykinin while not concentration-dependent, caused a small but significant shift to the right of substance P and also the selective agonist of tachykinin NK2 receptor induced contractions in the guinea pig trachea (El Sayah et al. 1998).

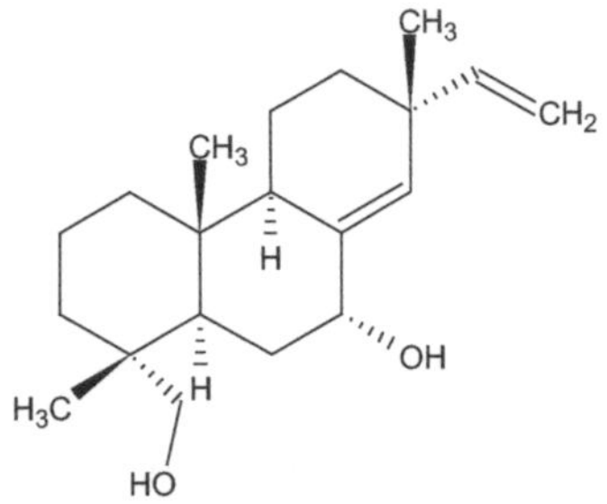

Fig. 2.67. Structure of 8(14),15-Sandaracopimaradiene-7α,18-diol.

Tetradenia riparia (Lamiaceae) is one of the most popular medicinal plants in Rwanda. Previously, several new substances were isolated from the leaves, including 8(14),15-Sandaracopimaradiene-7α,18-diol, Fig. 2.67. This diterpenediol exhibits a papaverine-like antispasmodic activity on the contractions of the guinea pig ileum provoked by methacholine, histamine, and BaCl$_2$ and on the noradrenaline-induced contractions of the rabbit aorta (Puyvelde et al. 1987).

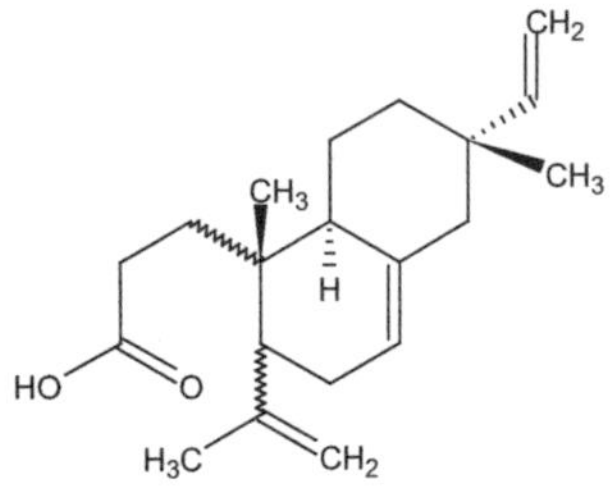

Fig. 2.68. Structure of 3,4-Secoisopimara-4(18),7,15-triene-3-oic acid.

The genus *Salvia* (Lamiaceae) includes over 900 species widespread all over the world. Various plants of this genus are widely used in folk medicine and some species are listed in the modern Pharmacopoeias. From the leaf surface exudate of the aerial parts of *Salvia cinnabarina* a secoisopimarane diterpenoid (Fig. 2.68) with a non-specific spasmolytic activity on histamine-, ACh-, and BaCl$_2$-induced contractions in the isolated guinea-pig ileum was obtained. The IC$_{50}$ value obtained was comparable with that of papaverine (Romussi et al. 2001).

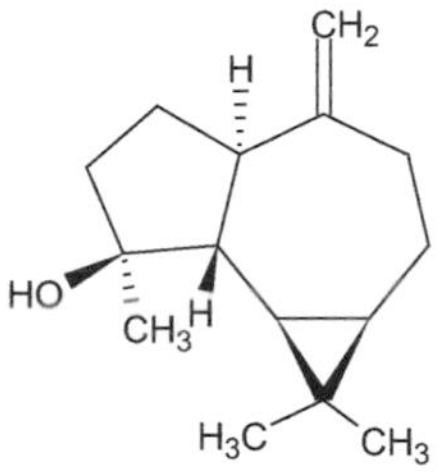

Fig. 2.69. Structure of Spathulenol.

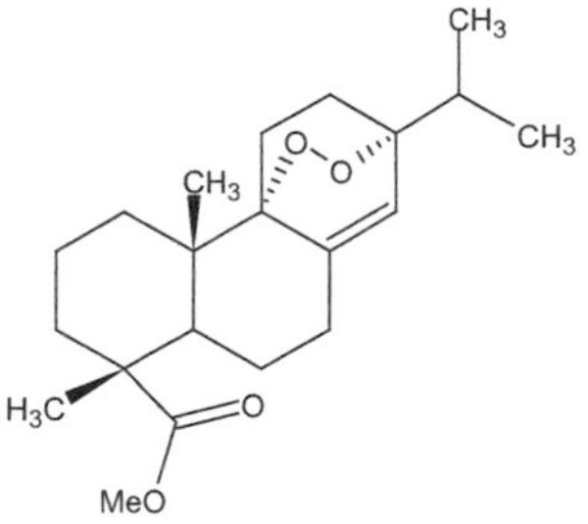

Fig. 2.70. Structure of 9α,13α-Epidioxyabiet-8(14)-en-18-oic acid methyl ester.

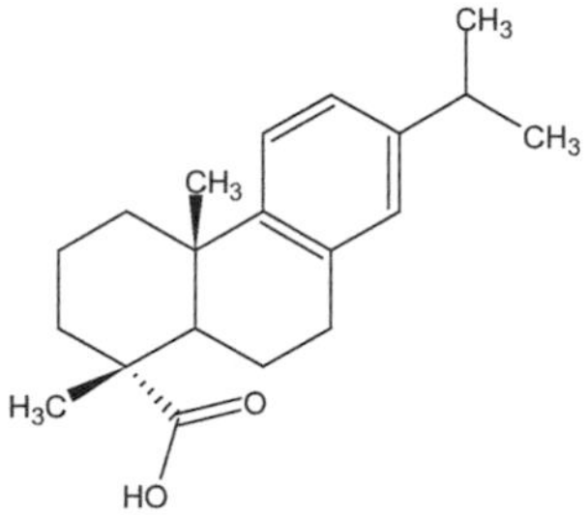

Fig. 2.71. Structure of Dehydroabietic acid.

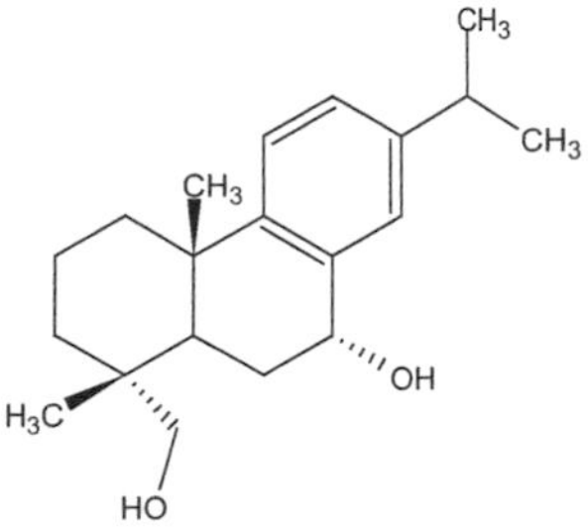

Fig. 2.72. Structure of 9α-Hydroxydehydro abietyl alcohol.

Lepechinia caulescens (Lamiaceae), popularly known as 'bretonica', is a perennial herb widely distributed in open areas of Quercus and Abies forests in central Mexico. An aqueous infusion prepared by decoction of the whole plant is used in folk medicine to treat diabetes, hypertension, gastrointestinal infections, dysmenorrhea and as abortifacient (Villavicencio-Nieto and Perez-Escandon 2006). Earlier pharmacological reports on *Lepechinia caulescens* showed that aqueous and MeOH extracts display hypoglycemic (Roman-Ramos et al. 2001) and vasorelaxant (Aguirre-Crespo et al. 2006) effects. In another study EtOAc extract of the leaves, which showed slight relaxing activity, led to 9α-hydroxydehydroabietyl alcohol and rosmarinic acid (Estrada-Soto et al. 2007).

Perez-Hernandez et al. (2008) isolated four terpenes (Figs. 2.69 to 2.72), from *Lepechinia caulescens*, responsible for the spasmolytic effect. These compounds had a direct relaxant effect on tonic contraction induced by depolarizing solutions in uterus smooth muscle in a concentration-dependent manner. Spathulenol (Fig. 2.69), the most potent terpene, inhibit the cumulative Ca²⁺-induced contraction in the depolarized uterus rings and its spasmolytic effect does not depend on NO or/and β-adrenergic receptors, and can be reverted by addition of Ca²⁺. These findings suggest that spathulenol, acts on voltage-sensitive calcium channels (Estrada-Soto et al. 2007).

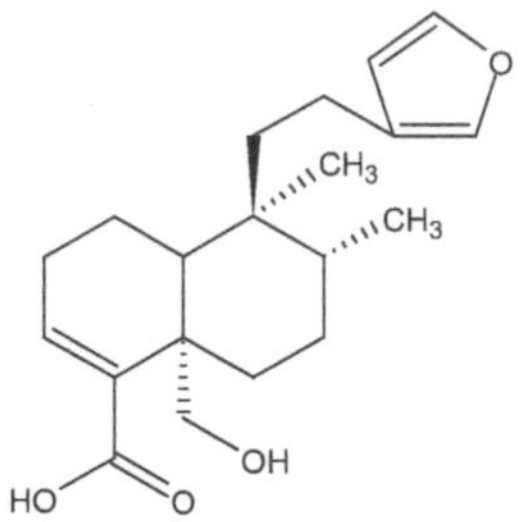

Fig. 2.73. Structure of Hautriwaic acid.

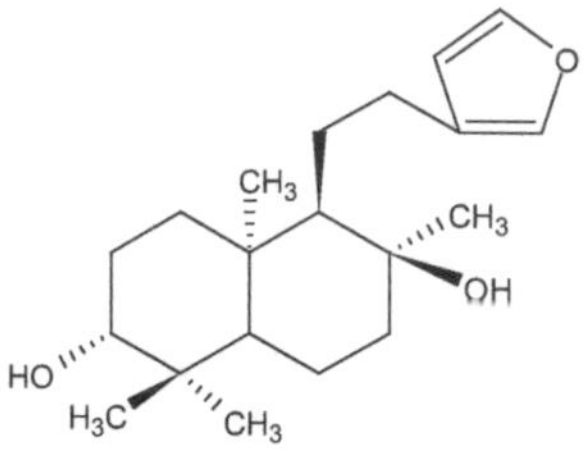

Fig. 2.74. Structure of ent-15, 16-epoxy-9 α H-labda-13(16)14-diene-3 β, 8 α-diol.

Found in *Dodonaea viscosa* (Sapindaceae). The ent-Labdane inhibited the ileum contractions evoked by ACh, histamine, and $BaCl_2$. In addition, both hautriwaic acid (Fig. 2.73) and ent-15,16-epoxy-9αH-labda-13(16)14-diene-3β,8α-diol (Fig. 2.74) were capable of relaxing contractions of the rat uterus induced by Ca^{2+} in K^+-depolarizing solution, displacing to the right the concentration-response curves to Ca^{2+}. These results suggest that ent-15, 16-epoxy-9 α H-labda-13(16)14-diene-3 β, 8 α-diol produce an interference with calcium metabolism in smooth muscle cells (Rojas et al. 1996).

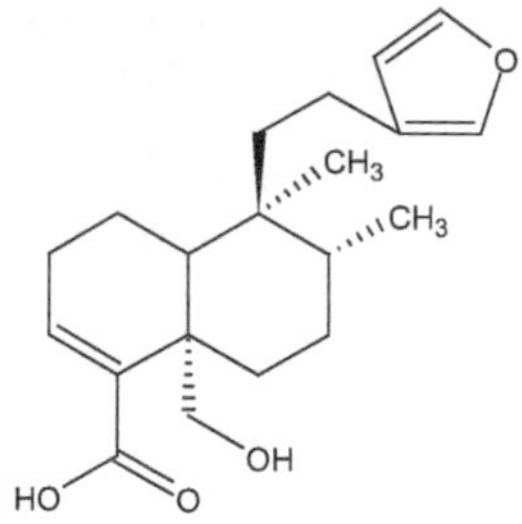

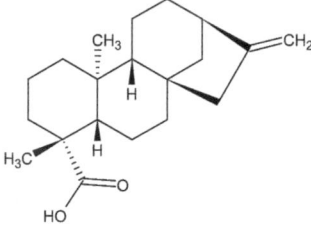

Fig. 2.75. Structure of ent-Beyer-15-en-19-oic acid.

Fig. 2.76. Structure of ent-Kaur-16-en-19-oic acid.

Viguiera hypargyrea (Asteraceae) is a perennial herb found in northern Mexico, locally known as 'plateada'. The underground parts of this plant have been used in Mexican traditional medicine for the treatment of gastrointestinal disorders. Two spasmolytic diterpene acids, Fig. 2.75 and 2.76, have been isolated from the roots of *Viguiera hypargyrea* by bioassay-guided fractionation. These compounds showed spasmolytic activity on histamine-, ACh-, and $BaCl_2$-induced contractions in the isolated guinea-pig ileum (Zamilpa et al. 2002).

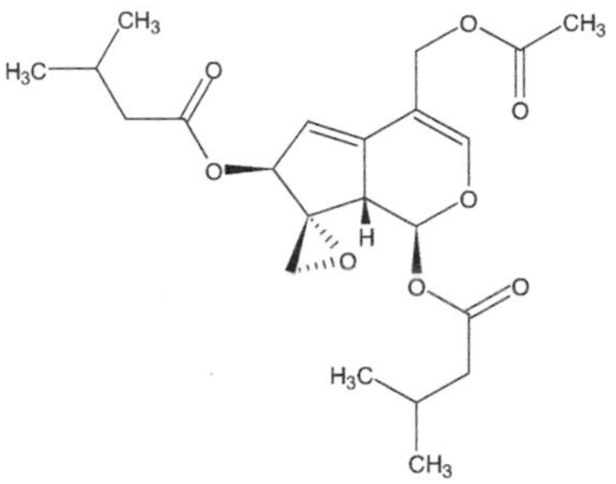

Fig. 2.77. Structure of Valtrate.

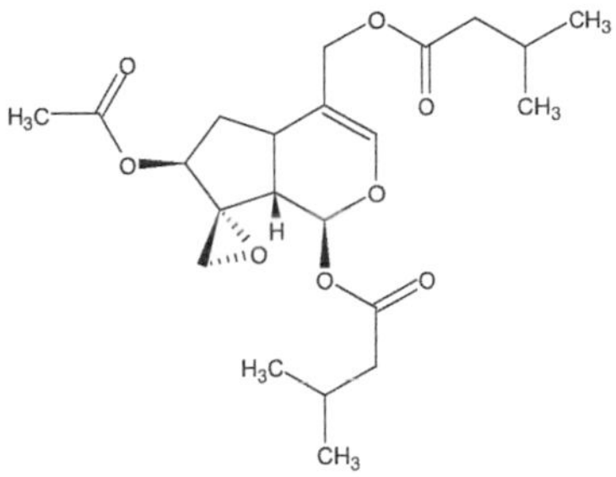

Fig. 2.78. Structure of Didrovaltrate.

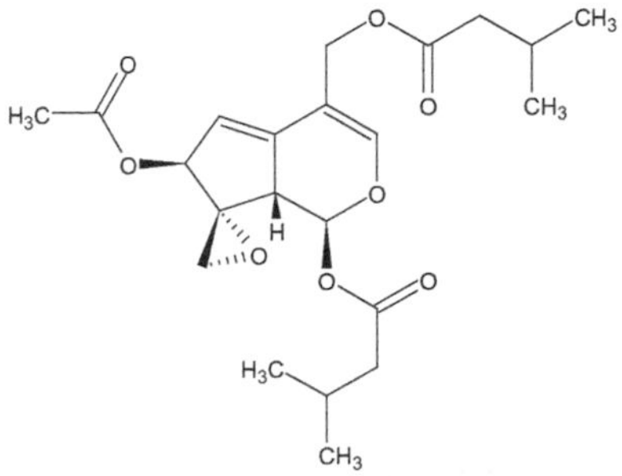

Fig. 2.79. Structure of Isovaltrate.

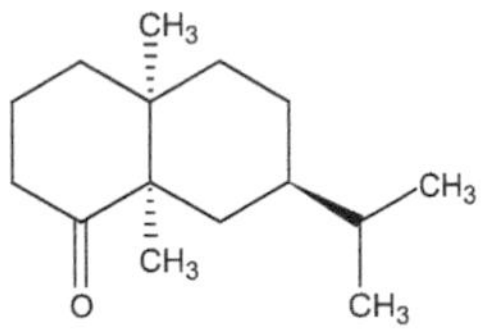

Fig. 2.80. Structure of Valeranone.

Valtrate, didrovaltrate, isovaltrate, and valeranone were isolated from subterranean parts of *Valeriana edulis* (Valerianaceas) and caused a suppression of rhythmic contractions in a closed part of the guinea-pig ileum *in-vivo*. These compounds relaxed potassium stimulated contractures and inhibited $BaCl_2$ contractions in guinea-pig ileum preparations *in vitro*. Guinea-pig stomach fundic strips stimulated by carbachol were also relaxed by these substances. Potassium stimulated smooth muscle cells were also relaxed by the valeriana compounds, even when autonomic receptors were blocked by appropriate antagonists. It is concluded that the valeriana compounds probably relax stimulated smooth muscle cells by acting as musculotropic agents and not by interacting with receptors of the autonomic nervous system (Hazelhoff et al. 1982).

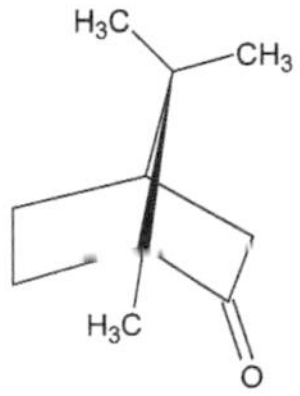

Fig. 2.81. Structure of Camphor.

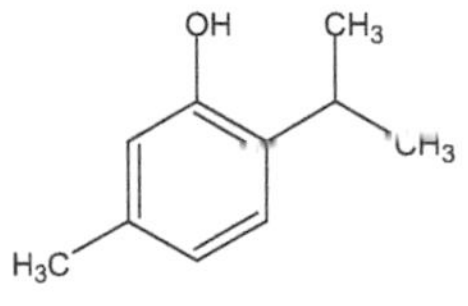

Fig. 2.82. Structure of Thymol.

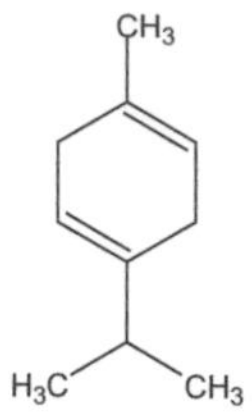

Fig. 2.83. Structure of γ-Terpinene.

The aerial parts of *Acalypha phleoides* (Euphorbiaceae) are usually prescribed in Mexican traditional medicine for a variety of gastrointestinal complaints. The essential oil, obtained from the aerial part of this plant, inhibits the spontaneous pendular movement of the rabbit jejunum. The monoterpenes camphor, thymol and γ-terpinene (Fig. 2.81 to 2.83) showed antispasmodic activity in the rabbit jejunum preparation, thymol was the most active compound, followed by camphor and γ-terpinene. However, thymol and camphor in high concentrations also showed tracheal relaxant properties, but γ-terpinene did not (Astudillo et al. 2004).

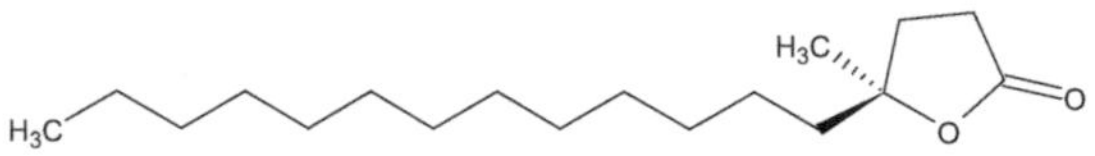

Fig. 2.84. Structure of 4-Methyl-tridecenylbutirolactone.

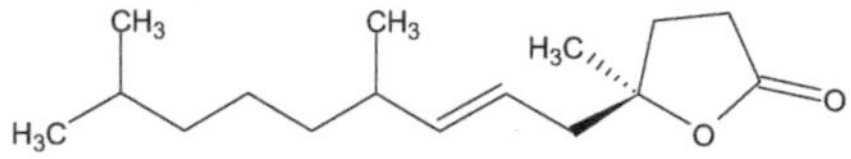

Fig. 2.85. Structure of 4,8R,13-Trimethyl,6(Z)ene-nonanebutirolactone.

Antispasmodic activity-guided fractionation of methanol extract from *Eryngium carlinae* (Umbelliferaceae) together with chemical analysis led to the isolation of 4-methyl-tridecenylbutirolactone (Fig. 2.84) and 4,8R,13-Trimethyl,6(Z)ene-nonanebutirolactone (Fig. 2.85). These isolated γ-lactones produce a significant antispasmodic effect on the contractions of the rat ileum induced by ACh, histamine and BaCl$_2$. 4-methyl-tridecenylbutirolactone is less active than 4,8R,13-trimethyl,6(Z) ene-nonanebutirolactone (Perez-Gutierrez et al. 2006).

Triterpenes

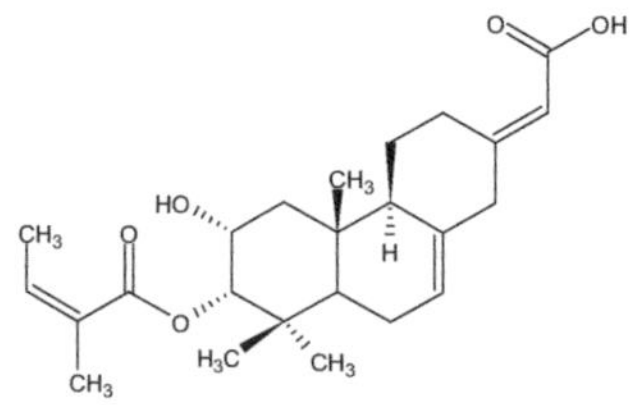

Fig. 2.86. Structure of 3α-Angeloyloxy-2α-hydroxy-13,14Z-dehydrocativic acid.

Brickellia paniculata (Asteraceae) has been used as spasmolytic in Mexican traditional medicine. 3α-Angeloyloxy-2α-hydroxy-13,14Z-dehydrocativic acid (Fig. 2.86), inhibited contractions on rat uterus precontracted by either KCl or oxytocin (Ponce-Monter et al. 2006).

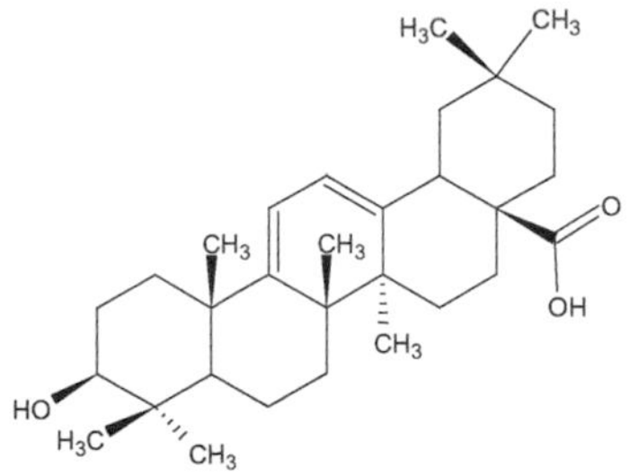

Fig. 2.87. Structure of Eucatyptanoic acid.

A triterpenoid acid named eucalyptanoic acid has been isolated from the fresh uncrushed leaves of *Eucalyptus camaldulensis* var. *obtusa*, which was found to be the spasmolytic on isolated rabbit jejunum, mediated through blockage of calcium influx at 1 mg/mL (Begum et al. 2002).

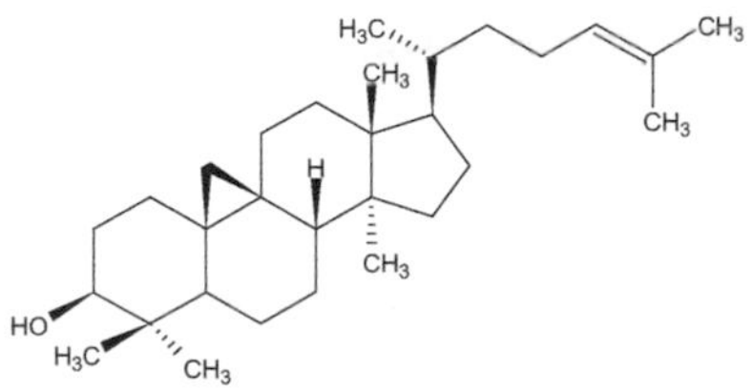

Fig. 2.88. Structure of Cycloartenol. 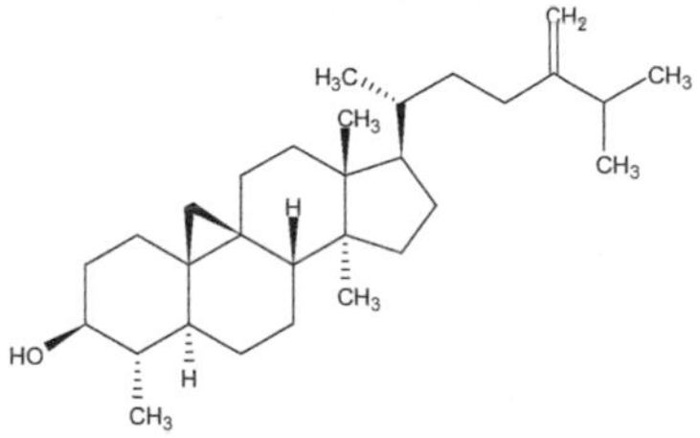   **Fig. 2.89.** Structure of Cycloeucalenol.

The mixture of two triterpenoids (CC), cycloartenol (Fig. 2.88) and cycloeucalenol (Fig. 2.89), obtained from *Herissanthia tiubae* (Malvaceae) a native plant from the northeast of Brazil, exhibited spasmolytic activity in isolated guinea-pig ileum. This relaxant activity was unspecific with regard to the contractile agent used: receptor agonists (carbachol and histamine) or membrane depolarization and the slopes of the Schild plot for CC were significantly different. In addition, CC was completely ineffective in relaxing other types of smooth muscles such as rat aorta and guinea-pig trachea even when the same agonists were used. These results suggest that these compounds are not acting as a receptor antagonist and that a common molecular mechanism for both stimulant agents specific for smooth muscle of ileum should be implicated in the spasmolytic action of CC. The inhibition of PKC is involved in spasmolytic effect of CC in the guinea-pig ileum (Gomes et al. 2005).

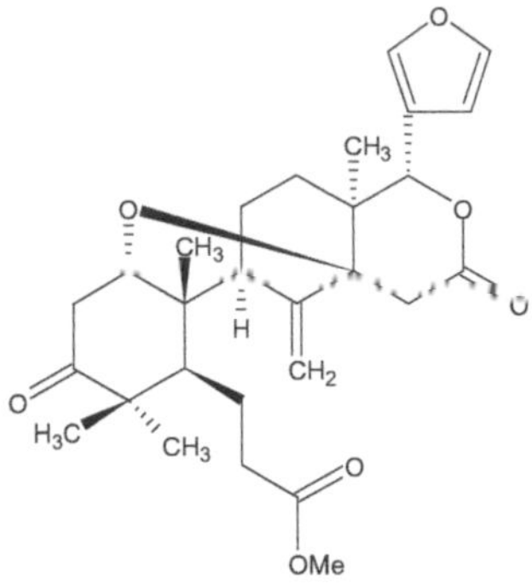

Fig. 2.90. Structure of Methyl angolensate.

Entandrophragma angolense (Meliaceae) is a popular plant commonly found on the West African coast. It is an opened crowned tree that resembles African mahogany. The stem bark is widely used in ethnomedical treatment of various gastrointestinal disorders including peptic ulcer in humans. Studies were done on the wood constituents of the plant (Orisadipe et al. 2001). Methyl angolensate (Fig. 2.90) isolated from *E. angolense*, produced a concentration dependent inhibition of the rat fundus strip, rabbit jejunum and was devoid of any contractile response on the guinea pig ileum. However, the inhibitory responses observed did not attenuate acetylcholine, a known muscarinic agonist, thus suggesting that cholinergic mechanisms may not be involved in the observed effects. The results suggest that methyl angolensate, might contribute possibly by forming a protective coat on the gastrointestinal tract and by inhibition of 5-HT_{2B}-receptors in rat fundus (Orisadipe et al. 2001).

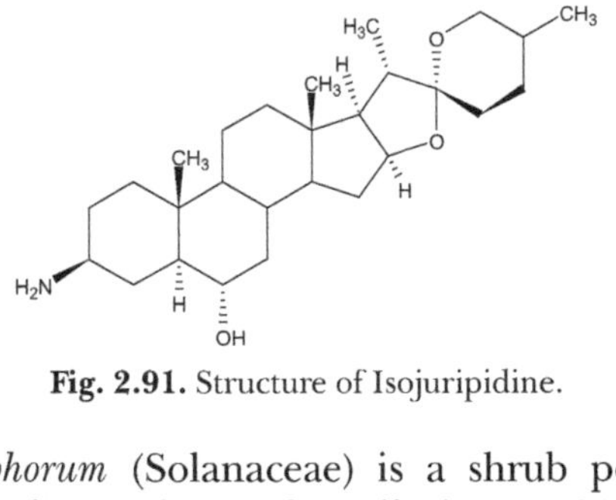

Fig. 2.91. Structure of Isojuripidine.

Solanum asterophorum (Solanaceae) is a shrub popularly known as 'jurubeba-defogo' in the northeast of Brazil. Figure 2.91, a steroidal alkaloid obtained from *S. asterophorum* leaves, inhibited phasic contractions induced by both histamine, ACh in guinea-pig ileum, This effect is probably due to inhibition of calcium influx through voltage-operated calcium channels (Oliveira et al. 2006).

Fig. 2.92. Structure of Erythrodiol.

Fig. 2.93. Structure of Betulinic acid.

Fig. 2.94. Structure of Maslinic acid.

Fig. 2.95. Structure of 2α,3α,23-Trihydroxyolean-12-en-28-oic acid.

Fig. 2.96. Structure of Bayogenin.

Fig. 2.97. Structure of Arjunilic acid.

Fig. 2.98. Structure of Methyl arjunolate.

Fig. 2.99. Structure of Arjungenin.

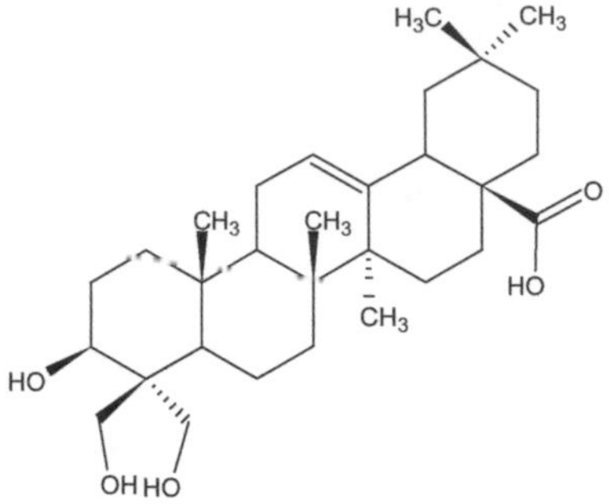

Fig. 2.100. Structure of 3β,23,24-Trihydroxyolean-12-en-28-oic acid.

Rhododendron collettianum (Ericaceae) is widely found in Kurram and in adjoining areas of Afghanistan. Nine pentacyclic triterpenes (Fig. 2.92 to 2.100) were isolated from the chloroform extract. Among the triterpenes tested, arjunilic acid was found to be most potent. Their structure-activity relationship showed that if the configuration of the -OH group at C-2 is changed from α to β the potency is decreased. In most of the compounds the position and configuration of the -OH group was found to be important for the inhibitory potency against the enzyme tyrosinase (Ullah et al. 2007).

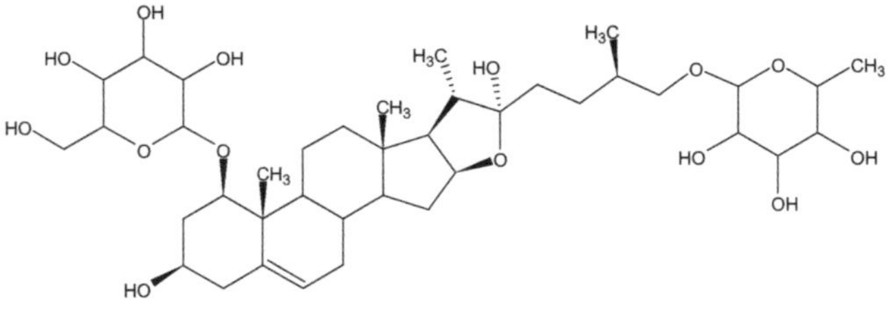

Fig. 2.101. Structure of Tropeoside A1.

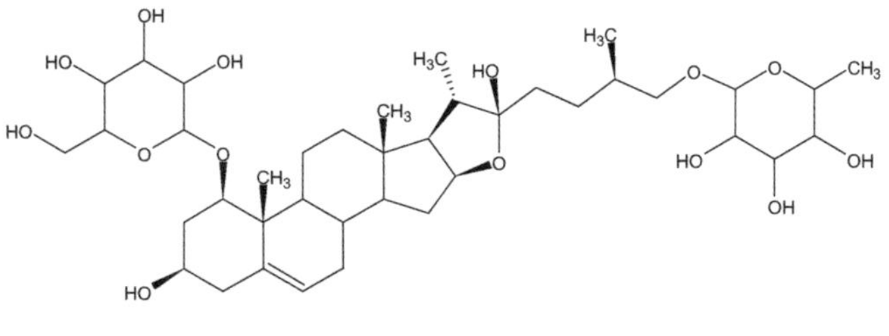

Fig. 2.102. Structure of Tropeoside A2.

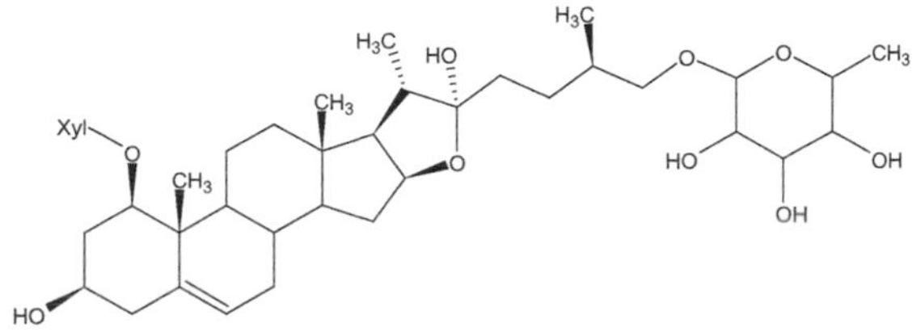

Fig. 2.103. Structure of Tropeoside B1.

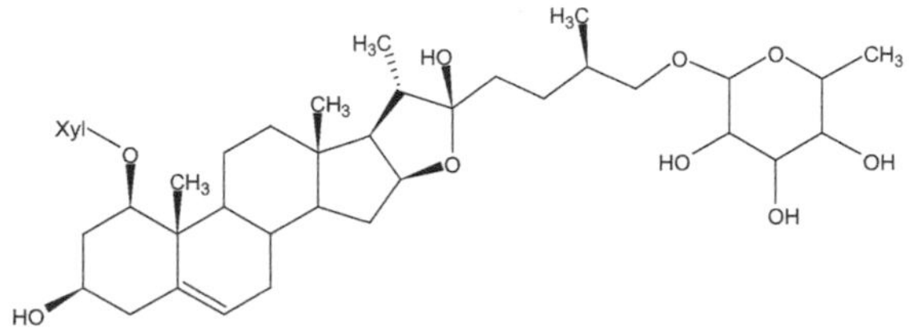

Fig. 2.104. Structure of Tropeoside B2.

The red onion, *Allium cepa* var. *tropea* (Alliaceae) is a typical Italian variety, cultivated in some areas of southern Italy (Calabria region). This variety, characterized by both white and purple flowers, is known for its distinctive bulb, lengthened or oval, red, and sweet. Analysis of the MeOH extract from the bulbs of Tropea onions revealed the presence of high concentrations of saponins. Among these, four saponins, named tropeosides A1/A2 (Figs. 2.101 and 2.102) and B1/B2 (Figs. 2.103 and 2.104), were found to possess antispasmodic activity in the guinea pig isolated ileum; such an effect might contribute to explaining the traditional use of onion in the treatment of disturbances of the gastrointestinal tract (Corea et al. 2005).

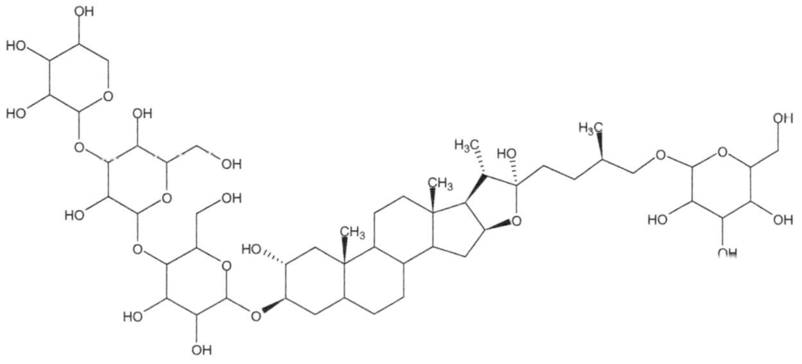

Fig. 2.105. Structure of Hirtifolioside A1/A2.

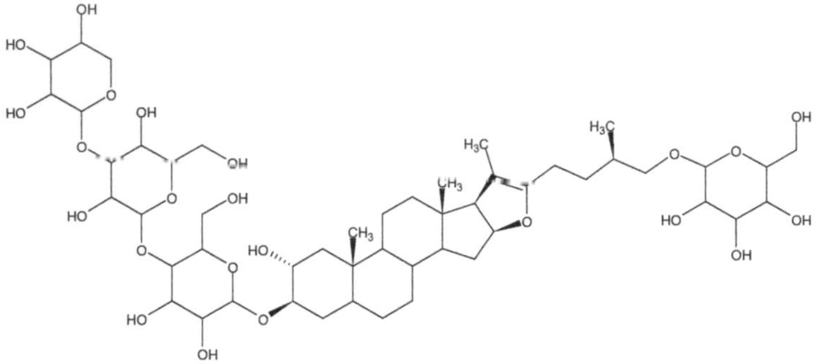

Fig. 2.106. Structure of Hirtifolioside B.

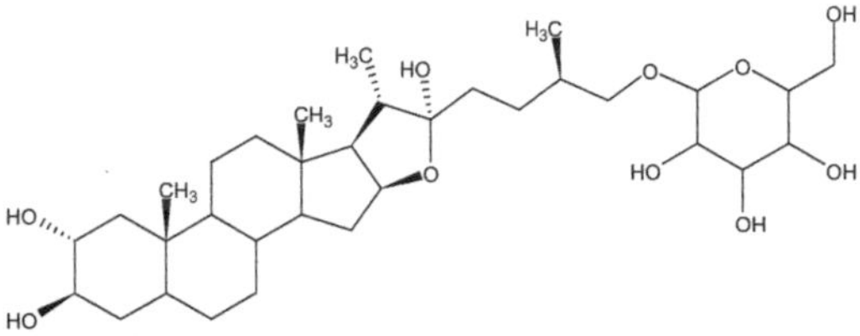

Fig. 2.107. Structure of Hirtifolioside C1/C2.

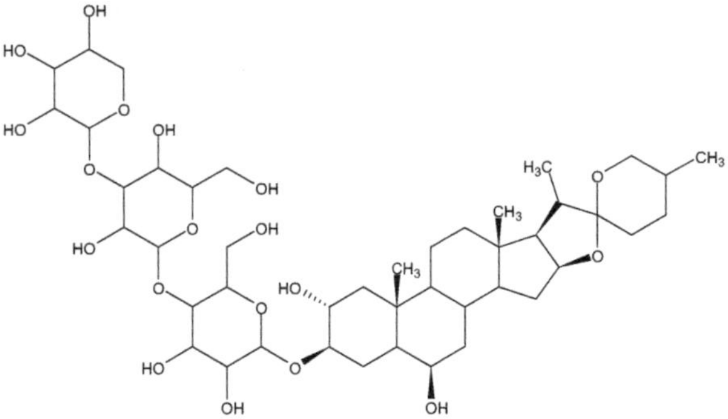

Fig. 2.108. Structure of Hirtifolioside D.

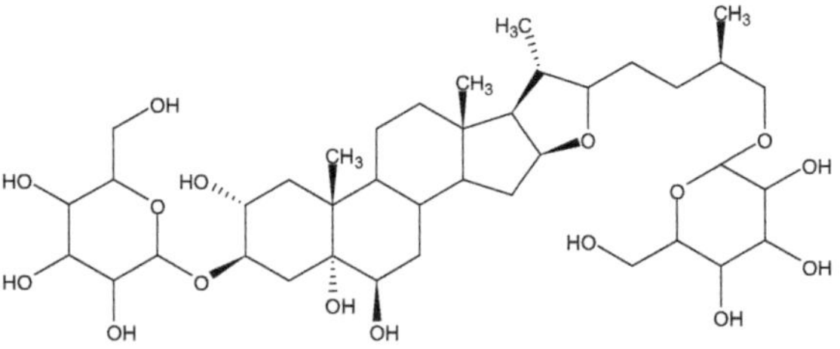

Fig. 2.109. Structure of Elburzensoside A1/A2.

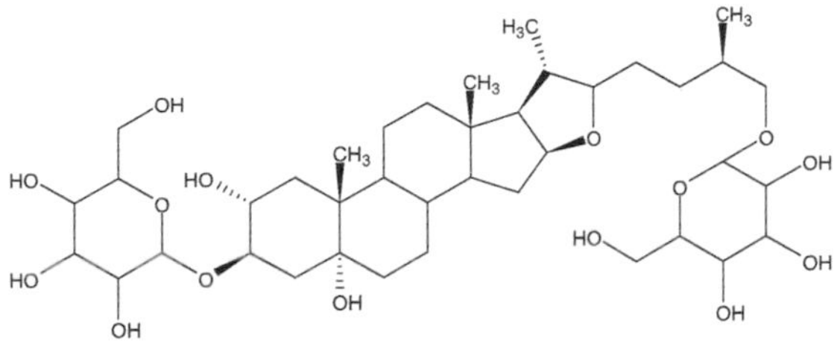

Fig. 2.110. Structure of Elburzensoside C1/C2.

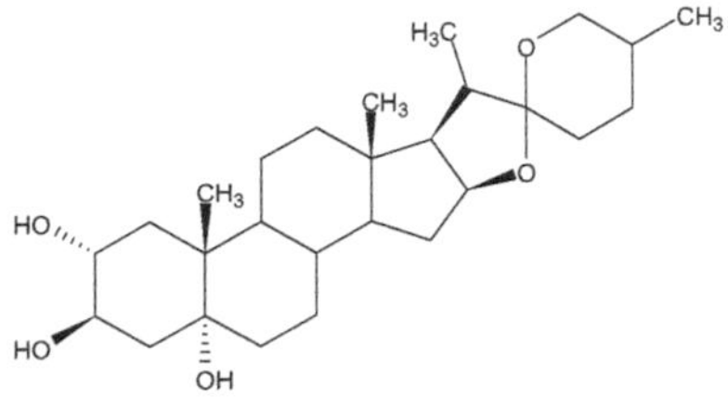

Fig. 2.111. Structure of Agapanthagenin.

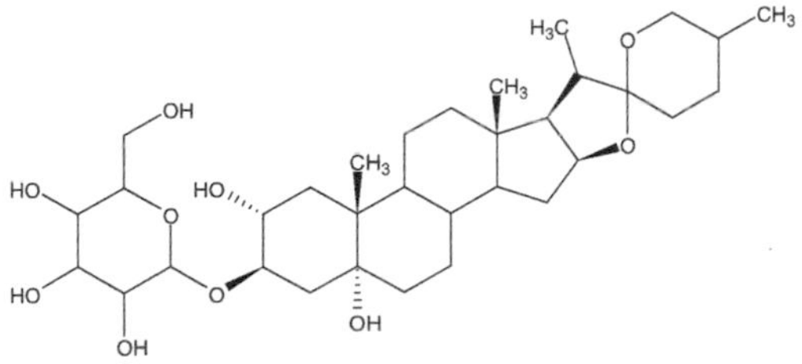

Fig. 2.112. Structure of Agapanthagenin glucoside.

A phytochemical study of *Allium hirtifolium* (Alliaceae) flowers has led to the isolation of saponins (Fig. 2.105 to 2.112), which were found to possess antispasmodic activity in the guinea-pig isolated ileum. They highlight the positive effects of a hydroxyl group at position 5 and of a glucose unit at position 26 and demonstrate the detrimental effects of both a hydroxyl group at position 6 and of a glucose unit at position 3. Among the tested compounds, elburzensoside C1/C2 (Fig. 2.110) and agapanthagenin (Fig. 2.111) showed the highest activity in reducing induced contractions as measured by the reduction of histamine release by about 50%. The observed effect therefore contributes to the explanation of the traditional use of onion and garlic in the treatment of disturbances of the gastrointestinal tract (Barile et al. 2005).

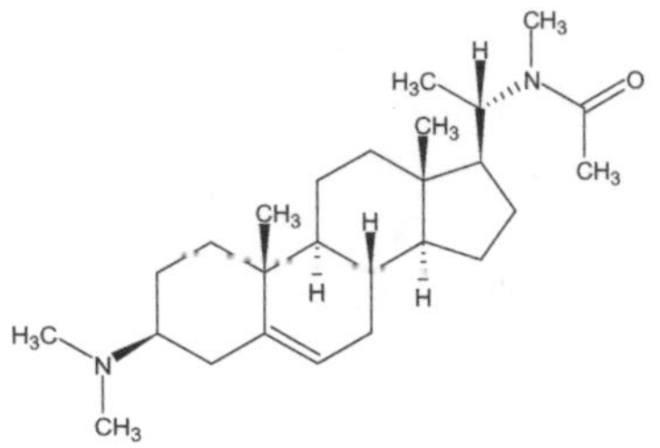

Fig. 2.113. Structure of Saracocine.

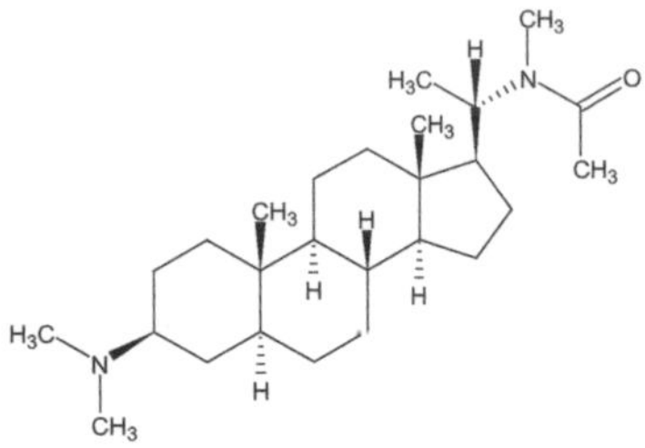

Fig. 2.114. Structure of Saracodine.

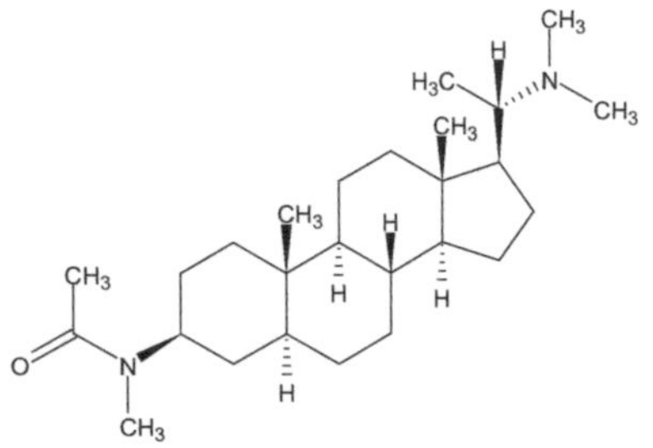

Fig. 2.115. Structure of Saracorine.

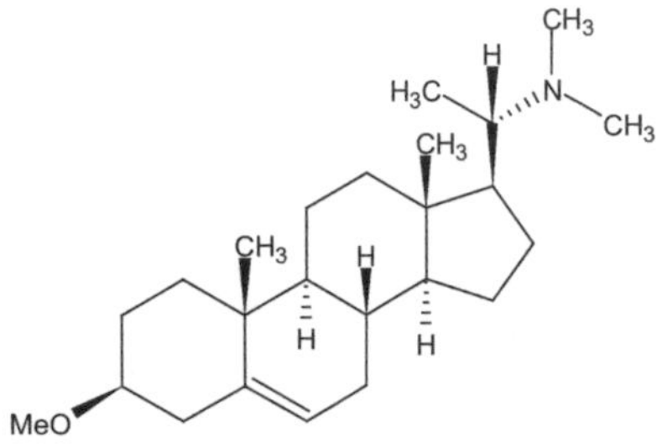

Fig. 2.116. Structure of Alkaloid-C.

Sarcococca saligna (Syn. *S. pruniformis*, *S. salicifolia*; Buxaceae) is a shrub that abundantly grows in the northern areas of Pakistan. The local population of northern Pakistan, call it 'ban sathra' while in the west, it is known as sweet box or Christmas box. People of Pakistan have been using it in disorders resulting from hyperactive states of the gastrointestinal tract. Characteristic steroidal compounds of this plant (Fig. 2.113 to 2.116), exhibited AChE inhibitory effect along with dominant calcium channel bloking (CCR) activity, possibly owing to which it showed spasmolytic, antidiarrhoeal and anti-secretory activities (Gilani et al. 2005).

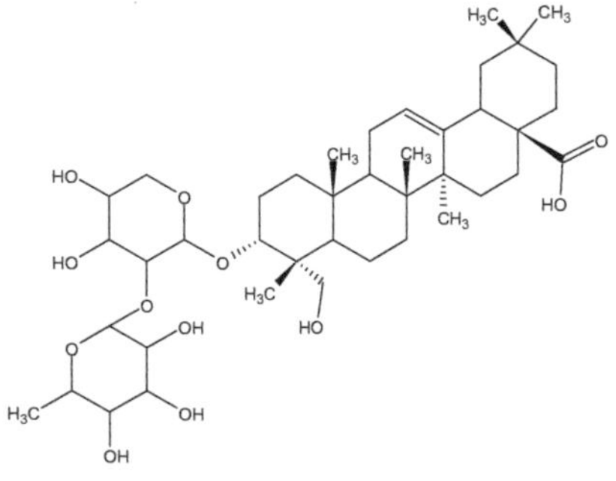

Fig. 2.117. Structure of α-Hederin.

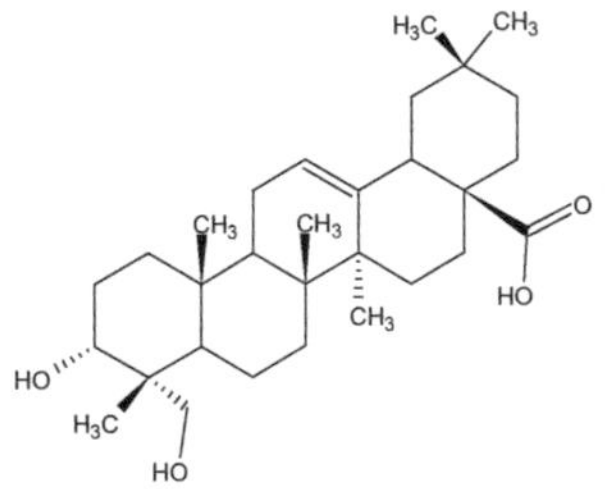

Fig. 2.118. Structure of Hederagenin.

Found in the leaves of ivy (*Hedera helix* Araliaceae). Hederin (Fig. 2.17) and hederagenin (Fig. 2.118) exhibited significant antispasmodic activity whereas the bidesmoside hederacoside C, the main saponin of ivy, was much less active (Trute et al. 1997).

Miscellaneous

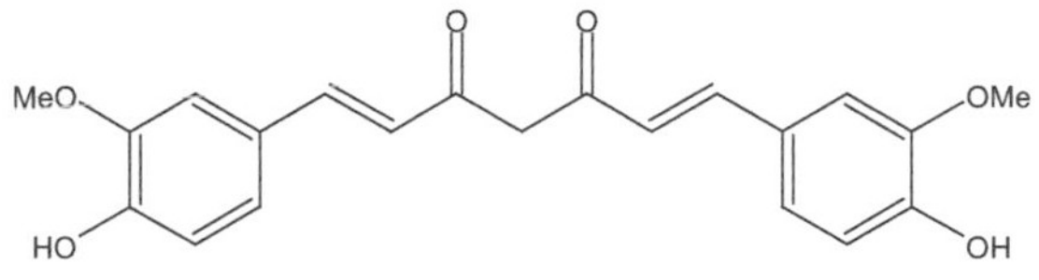

Fig. 2.119. Structure of Curcumin.

Curcumin isolated from *Curcuma longa* (Zingiberaceae) rhizomes, showed significantly inhibited the ileum pre-contracted with acetylcholine and histamine. In potassium depolarizing, curcumin reduced the contraction induced by calcium chloride. In rat uterus smooth muscle preparation, curcumin significantly reduced force and frequency of contraction induced by oxytocin. The results obtained from this study concluded that curcuminoids produced a smooth muscle, relaxing effect on isolated guinea-pig ileum and rat uterus by receptor-dependent and independent mechanism (Itthipanichpong et al. 2003).

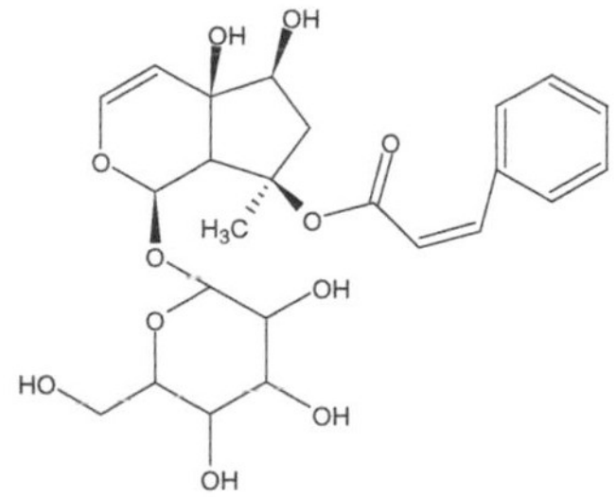

Fig. 2.120. Structure of Harpagoside.

Found in *Harpagophytum procumbens* (Pedaliacease). Harpagoside (Fig. 2.120) showed spasmolytic effect (Eichler and Koch 1970).

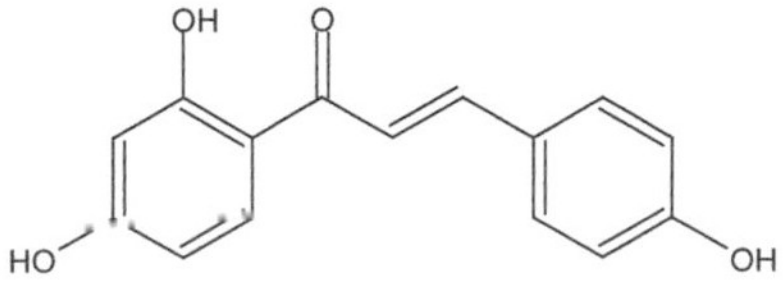

Fig. 2.121. Structure of Isoliquiritigenin.

Glycyrrhizae radix is used to treat abdominal pain as a component of shakuyakukanzoto (shaoyao-gancao-tang), a traditional Chinese medicine formulation. Isoliquiritigenin was isolated from an aqueous extract of liquorice as a potent relaxant, which inhibited the contraction induced by various types of stimulants, such as CCh, KCl, and $BaCl_2$ with IC_{50} values close to those of papaverine. Isoliquiritigenin (Fig. 2.121) also showed the most potent inhibition of mouse rectal contraction induced by CCh. These results suggest that isoliquiritigenin acts as a potent relaxant in the lower part of the intestine by transformation from its glycosides (Sato et al. 2007).

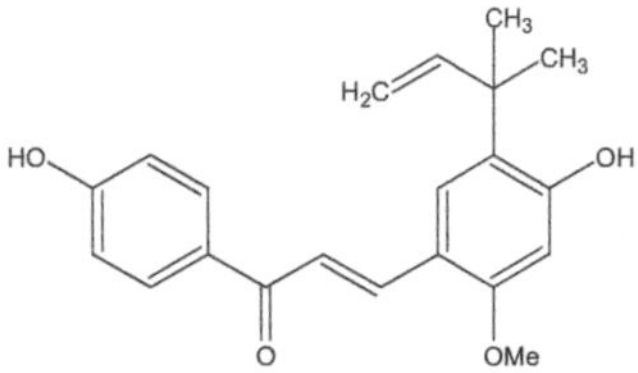

Fig. 2.122. Structure of Licochalcone A.

Liquorice is one of the most frequently used natural drugs prepared from *Glycyrrhiza* species roots and contains glycyrrhizin as an important pharmacologically active glycoside (Olukoga and Donaldson 2000). In Japan, liquorice prepared mainly from *G. uralensis* root is used in clinical practice to treat various abdominal spasmodic symptoms as a component of a traditional Chinese formulation ('Shakuyakukanwto' in Japanese) (Katsura 1995).

Licochalcone A (Fig. 2.122), a retrochalcone of *Glycyrrhiza inflata* (Fabaceae) root, has been known to possess muscle relaxants, shows a concentration-dependent relaxant effect on the contraction induced by carbachol. Pretreatment with licochalcone A enhances the relaxant effect of forskolin, an adenylyl cyclase activator, on the contraction in a similar manner to 3-isobutyl-1-methyl xanthine a phosphodiesterase inhibitor. These results indicate that licoch- alcone A may have been responsible for the relaxant activity of *G. inflata* root and acts through the inhibition of cAMP PDE (Nagai et al. 2007).

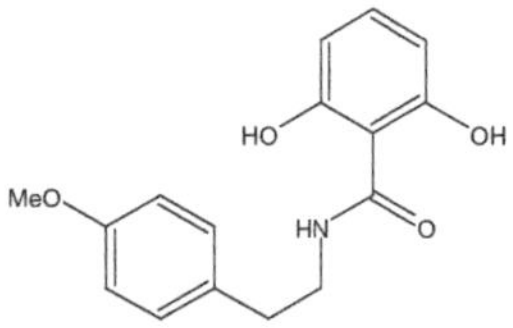

Fig. 2.123. Structure of Riparin.

Riparin (Fig. 2.123), is a constituent of *Aniba riparia* (Lauraceae). In rat depolarized uterus, riparin inhibited in a reversible and noncompetitive manner $CaCl_2$-induced contraction, a response mediated through voltage-dependent Ca^{2+} channels, and was also ineffective in suppressing noradrenaline-induced sustained contractions of rabbit aortic strips. However, in the aorta, the compound inhibited intracellular calcium-dependent transient contractions of noradrenaline and riparin was approximately two and a half times more potent than procaine a known inhibitor. In guinea-pig alveolar leucocytes, riparin inhibited intracellular Ca^{2+} accumulation induced by the calcium ionophore A23187. The results suggest that the inhibition of Ca^{2+} influx and of Ca^{2+} release from intracellular stores contribute to the spasmolytic effects, which may not involve cyclic AMP generation as the levels of this nucleotide were not increased in alveolar macrophages treated with this compound (Thomas et al. 1994).

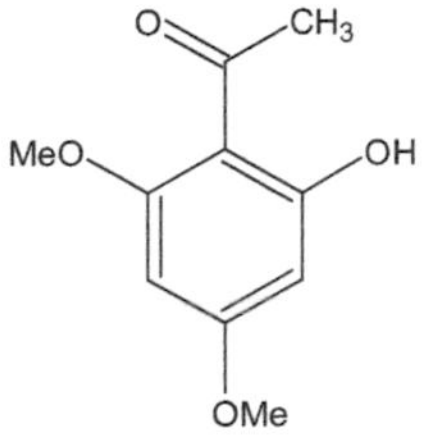

Fig. 2.124. Structure of Xanthoxyline.

Xanthoxyline (Fig. 2.124) is the major constituent isolated from the hexane extracts of *Sebastiania schottiana* (Euphorbiaceae), a native Brazilian medicinal plant used in folk medicine for the treatment of kidney disease. Xanthoxyline has also been isolated from other plants, such as *Fagara okinawaensis* (Rutaceae), *Pulicaria undulata* (Asteraceae), *Eucalyptus michaeliana* (Myrtaceae), *Sapium sebiferum* (Euphorbiaceae), and *Phyllanthus sellowianus* (Euphorbiaceae). Xanthoxyline caused a potent and concentration-dependent effect against several agonist mediated contractions of the ileum and urinary bladder from guinea pig and rat uterine smooth muscles *in vitro*. In addition, xanthoxyline also inhibited, in a graded manner, twitch responses evoked by electrical stimulation of strips of guinea pig longitudinal ileum, urinary bladder, dog ureter, and rat left atrium, with IC_{50} values of 50 to 480 μM (Filho et al. 1995). In another study Calixto et al. (1990) demonstrated that xanthoxyline induces a direct and non-selective inhibition of contractions triggered by agonists or electrical stimulation of smooth and cardiac muscle preparations.

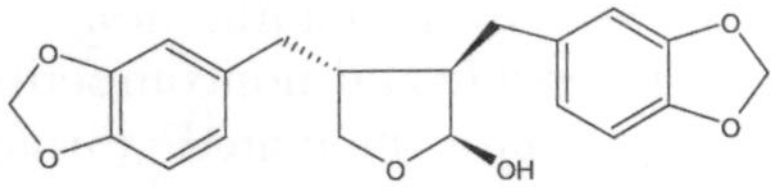

Fig. 2.125. Structure of Cubebin.

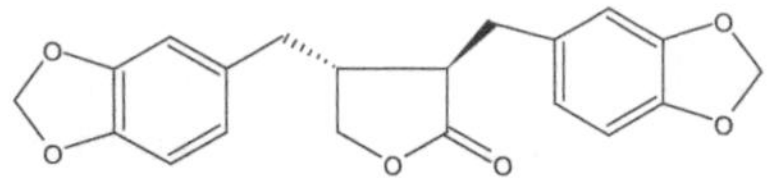

Fig. 2.126. Structure of Hinokinin.

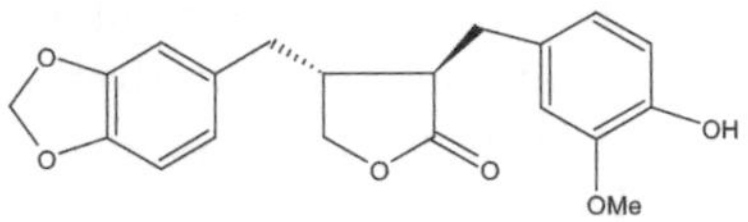

Fig. 2.127. Structure of Pluviatolide.

3,4-Dibenzyldihydrofuran-type lignans (Fig. 2.125 to 2.127) were isolated as antispasmodic principals from the n-hexane and chloroform extracts of *Aristolochia constricta* (Aristolochiaceae) a plant whose aerial parts have been used empirically in folk medicine for various purposes. These constituents showed antispasmodic effects on smooth muscle contraction in isolated guinea-pig ileum (Zhang et al. 2008).

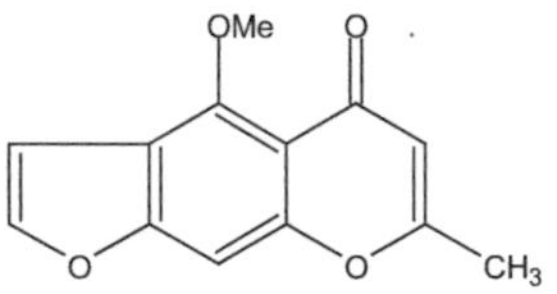

Fig. 2.128. Structure of Visnagin.

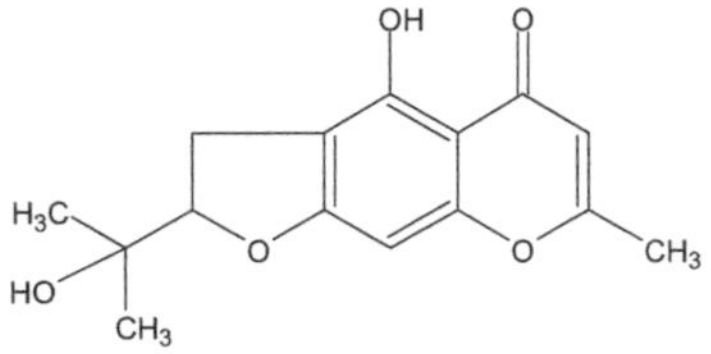

Fig. 2.129. Structure of Visamminol.

Cimicifuga dahurica (Ranunculaceae) with the guidance of the intestinal action resulted in the isolation of two active substances (Fig. 2.128 and 2.129). Pharmacological estimations using guinea pig jejunum revealed that the spasmolytic actions of visamminol and visnagin were one-third and one-tenth that of papaverine hydrochloride, respectively (Ito et al. 1976).

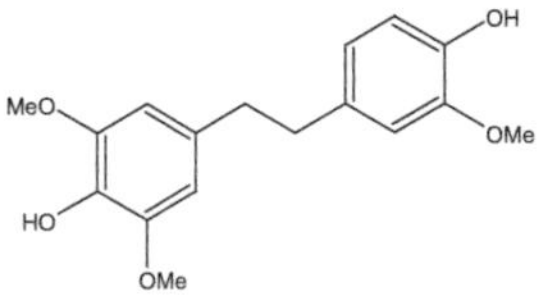

Fig. 2.130. Structure of Aloifol II.

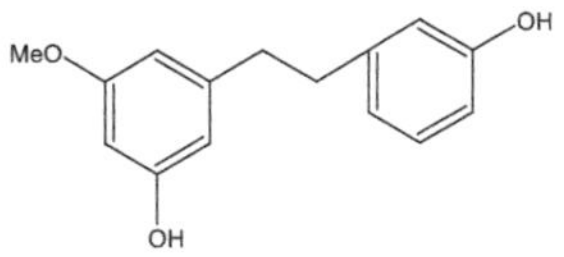

Fig. 2.131. Structure of Batatasin III.

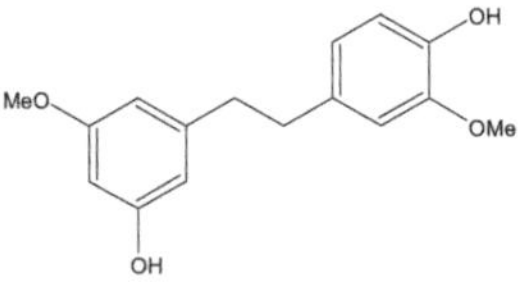

Fig. 2.132. Structure of Gigantol.

Nidema boothii (Orchidaceae), also known as *Epidendrum boothii*, is an orchid found from Mexico to Panama, Cuba, and Surinam, and its habitat is in tropical moist forests. *N. boothii* is not used as a traditional or alternative remedy. These compounds isolated from orchids induce a concentration-dependent inhibition of the spontaneous contractions of the rat ileum with a potency higher than or comparable to that of papaverine. It was also demonstrated that the isolated bibenzyls (Fig. 2.130 to 2.132), might exert their spasmolytic action not only by a nitrergic mechanism but also by inhibiting calmodulin-mediated processes (Hernandez-Romero et al. 2004).

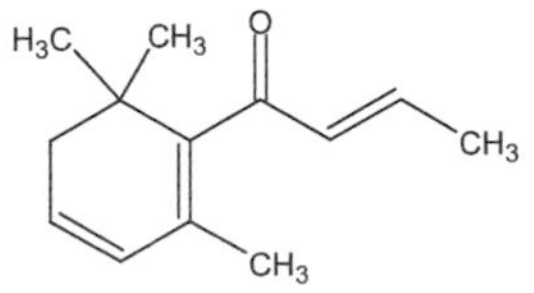

Fig. 2.133. Structure of β-Damascenone.

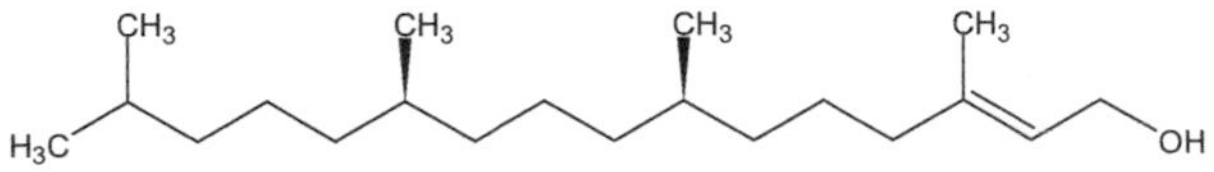

Fig. 2.134. Structure of Phytol.

The plant *Ipomoea pes-caprae* (Convolvulaceae) has previously been shown to antagonize smooth muscle contractions induced by several agonists via a non-specific mechanism. Bioassay-guided fractionation of this plant resulted in isolation of the antispasmodically acting isoprenoids. Their antispasmodic potencies were found to be in the same range as that of papaverine, a general spasmolytic agent. It is possible that β-damascenone (Fig. 2.133) and E-phytol (Fig. 2.134), by interfering with the contraction of vascular smooth muscle cells, are partly responsible for the previously reported effectiveness of *Ipomea* in the treatment of such dermatitis (Pongprayoon et al. 1992).

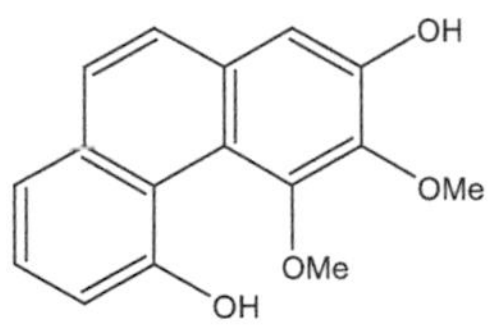

Fig. 2.135. Structure of 2,5-Dihydroxy-3,4-dimethoxyphenanthrene.

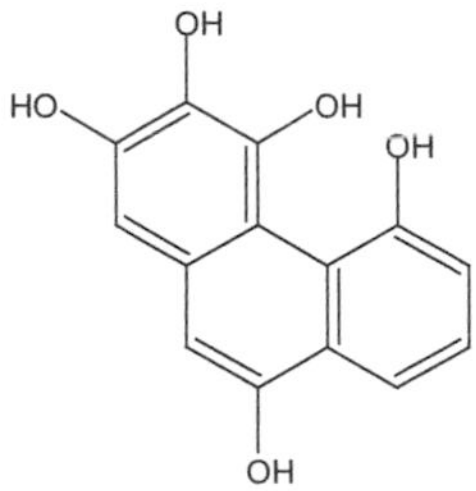

Fig. 2.136. Structure of Fimbriol A.

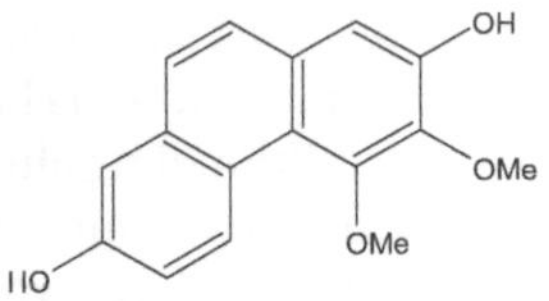

Fig. 2.137. Structure of Nudol.

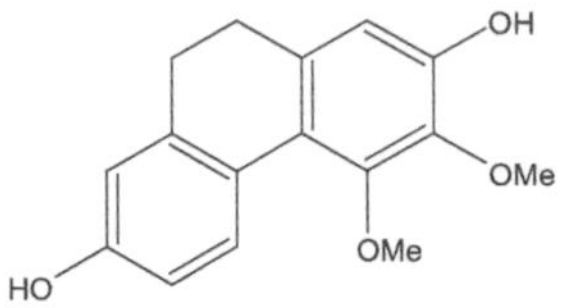

Fig. 2.138. Structure of Erianthridin.

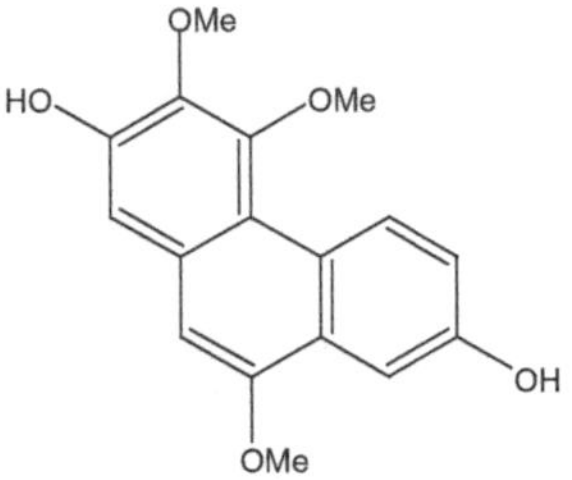

Fig. 2.139. Structure of Gymnopusin.

2,5-Dihydroxy-3,4-dimethoxyphenanthrene (Fig. 2.135) and fimbriol A (Fig. 2.136) occurring in *Maxillaria densa* (Orchidaceae) provoked a concentration-dependent inhibition of the spontaneous contractions of the rat ileum with potencies comparable to papaverine. The relaxant activity of the compounds does not involve a direct nitrergic or antihistaminergic mode of action or an interference with calcium influx into the smooth muscle cells (Estrada et al. 2004).

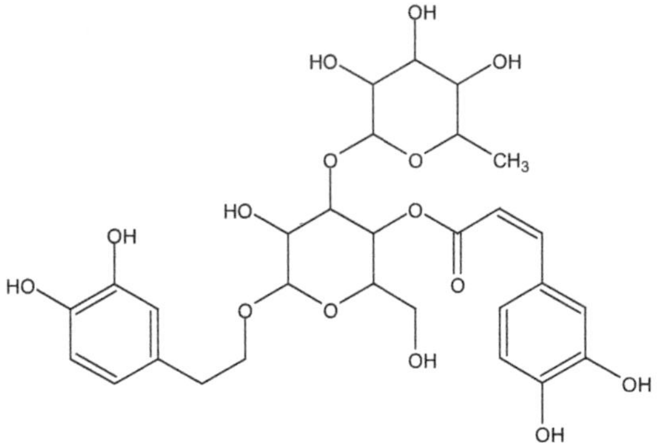

Fig. 2.140. Structure of Acteoside.

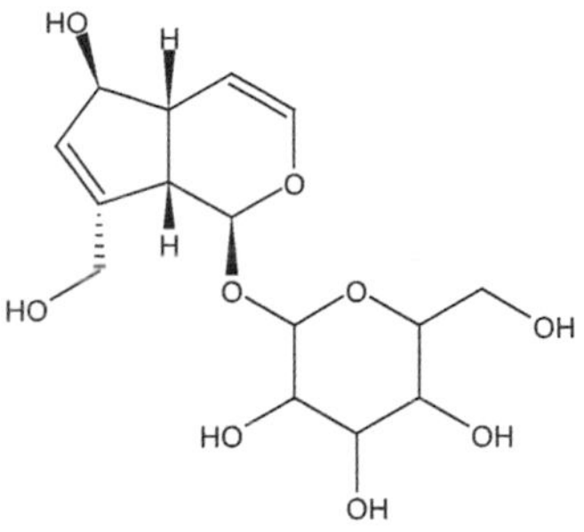

Fig. 2.141. Structure of Aucubin.

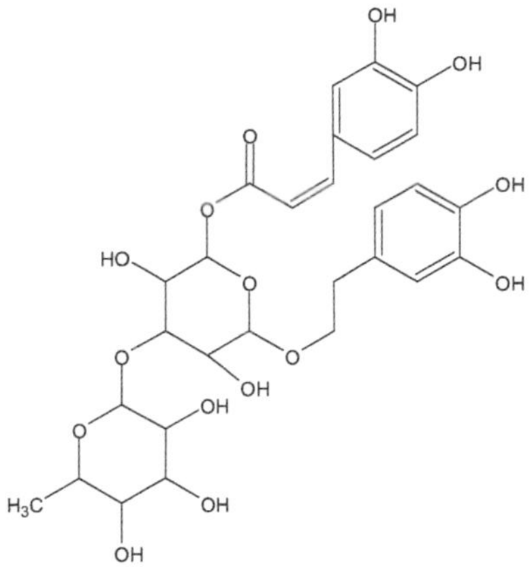

Fig. 2.142. Structure of Isoacteoside.

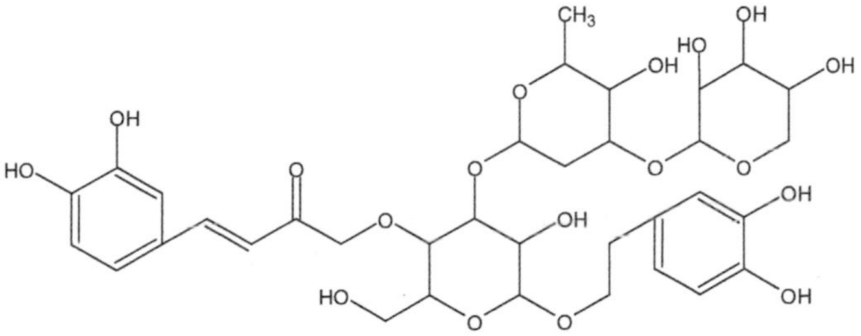

Fig. 2.143. Structure of Lavandulifolioside.

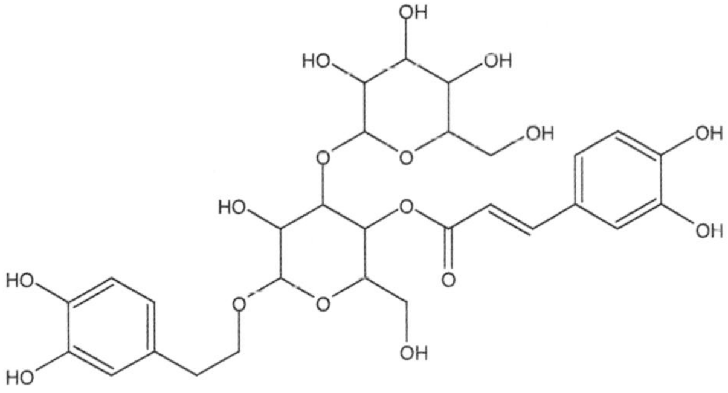

Fig. 2.144. Structure of Plantamajoside.

An ethanolic extract of the aerial parts of *Plantago lanceolata* (Plantaginaceae) inhibited the contractions of the guinea-pig ileum that were induced by various agonists such as ACh, histamine, potassium and barium ions. Additionally the trachea contractions induced by barium ions were also inhibited. Isolated compounds (Fig. 2.140 to 2.144) inhibited the ACh-induced contractions of the guinea-pig ileum. Luteolin and acteoside reduced the barium-induced contractions of the guinea-pig trachea (Fleer and Verspohl 2007).

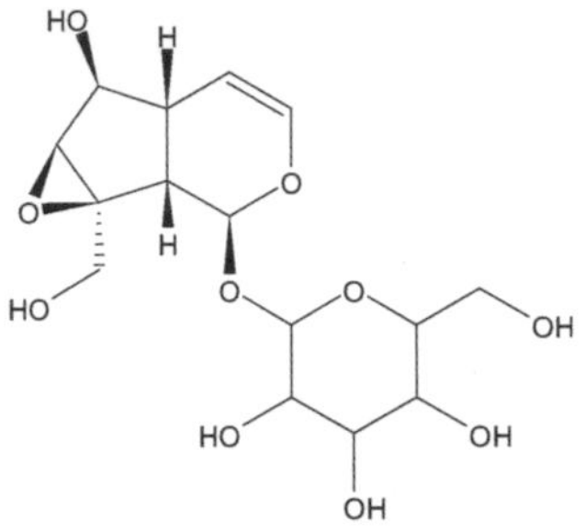

Fig. 2.145. Structure of Catalpol.

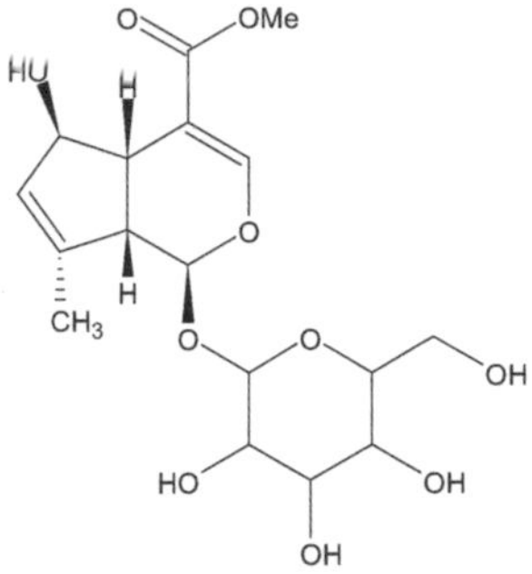

Fig. 2.146. Structure of Penstemonoside.

In vitro, the peracetates of penstemonoside, aucubin and catalpol, iridoids isolated from *Parentucellia latifolia* (Scrophulariaceae), antagonize the uterine muscular contractions induced by acetylcholine and calcium, in a similar way to papaverine. The antagonism is non-competitive against acetylcholine and competitive against calcium The two components, phasic and tonic, of the response of the vas deferens to potassium are reduced by the three iridoids. The reduction is similar for both phases. The antispasmodic activity of the three iridoids (Fig. 2.145 to 2.146), similar to papaverine, is related to an inhibiting effect of extracellular calcium, intracellular or both (Ortiz de Urbina et al. 1994).

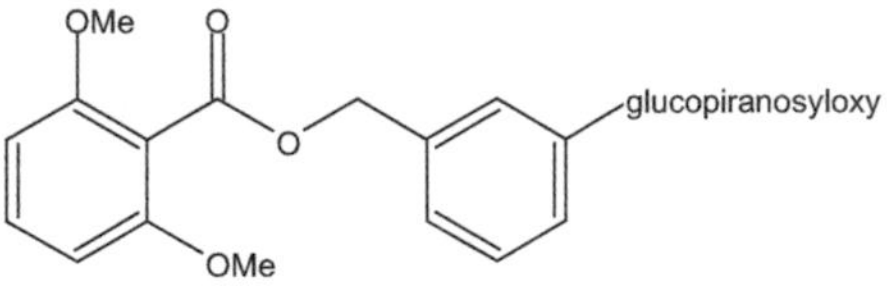

Fig. 2.147. Structure of 3'-(β-D-glucopyranosyloxy)benzyl-2,6-dimethoxybenzoate.

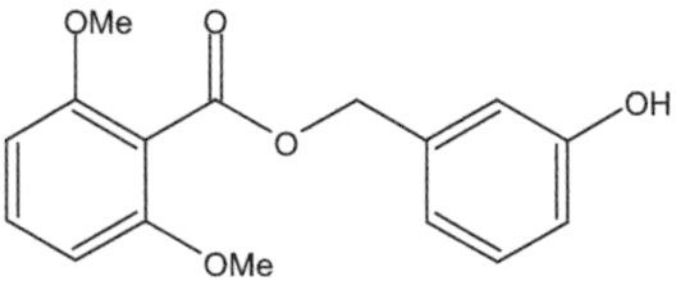

Fig. 2.148. Structure of 3'-Hydroxybenzyl-2,6-dimethoxybenzoate.

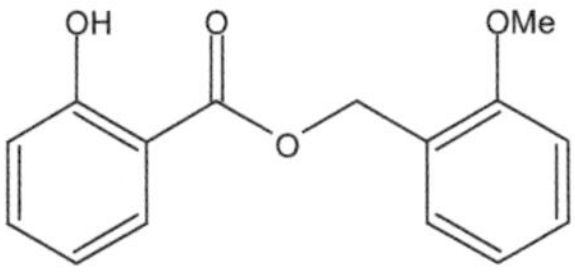

Fig. 2.149. Structure of 2'-Methoxybenzyl-2-hydroxybenzoate.

Fig. 2.150. Structure of Benzyl 2,6-dimethoxybenzoate.

Fig. 2.151. Structure of 3'-Methoxybenzyl 2-hydroxy-6-methoxybenzoate.

Fig. 2.152. Structure of Benzyl 2-hydroxy-6-methoxybenzoate.

Fig. 2.153. Structure of Benzyl 2,5,6-trimethoxybenzoate.

Fig. 2.154. Structure of Benzyl 2-hydroxy-5,6-dimethoxybenzoate.

Fig. 2.155. Structure of 3'-Methoxybenzyl 2,6-dimethoxybenzoate.

Nine benzoates (Fig. 2.147 to 2.155) were isolated from *Brickellia veronicifolia* (Asteraceae), (Rivero-Cruz et al. 2005). The most active inhibitors of the system calmodulin-PDEl were the first five benzyl benzoates, which inhibited the activity of PDEl in a concentration-dependent manner (Rivero-Cruz et al. 2007).

Fig. 2.156. Structure of Denthirsinin.

Fig. 2.157. Structure of Lusianthridin.

Bioactivity-guided fractionation of the active extract from *Scaphyglottis livida* (Orchidaceae) resulted in the isolation of denthirsinin (Fig. 2.156) and lusianthridin (Fig. 2.157) which were inactive and antagonized the histamine-induced contractions of the rat ileum. The contractions evoked by BaCl$_2$ were inhibited, lightly by denthirsinin, but lusianthridin enhanced the contractions elicited by BaCl$_2$ (Kovacs et al. 2008). In another study the relaxant response elicited by natural products is probably mediated by the neuronal release of NO. Compounds of *S. livida* induced the production of NO in ileal tissue, which in turn provoked relaxation of the ileal muscles by elevating the cyclic GMP content. Denthirsinin provoked the inhibition of noradrenaline-evoked contractions in the endothelium and appeared to induce a dose-response vasorelaxation by more than one mechanism (Estrada-Soto et al. 2006).

Conclusions

In the search for natural spasmolytic agents from plants a wide range of chemical compounds including alkaloids, coumarins, flavonoids, rotenoids, terpenoids, iridoids and triterpenes have been isolated with antispasmodic effect from several species in recent times. These compounds are able to relax isolated ileum and to inhibit contractions induced by receptor-dependent and -independent mechanisms. These data have provided evidence to support the use of traditional medicine as antispasmodic agents in the treatment of gastro-intestinal problems.

References

Abu-Niaaj, L., M. Abu-Zarga, S. Sabri, and S. Abdalla. 1993. Isolation and biological effects of 7-O-methyleriodictyol, a flavanone isolated from Artemisia monosperma, on rat isolated smooth muscles. Planta Med. 59: 42–45.

Abu-Niaaj, L., M. Abu-Zarga, S. Abadía, and S. Abdalla. 1996. Isolation and inhibitory effects of eupatilin, a flavone isolated from *Artemisia monosperma* Del., on rat isolated smooth muscle. Pharm. Biol. 34: 134–140.

Achem, S.R. 2004. Treatment of spastic esophageal motility disorders. Gastroenterol. Clin. North Am. 33: 107–124.

Afifi, F.U., E. Khalil, and S. Abdalla. 1999. Effect of isoorientin isolated from *Arum palaestinum* on uterine smooth muscle of rats and guinea pigs. J. Ethnopharmacol 65: 173–177.

Aguirre-Crespo, F., J. Vergara-Galicia, R. Villalobos-Molina, J.J. Lopez-Guerrero, G. Navarrete-Vazquez, and S. Estrada-Soto. 2006. Ursolic acid mediates the vasorelaxant activity of *Lepechinia caulescens* via NO release in isolated rat thoracic aorta. Life Sci. 79: 1062–1068.

Astudillo, A., E. Hong, R. Bye, and A. Navarrete. 2004. Antispasmodic activity of extracts and compounds of *Acalypha phleoides* Cav. Phytother. Res. 18: 102–106.

Barile, E., R. Capasso, A.A. Izzo, V. Lanzotti, S.E. Sajjadi, and B. Zalfaghari. 2005. Structure-activity relationships for saponins from *Allium hirtifolium* and *Allium elburzense* and their antispasmodic activity. Planta Med. 71: 1010–1018.

Begum, S., I. Sultana, B.S. Siddiqui, F. Shaheen, and A.H. Gilani. 2002. Structure and spasmolytic activity of eucatyptanoic acid from *Eucalyptus camaldulensis* var obtusa and synthesis of its active derivative from oleanolic acid. J. Nat. Prod. 65: 1939–1941.

Bergendorff , O., and O. Sternerl. 1995. Spasmolytic flavonols from *Artemisia abrotanum*. Planta Med. 61: 370–371.

Borrelli, F., N. Milic, V. Ascione, F. Capasso, A.A. Izzo, F. Petrucci, R.Valente, E. Fattorusso, and O. Taglialatela-Scafati. 2005. Isolation of new rotenoids from *Boerhaavia diffusa* and evaluation of their effect on intestinal motility. Planta Med. 71: 928–932.

Borrelli, F., V. Ascione, R. Capasso, A.A. Izzo, E. Fattorusso, and O. Taglialatela-Scafati. 2006. Spasmolytic effects of nonprenylated rotenoid constituents of *Boerhaavia diffusa* roots. J. Nat. Prod. 69: 903–906.

Calixto, J.B., O.G. Miguel, R.A. Yunes, and G.A. Rae. 1990. Action of 2-hydroxy-4,6-dimethoxyacetophenone isolated from *Sebastiania schottiana*. Planta Med. 56: 31–35.

Corea, G., E. Fattorusso, V. Lanzotti, R. Capasso, and A.A. Izzo. 2005. Antispasmodic saponins from bulbs of red onion, *Allium cepa* L. Var. Tropea. J. Agric. Food Chem. 53: 935–940.

Cos, P., M. Calomme, B. Sindambiwe, T. Bruyne, K. Cimanga, L. Pieters, A, Vlietinck, and DV. Berghe. 2001. Cytotoxicity and lipid peroxidation-inhibiting activity of flavonoids. Planta Med. 67: 515–519.

Cunha, K.M.A., L.A.F. Paival, F.A. Santos, N.V. Gramosa, E.R. Silveira, and V.S.N. Rao. 2003. Smooth muscle relaxant effect of kaurenoic acid, a diterpene from *Copaifera langsdorffii* on rat uterus *in vitro*. Phytother. Res. 17: 320–324.

da Silva, T.M., B.A. da Silva, and R. Mukherjee. 1999. The monoterpene alkaloid cantleyine from *Strychnos trinervis* root and its spasmolytic properties. *Phytomedicine* 6: 169–176.

Eichler, O. and C. Koch. 1970. Antiphlogistic, analgesic and spasmolytic effect of harpagoside, a glycoside from the root of *Harpagophytum procumbens* DC. Arzneimttel-Forsch. 20: 107–109.

El Sayah, M., V. Cechinel Filho, R.A. Yunes, T.R. Pinheiro, and J.B. Calixto. 1998. Action of polygodial, a sesquiterpene isolated from *Drymis winteri*, in the guinea-pig ileum and trachea *in vitro*. Eur. J. Pharmacol. 344: 215–221.

Emendorfer, F., F. Emendorfer, F. Bellato, V.F. Noldin, V. Cechinel-Filho, R.A. Yunes, F.D. Monache, and A.M. Cardozo. 2005. Antispasmodic activity of fractions and cynaropicrin from *Cynara scolymus* on guinea-pig ileum. Biol. Pharm Bull. 28: 902–904.

Estrada, S., J.J. López-Guerrero, and R. Villalobos-Molina. 2004. Spasmolytic stilbenoids from *Maxillaria densa*. Fitoterapia 75: 690–695.

Estrada-Soto, S., J.J. Lopez-Guerrero, and R. Villalobos-Molina. 2006. Endothelium-independent relaxation of aorta ring by two stilbenoids from the orchids *Scaphyglottis livida*. Fitoterapia 77: 236–239.

Estrada-Soto, S., A. Rodriguez-Avilez, C. Castañeda-Avila, P. Castillo-España, G. Navarrete-Vazquez, L. Hernandez, and F. Aguirre-Crespo. 2007. Spasmolytic action of *Lepechinia caulescens* is through calcium channel blockade and NO release. J. Ethnopharmacol. 114: 364–370.

Farhadi, A., K. Bruninga, J. Fields, and A. Keshavarzian. 2001. Irritable bowel syndrome: an update on therapeutic modalities. Expert Opin. Investig. Drugs 10: 1211–1222.

Filho, V.C., O.G. Miguel, R.J. Nunes, J.B. Calixto, and R.A. Yunes. 1995. Antispasmodic activity of xanthoxyline derivatives: Structure-activity Relationships. J. Pharm. Sci. 84: 473–475.

Fleer, H. and E.J. Verspohl. 2007. Antispasmodic activity of an extract from *Plantago lanceolata* L. and some isolated compounds. Phytomedicine 14: 409–415.

Fusi, F., F. Ferrara, C. Koorbanally, N.R. Crouch, D.A. Mulholland, and G. Sgaragli. 2008. Vascular myorelaxing activity of isolates from South African Hyacinthaceae partly mediated by activation of soluble guanylyl cyclase in rat aortic ring preparations. J. Pharm. Pharmacol. 60: 489–497.

Ghayur, M.N., H. Khan, and A.H. Gilani. 2007. Antispasmodic, bronchodilator and vasodilator activities of (+)-catechin, a naturally occurring flavonoid. Arch. Pharm. Res. 56: 970–975.

Gilani, A.H., M.N. Ghayur, A. Khalid, A. Zaheer-ul-Haq, M.I., and A. Choudhary. 2005. Presence of antispasmodic, antidiarrheal, antisecretory, calcium antagonist and acetilcholinesterase inhibitory alkaloids in *Sarcococca saligna*. Planta Med. 71: 120–125.

Gilani, A.H., A.U. Khan, M.N. Ghayur, S.F. Ali, and W. Herzig. 2006. Antispasmodic effects of Rooibos tea (*Aspalathus linearis*) is mediated predominantly through K$^+$ -channel activation. Basic Clin. Pharmacol. Toxicol. 99: 365–373.

Go, M.L., and S.L. Leung. 1984. The synthesis of 5-methyl-2-furylacetic acid. An alternative route. Acta. Pharm. Suec. 21: 77–80.

Gomes, A.Y.S., M.F.V. Souza, S.F. Cortes, and V.S. Lemos. 2005. Mechanism involved in the spasmolytic effect of a mixture of two triterpenes cycloartenol and cycloeucalenol, isolated from *Herissanthia tiubae* in the guinea-pig ileum. Planta Med. 71: 1025–1029.

Gomes, F.P. 1953. Antispasmodic activity of an active principle (visnagan) rom *Ammi visnaga*. C R Seances Soc. Biol. Fil. 147: 1836–1839.

Hasler, W.L. 2003. Pharmacotherapy for intestinal motor and sensory disorders. Gastroenterol. Clin. North Am. 32: 707–732.

Hazelhoff, B., and T.M. Malingre. 1982. Antispasmodic effects of Valeriana Compounds: An *in-Vivo* and *in-Vitro* Study on the guinea-pig ileum. Arch. Int. Pharmacodyn. 257: 274–287.

Heinrich, M., B. Heneka, A. Ankli, H. Rimpler, O. Sticher, and T. Kostiza. 2005. Spasmolytic and antidiarrhoeal properties of the Yucatec Mayan medicinal plant *Casimiroa tetrameria*. J. Pharm. Pharmacol. 57: 1081–1085.

Hernandez-Romero, Y., J. Rojas, R. Castillo, and A. Rojas. 2004. Spasmolytic effects, mode of action, and structure-activity relationships of stilbenoids from *Nidema boothii*. J. Nat. Prod. 67: 160–167.

Herrera, M.D., E. Marhuenda, and A. Gibson. 1992. Effects of genistein, an isoflavone isolated from *Genista ridentata*, on isolated guinea-pig ileum and guinea-pig ileal myenteric plexus. Planta Med. 58: 314–316.

Hiller, K.O., M. Ghorbani, and H. Schilcher. 1998. Antispasmodic and relaxant activity of chelidonine, protopine, coptisine, and *Chelidonium majus* extracts on isolated guinea-pig ileum. Planta Med. 64: 758–60.

Ito, M., Y. Kondo, and T. Takemoto. 1976. Spasmolytic substances from *Cimicifuga dahurica* (Maxim). Chem. Pharm Bull. 24: 580–583.

Itthipanichpong, C., N. Ruangrungsi, W. Kemsri, and A. Sawasdipanich. 2003. Antispasmodic effects of curcuminoids on isolated guinea-pig ileum and rat uterus. J. Med. Assoc .Thai. 86 Suppl. 2: S299–309.

Jailwala, J., T.F. Imperiale, and K. Kroenke. 2000. Pharmacologic treatment of the irritable bowel syndrome: a systematic review of randomized controlled trials. Ann. Intern. Med. 133: 136–147.

Kaneda, T., K. Shimizu, S. Nakajyo, and N. Urakami. 1998. The difference in the inhibitory mechanisms of papaverine on vascular and intestinal smooth muscles. Eur. J. Pharm. 355: 149–157.

Kar, K., V.N. Puri, G.K. Patnaik, R.N. Sur, B.N. Dhawan, D.K. Kulshrestha, and R.P. Rastogi. 1975. Spasmolytic constituents of *Cedrus deodara* (Roxb.) Loud: pharmacological evaluation of himachalol. J. Pharm. Sci. 64: 258–262.

Katsura, T. 1995. The remarkable effect of Kanzo-to and Shakuyakukanzo-to in the treatment of acute abdominal pain. Jpn. J. Orient. Med. 46: 293–299.

Kim, H.K., B.S. Cheon, Y.H. Kim, S.Y. Kim, and H.P. Kim. 1999. Effects of naturally occurring flavonoids on nitric oxide production in the macrophage cell line RAW 264.7 and their structure-activity relationships. Biochem. Pharmacol. 58: 759–765.

Ko, W.C. 1980. A newly isolated antispasmodic-butylidenephthalide. Jpn. J. Pharmacol. 30: 85–91.

Ko, W.C., S.W. Kuo, R. Sheuj, C.H. Lin, S.H., Tzeng, and C.M. Chen. 1999. Relaxant effects of quercetin methyl ether derivatives in isolated guinea pig trachea and their structure-activity relationships. Planta Med. 65: 273–275.

Ko, W.C., H.L. Wang, C.B. Lei, C.H. Shih, M.I. Chung, and C.N.Lin. 2002. Mechanisms of relaxant action of 3-O-methylquercetin in isolated guinea pig trachea. Planta Med. 68: 30–35.

Ko, W., P. Liu, J. Chen, I. Leu, and C. Shih. 2003. Relaxant effects of flavonoids in isolated guinea pig trachea and their structure-activity relationships. Planta Med. 69: 1086–1090.

Kovacs, A., A. Vasas, and J. Hohmann. 2008. Natural phenanthrenes and their biological activity. Phytochemistry 69: 1084–1110.

Lanata, C.F., and W. Mendoza. 2002. Improving diarrhoea estimates. Monitoring and Evaluation Team Child and Adolescent Health and Development. World Health Organization Geneva, Switzerland. URL: http://www.who.intlchild-adolescent–health New-Publications/CHILD-HEAL TH/EPI/ Improving-Diarrhoea-Estimates. pdf. (revised in May 2007).

Lemmens-Gruber, R., E. Marchart, P. Rawnduzi, N. Engel, B. Benedek, and B. Kopp. 2006. Investigation of the spasmolytic activity of the flavonoid fraction of *Achillea millefolium* on isolated guinea-pig ilea. Arzneimttel-Forsch. 56: 582–588.

Lin, C.H., F.N. Ko, Y.C. Wu, S.T. Lu, and C.M. Teng. 1993. The relaxant actions on guinea- pig trachealis of atherosperminine isolated from *Fissistigma glaucescens*. Eur. J. Pharmacol. 237: 109–116.

Lozoya, X., M. Meckes, M. Abou-Zaid, J. Tortoriello, C. Nozzolillo, and J.T. Arnason. 1994. Quercetin glycosides in *Psidium guajava* L. leaves and determination of a spasmolytic principle. Arch. Med. Res. 25: 11–15.

Maeda, T., K. Shinozuka, K. Baba, M. Hayashi, and E. Hayashi. 1983. Effect of Shakuyaku-kanzoh-toh a prescription composed of Shakuyaku (*Paeoniae Radix*) and Kanzoh (*Glycrryhizae Radix*) on guinea pig ileum. J. Pharmacobio-Dyn. 6: 153–160.

Mata, R., A. Rojas, L. Acevedo, S. Estrada, F. Calzada, I. Rojas, and E. Linares. 1997. Smooth muscle relaxing flavovonoids and terpenoids from *Conyza filaginoides*. Planta Med. 63: 31–35.

Moura, N.F., A.F. Morer, E.C., Dessoy, N. Zanatta, M.M. Burger, N. Ahlert, G.P. Porto, and B. Baldisserotto. 2002. Alkaloids Amides and Antispasmodic Activity of *Zanthoxylum hyemale*. Planta Med. 68: 534–538.

Nagai, H., J. He, T. Tani, and T. Akao. 2007. Antispasmodic activity of licochalcone A, a species-specific ingredient of *Glycyrrhiza inflata* roots. Pharm Pharmacol. 59: 1421–1426.

Namgoong, S.Y., K.H. Son, H.W. Chang, S.S. Kang, and H.P. Kim. 1994. Effects of naturally occurring flavonoids on mitogen-induced lymphocyte proliferation and mixed lymphocyte culture. Life Sci. 54: 313–320.

Nasiri, A., A. Holth, and L. Björk. 1993. Effects of the sesquiterpene capsidiol on isolated guinea-pig ileum and trachea, and on prostaglandin synthesis *in vitro*. Planta Med. 59: 203–206.

Oliveira, R.C., J.T. Lima, L.A. Ribeiro, J.L. Silva, F.S. Monteiro, T.S. Assis, M.F. Agra, T.M. Silva, F.R. Almeida, and B.A.Silva. 2006. Spasmolytic action of the methanol extract and isojuripidine from *Solanum asterophorum* Mart. (Solanaceae) leaves in guinea-pig ileum. Z. Naturforsch. [C]. 61: 799–805.

Olukoga, A. and D. Donaldson. 2000. Liquorice and its health implications. J.R. Soc. Health 120: 83–89.

Orallo, F. 2003. Pharmacological effects of (+)-nantenine, an alkaloid isolated from *Platycapnos spicata*, in several rat isolated tissues. Planta Med. 69. 135–142.

Orisadipe, A., S. Amos, A. Adesomoju, L. Binda, M. Emeje, J. Okogun, C. Wambebe, and K. Gamaniel. 2001. Spasmolytic activity of methyl angolensate: a triterpenoid isolated from *Entandrophragma angolense*. Biol. Pharm. Bull. 24: 364–367.

Ortiz de Urbina, A.V., M.L. Martín, B. Fernández, L. San Román, and L. Cubillo. 1994.

In Vitro antispasmodic activity of peracetylated penstemonoside, aucubin and catalpol. Planta Med. 60: 512–5.

Pasricha, P.J. 2001. Prokinetic agents, antiemetics. and agents used in irritable bowel syndrome. In: J.B. Hardman, L.E. Limbird, and A. Goodman Gilman (eds.). Goodman & Gilman's the pharmacological basis of therapeutics. McGraw Hill, New York. pp. 1021–1036.

Pasricha, P.J. 2006. Treatment of disorders of bowel motility and water flux; antiemetics; agents used in biliary and pancreatic disease. In: L.L. Brunton, J.S. Lazo, and K.L. Parker. (eds.). Goodman and Oilman's: The pharmacological basis of therapeutics. McGraw Hill, New York. pp. 983–1019.

Patnaik, G.K., and B.N. Dhawan. 1982. Evaluation of spasmolytic activity of Clausmarin A. a novel coumarin from *Clausena pentaphylla* (Roxb.). J. Ethnopharmacol 6: 127–137.

Patnaik, G.K., K.K. Banaudha, K.A. Khan, A. Shoeb, and B.N. Dhawan .1987. Spasmolytic Activity of Angelicin: A Coumarin from *Heracleum thomsoni*. Planta Med. 517–520.

Perez-Gutierrez, R.M., and R. Vargas-Solis. 2006. γ-Lactone isolated from metanol extract of the leaves of *Eryngium carlinae* and their antispasmodic effect on rat ileum. Bol. Lat. Caribe Plants Med. Arom. 5: 51–56.

Perez-Gutierrez R.M., F.M. Vargas-Solis R. Martinez and G.E. Baez. 2000. δ-Lactone from *Oedogonium capillare* and their effects on rat ileum. Nat. Prod. Res. 20: 305–310.

Perez-Hernandez, N., H. Ponce-Montera, and J.A. Medina. 2008. Spasmolytic effect of constituents from *Lepechinia caulescens* on rat uterus. J. Ethnopharmacol. 115: 30–35.

Ponce-Monter, H., M. Meckes, A. Macias, and M. Campos. 2006. Terpenoids relaxant effect of xanthomicrol and 3α-angeloyloxy-2α-hydroxy-13, 14Z-dehydrocativic acid from *Brickellia paniculata* on rat uterus. Biol. Pharm. Bull. 29: 1501–1503.

Pongprayoon, U., P. Baeckström, U. Jacobsson, M. Lindström, and L. Bohlin. 1992. Antispasmodic activity of beta-damascenone and E-phytol isolated from *Ipomoea pescaprae*. Planta Med. 58: 19–21.

Puyvelde, L.V., R. Lefebvre, P. Mugabo, N. Kimpe, and N. Schamp. 1987. Active principles of *Tetradenia riparia*; II. Antispasmodic activity of 8(14),15- sandaracopimaradiene-18-diol. Planta Med. 156–158.

Ragone, M.I., M. Sella, P. Conforti, M.G. Volonté, and A.E. Consolini. 2007. The spasmolytic effect of *Aloysia citriodora*, Palau (South American cedrón) is partially due to its vitexin but not isovitexin on rat duodenums. J. Ethnopharmacol. 113: 258–266.

Rojas, A., S. Cruz, and H. Ponce-Monter. 1996. Smooth muscle relaxing compounds from *Dodonae*. Planta Med. 62: 154–159.

Rojas, A., M. Bah, J.I. Rojas, and D.M. Gutiérrez. 2000. Smooth muscle relaxing activity of gentiopicroside isolated from *Gentiana spathacea*. Planta Med 66: 765–767.

Rivero-Cruz, B., I. Rivero-Cruz, and R. Rodriguez-Sotres. 2007. Effect of natural and synthetic benzyl benzoates on calmodulin. Phytochemistry 68: 1147–1155.

Roman-Ramos, R., C.C. Contreras-Weber, G. Nohpal-Grajeda, I.L. Flores-Saenz, and P.I. Alarcón-Aguilar. 2001. Blood glucose level decrease caused by extracts and fractions from *Lepechinia caulescens* in healthy and alloxan- diabetic mice. Pharm. Biol. 39: 317–321.

Romussi, G., G. Ciarallo, A, Bisio, N. Fontana, F. De simona, N. De Tommas, N. Mascolo and L. Pinto. 2001. A new diterpenoid with antispasmodic activity from *Salvia cinnabarina*. Planta Med. 67: 153–155.

Sato, Y., T. Akao, J. Heb, H. Nojima, Y. Kuraishic, T. Morotad, T. Asanod, and T. Tanib. 2006. Glycycoumarin from *Glycyrrhizae Radix* acts as a potent antispasmodic through inhibition of phosphodiesterase 3. J. Ethnopharmacol. 105: 409–414.

Sato, Y., J.X. He, H. Nagai, T. Tani, and T. Akao. 2007. Isoliquiritigenin, one of the antispasmodic principles of *Glycyrrhiza ularensis* roots, acts in the lower part of intestine. Biol. Pharm Bull. 30: 145–149.

Sotníková, R., V. Kettmann, D. Kostálová, and E. Táborská. 1997. Relaxant properties of some aporphine alkaloids from *Mahonia aquifolium*. Meth. Find. Exp. Clin. Pharmacol. 19: 589–597.

Thomas, G., U.J. Branco, J.M. Barbosa Filho, M. Bachelet, and B.B. Vargaftig. 1994. Studies on the mechanism of spasmolytic activity of (O-methyl-)-N-(2,6-dihydroxyben-zoyl) tyramine, a constituent of *Aniba riparia* (Nees) Mez. (Lauraceae), in rat uterus, rabbit aorta and guinea-pig alveolar leucocytes. J. Pharm Pharmacol. 46: 103–107.

Trute, A., J. Gross, E. Mutschler, and A. Nahrstedt 1997. *In vitro* antispasmodic compounds of the dry extract obtained from *Hedera helix*. Planta Med. 63: 125–129.

Ullah, F., H. Hussain, J. Hussain, I.A. Bukhari, M. Tareq, K. Khan, M.I. Choudhary, A.H. Gilani, and V.U. Ahmad. 2007. Tyrosinase inhibitory pentacyclic triterpenes and analgesic and spasmolytic activities of methanol extracts of *Rhododendron Collettianum*. Phytother. Res. 21: 1076–1081.

Villavicencio-Nieto, MA., and B.E. Perez-Escandon. 2006. Useful plants of the state of Hidalgo. Editorial Amalgama Arte Editorial. Pachuca de Soto. Hidalgo Mexico. p. 81.

Wu, X.L., Y.Y. Wang, J. Cheng, and Y.Y. Zhao. 2006. Calcium channel blocking activity of calycosin, a major active component of *Astragali Radix*, on rat aorta. Acta Pharmacol. Sin. A27: 1007–1012.

Zamilpa, A., J. Tortoriello, V. Navarro, G. Delgado, and L. Alvarez. 2002. Antispasmodic and antimicrobial diterpenic acids from *Viguiera hypargyrea* roots. Planta Med. 68: 281–283.

Zhang, G., S. Shimokawa, M. Mochizuki, T. Kumamoto, W. Nakanishi, T. Watanabe, T. Ishikawa, K. Matsumoto, K. Tashima, S. Horie, Y. Higuchi, and O.P.Dominguez. 2008. Chemical constituents of *Aristolochia constricta*: Antispasmodic effects of its constituents in guinea-pig ileum and isolation of a diterpeno-Lignan hybrid. J. Nat. Prod. 71: 1167–1172.

Chapter 3

Brazilian Ethnomedicinal Plants with Anti-inflammatory Action

José Carlos Tavares Carvalho

Introduction

The use of flora as a source of therapeutic material is as ancient as therapy itself. Most, if not all, botanical species appear to have a rudimentary immune system (Sarti 1995) that favored the development of means of chemical defense against the attack of bacteria, fungi, protozoa, insects, birds and other animals. The substances that constitute this barrier—the natural products, frequently evidence hormonal toxic activity against parasites, insects that transmit disease and other live organisms that afflict humans and could therefore constitute useful therapeutic and/or prophylactic agents.

In the beginning of history, when popular knowledge gradually emerged from darkness, the use of plants as a form of curing or preventing diseases became a subject of interest. The rebirth of interest in medicinal plants is not due to 'quackery' or to the passing fad of return to nature. The true motives of this phenomenon should be analyzed from the economic, temporal and social points of view. To this effect, one should consider the enormous progress of modern chemical industry, which favored greater speed and diversity in the production of new medicines, privileging therapy with a varied number of synthetic drugs. However, the heavy financial and temporal investments required for research and production of new medications have contributed toward the resurgence of medicinal plants as sources of pharmaceutical raw materials.

Laboratório de Pesquisa em Fármacos, Universidade Federal do Amapá, Rod. JK km 02 CEP 68902-280, Macapá-AP, Brazil.
E-mail: *farmacos@unifap.br*

Still within this context, we should take into account the fact that, with the elimination of the chemical synthesis stage, the cost of a drug of plant origin is reduced in view of the financial investment required for the development of drugs of synthetic origin. It is estimated that the development of a new synthetic drug involves funds amounting to 125 million US dollars, in at least 10 years of research, from synthesis through to its approval by the health authorities for sale on the market. Moreover, it is necessary to test close to 5,000 chemical compounds for this screening to result in a drug for therapeutic use (Korolkovas 1983, ABIFARMA 1989).

Hence, the need to seek inspiration in nature once more and to use the substances of innate defense present in medicinal plants. It is not a case of choosing between chemical substances extracted from nature and synthetic products. Far from these two groups excluding each other, as in the current stage of pharmacotherapy, it is not fitting to replace, by medicinal plants, the efficient medication available for the treatment of many diseases, instead it is suitable to recognize that, in a number of cases that are hardly negligible, medicinal plants offer undeniable advantages over synthetic drugs. It must also be acknowledged that one of the particularly relevant aspects of essential research, in the field of chemistry of medicinal plants, is the shorter time lapse that separates basic research from therapeutic use.

Consequently, some countries without resources and time to develop synthetic drugs required for the resolution of their health problems, have started to dedicate considerable efforts to the recovery and scientific exploration of their therapeutic traditions.

Such evolution in medicinal therapy requires a new approach in the evaluation and use of medicinal plants employed in popular medicine. The plant extracts henceforth began to be executed with well defined goals: to select substances endowed with specific actions. The forms of extraction and of fractionation were therefore conducted and guided through continuous biological tests of their various fractions until the pharmacologically active substance was isolated.

Thus plentiful and interesting compounds were discovered and, although few of them are of clinical utility in modern medicine, the action of some confers credibility to the use of plants in popular medicine. On the other hand, certain medicinal plants did not prove to have the effects that popular medicine attributed to them, and another even proved dangerous. Hence the researcher's duty is to oppose the indiscriminate use of medicines and methods employed in popular medicine. Each one of the plants should be submitted to a critical and thorough examination and no use should be made without medical recommendation.

In this spirit, the old Central de Medicamentos—CEME (Medications Center), a Brazilian Government Institution, in 1982, deployed Programa de Pesquisa de Plantas Medicinais—PPPM (Medicinal Plant Research Program), aimed at the development of alternative and complementary therapeutics, with scientific grounds, through the production of medication originating from the determination of the true pharmacological value of preparations of popular use on a so-called medicinal plant basis. These natural preparations would be submitted to various trials, whereby there would be confirmation or not of the therapeutic properties attributed thereto. Those preparations in which the medical action as well as the therapeutic effectiveness and the absence of harmful effects were proved, would be fit for inclusion in Relação Nacional de Medicamentos Essenciais do Brasil—RENAME (National List of Essential Medications from Brazil). A total 57 popular medicinal species were initially selected that formed the list of plants from the program (CEME 1983, 1985). These selected plants, according to their uses, were classified in nine groups, one of which was made up of plants with supposedly analgesic, anti-inflammatory, antipyretic and/or antispasmodic actions.

Necessity was, is and will continue to be the great lever to impel humankind. Pain caused man to seek the analgesic; disease, the medicine. It is, therefore, easy to infer that the use of plants to fight diseases is as old as humanity itself (Oliveira et al. 1991).

The study of plants, used empirically by witchdoctors, countrymen and villagers in the relief of infirmities, has contributed to the enrichment of the therapeutic arsenal. Ipecac roots, known by Brazilian witchdoctors, or *pajés*, and those of Rauvolfia, by the Veda physicians, constitute classic examples of plants used without a scientific basis in the old days, and today sanctioned by medicine (Jorge Neto 1973).

Possessing exuberant flora, Brazil these days constitutes one of the largest storage areas of plants considered to be of potential medicinal value. Scientific progress in recent times calls for knowledge and exploration of our plant resources. The study of plants, geared toward their use as a therapeutic weapon, should preferably be executed starting with those already consolidated in popular medicine. It constitutes an arduous and not very easy task to detect the active principles of these plants and to analyze them; nevertheless, without this, the medicinal use of these plants is not justified.

Brazil, which has around 120,000 species of higher plants, is a true vegetal empire. According to ethnobiological sources, several of these plants played an important role in Brazilian life, being used, by indigenous peoples, as a remedy for their maladies or as venom and anti-venom in their wars and in their hunting expeditions (Gottlieb and Mors 1979).

The first travelers were surprised by the large number of medicinal plants known by the Indians, and by the fact that the natives used simple remedies, unlike the Europeans who believed in semi-magical combinations of several plants. Actually, few primitive peoples acquired such complete knowledge of the physicochemical and therapeutic properties of their botanic environment as the South American indigenous tribes (Lévi-Strauss 1987).

Such knowledge was partly transmitted to explorers and also to slaves, when their ships docked off Brazilian coasts. These people and those who arrived here later, besides using these plants, also introduced others, further increasing our popular phytotherapic arsenal. Phytotherapy (treatment of diseases with preparations using a plant species base) was the essential item of the therapeutic arsenal until the mid-19th century. Since then, it has gradually given way to preparations made with pure molecules of active elements, isolated from medicinal plants endowed with more specific pharmacological action (Atisso 1979). Therefore, medicinal plants have lost their importance and started to be used only as an alternative therapy in Brazil and in other developing countries, where about 70–80% of the population does not have access to medical/pharmaceutical care. Thus phytotherapic substances were forgotten and were substituted by therapies based on the use of chemical substances (fully or partly synthetic).

Lately, unmistakable signs have appeared that phytotherapy is making a comeback and even in the First World, the market of natural products is expanding rapidly.

A medical plant means "Every plant being administered in any form and by some route to man or animal, that exercises some kind of pharmacological effect thereupon" (WHO 1992).

Through this definition plants that are pharmacologically inert, and those considered medicinal, but whose pharmacological effects are not well demonstrated, are excluded. In this case, popular preparations of the fruit of plant species *Pterodon emarginatus* have demonstrated significant effectiveness in terms of their use in popular medicine, where they are employed as anti-inflammatory substances. Therefore, this chapter aims to present a comprehensive review of studies of the plant species *P. emarginatus*.

The inflammatory process

The word inflammation is derived from the state of being inflamed. To inflame means to 'set on fire', which implies the color red, the possibility of heat and the generation of pain. The word inflammation is particularly appropriate as a term to describe the body's physiological response to a lesion, invasion by external factors and, occasionally, by self-aggression (Trowbridge and Emling 1996).

Inflammation can be defined as the response of the microcirculation and its components to injury. In recent years, considerable interest has been presented in the study of pathological etiology and pharmacological control of the inflammatory response. This interest originated from the discovery that in several existing diseases we can find tissue damage that resulted from an inflammatory response. There are conditions in which the stimulus that provokes the reaction would be innocuous, in other words, it does not have the capacity to activate an acute inflammatory response, as is the case in some nontoxic granulomatous reactions.

Inflammation is a tissue response that occurs in the connective tissue specifically; and is thus considered a mesenchymatic phenomenon.

Among mesenchymatous cells, we should emphasize the mast cells, the fibroblasts, and the phagocytic cells originating from the blood. Blood vessels are an integral and important part of the connective tissue. Fibers, such as those of collagen, especially the reticular and elastic fibers, form with the amorphous or essential substance, the non-cellular part of the connective tissue. We should add the important presence of nerve fibers, both sensory and motor, that, although not connective in terms of their nature, are an integral part of the physiology of the connective tissue.

In this manner, when the organism is attacked, the connective tissue reacts by determining the inflammatory response, in which practically all the connective tissue elements take part (Douglas 1994).

Inflammation can be considered a stereotyped response of the tissues, under the effects of harmful stimuli. The phlogogenic agents are those that produce inflammation, or phlogistic phenomenon (Douglas 1994).

The causes that lead to inflammation are multiple and of variable nature. The following types of phlogogen are recognized: a) Biological, in this case the factor that triggers inflammation is of a biological nature, in other words, live beings that, in the tissue, produce inflammation, like bacteria, viruses, protozoa, helminths etc. In general they are microorganisms that represent the most frequent cause of inflammation; b) Chemical, when the inflammation is determined by the action of chemical substances of a widely varied nature, such as acids, alkalis, caustic substances, croton oil, turpentine, silver nitrate, formaldehyde, polysaccharide (carragenin) as well as by substances released from cells, or tumors in a necrobiotic process. c) Physical, when the physical factors, such as excessive heat (burns), or exaggerated cold (freezing), ultraviolet rays, ionizing radiations, electricity or action of mechanical factors, like traumatisms, fractures or incisions, that are habitually observed in surgery, are important causes of inflammation. d) Immune, although immunological processes are of a chemical or cellular nature, they are also a very frequent cause of inflammation. They are grouped separately, as they constitute a group of phlogogenic factors that cause the inflammatory response in a complex manner; moreover many inflammations include in their pathogeny, a mechanism of immune nature, such as participation of the complement (Garcia Leme1977).

The inflammatory response is basically a series of complex events, not a single one separately, and it follows a direction that is essentially uniform. Different kinds of injuries and different degrees of damage can, however, lead to variations of intensity and duration as a particular aspect of the response.

The characteristics of inflammation were first described by Celsus (30 BC–38 AD). He described the four cardinal signs of inflammation: 'redness' (hyperemia), 'swelling' (edema), 'heat' (increase of local temperature) and pain. Afterwards Galeno (130–200 AD) and finally Hunter (1728–1793) added loss of function.

The phlogogenic agent has the property of acting in certain cells, injuring them, that is, provoking a reaction in them with results that are expressed as inflammation. There is a debate regarding the existence of specific target cells that initiate inflammation, or whether all cells can start it, when affected. Apparently, for each kind of phlogogenic agent there will be a response of specific cells, but among them the mast cells, monocytes and polymorphonuclear cells are the most important as cells affected by the inflammatory stimulus, and where metabolic and functional modifications are produced which will be expressed through the release of substances of phlogistic effect (mediators), that is, it will determine the reactive characteristic of inflammation (Douglas 1994).

Inflammation is a nonspecific response of the organism to etiopathogenic stimuli of different nature. In clinical practice, it is more frequently associated with traumatisms, infections, deposits of crystals or of antigen-antibody complexes (Calin 1984).

The clinical manifestations of the inflammatory process are pain, hyperalgesia, erythema and edema. At the histopathologic level, two successive pictures occur. An acute stage, characterized by arteriolar and venular vasodilation, increase of permeability in microvascularization, by cellular contraction in the venular endothelium (fenestration), with protein leakage that decreases the vascular oncotic pressure and increases the hydrostatic pressure in microcirculation with leakage of liquid to the interstice (edema); that favors the formation of inflammatory exudates within the tissue; migration of polymorphonuclear cells (chemotaxis), initially; and buildup of macrophages at the lesion site, about 24 hours later. Another, which is installed 36 to 48 hr after the inflammatory stimulus, in which continuous leucocitary migration, with predominance of monocytes, lymphocytes, plasmocytes and fibroblasts, and signs of regeneration and reconstruction of the connective matrix. At the molecular level, there is protein denaturation, from the action of lytic enzymes (proteases, stereases, collagenases), released by the rupture of the lysosome membrane, as a consequence of the phagocytic action. The stimulation of inflammatory cells is the starting point for the activation of a series of systems that synthesize and release intermediate lesion substances, like histamine, serotonin, bradykinin, prostaglandins, leukotrienes and several chemotactic factors. These substances are responsible for vasodilation, increase of vascular permeability, migration of leucocytes, besides other manifestations of the acute inflammatory process.

Prostaglandins are more consistently involved in this process. They are formed by arachidonic acid, released from the phospholipids of the membrane of the injured cells, by catalytic action of phospholipase A_2, phospholipase C and diglyceride lipase. The enzymes cyclooxygenase and hydroperoxidase catalyze the sequential stages of the synthesis of prostanoids (classical prostaglandins and thromboxanes), while lipoxygenases transform the arachidonic acid into leukotrienes.

Prostaglandins (PG) are vasodilators, with the exception of thromboxane A_2. This stimulates platelet aggregation, contrary to prostacyclin. PGD_2 is released by the mast cells activated by allergic stimuli or others. PGE_2 inhibits the action of lymphocytes and of other cells that participate in the allergic or inflammatory responses. Besides promoting vasodilatation, prostaglandins sensitize nociceptors (hyperalgesia) and stimulate the hypothalamic thermoregulatory centers. Leukotrienes increase vascular permeability and attract the leukocytes to the lesion site. Histamine and bradykinin increase capillary permeability and activate the nociceptors (Fuchs and Wannmacher 1992).

The inflammatory reaction is present in almost all lesions produced in the human organism. Traumas, infections, immune reactions to external agents and autoimmune processes are accompanied, to a greater or lesser extent, by inflammatory reactions (Smith 1988).

Anti-inflammatory drugs and analgesics of plant origin

The capacity to alleviate pain through the use of plants or plant-based preparations was probably the first great victory of medicine and preceded by far the capacity to cure. Plants like the henbane, or mandrake, asafoetida and the poppy in the first Pharmacopeias of Babylon and of Sumeria (Pelt 1979).

As regards analgesics, it can be said that they are the best examples of the close relationship between primitive processes of medicinal plant selection and pharmacology and modern chemistry. Aspirin has its origin in the bark of the willow, a tree of the genus *Salix* that grows at the water's edge and only feels good 'with its feet wet'. According to the theory of signatures ('everything that nature creates', Swiss physician Paracelsus wrote in the 16th century, "receives the image of the virtue that it intends to hide there"), this indicates that the willow does not feel the cold, and therefore cures colds, the flu, fever, rheumatisms, etc., and, it is the bark that keeps the willow protected and warm, it is in this bark that one hopes to find the active principle in question. In the 18th century, it was discovered that the willow bark is as bitter as that of a tree from Peru, the quina, whose extract, quinine, was considered a popular medicine against malaria. Hence, an infusion of willow bark began to be used as antithermal medication by the Europeans. In 1828, the French citizen Leurox managed to extract from the willow bark a substance that he called salicin (from the Latin name for the plant). Soon afterwards, a Swiss pharmacist, called Pagensthcher, distilled from 'meadowsweet', a plant of the genus *Spiraea*, which also likes having its 'feet wet', a substance that closely resembles salicin—methyl salicylate. A few years later, inspired by these discoveries, a German researcher artificially produced salicylic acid. From this substance a derivative was obtained: acetylsalicylic acid, which is none other than the official name of aspirin, a universal medication for pain, whose 'spir' syllable brings to mind its plant origin: *Spiraea* (Pelt 1979).

Similar processes gave rise to many modern medications, whose principles derive from plants known by man since time immemorial. However, it is necessary to elucidate for once and for all the relations between science and empiricism, between scientists and those people that are in possession of secular knowledge, sometimes beneficial.

Constituents of plants with anti-inflammatory activity

There has been great progress in understanding the biochemistry of the inflammatory pathophysiology, and this favors the establishment of new *in vitro* test systems for the screening of several substances, which stimulates the discovery of new anti-inflammatory agents.

The term 'anti-inflammatory drugs' includes agents that intervene in an acute and chronic inflammatory process, such as rheumatic diseases or arthritis, and in which the metabolism of arachidonic acid occupies an important role in this process.

In analyzing the action of phytodrugs on the inflammatory process, we should consider the generation of prostaglandins, main mediators of inflammation. Depending on the metabolic pathway of the enzymatic system, arachidonic acid is converted in a variety of highly active metabolites. By the cyclooxygenase pathway it results in the formation of stable prostaglandins, PGE_2, PGD_2, and $PGF_{2\alpha}$ as well as prostacyclin (PGI_2) and thromboxane B_2, unstable intermediate cyclic endoperoxide pathway, PGG_2 and PGH_2. By the 5-lipoxygenase pathway it produces leukotriene B_4 and sulphopeptide leukotriencs (LT), LTC_4, LTD_4, and LTE_4, as well as 5-hydroxyeicosatetraenoic acid (5-HETE). These metabolites play important roles in inflammation, associated with vasodilatation, increase of capillary permeability, pain and chemotaxis (Schrör 1984). Consequently, the following classes of compounds of vegetal origin that act on the inflammatory process are:

1) Phenolic carboxylic acids, simple phenols, flavonoids, tannins and phenylpropane derivatives

After the introduction of aspirin, a classic non-steroidal anti-rheumatic agent, countless similar structures have been synthesized, but not essentially with the same activity. Salicylic acid derivatives commonly occur in many plants, but only plants that contain them (salicylic acid and/or salicylic alcohol), are used in phytotherapy. Some of the most important are: *Salix* spp., *Populus tremula, Filipendula ulmaria* (syn. *Spiraea ulmaria*), *Gaultheria procumbens, Betula* spp., *Viola tricolor,* and *Primula elatior.* During *in vitro* tests, in which cyclooxygenase, *Salix* extracts and all compounds present in willow bark were used, D-salicin, salicostine, salireposide, and chalconic glycosides, isosalipurposide were inactive in the concentration of 250 μM (Wagner and Reger1987). Some negative results were described by Flower and Vane (1974) for salicylic acid and its main metabolite, salicuric acid. However, free salicylic acid, formed *in vivo* in the liver by oxidation of salicylic alcohol, exercises a similar anti-inflammatory activity to aspirin (Whitehouse et al. 1977, Ferreira and Vane 1979). Esters were isolated from 4 and 5-methoxy-salicylic acid of *Primula* spp., and tested on a cyclooxygenase system (Wagner et al. 1987a). The most active inhibitor was the ester of methyl-4-methoxy salicylic acid, which presented inhibition of 50% in the concentration of 250 μM.

The phenolic compounds, eugenol, thymol, and carvacrol from the essential oil of *Syzygium aromaticum, Thymus vulgaris* and *Ledum palustre*, respectively, are considered structural analogues of salicylic derivatives. The IC_{50} value of these compounds was considered similar to that of indomethacin, in other words, 1.2 μM (Wagner et al. 1987b). The fact that acetyl eugenol presents a greater effect than eugenol suggests that the inhibition mechanism is similar to of aspirin, which inhibits the cyclooxygenase enzyme through an irreversible bond by transfer from the acetyl group to the protein.

When phenolic compounds were tested, it was noted that carvacrol presented inhibitory activity on cyclooxygenase only in the presence of adrenalin, which is commonly used in test systems as a cofactor. In the absence of this cofactor, carvacrol demonstrated stimulant activity on the enzymatic system, above the concentration of 10 μM. Detailed experiments, including a system of cells (platelets), revealed that carvacrol and other similar phenolic compounds, with certain substitution patterns, act as electron donors and are cooxidated by hydroperoxidase in prostaglandin biosynthesis. Moreover, similar observations were obtained with paracetamol by Robak et al. (1978) and with guajacol by Egan et al. (1980).

Flavonoids, like salicylic acid, were introduced into therapeutics before being tested on the metabolic activity of prostaglandins. Some flavonoids were investigated by Baumann et al. (1980) and by Michel et al. (1985). 5.7-dihydroxyflavone galangin (IC_{50}:5.5 µM) was proven to be the most active cyclooxygenase inhibitor.

Flavonoids have chemical structures with a C_6-C_3-C_6 carbon skeleton, represented by two aromatic rings A and B (Robbers et al. 1997). Those that exhibit an orthodihydroxy structure in the A or B ring were powerful inhibitors, more than with a free 3-hydroxyl group. Certain pre-ringed flavonoids, in general, morusin, from *Morus alba*, were also active, hypothetically due to their high liposolubility (Kimura et al. 1986, Evans et al. 1987). Catechin also presents cyclooxygenase inhibition activity (IC_{50}:130 µM) (Michel et al. 1985).

Several tannins have been repeatedly described, regarding the fact that they have antiphlogistic activity (Kimura et al. 1986), and some of them have presented activity in tests *in vitro*, involving cyclooxygenase. The most active cyclooxygenase inhibitors of this class, are the lichen substance, 4-0-methyl cryptochlorocaffeic acid, (IC_{50}: approximately 10 µM) and a mixture of gallotannin consisting of tetra-, penta-, and oligo-galloylglucose (IC_{50}: approximately 10 µM). Ellagitannins are great inhibitors, which indicates that cyclooxygenase inhibition depends on the number of galloil residues. Similar dependence for this radical was observed by Nishizawa et al. (1983), who used different enzyme systems. In view of these observations, tannins should be removed from the plant extracts before their evaluation, on cyclooxygenase, to detect the presence of other powerful cyclooxygenase inhibitors. As the mechanism of action of gallotannins is important, it was suggested that they act by destroying oxygen radicals, in the same manner as it has been established for antioxidant agents, like gallic acid ethyl ester. In comparison, all the other phenolic compounds probably act by direct reversible competitive inhibition of the enzyme.

The structure-activity relationship of the class of cyclooxygenase inhibitors of plant origin is interesting, as it is believed that all compounds have at least one phenolic group or one equivalent substitute, like the sulfoxide group in thiosulfinates. Masking these groups leads to a reduction of the significant inhibitory effect. The cathecolic structure does not appear essential. Lipophilic substituents increase the inhibitory activity, and several additional polar groups are found in glycosides, decreasing or even suppressing their effects.

Numerous investigations have also been described involving the inhibitory activity of phenols, phenol carboxylic acids, coumarins, and flavonoids in the 5-lipoxygenase system test. The most powerful inhibitors of 5-lipoxygenase are the flavonoids: quercetin, 7-hydroxyethyl-quercetin, crisiol and baicalein, and the coumarin esculetin, which have IC_{50} values between 0.1 and 5 µM (Yoshimoto et al. 1983, Hope et al. 1983, Furukawa et al. 1984).

The presence of a cathecolic structure appears essential for the inhibitory activity of 5 lipoxygenase, which allows us to suggest that such inhibitors act by free oxygen radical sweeping mechanism. However, the IC_{50} value of cinnamic acid is very high (46 to 100µM), depending on the test system used. The high polarity of esters of caffeic acid, chlorogenic acid, cichoric acid, and rosmarinic acid, is completely inactive. Therefore, a certain degree of liposolubility seems to be a necessary requisite for a lipoxygenase inhibitor (Furukawa et al. 1984).

These hypotheses are in accordance with the results obtained for wedelolactone (IC_{50}: 2.5 µM), isolated from *Eclipta alba* and *Wedelia calendulacea*, which presents inhibitory activity similar to the most potent lipoxygenase inhibitor, which has been described, nordihydroguaiaretic acid (NDGA; IC_{50}: 1.5 µM) (Wagner 1996). Chemiluminescence, due to the release of oxygen radicals, is inhibited in a dose-dependent manner by wedelolactone, in concentrations that range between 3×10^{-5} and 3×10^{-9} µM. Similar inhibitory activity is presented by NDGA, and the scavenger property of this class of compounds was confirmed by Wagner in 1996. Investigations on the structure-activity relationship of a series of coumarins and flavonoids demonstrated that the high inhibitory property is not only due to derivatives with cathecol type substitution in ring B and an OH group or OH together with a second lipophilic group (OCH_3, OC_2H_5) in ring A (Wong et al. 1988).

There have been descriptions of the topic application of pure phenols (eugenol and thymol) and flavonoids, or corresponding drug extracts, for the treatment of diseases of the respiratory tract, gastric inflammation, or rheumatic diseases. However, when used systemically, the ideal plasma concentration for triggering of the therapeutic response, is only obtained in exceptional circumstances.

2) Triterpenes, steroids, and sesquiterpenes

These types of substances exercise their antiphlogistic effects by intervening in mechanism of the immunological reaction. For example, a hyperactivity of the complement system, characterized by formation of immune complexes and inflammatory humoral factors, is a factor that contributes to the clinical manifestation of rheumatoid arthritis, acute glomerulonephritis, and systemic lupus erythematosus. In other autoimmune processes, the main part is presented by hyperactivity of the cell defense mechanism (macrophages, lymphocytes) with excessive secretion of interleukin 1 and 2 (Otterness and Bliven 1982).

Inhibition of the classic complement pathway by phenolics such as rosmarinic acid has been described by some groups of researchers (US Pat 1981, Garcza et al. 1985). Other esters of caffeic acid, such as, chlorogenic acid, isochlorogenic acid, and cinarin, also inhibit the complement (Wagner 1996). However, results obtained with several triterpenes, demonstrated their relevance for therapeutic application, with respect to inhibition of the complement system (Wagner and Reger 1987). The most powerful complement system inhibitor demonstrated was α-, β-boswellic acids, isolated from *Boswellia serrata* (incense), which provoked practically 100% of inhibition in the concentration of 0.1 µM. Glycyrrhetinic acid, which is used as succinic ester (Carbenoxolon, Bigastrone) for ulceration treatment, has very intense inhibitory activity. As an inhibitor of the classic complement pathway, boswellic acid is indicated, in some cases (Wagner et al. 1996). However, this has a corticomimetic activity, inhibiting corticoid degradation, and also intervenes with the metabolism of arachidonic acid (Tamura et al. 1979, Inoue et al. 1986). The mechanism of the antiphlogistic action of aescine (Hiai et al. 1981), saikosaponins of *Bupleurum falcatum* (Yamamoto et al. 1975), and saponins of *Dodonea viscosa* (Wagner et al. 1987b) is probably via action resembling that of glycyrrhetinic acid, in other words, inhibitor of the complement system. Numerous antiphlogistic agents act like helenalin (sesquiterpene lactone) and many other structural derivatives that occur in the species of *Arnica, Eupatorium, Tanacetum, Parthenium*, and others from the family Asteraceae. These terpenes have been studied in several animal models by Hall et al. (1979), but their mechanisms of action are still unknown. A tanacetum extract (*Chrysanthemum parthenium*) was investigated that is popularly employed in England for the treatment of rheumatism. Using the chemiluminescence method, it was discovered that this extract has the capacity to inhibit the release of platelet activating factor (PAF) induced by granulocytosis (Capasso 1986). This served as a basis for the establishment of a similar activity for the active principle of the drug extract, parthenolide (Capasso 1986), whereas it is possible that this sesquiterpene lactone acts on protein kinase C of granulocytes. The increase in activity of the phospholipase A_2 enzyme, by stimulus of the lipocortin bond, leads to the increase of the arachidonic acid metabolism. However, inhibition of protein kinase C indirectly leads to inhibition of the arachidonic acid metabolism. This mechanism of action is in accordance with the observations of Hayes and Foreman (1987).

The anti-oxygen-free-radical-effect can be measured by the chemoluminescence method, using the system of purin xantinoxidase of free cells. Based on this fact, it is interesting to note an antiphlogistic principle isolated from the fruit juice of *Ecballium elaterium*, which is applied externally in diluted form for the treatment of sinusitis in Turkey, and that was identified as being curcurbitacin B (Yesilada et al. 1988). After oral administration of this compound in mice, vascular permeability inhibition was measured by the method of Mustard et al. (1965). In the dose of 200 mg/kg, the inhibition was 70%, which is significantly higher than that presented by aspirin. The mechanism of action is still unknown, but in view of the irritant action of curcubitacin, we have not arrived at its definition. Based on the fact that cucurbitacins are characteristic constituents of *Bryonia dioica*, which has been commonly used to treat rheumatism and gout, it is suggested that this substance acts by a mechanism similar to that of curcurbitacin. However, the antiphlogistic and antirheumatic action of steroids from *Withania somnifera* and *Lycium chinensis* appears to be via suppression of T lymphocytes (Bähr and Hansel 1982, Haruna et al. 1985, Choi and Woo 1986).

Studies of plants of the genus *Pterodon*: a brazilian medicinal species

The fruits of the Sucupira Branca or Faveira, common names of the species *Pterodon emarginatus* Vog. (Fabaceae) are used in the form of hydroalcoholic macerate in popular medicine in the treatment of rheumatism, sore throat, spinal problems and even as a depurative and fortifier, as indicated by the ethnobotanical study conducted in 30 cities from the interior of Minas Gerais (Leite de Almeida and Gottlieb 1975).

The genus *Pterodon* includes four native species in Brazil: *P. abruptus* Benth, *P. apparicori* Pedersoli, *P. emarginatus* Vog. (synonymy -*P. pubenscens* Benth) and *P. polygalaeflorus* Benth. The investigation of the four species was motivated by the discovery of the anti-cercarian action of the oil from the fruit of *P. pubenscens* Benth (Mors et al. 1967), and from the other three species (Fascio et al. 1975) besides *in vitro* antimicrobial activity of the oil obtained from *P. pubenscens* Benth (Jorge Neto1973).

In 1967, Mors et al. isolated from the essential oil of the fruits of *P. pubenscens* the substance 14,15-epoxigeranilgeraniol, which proved effective as a chemoprophylactic agent in schistosomiasis. In 1970, new diterpenoids were obtained from *P. emarginatus* by Mahjan and Monteiro. In the same year, Fascio et al. (1970) isolated two new terpenes from the oil of the fruit of *P. pubenscens*.

Based on a chemical study of Brazilian leguminous plants, Bras Filho et al. (1971) managed to isolate isoflavones from *P. pubenscens*. Proceeding with the study of the species *P. emarginatus*, Mahjan and Monteiro (1972) isolated more new diterpenoids from this species. Diterpenes, found in the species *P. pubenscens*, were effective in the prophylaxis of schistosomiasis (Dos Santos Filho et al. 1972). Isoflavones were also found in the species *P. apparicori* (Galira and Gottlieb 1974).

Fourteen furane diterpenes, among other substances, were described and isolated from the fruits of the genus *Pterodon*. Four of these diterpenes belong to the species *P. polygaeflorus* Benth (Fascio et al. 1975).

Nunan (1985) demonstrated the anti-inflammatory activity of the furane-diterpene 6,7- dihydroxyvouacapan-17 sodium oate (PpDDE-81), isolated from the oil of the fruits of *P. polygalaeflorus* Benth, in carragenin-induced paw edema.

From the hexane extract of the seeds of *P. polygalaeflorus* Benth, by steam stripping, Campos et al. (1990) obtained essential oil in which they identified the following constituents: ylangene, α-capaene, β-caryophyllene, α-humulene, γ-elemene and δ-cadinene. Teixeira et al. (1991), from the unsaponifiable fraction of the ethanolic extract of the seeds, isolated and characterized 6α, 7β-dihydroxy vouocapan-17β-oic acid. Campos et al. (1992), continuing with the phytochemical study of the said species, recorded the isolation of another two vouocapanic diterpenes that were: 6α-hydroxy-7β-acetoxy vouocapan-17β-methyloate and 6α-hydroxy-7β-acetoxy vouocapan-14(17)-ene.

Natural diterpenoids isolated from the fruit of species of *Pterodon:*

P. pubescens—diterpene methyl 6α,7β-diacetoxy-14-hydroxy vinhaticoate.

P. apparicioi, and *P. emarginatus*—Diterpene 7β-acetoxy vouacapan.

P. apparicioi, and *P. emarginatus* Diterpene 6α,7β-diacetoxy vouacapan.

P. emarginatus, and *P. polygalaeflorus* Diterpene Vouacapan-6α,7β,14β-triol.

P. pubescens Methyl 6α, 7β-diacetoxy-12-16-dihydro-12,14-dihydroxy-16-oxo-vinhaticoate.

P. pubescens Diterpene 6α,7β-diacetoxy vouacap-14β-al.

P. emarginatus, P. polygalaeflorus, P. apparicioti, and *P. pubescens* 6α, 7β-diacetoxivouacapânico-14β-oato.

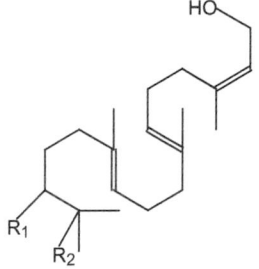

P. pubescens, P. emarginatus, P. polygalaeflorus and *P. apparicieri* Geranilgeraniol linear diterpenoid.

Botanical description of the *Pterodon emarginatus* Vog. species

Systematic position of the species according to Cronquist (Gemtchújnicov, 1976):

Division	:	Angiospermae
Class	:	Dicotyledoneae
Subclass	:	Archichlamydeae
Order	:	Rosales
Suborder	:	Leguminosineae
Family	:	Fabaceae
Subfamily	:	Faboideae (Papilionoideae)
Tribe	:	Dalbergieae
Subtribe	:	Geoffraeinae
Genus	:	*Pterodon*
Species	:	*P. emarginatus* Vog.
Common synonymy	:	Sucupira-branca, faveira.
Foreign common synonymy	:	Alcornoque (Venezuela and Colombia)

The common name sucupira-branca is employed for several vegetal species of the family Fabaceae, including *Pterodon emarginatus* Vog. This species is characterized by its presentation in the form of a tree with a height of 5–10 m and 50–70 diameter cm; grayish ritidome coming loose in rounded sheets with a diameter of 2–3 cm and leaving scars; alternate, compound leaves, 8–20 cm in length, between the rachis and petiole; angular and slightly pilous rachis; folioles with a width of 3 cm, numbering from 10 to 20, alternate and underlapping, of emarginated apex, with small hairs; inflorescence in terminal panicule, shorter than the leaves; acute caduceus membranaceous bracteoles, measuring 2 mm; flowers with 6–8 mm pedicel, with one only flower; calyx with 5 segments, 2 of which are large, petaloid, puberulous, with internal and external glandular punctuations and the other 3 exhibiting sepaloid denticles; corolla with violaceous petals, glabrous and with glandular punctuations; oval vexilli with emarginated apex, with two mucrons; wings oboval, bifid and mucronate, glabrous ovals, stipitate; full style with single stigma; monadelphous stamens, glabrous oval anthers, dorsifixed, with a gland at the apex; drupaceous-woody orbicular legume, with surrounding wing, monospermous, 3-4 cm in length and 2-2.5 cm in width (Corrêa 1984).

Fig. 3.1. *Pterodon emarginatus* Vog. Fruits without the outside layer (5 cm), and with dark outside layer (epicarp–5.5 cm).

Color image of this figure appears in the color plate section at the end of the book.

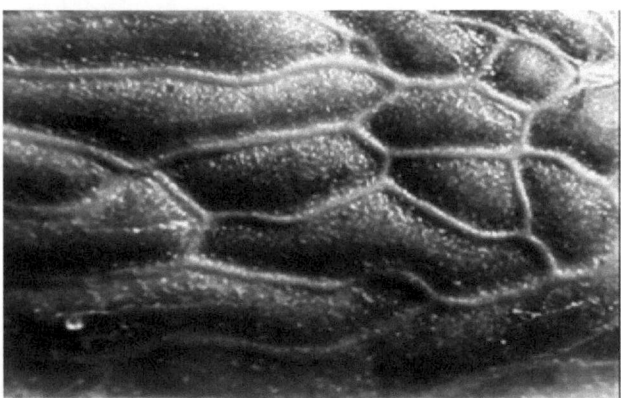

Fig. 3.2. Veins of the fruits of *Pterodon emarginatus* outlining small salient and elongated regions that correspond to the oil producing glands.

Color image of this figure appears in the color plate section at the end of the book.

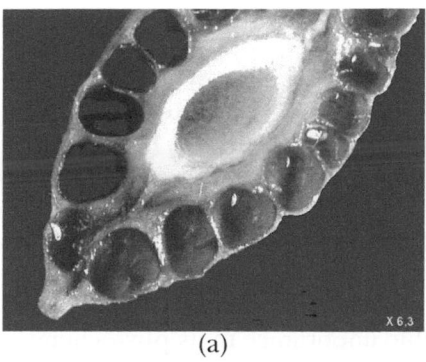
(a)

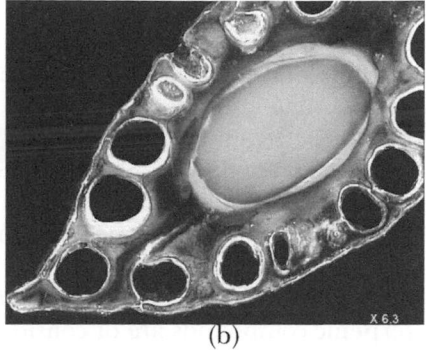
(b)

Fig. 3.3. *Pterodon emarginatus* Vog. (fruit) (a) Cross section (6.3 x), showing details of the transversal cut of the schizogenous glands. (b) Cross section (6.3 x) of the schizogenous glands, with the seed observed at the center. Coloration of aqueous methyl green solution 1:20.

Color image of this figure appears in the color plate section at the end of the book.

The genus *Pterodon* previously included five native species from Brazil, *P. pubescens* Benth., *P. emarginatus* Vog., *P. polygalaeflorus* Benth., *P. apparicioi* Perdesoli, and *P. abruptus* Benth., while today it is known that the species *P. emarginatus* Vog. and *P. pubescens* Benth. are botanical synonyms (Lorenzi 1992). The chemical investigation of these species started when it was demonstrated that the oil of *Pterodon pubescens* fruits inhibited the penetration of cercaries through the skin, and that the elements responsible for this activity were 14,15-epoxi geranilgeraniol and linear diterpenoid 14,15-dihydroxy-14,15-dihydroxy geranilgeraniol (Mors et al. 1967). A series of furane diterpenes was isolated from all the species of *Pterodon*, and in the case of the species *P. emarginatus*, 6α,7β-dihydroxy vouacapan-17β-oic acid appears to be the majority compound, although linear diterpenes are also found (Fascio et al. 1975).

Fig. 3.4. 6,7-dihydroxy vouacapan-17-oic acid, isolated from *Pterodon emarginatus* Vog.

Live organisms can be considered a biosynthetic laboratory not only as regards to the chemical compounds of the primary metabolism (carbohydrates, proteins, fats) used as food by human beings and by animals, but also refers to those of the secondary metabolism (glycosides, alkaloids, terpenoids) that exercise physiologic effects. It is the latter that confer therapeutic properties to active principles extracted from plants and animals. Drugs can be used as such in a natural form or in the form of extracts, whose principles are employed as therapeutic agents (Robbers et al. 1997).

In the vegetal species *Pterodon emarginatus*, it was verified that terpenic compounds are of considerable importance in its phytochemical constitution, with the detection of both monoterpenes in the essential oil and tricyclic diterpenes in the crude hexane extract.

Terpenoids are widely distributed in nature and are found in abundance in higher plants. Moreover, some fungi are capable of producing several interesting terpenoids and marine organisms are a prolific source of somewhat uncommon terpenoids. These compounds are also found in pheromones of insects and in their defense secretions. Terpenoids are defined as natural products whose structure can be divided into isoprene units; and for this reason, these compounds are also called isoprenoids. The isoprene units originate biogenetically from acetate through the mevalonic acid pathway and have ramified chains, with five units of carbon that contains two double bonds (Harborne et al. 1981).

The number of different terpenoids isolated from natural sources is approximately 20 thousand, far higher than that of any other group of natural products. Vegetal terpenoids play a prominent role in discussions about chemical ecology, as they perform important duties as phytoalexins, insectifuges, insect repellents, polinic attraction agents, agents of defense against herbivores, pheromones, allelochemicals, vegetal hormones and signaling molecules. Terpenoids are involved in almost all the possible interactions among plants, between plants and animals and between plants and microorganisms (Teranishi et al. 1993).

The extensive information about the pharmacological action of terpenic compounds is illustrated by the existence of countless published works, and as they have characteristic chemical structures, which have been of considerable interest to medical chemistry laboratories. Through the analysis of the essential oil of *Pterodon emarginatus* fruits it was possible to identify the following compounds: β-caryophyllene (bicyclic sesquiterpene), α-pinene (bicyclic monoterpene), myrcene (monoterpene), ethyl-eugenol (phenylpropanoid), eugenol (phenylpropanoid), geraniol (acyclic monoterpene) and methyl eugenol (phenylpropanoid), and from the crude hexane extract they isolated 6α,7β-dihydroxy vouacapan-17β-oic acid (furanic tricyclic diterpene). Some of the physical chemical properties of these compounds, such as liposolubility, are relevant characteristics for their potent biological activities.

There are various references in literature to studies that emphasize activities of terpenic compounds, relating discussions of results that correlate the heterogeneity of such compounds with various pharmacological actions (Hall et al. 1979, Hall et al. 1987, Russin et al. 1989, Giordano et al. 1990, Zitterl-Eglseer et al. 1991, Zheng et al. 1992, Kaneda et al. 1992, Giordano et al. 1992, Martin et al. 1993, Valencia et al. 1994, Ulubelen et al. 1994, Guardia et al. 1994, Thebtaranonth et al. 1995, Tambe et al.1996, Williams et al.1996, Kubo et al. 1996, Mata et al. 1997).

Due to the broad spectrum of biological action, terpenic compounds favored the performance of countless studies with the intention of elucidating the possible metabolites formed in mammals. Therefore, the metabolism of caryophyllene and caryophyllene oxide were studied in rabbits. Each one of these sesquiterpenes was converted predominantly into primary, secondary or tertiary alcohols, whereas primary alcohol was found in higher concentration. In many cases, methyl vinyl and exo-methylene groups were hydroxylated and converted into glycol via epoxide. Eight new metabolites were determined by chemical and spectrophotometric methods (Asakawa et al. 1986). The renal excretion of metabolites of α-pinene was studied in humans, in individuals exposed to inhalation of this substance, in which there were 60% of pulmonary absorption, around 1–4% of the total was excreted in the *cis*- and *trans*-verbenol forms. Verbenols were eliminated 20 hr after exposure for 2 hr. The renal excretion of α-pinene was below 0.001% (Levin et al. 1992). A similar result was obtained in another study, with demonstration of the high affinity of this terpenic compound with the adipose tissue (Falk et al. 1990).

Diterpenoids, such as the 6α,7β-dihydroxy vouacapan-17β-oic acid isolated from the crude hexane extract of fruits, involve a large group of nonvolatile C_{20} compounds, derivatives of geranylgeranyl pyrophosphate. Although they are mostly of a vegetal or fungic origin, they can also be biosynthesized by some marine organisms and by insects. Some acyclic diterpenoids are known but the vast majority is of carbocyclic compounds that contain up to five rings. In the case of 6α,7β-dihydroxy vouacapan-17β-oic acid, this is a furanic tricyclic diterpene, in other words, with three rings and a furanic ring (Fig. 3.4). Contrary to the triterpenoids, diterpenoids rarely combine with sugars to form glycosides; with the exception of stevioside. Diterpenoids can present a large number of different biological activities (Dev 1986), and 6α,7β-dihydroxy vouacapan-17β-oic acid was initially isolated in the attempt to prove the active principle responsible for anti-cercarian activity (Fascio et al. 1975).

The participants of a study investigated the mode of action of crude hexane extract obtained from the fruits of the vegetal species *Pterodon emarginatus* as an anti-inflammatory agent and its effects on the various mechanisms of inflammation and on the gastric mucosa.

Redness, swelling, heat and pain are always present in the inflammatory process. In spite of this, when different inflammatory agents are introduced in the tissues, the phenomena are slightly different in extent and duration. Therefore, for a considerable number of agents, certain aspects appear common, while others are specific (Van Arman et al. 1965).

The edematogenic response is one of the signs of the inflammatory response arising from the increase of vascular permeability, which occurs in microcirculation, due to the action of the mediators released. The release of mediators in the inflammatory process is highly dependent on the inflammatory stimulus.

Many of the methods commonly used for screening of anti-inflammatory drugs are based on the ability of such test agents to inhibit the edema produced in the rat's rear paw by the injection of inflammatory agents, such as carragenin, dextran, histamine and other inflammatory mediators.

Carragenin is a mucopolysaccharid obtained from Irish sea algae, *Chondrus crispus*. Gardner (1960) employed carragenin to produce experimental arthritis in rabbits and guinea pigs, and showed that the response of this material depends entirely on local stimulus for the inflammatory process, not being antiogenic, not producing systemic effects and presenting a high degree of reproducibility.

Carragenin has been used traditionally in the laboratory to experimentally induce acute paw edema in animals (Winter et al. 1962). Many mediators are involved in the carragenin-induced edema, including histamine, serotonin, kinins and PGs (Willis 1969a and 1969b). It is necessary to consider the role of these mediators in the carragenin-induced edema, obtaining information about the nature of the irritation, inflammatory response site and the standard of the pharmacological mediators released (Di Rosa and Sorrentino 1970). The increase in paw size resulting from the subplantar injection of carragenin was found to represent 3% of the edema volume. Using techniques based on the quantity of dislocated fluid (Van Arman et al. 1965, Vinegar 1968), volumetric measurements could be taken repeatedly in normal non-anesthetized animals and the course of time of the edema formation determined (Vinegar 1968).

This inflammation model (Di Rosa 1972) is identified by three different phases of release of mediators during edema development, namely: starting from experiments, using histamine and serotonin (5-HT) antagonists, an initial phase was determined in which there is release of histamine and serotonin (from 0 to 1.5 hr), whose mediators are released at the same time and in a satisfactory concentration to produce effects on vascular permeability. This explains the failure of antihistamines and of serotonin antagonists used to modify early paw edema.

The second phase of the inflammatory response (from 1.5 to 2.5 hr) is mediated by kinins, with the transitory role of increasing vascular permeability after the carragenin injection (Di Rosa and Sorrentino 1970).

The last phase of the inflammatory response (from 2.5 to 6.0 hr) is mediated by a substance that is not a histamine, serotonin or kinins, but it is demonstrated that prostaglandin is present in carragenin-induced exudates (Willis, 1969a and b).

A time-edema curve of two periods is obtained as a result, in which there is accelerated edema formation (biphasic). The development of the first phase starts immediately after the carragenin injection, quickly accelerating and then diminishing between 20 mins and 1 hr. The second phase starts 1 hr after the injection of the inflammatory agent. The accelerated production of the edema is between 1 to 3 hr and diminishes from 3 to 7 hr.

The importance of the biphasic nature of the time-carragenin-induced paw edema curve is recognized by effects of standard non-steroidal and steroidal anti-inflammatory drugs (Katsuhiko and Takashi 1982).

Hydrocortisone produces dose-dependent inhibition in the second phase, with little effect in the first phase. The second phase is sensitive to both drugs, steroidal and non-steroidal. The high sensibility to steroids induces evaluation of the importance of stress, generated by different causes, such as disease, noise or even handling of the animal, on edema formation when studying drugs with possible inhibitory actions, as endogenous adrenocorticoids are released in stress situations, and act by decreasing the volume of the edema sensitive to exogenous drugs (Vinegar et al. 1974).

The differences in the mechanisms of action of steroidal and non-steroidal drugs on carragenin-induced inflammation are evident. Steroidal anti-inflammatory drugs prevent neutrophils from taking part in the acute inflammatory response and, thus, decrease the quantity of enzymes, such as phospholipase A_2 for prostaglandin synthesis (Vinegar et al. 1982).

Since VANE's original hypothesis in 1971, cyclooxygenase inhibition proves to be the unifying hypothesis, which can explain all the anti-inflammatory actions of NSAIDs. Vane (1971) demonstrated that standard non-steroidal anti-inflammatory drugs inhibit cyclooxygenase, showing a correlation between *in vitro* inhibition of this synthetase and *in vivo* anti-inflammatory activity.

Studies indicate that prostaglandins are generated during the period in which the edema is formed (Vinegar et al. 1976). Carragenin phagocytosis by neutrophils occurs at the same time as the prostaglandins are synthesized by neutrophils.

Through these observations involved in the generation of the inflammatory process triggered by carragenin, it was verified that the crude hexane extract acts on these mechanisms, in which this extract behaved by determining a dose-response effect whose DE_{50} evaluated at the maximum peak of this process was 108 mg/kg. Knowing the phases involved in the generation of carragenin-induced inflammation, the crude hexane extract, when administered orally after the application of this phlogistic agent, significantly inhibited this process, with kinetics resembling the standard substance used, which was indomethacin.

The hypothesis of involvement of the crude hexane extract and the substance isolated from the extract in inhibition of the synthesis of prostaglandins via cyclooxygenase is suggested in this study, as according to the results obtained by the administration of these test substances, significant action was observed on the process triggered by carragenin, and did not inhibit that developed by the injection of dextran and histamine in the animals' paws. Action on the inflammation triggered by the injection of PGE_2 was observed as well.

In the experimental models in which there is participation of histamine, such as moderate thermal injuries, ultraviolet irradiation and injection of known histamine releasers, such as dextranes and the 48/80 compound, these are established through an immediate response. After immersing the rat's paw in lukewarm water, the quantity and release time of histamine are correlated with the high early edematogenic response. Once the edema is fully developed, there is no additional release of histamine (Horakova and Beaven 1974). Similar correlation was found by Nishida and Tomizawa (1980), after the injection of dextranes of different molecular weights in the rat's paw.

The response to dextrane is particularly suppressed by a pre-treatment with a combination of antagonists of H_1 and H_2 receptors (Nishida and Tomizawa 1980). These treatments were not effective in suppressing the late response in the injured tissue or the increase of vascular permeability induced by the 48/80 compound (Owen and Woodward 1980). Histamine has a dominant function in vasodilatation induction (when it interacts with its receptors), and functions in the mediation of the increase of vascular permeability.

Regardless of the role that it plays in these reactions, the dose of histamine released is a useful indicator of the degranulation of mast cells, which occurs during rapid accumulation after the dextrane injection, but not for late inflammatory responses as for carragenin.

The accumulation of free proteins, induced by dextrane, is not affected by treatment with indomethacin, while the infiltration of plasma proteins and neutrophils in the response to carragenin, is inhibited.

In view of all these observations, the participation of histamine in edema caused by dextrane is considered important, having maximum peak between the first and second hour, which distinguishes it from the edema provoked by carragenin (Carvalho et al. 1993).

The administration of crude hexane extract and the substance isolated from this extract in edema by dextrane and histamine did not produce its inhibition, as is the case of vascular permeability induced by histamine, thus demonstrating negative anti-histaminic activity, serving to reinforce the involvement with the mechanisms triggered by carragenin.

Both in dextrane-induced edema, and in that caused by histamine, the standard drug used was cyphroeptadine, a substance that blocks H_1 receptors, which significantly inhibited these edematogenic processes, besides vascular permeability.

Rat paw arthritis, induced by the application of Freund's adjuvant, has characteristics of rheumatoid arthritis. Arthritis is one of the most stressful and incapacitating syndromes found in the medical practice and, although it is one of the oldest diseases, there is no drug that leads to a permanent cure. In fact, steroidal and non-steroidal anti-inflammatory drugs are used for the improvement of symptoms, but they produce temporary relief and also produce severe side effects (Geetha and Varalakshmi 1998). So traditional phytotherapic medications, used for this pathology, represent possible sources of new drugs that can be employed, and through ethnobotanical studies it was demonstrated that the species *Pterodon emarginatus* is employed for this purpose in the form of cataplasm (Leite de Almeida and Gottlieb 1975).

Freund's complete adjuvant is comprised of mineral oil that contains dead tuberculosis bacilli (Freund et al. 1942). It is the most powerful adjuvant in terms of its capacity to stimulate cellular and humoral response, but is highly toxic since it contains non-metabolizable mineral oil and produces severe local reactions. It acts as deposition adjuvant, with the consequent slow release of the antigen. The mycobacterium acts directly on the immunocompetent cells.

Adjuvant arthritis, which occurs exclusively in the rat, was described for the first time by Stoerck et al. (1954), who obtained it by inoculation of homologous spleen cells with Freund's adjuvant in rats, that is, suspension of *M. tuberculosis* in pure mineral oil. Subsequently, Pearson (1956) rediscovered the phenomenon, obtaining arthritis compatible with the previous case by inoculation of Freund's complete adjuvant and homologous muscular tissue. It soon became evident that vigorous polyarthritis could be obtained just by the inoculation of Freund's complete adjuvant, without the addition of any other antigen (Pearson and Wood, 1959).

Arthritis thus obtained was called adjuvant arthritis. This experimental syndrome has been used for the study of pathogeny, pathology and therapeutics of conditions of an inflammatory nature, due to its similarities with some human chronic and evolutive affections, including rheumatoid arthritis (Sharp et al. 1961, Pearson et al. 1961, Newbould 1963, Pearson and Wood 1963, Pearson 1964).

Adjuvant arthritis does not develop in other species like the rabbit, guinea pig, hamster and mouse, and even in the rat it develops in only a few bloodlines.

Amado et al. (1986) demonstrated the development of the inflammatory response induced by Freund's complete adjuvant in two rat bloodlines: Holtzman and Lou/M isogenics, and concluded that, regardless of the mechanism through which some animals present greater resistance to inoculation of Freund's complete adjuvant, there is considerable variability in the frequency and severity of adjuvant arthritis in the bloodlines studied.

The injection of Freund's complete adjuvant (dead *Mycobacterium tuberculosis* suspended in mineral oil) in rat paws produces inflammation (primary reaction) that reaches maximum peak in 4 to 5 d; after this it remains temporarily stationary. Approximately 10 d after the injection, inflammatory lesions are discovered in the left paw (injected with saline only) and in the forepaws, ears and tail (secondary reaction).

The syndrome of the adjuvant was studied extensively and was approved as a convenient experimental model for evaluation of anti-inflammatory and/or immunosuppressive drugs. The inflammation associated with the adjuvant can be reduced through treatment with anti-inflammatory drugs like hydrocortisone and related substances as well as aspirin and related compounds (Di Pasquale et al. 1975, Kaiser and Gleen 1974, Sofia et al. 1976).

The inflammatory reaction in the injected paw was significantly greater in animals from the control group when compared with that of the groups treated with crude hexane extract and indomethacin, respectively. Likewise, the inflammatory reaction in the contralateral, non-injected paw appeared significantly reduced in these latter animals, indicating a certain resistance to the effects of the adjuvant.

Secondary lesions, characterized by the appearance of nodules in the ears, tail and edema in the forepaws, were observed as from the 10th day of the experimental period only in the control and indomethacin-treated groups (with lesser intensity).

Based on this data, it can be suggested that the rats from the control group were highly susceptible to inoculation of Freund's complete adjuvant, developing very severe arthritis, while the rats treated with crude hexane extract and indomethacin were more resistant.

Studies carried out with different rat bloodlines show that the frequency and the severity of adjuvant-induced arthritis vary a lot (Glenn and Gray

1965) and that genetic susceptibility is necessary for adjuvant arthritis to present all its peculiar clinical manifestations (Swingle et al. 1969, Rosenthale 1970), suggesting that the apparent resistance of some rat bloodlines is caused by insufficient drainage at the adjuvant injection site.

Several studies also showed the involvement of immune mechanisms in the development of secondary lesions of adjuvant arthritis (Whitehouse and Whitehouse 1968, Wegelius et al. 1970, Paulus et al. 1973, Pearson et al. 1975, Paulus et al. 1977, Mackenzie et al. 1978).

All the observations described, along with the result obtained with the oral administration of crude hexane extract, lead us to believe in the existence of a modulator effect of this extract and its components, on the mechanisms triggered by the intraplantar injection of Freund's complete adjuvant.

Antinociceptive drugs are generally classified according to their mode of action, in other words, they act centrally or peripherally depending on their site of action (Matsuda et al. 1998).

The model of abdominal contortion, induced by irritant substances, such as acetic acid for instance, injected in the peritoneal cavity of mice, is intended to test drugs that could present some analgesic activity, associated with anti-inflammatory action.

Through the intermittent contortions of the abdomen, torsion of the trunk and extension of the rear paws, it is possible to quantify the algesia induced by the intraperitoneal injection of the irritant agent, and thus to evidence whether a drug can or cannot present analgesic action.

In pain relief, opiates are generally considered analgesics, as they act on the central nervous system, while moderate analgesic-antipyretic compounds have largely peripheral effects (Winter and Flataker 1965, Matsuda et al. 1998).

Following the discovery of drugs resembling aspirin, Vane (1971) proposed that its analgesic effects result from the reestablishment of the nociceptive threshold by prostaglandin synthesis inhibition (Ferreira et al. 1990).

Hyperalgesia, induced by prostaglandins (nociceptive sensitization), results from the increase of $AMPc/Ca^{++}$ concentration in the nociceptors. This suggestion was particularly based on the fact that neuronal stimulation for activation of adenylate cyclase, by sympathomimetic amines, was capable of causing hyperalgesia in the rat paw. Several pharmacological and physiologic effects suggest the existence of the somatosensory influence (Wiesenfeld-Hallw and Hallin 1984).

Non-steroidal anti-inflammatory drugs have the common characteristic of not increasing the threshold stimulus, to evoke pain response in normal tissues, as is the case of narcotic analgesics. It is known that they restore the threshold for pain only in inflamed tissues and therefore, the effect of pain relief of this group is considered antialgic and non-analgesic (Ferreira et al. 1978).

As mentioned previously, Duarte et al. (1988) demonstrated that the administration of acetic acid (i.p.) leads to prostaglandin synthesis, as well as the release of mediators of the sympathetic system, perhaps by pathways similar to those observed by generation of $AMPc/Ca^{++}$second messengers. In accordance with this and plenty of other evidence, we can affirm that anti-inflammatory substances can also be involved in the analgesic activity at the peripheral level (Carvalho 1993).

Duarte et al. (1992) showed that $6\alpha,7\beta$-dihydroxy vouacapan-17β-oic acid isolated from the species *Pterodon polygalaeflorus* has an anti-nociceptive activity that is partially blocked by opioid antagonists, and the sodium salt of this acid has an anti-nociceptive activity with involvement of dopamine and its specific receptor D_2 and catecholamines (Duarte et al. 1996).

In this case the crude hexane extract demonstrated a dose-response effect when administered orally in mice, submitted to the contortions test, as did the substance isolated from the extract. These results corroborate that the anti-nociceptive effect of the crude hexane extract and substance is through a peripheral action accompanied by the anti-inflammatory action, since they were effective in inhibiting the inflammatory processes triggered by carragenin, nystatin and Freund's adjuvant, and did not significantly increase the latency time of animals in the hot plate test.

Vane (1971) demonstrated that many of the effects provoked by NSAIDs on the gastric mucosa, are results of the inhibition of prostaglandin synthesis. The initial injury is due to the direct damage mediated by NSAIDs, followed by a systemic effect, in which prostaglandin synthesis is inhibited. The specific synthesis of prostaglandin is tissue-specific, and those most prevalent in the gastrointestinal mucosa are E_2 (PGE_2), PGI_2 and PGF_2, considered cellular protectors of the gastric mucosa, as they act through increase of bicarbonate transportation and regulate the blood flow, avoiding ischemia and stimulating the longitudinal muscular contraction of the fundic region (Carson and Stron 1993). Prostaglandins inhibit the secretion of gastric acid endogenously, and, when administered exogenously, inhibit through reduction of intracellular AMPc synthesis. They are also critical components of defense of the gastrointestinal mucosa, protecting it against damage by a variety of agents (Shorrock and Rees 1988).

Local injury can also occur as a result of biliary excretion with active metabolites and subsequent duodenogastric reflux. The effects are cumulative, some local and others by systemic mechanisms that, alone, are sufficient to produce damage in the gastrointestinal mucosa. Much knowledge, involving damage induced by NSAIDs, is derived from previous studies, using salicylates, such as aspirin, which produces a pH-dependent local damaging effect (Guslandi 1997). At pH < 3.5, an acidity level commonly found in the stomach, aspirin predominates in its lipophilic non-ionized form. This condition favors transportation through the plasma into the epithelial cells membranes of the mucosa and the 'entrapment of ions'. The dissociated ions are imprisoned outside the cell. Within a few minutes after aspirin ingestion, the breakage of epithelial surface protection occurs, resulting in mucosa permeability increase. Na^+ and K^+ enter the luminal fluid, and the H^+ ions diffuse back inside the gastric lumen. The rediffusion of H^+ ions leads to additional damage to the mucosa.

The spectrum of injury by NSAIDs includes accentuated combination of subepithelial hemorrhage, erosions and ulcerations. The distinction between erosions and ulcerations depends on the pathological definition of the lesion penetration, for example, at the mucosa level for ulcers and limited superficial involvement of the mucosa for erosions (Lichtenstein et al. 1995). NSAID administration studies of short duration show dose-dependent injury in the gastrointestinal mucosa, designated NSAID gastropathy, with all the NSAIDs showing the potential to produce superficial injury (Lanas et al. 1995). The short–lasting administration of NSAIDs can cause epithelial damage of the ultrastructural gastric surface in a matter of minutes, and severe endoscopic gastrointestinal subepithelial hemorrhages and erosions within several hours after ingestion (Grahan and Smith 1986).

Prostaglandins of groups A, E and I inhibit the release of acid from the parietal cells. The theories for acid reduction include inhibition of biosynthesis of antisecretory prostanoids, and another possibility is stimulation of cyclic AMP generation, which reduces the secretion of hydrochloric acid (Di Bella et al. 1995).

Evidence accumulated during the last decade to support the view of the pharmacokinetic behavior of NSAIDs, not only contributes decisively to the therapeutic effects, but also to the type and incidence of their side effects. For this purpose it was shown that NSAIDs reach particularly high concentrations in the compartments in which they cause effects and side effects. Specifically, the data relating to these aspects indicates that NSAID accumulation inside the gastric mucosa cells, *a priori*, is a chief factor associated with the intracellular intervention of biochemical events, thus resulting in damage to the gastric mucosa (McCormack and Brune 1987).

Gastric cytoprotection is a property of certain substances, particularly prostaglandins, when used in non-antisecretory doses, to protect the gastric mucosa from inflammations and necrosis caused by exposure to harmful agents. The association between alterations of endogenous prostaglandins and damage in the gastric mucosa induced by several drugs was observed. The adaptable cytoprotection process, in which moderate irritant substances protect the gastric mucosa against the harmful effects of several necrotizing agents, appears to be mediated by prostaglandins. The exact mechanisms that are involved in cytoprotection activity have not yet been elucidated, although several theories were proposed. The involvement of thromboxanes, leukotrienes and endogenous sulphidrils was suggested in the pathogenesis of the damage to the gastric mucosa induced by several necrotizing agents (Parmar et al. 1987).

The structural requirements for the gastric cytoprotection activity of several sesquiterpene lactones were investigated. Theoretical-experimental studies of the potential of active centers of these substances were demonstrated. The biological evaluation of reduced analogues, and the simulation of molecular interactions between these compounds and an endogenous cysteine residue suggest that the presence of a non-steric Michael acceptor seems to be the essential structural requirement for cytoprotection activity in this family of substances. This observation suggests that cytoprotection is mediated by a Michael reaction between the mucosa peptides containing sulphidrils and Michael acceptors that are in the lactonic molecules. Besides this mechanism of action, it is suggested that sesquiterpenes stimulate the synthesis of endogenous prostaglandins (Giordano et al. 1992).

Gastric cytoprotection of β-caryophyllene, an anti-inflammatory sesquiterpene, was investigated in rats. Oral administration of β-caryophyllene in rats significantly inhibited damage to the gastric mucosa induced by necrotizing agents like absolute ethanol and HCl 0.6 N, although it did not prevent the gastric lesions induced by stress (water immersion) and indomethacine. This compound merely affected the secretion of gastric acid and pepsin. Accordingly, it can be affirmed that β-caryophyllene produces anti-inflammatory effects without producing any damage in the gastric mucosa, typical of NSAIDs. Moreover, this compound manifested cytoprotective effects and effectiveness in two dimensions, being clinically safe and potentially effective (Tambe et al. 1996).

Caryophyllene was identified in the essential oil of *P. emarginatus* fruits and this substance is also present in crude hexane extract, in which diterpene 6,7-dihydroxy vouacapan-17-oic acid (Fig. 3.4) was also found. It can be observed that the crude hexane extract produced a significant cytoprotection effect on the incidence of acute gastric lesions, produced by stress during a 17-h period, and in the evaluation of the anti-inflammatory power and occurrence of gastric ulcers a significant anti-inflammatory effect was observed together with the occurrence of gastric damage, with a zero rate of ulceration. Therefore, owing to the fact that all the components found in the fruit extract (crude hexane extract) are related to terpenic compounds, and as demonstrated above, since there are several studies demonstrating the effectiveness of gastric protection for this class of active principles, it is suggested that the low degree of interference of this substance on the gastric mucosa is due to the presence of these compounds.

Several factors are related to induction of hepatic injury. Starting with endotoxemia, which is commonly observed in patients with some liver diseases and in animals with experimental liver injuries (Leng-Peschlow 1996).

It was recently demonstrated that lipidic peroxidation in the livers of rats treated with CCl_4 increases with the progression of the lesion (Aguilar-Delfin et al. 1996, Ohta et al. 1997), and that lipidic peroxidation levels in the hepatic microsomal fraction also increases with the progression of the liver injury, besides which the decrease of levels occurs the faster the lesion is treated (Aguilar-Delfin et al. 1996).

Gonzáles Padrón et al. (1996) demonstrated the possibility that depression of hepatic microsomal glucose-6-phosphatase activity in rats intoxicated with CCl_4 is maintained as a result of the effect of lipidic peroxidation on the cellular components present in the phase of the G6 system. It was also demonstrated that in the exposure of the hepatic microsome, ascorbic acid/Fe^{+2}, product of lipidic peroxidation, contributes toward the oxidative inactivation of the G6 phase (Ohyashiki et al. 1995).

Therefore, it is suggested that for protection against liver injury produced by CCl_4 it is necessary for the drug to exhibit lipidic antiperoxidative action. In the case of the substance used as standard in this trial, silymarin (flavonolignan), acts through incorporation in the cell membrane, thereby increasing the membrane's resistance to injury, probably by changing the physico chemical properties, besides protecting the physiological antioxidants (glutation, superoxide dismutase) and preventing their depletion and consequently the function and the structure of the cell (Leng-Peschlow 1996). However, the animals that were treated with crude hexane extract and submitted to CCl_4 intoxication did not present injury suppression, on the contrary, its potentialization occurred. Austin et al. (1988) demonstrated that terpenic compounds stimulate the activity of cytochrome P-450, similar to phenobarbital. Therefore, the effect of crude hexane extract on CCl_4 intoxication can be justified by the major participation of terpenic compounds on the hepatic metabolism, requiring its integrity in terms of enzymatic activity, for its metabolites to be eliminated, since for some terpenes found in this extract, like caryophyllene, one of its metabolites is a compound with epoxide characteristic, for which the occurrence of irreversible stable bonds is suggested in the microsomal system, possibly leading to cases of hepatic intoxication (Asakawa et al. 1986).

In the histopathological analysis of the animals treated with crude hexane extract, during 30 d with the effective dose for the anti-inflammatory activity the occurrence of mild liver lesions was observed, to which it is possible to attribute the action of the components present in this extract; however, in all the trials for evaluation of the anti-inflammatory activity this proved effective, and there was no occurrence of damage in the gastrointestinal tract, from the esophagus to the large intestine, and no alterations were observed at the level of mucosa and its cells.

Conclusion

Through the studies related here, only one Brazilian vegetal species was presented that is widely used as an anti-inflammatory drug. There are hundreds of other species with this effect which have already been proven scientifically. Consequently, going by the results presented, and based on mechanisms already described in literature, and others that might come to light in this study, it is possible to suggest the anti-inflammatory activity of products of the fruit of the plant species *Pterodon emarginatus*.

References

Amado, C.A.B. and S. Jancar. 1986. Resposta inflamatória ao adjuvante completo de Freund —Comparação entre duas linhagens de ratos. Arquivos de Biologia e Tecnologia 29(4): 611–619.

Abifarma—Associação Brasileira Da Indústria Farmacêutica. 1989. A indústria farmacêutica no Brasil: a realidade. Rio de Janeiro, ABIFARMA.

Aguilar-Delfin, I., F. López-Barrera, and R. Hernández-Muñoz. 1996. Selective enhancement of lipid peroxidation in plasma membrane in two experimental models of liver regeneration, partial hepatectomy and acute CCl_4 administration. Hepatology 24: 657–662.

Asakawa, Y., T. Ishida, M. Toyota, and T. Takemoto. 1986. Terpenoid biotrasformation in mammals. IV. Biotransformation of (+)-longifolene, (-)-caryophyllene, (-)-caryophyllene oxide, (-)-cyclocolorenone, (+)-nootkatone, (-)-elemol, (-)-abietic acid and (+)-dehydroabietic acid in rabbits. Xenobiotica. 16(8): 753–767.

Atisso, M.A. 1979. As plantas medicinais voltam a florescer. vol. 9.O correio da Unesco. pp. 6–7.

Austin, C.A., E.A.Shephard, S.F. Pike, B.R. Rabin, and I.R. Phillips. 1988. The effects of terpenoid compounds on cytochrome P-450 levels in rat liver. Biochemical Pharmacology 37(11): 2223–2229.

Bähr, V., and R. Hänsel. 1982. Immunomodulating properties of 5, 20α(R)– Dihydroxy-6α,7α-epoxy-1-oxo-(5α)-witha-2,24-dienolide and solasodine. Planta Medica 44: 32–33.

Baumann, J., G.Wurm, and F.V. Bruchhausen. 1980. Hemmung der prostaglandinsynthetase durch flavonoide und phenolderivate im vergleich mit deren O_2^- -Radikalfänger-eigenschaften. Archiv der Pharmazie 313: 330–337.

Bras Filho, R., O.R. Gottlieb, and R.M.V. Assumpção. 1971. Chemistry of Brazilian Leguminosae. XXXIV Isoflavones of *Pterodon pubenscens*. Phytochemistry 10 (11):2835–2836.

Calin, A.A. 1984. Pain and inflammation. American Journal of Medicine 77 (3A): 9–16.

Campos, A.M., A.A.Craveiro, and T.C. Teixeira. 1990. Óleo essencial das sementes de *Pterodon polygalaeflorus* Bent. In: Resumos da Reunião da Sociedade Brasileira de Química, PN-004.

Campos, A.M., E.R. Silveira, and T.C. Teixeira. 1992. Análise química do óleo da semente de *Pterodon polygalaeflorus*. In: Resumos da Reunião da Sociedade Brasileira de Química, PN-061.

Capasso, F. 1986. The effect of an aqueous extract of *Tanacetum parthenium* L. on arachidonic acid metabolism by rat peritoneal leucocytes. Journal of Pharmacy and Pharmacology 38: 71–72.

Carson, J.L. and B.L. Stron. 1993. The gastrointestinal toxicity of the non-steroidal antiinflammatory drugs. In: Side-effects of Antiinflammatory Drugs. Kluwer Academic Dordrecht, Boston, Baltimore. 3: 1–8.

Carvalho, J.C.T. 1993. *Caesalpinia ferrea* (Pau-ferro): Avaliação da atividade antiinflamatória e analgésica. Dissertação de Mestrado. FCF/USP/RP CEME-COPESP (1983/1985)-Relatório da Comissão de Seleção de Plantas instituídas pela Portaria CEME No. 093.

Choi, J.S. and W.S. Woo. 1986. Triterpenoid glycosides from the roots of *Patrinia scabiosaefolia*. Planta medica 53: 62–65.

Corrêa, M.P. 1984. Dicionário das plantas úteis do Brasil e das exóticas cultivadas. Ministério da Agricultura (IBDF), Rio de Janeiro, p. 153.

Dev, S. 1986. Handbook of terpenoids: Diterpenoids, vol. 1–4. CRC Press, Inc. Boca Raton, Florida.

Di Bella, M., D.Braghiroli, and D.T. Witiak.1995. Antiallergic and antiulcer drugs. In: W.O.Foye, T.L. Lemke, and D.A. Williams.(eds.). Medicinal Chemistry, Williams & Wilkins, New York. pp. 416–443.

Di Rosa, M. 1972. Biological properties of carrageenan. Journal of Pharmacy and Pharmacology 24: 89–102.

Di Rosa, M. and L. Sorrentino. 1970. Some pharmacodynamic properties of carrageenin in the rat. British Journal of Pharmacology 38: 214.

Di Pasquale, G.,C. Rassart, P. Welas, and J. Gingold. 1975. Effect of single and multiple treaments with phenilbutazone in normal and adjuvant induced polyarthritic rats. Agents and Actions 5: 52–56.

Dos Santos Filho, D., W. Vichnewski, P.M. Baker,and B. Gilbert. 1972. Prophylaxis of Schistosomiaris. Diterpenes from *Pterodon pubenscens*. Anais da Acadêmia Brasileira de Ciências 44(1). 45–49.

Douglas, C.R. 1994. Fisiologia da inflamação. Tratado de Fisiologia aplicada às ciências da Saúde. 1ª edição, Robe editorial, S. Paulo. pp. 1315–1355.

Duarte, I.D.G., M.Nakamura, and S.H. Ferreira. 1988. Participation of the sympathetic system in acetic acid induced writhing in mice. Brazilian Journal of Medical and Biological Research 21: 341–343.

Duarte, I.D.G., M. Nakamura-Craig, and D.L. Ferreira-Alves. 1992. Possible participation of endogenous opioid peptides on the mechanism involved in analgesia induced by vouacapan. Life Sciences 50: 891–897.

Duarte, I.D.G., D.L. Ferreira-Alves, D.P. Veloso, and M. Nakamura-Craig. 1996. Evidence of the involvement of biogenic amines in the antinociceptive effect of a vouacapan extracted from *Pterodon polygalaeflorus* Benth. Journal of Ethnopharmacology 55: 13–18.

Egan, R.W., P.H.Gale, G.C. Beveridge, L.J. Marnett, and F.A. Kuehl. 1980. Advances in Prostaglandin and Thromboxane Rescarch.vol.6. In: B.Samuelsson, P.W. Ramwell, and R. Paoletti (eds.). Raven Press, New York. pp. 153–155.

Evans, A.T., E. Formukong, and FJ. Evans. 1987. Action of *Cannabis* constituents on enzymes of arachidonate metabolism: antiinflammatory potential. Biochemical Pharmacology 36: 2035–2037.

Falk, A.A., M.T. Hagberg, A.E. Lof, E.M. Wigaeus Hjelm, and Z.P. Wang. 1990. Uptake, distribution and elimination of alpha-pinene in man after exposure by inhalation. Scandinavian Journal of Work and Environmental Health. 16(5): 372–378.

Fascio, M., B.Gilbert, W.B. Mors, and T. Nishida. 1970. Two new diterpenes from *Pterodon pubenscens*. Anais da Academia Brasileira de Ciências 42 (Supl.): 97–101.

Fascio, M., W. Mors, B. Gilbert, M.B. Mahajan, M.B. Monteiro, D. Dos Santos Filho, and W. Vichnewski. 1975. Diterpenoid Furans from *Pterodon* species. Phytochemistry 15: 201–203.

Ferreira, S.H. 1979. A new method for measuring variations of rats paw volumes. Journal of Pharmacy and Pharmacology 31: 648.

Ferreira, S.H., and J.R. Vane. 1979. Handbook of Experimental Pharmacology. vol. 50/ II.In: J.R Vane and S.H. Ferreira (eds.). Springer-Verlag, Heidelberg-Berlin-New York. p. 371.

Ferreira, S.H., and J.R. Vane. 1979. Mode of action of anti-inflammatory agents which are prostaglandin syntetase inhibitors. In: J.R.Vane, and S.H.Ferreira. (eds.). Anti-inflammatory Drugs. Springer-Verlag,Berlin, Heidelberg, New York. pp. 348–398.

Ferreira, S.H., S.Moncada, and J.R. Vane. 1973. Some effects of inhibiting endogenous prostaglandin formation on the responses of the cat spleen. British Journal of Pharmacology 47: 48–58.

Ferreira, S. H., B.B.Lorenzetti, and F.M.A. Correa. 1978. Central and peripheral antialgesic action of aspirin-like drugs. European Journal of Pharmacology 53: 39.

Ferreira,S.H., B.B. Lorenzetti, M.S.A. Castro, and F.M.A.Correa.1978. Antialgic effect of aspirin-like drugs and the inhibition of prostaglandin synthesis. In: D.C.Dumonde, and M.K. Jasani (eds.). The Recognition on Anti-rheumatic Drugs. MTP Press Limited, St. Leonard, House, Lancaster, UK. pp. 25–37.

Ferreira, S.H., B.B. Lorenzetti, and D.I.De Campos. 1990. Induction, blockades and restoration of a persistent hypersensitive state. Pain 42: 365–371.

Flower, R.J. and J.R.Vane. 1974. Prostaglandin Synthetase Inhibitors In: H.J.Robinson and J.R. Vane (eds.). Inflammopharmacology. Raven Press, New York. pp. 9–18.

Freund, J., and K. Mac Dermott. 1942. Sensitization to horse serum by means of adjuvants. Proceedings of the Society of Experimental Biology and Medicine 49: 548–553.

Fuchs, F.D. and L. Wannmacher. 1992. Farmacologia clínica da inflamação. Farmacologia clínica—Fundamentos da terapêutica racional. Guanabara Koogan, Rio de Janeiro. pp. 149–171.

Furukawa, M., T. Joshimoto, K. Ochi, and S.Yamamoto. 1984. Studies on arachidonate 5-lipoxygenase of rat basophilic leukemia cell. Biochemica Et Biophysica Acta-Molecular Basis of Disease 795: 458–465.

Galira, E. and O.R. Gottlieb. 1974. Chemistry of Brazilian Leguminosae. XLVI Isoflavones from *Pterodon appariciori*. Phytochemistry 13 (11): 2593–2595..

Garcia Leme, J. 1977. Mecanismos de regulação do desenvolvimento de reações inflamatórias. Medicina. 9: 10–71.

Garcza, L., H. Koch, and E. Löffler. 1985. Peroxide als pflanzeninhaltsstoffe, 3. Mitt. 1 β-hydroperoxyisonobilin, ein neues hydroperoxid aus den blüten der römischen kamiliie (*Arthemis nobilis*). Archiv der Pharmazie 318, 1090.

Gardner, D.L. 1960. Adjuvant induced arthritis in the rats. Annual of Rheumatism Disease 19: 369.

Geetha, T. and P. Varalakshmi. 1998. Anti-inflammatory activity of lupeol and lupeol linoleate in adjuvant-induced arthritis. Fitoterapia 59(1): 13–19.

Gemtchújnicov, I.D. 1976. Manual de taxonomia vegetal. Editora Agronômica Ceres, S. Paulo, pp. 156–167.

Giordano, O.S., E. Guerreiro, M.J. Pestchanker, J. Guzman, D. Pastor, and T. Guardia. 1990. The gastric cytoprotective effects of several sesquiterpene. Journal of Natural Products 53(4): 803–809.

Giordano, O.S., M.J. Pestchanker, E. Guerreiro, J.R. Saad, R.D. Enriz, A.M. Rodriguez, E.A. Jauregui, J. Guzman, A.O. Maria, and G.H. Wendel. 1992. Structure-activity relationship in the gastric cytoprotective effect of several sesquiterpene lactones. Journal of Medicinal Chemistry 35(13): 2452–2458.

Glenn, E.M. and I. Gray. 1965. Chemical in adjuvant induced polyarthritis of rats. American Journal of Veterinary Research 26: 1195.

González Padrón, A., E.C.D. De Toranzo, and J.A. Castro. 1996. Depression of liver microssomal glucose-6-phosphatase activity in carbon tetrachloride-poisoned rats. Potential synergistic effects of lipid peroxidation and of covalent binding of haloalkane-derived free radicals to cellular components in the process. Free Radical Biology and Medicine 21: 81–87.

Gottlieb, O.R. and W. Mors.1979. A Floresta brasileira: Fabulosa Reserva Fitoquímica. O Correio 9: 35–38.

Grahan, D.Y., and J.L. Smith. 1986. Aspirin and the stomach. Gastroenterology 104: 390–398.

Guardia, T., J.A. Guzman, M.J. Pestchanker, E. Guerreiro, and O.S. Giordano. 1994. Mucus synthesis and sulfhydryl groups in cytoprotection mediated by dehydroleucodine, a sesquiterpene lactone. Journal of Natural Products 57(4): 507–509.

Guslandi, M. 1997. Gastric toxicty of antiplatelet therapy with low-dose aspirin. Drugs 53(1): 1–5.

Hall, I.H., K.H. Lee, C.O. Starnes, Y. Sumida, R.Y. Wu, T.G. Waddell, J.W. Cochran, and K.G.Gerhart. 1979. Anti-inflammatory activity of sesquiterpene lactones and related compounds. Journal of Pharmaceutical Science 68(5): 537–542.

Hall, I.H., K.H. Lee, and H.C. Sykes. 1987. Anti-inflammatory agents; IV. Structure activity realationships of sesquiterpene lactone esters derived from helenalin. Planta Medica 53(2): 153–156.

Harborne, J.B. 1981. Phytochemical Methods. A guide to modern techniques of plant analysis. 2nd edit. Chapman and Hall, London.

Haruna, M., R. Gregory, and K. Lee. 1985. Antitumor agents, 71. Nudaphantin, a new cytotoxic germacranolide, and elephantopin from Elephantopus nudatus 48: 93–96.

Hayes, N.A. and J.C. Foreman.1987. The activity of compounds extracted from feverfew on histamine release from rat mast cells. Journal of Pharmacy and Pharmacology 39: 466–470.

Hiai, S., H. Yokoyama, and H. Oura. 1981. Effect of escin on adrenocorticotropin and corticosterone levels in rat plasma. Chemical and Pharmacology Bulletin 29: 490–494.

Hope, W.C., A.F. Welton, C. Fiedler-Nagy, C. Batula-Bernard, and J.W. Coffey. 1983. *In vitro* inhibition of the biosynthesis of slow reacting substance of anaphylaxis (SRS-A) and lipoxygenase activity by quercetin. Biochemical Pharmacology 32: 367–371.

Horakova, Z. and M.A. Beaven.1974. Time course of histamine release and edema formation in the rat paw after thermal injury. European Journal of Pharmacology 27: 305–312.

Horakova, Z., B.M. Bayer, A.P. Almeida, and M.A. Beaven.1980. Evidence that histamine does not participate in carrageenan-induced pleurisy in rat. European Journal of Pharmacology 62: 17–25.

Inoue, H., K. Saito, Y. Koshihara, and S. Muroto. 1986. Inhibitory effect of glycyrrhetinic acid derivatives on lipoxygenase and prostaglandin synthetase. Chemical and Pharmacology Bulletin 34: 897–901.

Inoue, H., T. Mori, S. Shibata, and Y. Koshikara. 1989. Modulation by glycerrhetinic acid derivates of TPA-induced mouse ear oedema. British Journal of Pharmacology 96: 204–210.

Jorge Neto, J. 1973. Contribuição ao Estudo Farmacognóstico dos Frutos de *Pterodon pubescens* (Bentham) Bentham. Tese de Doutorado. Faculdade de Ciências Farmacêuticas da Universidade Estadual Paulista.

Jorge Neto, J. 1989. Contribuição ao estudo farmacognóstico da folha do Pseudocaryophyllus guili (Speg.) Burr. Araraquara, FCF/Unesp, Tese de Livre Docência.

Kaiser, D. G., and E.M. Gleen. 1974. Aspirin-flubiprofen Interaction in the adjvant—induced polyarthritic rat. Research Communications in Molecular Pathology and Pharmacology 8: 731–734.

Kaneda, N., J.M. Pezzuto, A.D. Kinghorn, N.R. Farnsworth, T. Santisuk, P. Tuchinda, J.Udchachon, and V. Reutrakul. 1992. Plant anticancer agents, L. cytotoxic triterpenes from *Sandoricum koetjape* stems. Journal of Natural Products 55(5): 654–659.

Katsuhiko, M. and M. Takashi. 1982. Comparison of the anti-inflammatory effects of dexametasone, indometacin and 8W755 on carrageenin-induced pleurisy in rats. European Journal of Pharmacology 77: 229–236.

Kimura, Y., H. Okuda, T. Nomura, T. Fukai, and S. Aichi. 1986. Pharmacological study on *Panax ginseng* C.A. Meyer.III. Effects of red ginseng on experimental disseminated intravascular coagulation (?). Effects of ginsenosides on blood coagulative and fibrinolytic systens. Chemical and Pharmaceutical Bulletin 34· 1153–1157.

Korolkovas, A. 1983. Modificação molecular na obtenção de novos fármacos. In: 2°. Simpósio Nacional de Produtos Naturais. Anais 73–121.

Kubo, I., S.K. Chaudhuri, Y. Kubo, Y. Sanchez, T. Ogura, T. Saito, H. Ishikawa, and H. Haraguchi. 1996. Cytotoxic and antioxidative sesquiterpenoids from *Heterotheca inuloides*. Planta Medica 62(5): 427–430.

Lanas, A.I., J. Nerin, F. Esteva, and R. Sainz. 1995. Non-steroidal antiiflammatory drugs and prostaglandin effects on pepsinogen secretion by dispersed human peptic cells. Gout 36: 657–663.

Leite de Almeida, M.E., and O.R. Gottlieb. 1975. The chemistry of Brazilian leguminosae. Further Isoflavones from *Pterodon apparicia*. Phytochemistry 14(12): 2716.

Leng-Peschlow, E. 1996. Properties and medical use of flavonolignans (Silymarin) from *Silybum marianum*. Phytotherapy Research 10: 25–26.

Levin, J.O., K. Eriksson, A. Falk, and A. Lof. 1992. Renal elimination of verbenols in man following experimental alpha-pinene inhalation exposure. International Archives of Occupational and Environmental Health 63(8): 571–573.

Lévi-Strauss, C. 1987. Physical Anthropology, Linguistics and Cultural Geography of South American Indians. vol. 6. Handbook of South American Indians 2a ed. fac similar, Cooper Square. Publ. Inc., New York. p. 715.

Lichtenstein, D.R., S. Syngal, and M.M. Wolpe. 1995. Non-steroidal antiinflammatory drugs and the gastrointestinal tract. Arthritis and Rheumatism 38: 5–18.

Lorenzi, H. 1992. Manual de identificação e cultivo de plantas arbóreas nativas do Brasil. Árvores Brasileiras. Editora Plantarum Ltda. Campinas-SP, p. 227.

Mackenzie, A.R., C.R. Pick, P.R. Sibley, and B.P. White. 1978. Suppression of rat adjuvant disease by cyclophosphamide pretreatment: evidence: for an antibody mediated component in the pathogenesis of the disease. Clinical Experimental and Immunology 32: 86.

Mahjan, J.R. and M.B. Monteiro. 1970. New diterpenoids from *Pterodon emarginatus*. Anais da Academia Brasileira de Ciências. 42 (supl.). 103–107.

Mahjan, J.R. and M.B. Monteiro. 1972. New diterpenoids from *Pterodon emarginatus*. Anais da Academia Brasileira de Ciências 44(1): 51–53.

Mahjan, J.R. and M.B. Monteiro. 1973. New diterpenoids from *Pterodon emarginatus*. Journal of The Chemical Society Perkin Transactions 1: 520–525.

Martin, S., E. Padilla, M.A., Ocete, J., Galvez, J. Jimenez, and A. Zarzuelo. 1993. Anti-inflammatory activity of the essential oil of *Bupleurum fruticescens*. Planta Medica 59(6): 533–536.

Mata, R., A. Rojas, L. Acevedo, S. Estrada, F. Calzada, I. Rojas, R. Bye, and E. Linares. 1997. Smooth muscle relaxing flavonoids and triterpenoids from *Conyza filaginoides*. Planta Medica 63(1): 31–35.

Matsuda, H., M. Yoshikawa, M. Linuma, and M. Kubo. 1998. Antinociceptive and anti-inflammatory activities of limonin isolated from the fruits of *Evodia rutaecarpa* var. *bodinieri*. Planta Medica 64: 339–342.

McCormack, K. and K. Brune. 1987. Classical absortion theory and the development of gastric mucosal damage associated with the non-steroidal anti-inflammatory drugs. Archives of Toxicology 60(4): 261–269.

Michel, F., L. Mercklein, R. Rey, and A. Crastes De Paulet. 1985. In: L. Farkas, M. Gabor, and F. Kallay (eds.). Comparative effects of some flavonoids on cyclooxygenase and lipoxygenase activities in different cells systems or subfractions, Flavonoids and Bioflavonoids. Elsevier, Amsterdam pp. 389–401.

Mors, W.B., M.F. Dos Santos, and H.B. Monteiro. 1967. Chemoprophyloactic agent in Schistosomiasis: 14, 15-epoxygeranylgeraniol. Science 157 (3791): 950–951.

Mustard, J.F., J.Z. Movat, D.R.L. Macmorine, and A. Senyi. 1965. Release of permeability factors from blood platelet. Proceedings of the Society Experimental Biology and Medicine 119: 981–988.

Newbould, B.B. 1963. Chemotherapy of arthritis induced in rats by mycobacterial adjuvant. British Journal of Pharmacology 21: 127.

Nishida, S., and S. Tomizawa. 1980. Effects of compounds 48/80 on dextran-induced paw edema and histamine content of inflammatory exudate. Biochemical Pharmacology 29: 2641–2646.

Nishizawa, M., T. Yamagishi, and T. Ohyama. 1983. Inhibitory effect of gallotannins on the respiration of rat liver mitochondria. Chemical and Pharmaceutical Bulletin 31: 2150–2152.

Nunan, E.A. 1985. Estudo da atividade antiinflamatória de furano-diterpeno isolados de *Pterodon polygalaeflorus* Benth e de alguns de seus derivados. Tese de doutorado. Departamento de Química. UFMG, Belo Horizonte.

Ohta, Y., K. Nishida, E. Sasaki, M. Kongo, and I. Ishiguro. 1997. Attenuation of disrupted hepatic active oxygen metabolism with the recovery of acute liver injury in rats intoxicated with carbon tetrachloride. Research Communications in Molecular Pathology and Pharmacology 95: 191–208.

Ohyashiki, T., K. Kamata, M. Takeuchi, and K. Matsui. 1995. Contribuition of peroxidation products to oxidative inactivation of rat liver microsomal glucose-6-phosphatase. Journal of Biochemistry 118: 508–514.

Oliveira, F., G. Akisue, and M.K. Akisue. 1991. Farmacognosia, S. Paulo. Ed. Atheneu 62: 412.

Otterness, I.G. and M.L. Bliven. 1982. In: I.Otterness, R. Capetols, and S.T. Wong (eds.). therapeutic control of inflammation diseases, Advances in Inflammation Research, vol. 7 Raven Press, New York. p. 185.

Owen, D.A.A. and D.F. Woodward. 1980. Histamine and histamine H_1 and H_2-receptor antagonists in acute inflammation. Biochemical Society Transactions 8: 150–156.

Parmar, N.S., M. Tariq, and A.M. Ageel. 1987. Gastric cytoprotection: A critical appraisal of the concept, methodology, implications, mechanisms and future research prospects. Agents Actions 22: 114–122.

Paulus, H.E., H.I. Machleder, R. Bangert, J.A. Stratton, L. Goldberg, M. Whitehouse, D.T.Y. Yu, and L.M. Pearson. 1973. A case report thoracic duct lymphocyte drainage in rheumatoid arthritis. Clinical Immunology and Immunopathology 1: 173.

Paulus, H.E., H.I. Machleder, S. Levine, D. Yu, and N.S. MacDonald. 1977 Lymphocyte involvement in rheumatoid arthritis. Arthritis and Rheumatism 20: 1249.

Pearson, C.M. 1956. Development of arthritis, periarthritis and periotitis in rats given adjuvants. Proceedings of the Society of Biology and Medicine 91: 95.

Pearson, C.M. 1964. Experimental models in rheumatoid disease. Arthritis and Rheumatism 7: 80.

Pearson, C.M. and F.D. Wood. 1959. Studies of polyarthritis and other lesions induced in rats by injection of micobacterial adjuvant. I. General clinical and pathologic characteristics and some modifying factors. Arthritis and Rheumatism 2: 440.

Pearson, C.M., B.H. Waksman, and J.T. Sharp. 1961. Studies of arthritis and other lesions induced in rats by injection of micobacterial adjuvant. Journal of Experimental Medicine 113: 485.

Pearson, C.M. and F.D. Wood. 1963. Studies of arthritis and other lesions induced in rats by the injection of mycobacterial adjuvant. VII—Pathologic details of the arthritis and spondylitis. American Journal of Pathology 42: 73.

Pearson, C.M., H.E. Paulus, and J.I. Machleder. 1975. The role of the lymphocyte and its products in the propagation of joint disease. Annual Review of Science 256: 150.

Pelt, J.M. 1979. A revolução verde da medicina. O Correio da UNESP, Ano 7, No. 9.

Robak, J., A. Wieckowski, and R.J. Gryglewski. 1978. The effect of 4-acetamidophenol on prostaglandin synthetase activity bovine and ram seminal vesicle microsomes. Biochemical Pharmacology 27: 393–396.

Robbers, J.E., M.K. Speedie, and V.E. Tyler. 1997. Terpenóides. Farmacognosia e Farmacobiotecnologia. Editorial Premier. S. Paulo pp. 91–121.

Rosenthale, M.E. 1970. A comparative study of the Lewis and Sprague Dawley rat in adjuvant arthritis. Archives International of Pharmacodynamic 188: 14.

Russin, W.A., J.D. Hoesly, C.E. Elson, M.A. Tanner, and M.N. Gould. 1989. Inhibition of rat mammary carcinogenesis by monoterpenoids. Carcinogenesis 10(11): 2161–2164.

Sarti, S.J. 1995. Uso e importância das plantas medicinais através dos tempos. Apostila do Curso de Extensão Universitária-UEPb. Campina Grande—Pb.

Schrör, K. 1984. Prostaglandine und Verwandte Verbindungen, G. Thieme Verlag, Stuttgart, New York.

Sharp, J.T., B.H. Waksnan, C.M. Pearson, and S. Madoff. 1961. Studies of arthritis and other lesions induced in rats by injection of mycobacterial adjuvant. IV Examination of tissues and fluids for injections agents. Arthritis and Rheumatism 4: 169.

Shorrock, C.J. and W.D. W.Rees. 1988. Overview of gastric mucosa protection. American Journal of Medicine, vol. 84 (suppl. 2), 25–34.

Smith, G.N. 1988. The treatment and management of soft tissue injuries. Practitioner 232: 1245–1248.

Sofia, R.D., C.L. Rnobloch, and B.H. Vassar. 1975. Inhibition of primary lesions of adjuvant-induced polyarthritis in rats (18 hours arthritis test) for specific detection of clinically effective anti-arthritis drugs. Journal of Pharmacology and Experimental Therapeutic. 193: 918–931.

Sofia, R.D., L.C. Knobloch, and J.L. Douglas. 1976. Effect of concurrent administration of aspirin, indoethacin or hydrcortisone with gold sodium thiomalate against adjuvant induced arthritis in the rat. Agents and Actions 6: 728–734.

Stoerck, H.C., T.C. Bielinski, and T. Budzilovich. 1954. Chronic polyarthritis in rats injected with spleen in adjuvants. American Journal of Pathology 30: 616.

Swingle, K.F. and F.E. Shideman. 1972. Phases of the inflammatory response to subcutaneous implantation of a cotton pellet and their modification by certain anti-inflammatory agents. The Journal of Pharmacology and Experimental Therapeutics 13: 226–234.

Swingle, K.F., L.W. Jaques, and D.C. Kvam. 1969. Differences in the severity of arthritis in four strains of rats. Proceedings of the Society of Experimental Biology and Medicine 132: 608.

Swingle, K.F., M.J. Reiter, and M. Schwartzmiller. 1981. Comparison of croton oil and cantharidin induced inflammation of the mouse ear and their modification by topically applied drugs. Archives International of Pharmacodynamic 254: 168–176.

Tambe, Y., H. Tsujiuchi, G. Honda, Y. Ikeshiro, and S. Tanaka. 1996. Gastric cytoprotection of the non-steroidal anti-inflammatory sesquiterpene, beta-caryophyllene. Planta Medica 62(5): 469–470.

Tamura, Y., T. Nishikawa, K. Yamada, M. Yamamoto, and A. Kumagai. 1979. Effects of glycyrrhetinic acid and its derivatives on λ^4-5α- and 5β-reductase in rat liver. Arzneimittel-Forschung/Drug Research 29: 647–649.

Teixeira, T.C., A.M. Campos, and E.R. Silveira. 1991. Isolamento do ácido 6α, 7β-dihidroxivouocapan-17-β-óico das sementes de *Pterodon polygalaeflorus*. In: Resumos da Reunião da Sociedade Brasileira de Química, PN-048.

Teranishi, R., R.G. Buttery, and H. Sugisawa. 1993. Bioactive volatile compounds from plants. American Chemical Society, Washington, DC.

Thebtaranonth, C., Y. Thebtaranonth, S. Wanauppatamkul, and Y. Yuthavong. 1995. Antimalarial sesquiterpenes from tubers of *Cyperus rotundus*: Structure of 10,12-peroxycalamenene, a sesquiterpene endoperoxide. Phytochemistry 40(1): 125–128.

Trowbridge, H.O. and R.C. Emling. 1996. Mediadores químicos da resposta vascular. Inflamação uma revisão do processo. Quitessence Publishing Co. Inc., S. Paulo. pp. 27–42.

Ulubelen, A., G. Topcu, C. Eris, U. Sonmez, M. Kartal, S. Kurucu, and C. Bozok-Johansson. 1994. Terpenoids from *Salvia sclarea*. Phytochemistry 36(4): 971–974.

US Pat. 4.379361, 11, May, Europ. Patent Appl., No. 80107752.0, Ball 81/4 (1981).

Valencia, E., M. Feria, J.G. Diaz, A. Gonzales, and J. Bermejo. 1994. Antinociceptive, anti-inflammatory and antipyretic effects of lapidin, a bicyclic sesquiterpene. Planta Medica 60(5): 395–399.

Van Arman, C.G., A.J. Begany, L.M. Miller, and H.L. Pless. 1965. Some details of the inflammation caused by yeast and carrageenan. Journal of Pharmacology Experimental and Therapeutic 150: 328–334.

Vane, J.R. 1971. Inhibition of prostaglandin synthesis as a mechanism of action for aspirin-like drugs. Nature New Biology 231: 232–235.

Vane, J.R. 1979 The mode of action of aspirin ande similar compounds. Journal of Allergy Clinical and Immunology 58: 691–712.

Vinegar, R. 1968. Quantitative studies concerning kaolin-edema formation in rats. Journal of Pharmacology Experimental and Therapeutic161: 389–395.

Vinegar, R., W. Schreiber, and R. Hugo. 1969. Biphasic development of carrageenin edema in rats. Journal of Pharmacology and Experimental Therapeutic 166: 96–103.

Vinegar, R., A.W. Macklw, J.F. Truax, and J.L. Selph. 1974. Formation of pedal edema in normal and granulocytopenic rats. In: G. Van Arman (ed.). White Cells in inflammation, by Springfield, New York. 111:Thomas pp. 111–138.

Vinegar, R., J.F. Truax, and J.L. Selph. 1976. Quantitative studies of the pathway to acute carrageenan inflammation. Federation Proceedings 35: 2447–2456.

Vinegar, R., J.F. Truax, J.L. Selph, and F.A. Voelker. 1982. Pathway of onset, development, and decay of carrageenan pleurisy in the rat. Federation Proceedings 41: 2588–2594.

Wagner, H. 1996. Immunosuppressive effects of hypericin on stimulated human leukocytes: inhibition of the arachidonic acid release, leukotriene B$_4$ and interleukin-1α production, and activation of nitric oxide formation. Phytomedicine 3: 19–28.

Wagner, H. and H. Reger. 1987. Moderne phytotherapie. Deutsche Apotheke Zvestig, 126: 159–160.

Wagner, H., M. Wierer, and B. Fessler. 1987a. Effects of garlic constituents on arachidonate metabolism. Planta Medica 53: 305–306.

Wagner, H., B. Fessler, H. Lotter, and V. Wray. 1987b. New chlorine-containing sesquiterpene lactones from *Chrysanthemun parthenium*. Planta Medica 54: 171–172.

Wagner, H., W. Knaus, and E. Jordan. 1996. Cirsiliol and caffeic acid ethyl esther, isolated from *Salvia guaranitica*, are competitive ligands for the central benzodiazepine receptors. Phytomedicine 3: 29–31.

Wegelius, O., V. Laine, B. Lidstron, and M. Klockars. 1970. Fistula of the thoracic duct as an immunossuppressive treatment in rheumatoid arthritis. Acta Medica Scandinavica 187: 539.

Whitehouse, M.W., K.D. Rainsford, N.G. Ardlie, I.G. Young, and K. Brune. 1977. In: Proceedings of the Symposium on Aspirin and Related Drugs: Their Actions and Uses K.D. Rainsford, K. Brune, and M.W. Whitehouse (eds.). Birkhäuser Verlag, Basel -Stuttgart. pp. 43–57.

Whitehouse, N.V. and D.J. Whitehouse.1968. Inhibition of rat adjuvant arthritis without drugs or antibody therapy. Arthritis and Rheumatism 11: 519.

WHO—World Health Organization. 1992. Quality Control Methods for Medicinal Plant Materials. 88p.

Wiesenfeld-Hallw, Z. and R.G. Hallin. 1984. The influence of the sympathetic system on mechanoreception and nociception. Human Neurobiology 3: 41.

Wiiliams, H.J., G. Moyna, S.B. Vinson, A.I. Scott, A.A. Bell, and R.D. Stipanovic. 1996. Beta-caryophyllene derivatives from the wild cottons *Gossypium armourianum*, *Gossypium herknessii*, and *Gossypium turneri*. Agricultural Research Service. 77, 78.

Willis, A. L. 1969a. Parallel assay of prostaglandin-like activity in rat inflammatory exudate by means of cascade superfusion. Journal of Pharmacy and Pharmacology 21: 126.

Willis, A.L. 1969b. Release of histamine, kinin and prostaglandin during carrageenan-induced inflammation in the rat. In: P. Mantegazza and E.W. Horton (eds.). Prostaglandins, Peptides and Amines. London and New York. 31.

Winter, C.A. and L. Flataker. 1965. Reaction thresholds to pressure in edematous hindapaws of rats and responses to analgesic drugs. Journal of Pharmacology and Experimental Therapeutics 150: 165–171.

Winter, C.A., E.A. Risley, and G.W. Nuss. 1962. Carrageenin-induced oedema in the hind paw of the rat as an assay for anti-inflammatory drugs. Proceeding of the Society of Experimental Biology and Medicine 111: 544.

Wong, S.M., S. Antus, A. Gottsegen, B. Fessler, G.S. Rao, J. Sonnenbichler, and H. Wagner. 1988. Wedelolactone and coumestan derivatives as new antihepatotoxic and antiphlogistic principles. Arzneimittel-Forschung/Drug Research 38: 661–665.

Yamamoto, M., A. Kumagai, and Y. Yamamura.1975. Effekt von bencyclan auf die scherungsinduzierte plättchenaggregation. Arzneimittel-Forschung/Drug Research 25: 1021–1025.

Yesilada, E., S. Tanaka, E. Sezik, and M. Tabata. 1988. Isolation of ana anti-inflammatory principle from the fruit juice of *Ecballium elaterium*. Journal of Natural Products 51: 504–508.

Yoshimoto, T., M. Furukawa, S. Yamamoto,T. Horie, and S. Watanabe-Kohno.1983. Flavonoids: potent inhibitors of arachidonate 5-lipoxygenase. Biochemica Biophysica Research Communication 116: 612–618.

Zheng, G.Q., P.M. Kenney, and L.K. Lam.1992. Sesquiterpenes from clove (*Eugenia caryophyllata*) as potential anticarcinogenic agents. Journal of Natural Products 55(7): 999–1003.

Zitterl-Eglseer, K., J. Jurenitsch, S. Korhammer, E. Haslinger, S. Sosa, R. Della Loggia, W.Kubelka, and C. Franz. 1991. Sesquiterpenelactones of *Achillea setacea* with antiphlogistic activity. Planta Medica 57(5): 444–446.

Chapter 4

Women's Knowledge of Herbs used in Reproduction in Trinidad and Tobago

Cheryl Lans[1] and *Karla Georges[2]*

Introduction

The potential usefulness of certain plants is assessed from a non-experimental validation method. The reproductive uses of *Ageratum conyzoides* and *Cassia occidentalis* were common across all the three research studies. Plants described similarly for reproductive health in Moodie and Lans were *Abelmoschus moschatus*, *Achyranthes indica*, *Ambrosia cumanensis*, *Eryngium foetidum*, *Mimosa pudica* and *Stachytarpheta jamaicensis*. Wong's study was limited geographically but reflected reproductive health uses for the following plants recorded by Lans: *Brownea latifolia*, *Capraria biflora*, *Coleus aromaticus*, *Eupatorium macrophyllum*, *Fleurya aestuans*, *Hibiscus rosa-sinensis*, *Justicia secunda*, *Leonotis nepetaefolia*, *Petiveria alliacea* and *Wedelia trilobata*. *Achyranthes aspera*, and *Hibiscus rosa-sinensis* have antifertility activity which is not permanent. *Ageratum conyzoides*, *Cola nitida*, *Coleus barbatus*, *Mimosa pudica* and *Piper auritum* exert even stronger activity on the reproductive process.

The plants used for reproductive reasons in Trinidad and Tobago collected by three researchers: Wesley Wong (Wong 1976), Sylvia Moodie-Kublalsingh (Moodie 1994) and Cheryl Lans (Lans 2007) are discussed here (Table 4.1). Wong conducted his research in Blanchisscuse on the north coast of Trinidad in 1966 and 1967. Moodie conducted her research

[1]PO Box 72045 Sasamat, Vancouver, V6R4P2.
[2]Veterinary Public Health, The University of the West Indies, Faculty of Medical Sciences, School of Veterinary Medicine, EWMSC Mount Hope, Trinidad and Tobago West Indies.

with Spanish-speakers over several years from 1970 to 1986, with the majority of the interviews conducted between 1977–1981. The interviews covered most of Trinidad. Lans research was conducted from 1995–2000 and covered both islands. The major focus was on animal health but some of the interviewees became offended if human health information was not recorded as well. The data collection methods used by the three researchers are documented in other publications. This chapter is mainly on women's experiences seen indirectly through the plants that they and their female ancestors have used for childbirth. This allows their experiences to be shared without their identities being decipherable by persons in the small population nation. One interview was conducted by the second author in July 2008 and this is provided in Fig. 4.1.

Tisane (drink a bottle) to clean out the womb and assist in conceiving.
Grandmothers grave: To clean the uterus
Pumpkin for cooling: Just before giving birth
Castor oil, lamp oil, olive oil after giving birth: Use while bathing, you can also go outside and bathe in the rain and you would not catch a chill.

Okras: Boil and eat them if you are having a difficult labour. (Cows are given bamboo leaves but not people, okra works for people).

If the baby is not latching on to the breast use St. John's bush (Justicia secunda).

Once this interviewee was in the hospital and a woman was having trouble to pass the afterbirth—the midwife told her to blow into a rum bottle, and that worked.

Fig. 4.1 2008 interview with a folk medicine expert of Trinidad.

Traditional human health care for reproductive purposes focuses equally on the pregnancy, parturition and the postpartum period (Sobo 1996). In Jamaica of the125 pregnant women in a 1983 study, 82% reported drinking bush teas during their pregnancy. It is difficult to distinguish between amenorrhoea (absence or suppression of menstruation due to illness, depression or malnutrition) and early pregnancy. Knowing this difficulty, women in some Latin American and Caribbean countries deliberately or unconsciously blur the differences between abortifacient and menses-stimulating effects. This gives them some control over reproduction in Christian countries that have strict social or legal restriction against abortion. Women of Spanish and Mexican descent in New Mexico have used several plants as emmenagogues and abortifacients. Of the plants used, at least three are the same as those used in Trinidad and Tobago: ruda (*Ruta graveolens*), wormgrass (*Chenopodium ambrosioides*), and *Artemisia* species.

Research conducted in Jamaican by Sobo in 1996 shows that birth, defecation and menstruation are defined traditionally as cleansing processes. Both Morton (1981) and Sobo (1996) documented the habitual and widespread use of emmenagogues and bush teas in Latin America, to 'clean out' the womb and restore vitality after pregnancy or to 'bring back' or 'bring down the menses'. Plants were given to speed delivery after the onset of labor in home births. After births or miscarriages, mild purgatives were given to induce the quick delivery of the placenta and pregnancy-related waste matter, out through the vagina.

Moodie-Kublalsingh documented the Spanish tradition in Trinidad (Moodie 1994). One concoction she documents was called a *bebedizo* (*bebeyis*), this was a tisane or laxative given to new mothers after childbirth. It contained powdered and toasted clove (*Eugenia caryophyllata*), nutmeg (*Myristica fragrans*), cinnamon (*Cinnamomum verum*) and ginger (*Zingiber officinale*) in rum (*Cogian el clavo de especias, la memoscá, la canela, el genjibre*).

Squatting over a pot of hot water, in a utensil called ('tencil) in Tobago ensures the ejection of all waste. The steam enters the body and 'melts' all recalcitrant matter, which then slides out. In earlier times expectant mothers were given tea made of milk bush roots for five or nine successive days to 'cool down the body'. The scissors used to cut the umbilical cord are placed beneath the place where the baby's head is to lie, and left there for 9 d when the mother and child first emerge from the house. The new mother can assume all household duties after 9 d. When the baby and mother emerge from the house 9 d after the birth, a ceremony is held to present the new member of the family to relatives and the family ancestors (Herskovits and Herskovits 1947). The Caribs in Dominica had a similar set of practices for new mothers. A special bath was given 8 or 9 d after childbirth to the mother and to the newborn infant. Plants used in the mother's bath were framboisin (*Ocimum micranthum*), coton noir (*Gossypium vitifolium*), roucou (*Bixa orellana*), verveine (*Stachytarpheta* species), semen contra (*Chenopodium ambrosioides*), sou marque (*Cassia bicapsularis*), pistache (*Arachis hypogea*), and bouton blanc (*Egletes prostrata*) (Hodge and Taylor 1957).

In the past women who were pregnant or menstruating were considered 'heaty'. Pregnant or menstruating women were not to climb over a rope tethering an animal or sit cross-legged or on a doorstep where people would cross over them. Women also could not untie or tie hunting dogs, step over them or could not climb trees during their menstrual cycle.

There are still women who practice being informal midwives. *Mimosa pudica* was used by one midwife to unwrap the cord from around a baby's neck. Two plant tops were tied crossways, put in a pot and drawn. She claimed that 15 min after the pregnant woman drank the tisane the baby gave a flip. However, a caesarean was still needed because the baby's due date had past.

Validation of women's reproductive herbal health knowledge

There were many variables that Latin American and Caribbean women used when they chose plants for reproductive conditions: the hot-cold valence, irritating action, emmenagogic, oxytocic, anti-implantation and/or abortifacient effects. Uteroactive plants used in Latin America and the

Table 4.1. Comparison of women's traditional plant use in three Trinidad and Tobago research projects

	Reference	Reference	Reference
Latin name, local names	Moodie 1970s–1980s	Wong 1966, 1967	Lans 1995–2000
Abelmoschus moschatus, Gumbo musque			Leaves, seeds female complaints, remove placenta
Achyranthes indica Man better man			Female complaints
Ageratum conyzoides, Calero macho, Calero hembra, Z'herbe á femme	Herb teas for women's disorders and root decoctions postpartum clotting	Herb teas for dysmenorrhea, menorrhagia, root decoctions for postpartum clotting	Female complaints
Allium sativum, garlic		Clove rubbed on belly for parturition	
Aloe barbadensis, Sábila, aloes	To regulate menstrual cycle		
Ambrosia cumanensis, Altamis, Altamisa	Herb tea for menorrhagia, postpartum depurants. Good for women if they are not looking after their health		3-inch leafy branch Inflammation, abortion, menstrual pain
Artemisia absinthium, wormwood			Female complaints
Bauhinia excisa, liane tassajo		Root vine decoctions and infusions for amenorrhea	
Bixa orellana, Roucou, dormidero, Annatto	Bottles, 9 each time. If woman does not conceive there is no hope		
Borreria verticillata, white head broom, balai tête blanc		Herb decoction for dysmenorrhea	
Brownea latifolia, cooperhoop, cupajupa, bois de rose		Teas and infusions of flowers for amenorrhea	Flower, leaves female complaints
Capraria biflora, du thé pays		Herb teas for dysmenorrhea, as postpartum depurants	Leaves for menstrual pain

	Reference	Reference	Reference
Cassia occidentalis, Z'herbe puante Brusca, wild coffee, herbe puante	Used with root decoctions and infusions for womb inflammation, abortifacients, purgative, postpartum depurants.	Used with root decoctions and infusions for womb inflammation, abortifacients, purgative, postpartum depurants	Used to bring down the 'clot blood'. Cut it big, mostly roots, use 3 roots, boil them for 15 minutes, and throw it in a 'tensil'. Sit down on the 'tensil', as hot as you can bear.
Chaptalia nutans, dos blanc		Leaf tea for amenorrhea	
Chenopodium ambrosioides, worm grass, semence contre les vers		Herb teas and infusions in postpartum depurants	
Cocos nucifera, coconut		Root tea for amenorrhea. Husk tea for dysmenorrhea, amenorrhea, menorrhagia	
Coleus aromaticus, Spanish thyme, diten		Root infusion for menorrhagia	Leaves shorten ?labour
Cordia curassavica, Black sage			Leaves used for menstrual pain
Cordyline terminalis, rayo		Leaf in teas for amenorrhea	
Croton flavens, cancanapire		Leaves rubbed on belly for postpartum pain	
Curcuma longa, turmeric			Women boil the ground saffron with massala, ginger and salted butter, to 'bring down everything' after parturition.
Eleutherine bulbosa, Dragon blood			Bulb female complaints
Eryngium foetidum, Chadron bénée Culantro	Extract juice from leaves and drink for abdominal pains		Leaves menstrual pain, remove placenta, shorten labur

Table 4.1. contd...

Table 4.1. contd...

	Reference	Reference	Reference
Eupatorium macrophyllum, z' herbe chatte		Leaf teas for amenorrhea, dysmenorrhea, root tisane as postpartum depurant, leaf poultice is vaginal suppository for womb inflammation and prolapse	Female complaints
Euphorbia thymifolia, malômē		Herb decoction used as lactogogue, postpartum depurant	
Datura stramonium, niungué		Leaf poultice on belly for prolapse of womb	
Fleurya aestuans, Laportea aestuans, ortie blanche/rouge, white/red stinging nettle		Decoctions of plant and root to facilitate delivery of afterbirth	Leaves to shorten labor
Hibiscus rosa-sinensis, hibiscus, fleur barrière		Infusions for amenorrhea, flowers	Female complaints
Justicia secunda, chanchamunchin, Saint John bush		Herb teas baths for amenorrhea	Female complaints
Kalanchoe pinnata, wonder of the world, caractère des hommes		Leaf poultice for dysmenorrhea	
Leonotis nepetaefolia, chandelier		Leaf tea in vaginal suppository for womb prolapse	Menstrual pain
Mimosa pudica, Adolmidero, marie honte, Mese marie	Mimosaceae Promote fertility in women. Use root with other bush for tisane. Take tea for 9 days to take out 'cold' from system. After tisane take a purge and the woman should conceive after that.		Childbirth
Myristica fragrans, nutmeg, muscade		Decoction infusions for dysmenorrhea, in postpartum depurants	
Neurolaena lobata, Hoja de lanza	Leaf tea for women's disorders	Leaf infusion for dysmenorrhea	

	Reference	Reference	Reference
Nicotiana tabacum, tobacco		Cured leaf poultice for postpartum pain	
Ocimum gratissimum, diten jaraba		fresh leaf tea for dysmenorrhea	
Parthenium hysterophorus, White head broom			Female complaints
Petiveria alliacea, gully root, kojo root, mapurite		Root teas and infusions for dysmenorrhea, womb inflammation in abortifacients	Menstrual trouble.
Pilea microphylla, Du thé bethelmay			Leaves inflammation, womb cleanser
Pimpinella anisum, Hoja de gato	For women's disorders, leaf. Root par la barrija. La hoja cuando una mujer hace un muchachito y la sangre la para		
Piper marginatum, l'anis bois		Drink to ease parturition	
Pityrogramma calomelanos, white back fern		Fern frond infusions for menorrhagia	
Rauvolfia ligustrina, raíz de paraeria, reydeparél	Clean out the womb		
Ricinus communis, castor oil bush, palma Christi		Leaf poultice for inflammation of womb. Seed oil consumed during pregnancy, postpartum period for empacho—indigestion	
Rolandra fruticosa, lanve blan		Leaf tea for amenorrhea	
Ruta graveolens, Ruda			Leaves childbirth, carminative, menstrual pain, cold in womb
Sataria species, Gamelote	Use root for hemorrhage in women. Poaceae		

Table 4.1. contd...

	Reference	Reference	Reference
Spondias mombin, Hog plum			Used to bring down the 'clot blood'. Take a few branches boil it put it in a tub and sit down over it on a stool.
Stachytarpheta jamaicensis, Verbena	Verbenaceae, Cooling, increase milk in nursing mothers		Ground saffron with massala, ginger and salted butter, is drawn with Vervine to make a tea and is used to 'clean out' the body.
Trimezia martinicensis, dragon's blood		Corm in teas for amenorrhea	
Urena sinuata, patte chien		Root infusion is postpartum depurant	
Vetiveria zizanioides, Vetivert			Plants shorten labor
Wedelia trilobata, Venven caribe		Leaf and flower head tea for amenorrhea	Leaves for female complaints
Zea mays, corn		Husk tea for amenorhea, menorrhagia	

Table 4.2. Ethnomedicinal plants used for reproductive problems in Trinidad and Tobago (source Lans 2007)

Scientific name	Family	Common name	Part used	Use
Abelmoschus moschatus	Malvaceae	Gumbo musque	Leaves, seeds	Female complaints, remove placenta
Achyranthes indica	Amaranthaceae	Man better man		Female complaints
Ageratum conyzoides	Asteraceae	Z'herbe á femme		Female complaints
Ambrosia cumanenesis	Asteraceae	Altamis		Inflammation, abortion, Menstrual pain
Aristolochia rugosa, *Aristolochia trilobata*	Aristolochiaceae	Mat root, anico	Root	Remove placenta, abortion, menstrual pain
Artemisia absinthium	Asteraceae	Wormwood		Female complaints

Scientific name	Family	Common name	Part used	Use
Brownea latifolia	Fabaceae	Cooper hoop	Flower, leaves	Female complaints
Capraria biflora	Scrophulariaceae	Du thé pays	Leaves	Menstrual pain
Chamaesyce hirta	Euphorbiaceae	Malomay		Infertility
Cocos nucifera	Arecaceae	Coconut	Shell	Abortion
Cola nitida	Sterculiaceae	Obie seed	Seed	Infertility
Coleus aromaticus	Lamiaceae	Spanish thyme	Leaves	Shorten labour
Cordia curassavica	Boraginaceae	Black sage	Leaves	Menstrual pain
Croton gossypifolius	Euphorbiaceae	Bois sang	Leaves	Menstrual pain
Eleutherine bulbosa	Iridaceae	Dragon blood	Bulb	Female complaints
Entada polystachya	Fabaceae	Mayoc chapelle	Twigs	Menstrual pain
Eryngium foetidum	Apiaceae	Chadron bénée	Leaves	Menstrual pain, remove placenta, shorten labour
Eupatorium macrophyllum	Asteraceae	Z'herbe chatte		Female complaints
Hibiscus rosa-sinensis	Malvaceae	Hibiscus	Flowers	Female complaints
Justicia secunda	Acanthaceae	St. John's bush		Female complaints
Laportea aestuans	Urticaceae	Red stinging nettle	Leaves	Shorten labour
Leonotis nepetaefolia	Lamiaceae	Shandileer		Menstrual pain
Mimosa pudica	Fabaceae	Mesc marie		Childbirth
Parthenium hysterophorus	Asteraceae	White head broom		Female complaints
Pilea microphylla	Urticaceae	Du thé bethelmay	Leaves	Inflammation, womb cleanser
Ruta graveolens	Rutaceae	Ruda	Leaves	Childbirth, carminative, menstrual pain, cold in womb
Vetiveria zizanioides	Poaceae	Vetivert	Plant	Shorten labor
Wedelia trilobata	Asteraceae	Venven caribe	Leaves	Female complaints

Caribbean were defined by Browner et al. (1998) as those that cause the uterus to expel its contents. In Trinidad rice paddy is called a 'heated substance' and it is used for retained placenta because the 'heat' of the rice paddy would help break down the uterine lining and therefore assist in cases of retained placenta. This explanation may be similar to that used by Browner and colleagues who describe uteroactive plants used in Mexico in metaphorical terms of 'warming' or 'irritating'. 'Warming' the body, blood and womb, causes the womb to 'open' to release detained menstrual flow or expel a full-term fetus. 'Irritating' plants 'open' the uterus and stimulate contractions that will release blocked menstrual blood or push out a full-term fetus.

Non-experimental validation

Chemical constituents that correspond to 'warming' are perhaps those that cause *in vivo* or *in vitro* uterine contractions. For example, hogplum acts as a uterine stimulant perhaps because it contains saponins and glycosides. Irritating chemical constituents are oils, tannins, limonene. Uterotonic agents are: oxytocin > prostaglandin $F_2\alpha$ ($PGF_2\alpha$) > serotonin > acetylcholine > ergometrine. Oleanonic acid and 3-epioleanolic acid, also have uterotonic activity (Sewram et al. 2000, 2001). Several plants exert anti-fertility activity by acting on the hypothalamus, pituitary glands and gonads or have direct hormonal effects on reproductive organs that result in a reduced production of steroids by the ovary (Shibeshi et al. 2006).

The estrogenic potential of certain plants may be related to the direct activity of their 'estrogen-like' compounds on target organs or to an indirect effect through stimulation of Follicle Stimulating Hormone (FSH) production or through the activity of 'FSH-like' compounds in the extract of plants. There are flavanoids with proven estrogenic potential. Dried leaves of four medicinal plants *Aloe buettneri* (Liliaceae), *Justicia insularis* (Acanthaceae), *Dicliptera verticillata* G.J.H. Amshoff (Acanthaceae) and *Hibiscus macranthus* (Malvaceae) were tested on pieces of rat ovary (Telefo et al. 2004). High concentrations of the plant extract increased the amount of progesterone secreted. Both low and high concentrations of the plant extract increased the production of estradiol.

Many plant-derived compounds (flavonoids, lignans, coumestans) can act like endogenous hormones, they can stimulate the hypothalamus-pituitary complex with an resulting increase in FSH, which then increases the production of steroids by the ovary and estradiol synthesis (Oliver-Bever 1986, Kurzer and Xu 1997).

Table 4.3 is a preliminary evaluation of the efficacy of Trinidad and Tobago's ethnoveterinary medicines. This non-experimental method consists of a review of the published phytotherapeutic research to assess the concordance or discordance of the research results with the folk medicinal claims made by the key respondents.

Table 4.3. Evaluation of the plant remedies used for reproductive purposes in Trinidad and Tobago

Latin Name	Irritating Chemicals	Useful Medicinal Properties	Oxytocic/Uteroactive Chemicals
Abelmoschus moschatus L, Gumbo musque, Malvaceae		Myricetin is anti inflammatory	
Achyranthes indica (L.) Mill. *Achyranthes aspera* var *indica*, Man better man Amaranthaceae			Unspecified estrogenic activity
Ageratum conyzoides L, Calero macho, Calero hembra, Z'herbe á femme, Asteraceae	Oleic acid		Coumarin, glucosides, stigmasterol, linoleic acid
Aloe buettneri A. Berger, Liliaceae (similar species to reported one)			Coumarins, glycosides, quinones
Capraria biflora L, du thé pays, *Capraria lanceolata* Vahl, Scrophulariaceae		Anti-inflammatory activity	
Cassia occidentalis L,, Z'herbe puante Brusca, wild coffee, herbe puante, Fabaceae		Fungicidal, Anti-inflammatory, Antibiotic, Anti-hepatotoxic	Anthracenic heterosides, laxative activity
Chamaesyce hirta (L.) Millsp. Euphorbiaceae		Anti-stress activity	
Chenopodium ambrosioides L, worm grass, semence contre les vers, Chenopodiaceae		Seeds may be spermicidal	Antiinflammatory activity, calcium and ascorbic acid
Cola nitia (Vent.) Schott & Endl, cola nut, Sterculiaceae			antiestrogens and progestin-like substances
Cordia curassavica Roemer and Schultes, black sage Boraginaceae			spathulenol is spasmolytic

Table 4.3 contd...

Table 4.3. contd...

Latin Name	Irritating Chemicals	Useful Medicinal Properties	Oxytocic/Uteroactive Chemicals
Croton flavens L, Cancanapire, Euphorbiaceae	gamma-bisabolene	Anti-inflammatory activity	Camphor, terpenene, borneol
Curcuma longa L, turmeric Zingiberaceae,		Antiinflammatory, Antioxidative, Antitumor, Antifungal	
Datura stramonium L, niungué, Solanaceae	Citric acid		Scopoletin and ferulic acid are uterosedative
Eryngium foetidum L, Chadron bénée Culantro, Apiaceae	Capric acid, furfural		
Eupatorium macrophyllum L, z' herbe chatte, Asteraceae	limoene		
Euphorbia thymifolia L, malômē Euphorbiaceae	tannins		
Fleurya aestuans (L.) Gaud, *Laportea aestuans*, ortie blanche/rouge, white/red stinging nettle, Urticaceae	Irritatating chemicals		
Hibiscus macranthus Hochst ex A. Rich Malvaceae, similar species to that reported			Alkaloids, coumarins, glycosides, flavonoids
Justicia insularis T. Anders, Acanthaceae (similar species to that reported)			Alkaloids, glycosides, flavonoids
Leonotis nepetaefolia (R. Br.) W. T. Aiton, chandelier, Lamiaceae			Acts on smooth muscle
Myristica fragrans Houtt, nutmeg, muscade, nuez moscada, Myristicaceae		Possible aphrodisiac	
Nicotiana tabacum L, Tobacco, Solanaceae		Fertility disruptant	

Latin Name	Irritating Chemicals	Useful Medicinal Properties	Oxytocic/Uteroactive Chemicals
Ocimum gratissimum L, diten jaraba, Lamiaceae,			
Petiveria alliacea L, gully root, kojo root, mapurite Phytolaccaceae		Anticonvulsant, Analgesic	Induces smooth muscle contractions
Pilea microphylla L, Du thé bethelmay Urticaceae		Antioxidant, radioprotective	
Pimpinella anisum L, Hoja de gato, Umbelliferae		Anti inflammatory	
Piper marginatum Jacq, l'anis bois, Piperaceae			Related plant is stimulant
Ricinus communis L, castor oil bush, palma Christi, Euphorbiaceae,		Anti inflammatory	
Ruellia tuberosa L, minny root, Acanthaceae			Purgative and causes perspiration
Ruta graveolens, ruda, Rutaceae,			Has side effects
Spondias mombin L, hog plum, Anacardiaceae	Tannins	Antibacterial, Antiviral	Saponins, glycosides
Stachytarpheta jamaicensis (L.) Vahl, verbena, vervine Verbenaceae,	Tannins	Febrifuge, Analgesic	Glycosides
Urena sinuata L, *Urena aculeata* Mill, *Urena lobata* ssp. *sinuata* (L.) Borss. Waalk, *Urena lobata* var. *sinuata* (L.) Hochr, patte chien, Malvaceae		Anti inflammatory	

Non-experimental validation of plants used for reproductive problems

Abelmoschus manihot consumption may prevent bone loss perhaps due to its high content in calcium and lutein (84.5 mg/100 g dry leaves). It did not show any estrogenic effect. Myricetin from the aerial portion of the plant *Abelmoschus moschatus* improved insulin sensitivity in obese rats and is anti inflammatory (Hiermann et al. 1998, Puel et al. 2005, Liu et al. 2007).

Achyranthes indica has several synonyms: *Achyranthes aspera* var. *indica*, *Achyranthes aspera* var. *obtusifolia*, and *Achyranthes obtusifolia*. *Achyranthes aspera* is used in rural Ethiopia for fertility control. *Achyranthes bidentata* ('Niu Xi' in Chinese medicine, Radix Achyranthes Bidentatae) is used in traditional Chinese medicine as a tonic, emmenagogue, diuretic, and antifertility agent. An aqueous extract of powdered *Achyranthes bidentata* can accelerate the regeneration of a crushed common peroneal nerve in rabbits. The ethanol extract of the root of *Achyranthes aspera* has antifertility activity which is not permanent. The ethanol extract has estrogenic activity demonstrated by the significant increase in uterine weight, diameter of the uterus, thickness of endometrium and vaginal epithelial cornification in immature ovariectomized female albino rats (Vasudeva and Sharma 2006, Shibeshi et al. 2006, Sun 2006, Ding et al. 2008).

Ageratum conyzoides was found to reduce acute pain in the mammalian central nervous system, but this was contradicted by a 1991 study. *Ageratum conyzoides* plant extract inhibited uterine contractions induced by 5-hydroxytryptamine but had no effect on uterine contractions induced by acetylcholine (Vyas and Mulchandani 1986, Yamamoto et al. 1991, Sampson et al. 2000, Achola and Munenge 1998, Silva et al. 2000).

Artemisin extracted from *Artemisia annua*, reduced the fertility of rats, especially at high doses (5 to 10 times the therapeutic dose used in humans). The rats had lower maternal progestagens and testosterone. The doses given to the rats were 5 to 10 times the therapeutic dose used in humans (Wenqiang et al. 2006, Boareto et al. 2008).

Annatto from *Bixa orellana* seeds, is used as a food color additive, in pharmaceuticals and in the cosmetics industry (Paumgartten et al. 2002). The carotenoid pigment bixin is a major component of annatto (28%). Annatto had no effect on the fertility of rats.

The aqueous extract (50–200 mg/kg) of dried leaves of *Capraria biflora* has moderate pain relieving properties, and antimicrobial and anti-inflammatory activities, some of which are attributed to biflorin (Acosta et al. 2003, Vasconcellos et al. 2007).

Crude stem bark hydro-alcohol extracts of *Cola nitida* (20 µg/ml) inhibited the release of luteinizing hormone (LH) induced by luteinizing hormone-releasing hormone (LHRH), but had no effect on FSH release. The active compounds may be phenols: catechin, quinic acid, tannic acid, chlorogenic acid and flavonoids (Benie and Thieulant 2004).

Coleus barbatus is used to abort pregnancy in Brazil and as an emmenagogue in other countries (Almeida and Lemonica 2000). *Coleus barbatus* showed an anti-implantation effect in the pre-implantation period in rats, but after embryo implantation the extract had little effect.

The hexane, chloroform and methanol extracts from the aerial parts of *Cordia curassavica* essential oil, was evaluated against 13 bacteria and

5 fungal strains (Hernandez et al. 2007). The oil and extracts exhibited antimicrobial activity against Gram-positive and Gram-negative bacteria and five fungal strains. *Sarcina lutea* and *Vibrio cholerae* were more sensitive to the essential oil and *Vibrio cholerae* to the hexane extract. *Rhizoctonia solani* was more sensitive to the essential oil and *Trichophyton mentagrophytes* to the hexane extract.

The diterpenes isolated from the leaves of *Croton zambesicus* have vasorelaxant activity (Sylvestre et al. 2006, Baccelli et al. 2007).

Curcuma comosa has traditionally been used in Thailand for treatment of inflammation in postpartum uterine bleeding (Sodsai et al. 2007). Its anti-inflammatory effects were established.

The oil of *Eupatorium macrophyllum* contains mainly monoterpenes (sabinene and limonene) (Maia et al. 2002), which Latin American respondents refer to as 'irritating' chemicals that are used to stimulate contractions.

Hibiscus rosa-sinensis has anti-fertility activity (Nivsarkar et al. 2005). *Hibiscus rosa-sinensis* flower decoctions are used in folklore medicine in India and Vanuatu as aphrodisiacs, for menorrhagia, uterine hemorrhage and for fertility control (Bourdy and Walter 1992, Kasture et al. 2000). Flower extracts produced an irregular estrus cycle in mice with prolonged estrus and metoestrus and other indications of antiovulatory effects, androgenicity and estrogenic activity (Prakash et al. 1990, Murthy et al. 1997).

Mimosa pudica root powder (150 mg/kg body weight) when administered intragastrically, altered the estrus cycle pattern in female *Rattus norvegicus* (Valsala and Karpagaganapathy 2002). Dioestrus lasted for 2 wk. There was a significant reduction in the number of normal ova and a significant increase in the number of degenerated ova. The dried methanol extract of the powdered roots of *Mimosa pudica* (300 mg/kg body weight/day) were given to female mice for 21 d and caused a prolonged estrus cycle with a significant increase in the duration of the diestrous phase (Ganguly et al. 2007). No significant changes were observed in the level of progesterone and LH. A significant increase in the estradiol level in the diestrus stage was found in the treated animals, along with a significant decrease in the secretion of FSH in proestrus and estrus stages. The treated animals had smaller litters, but the pups were normal and the treatment had no effect on subsequent untreated litters. *Mimosa tenuiflora* seeds (10% of the ration) given to pregnant rats from days 6–21 of pregnancy caused bone malformations in the fetal pups (Medeiros et al. 2008).

Ocimum gratissimum var. *macrophyllum* accessions contained thymol as the major volatile oil constituent, and xantomicrol as the major flavone. The accessions could be divided based on volatile oil constituents into six groups: (1) thymol: α-copaene; (2) eugenol:spathulenol;

(3) thymol:p-cymene; (4) eugenol:γ-muurolene; (5) eugenol:thymol: spathulenol; and (6) geraniol (Viera et al. 2001).

Parthenium hysterophorus interferes in the functioning of the hypothalamo-hypophyseal axis in the mouse brain (release of pituitary hormones) which may in turn affect the physiology of the peripheral endocrine glands which (if the impact is negative) can in turn can lead to polycystic ovaries, hyperandrogenaemia, hypothalamic amenorrhoea, and functional hyperprolactinaemia (Verma et al. 2007).

Aqueous and ethanol extracts of aerial parts of *Piper auritum* have produced spasmogenic uterine stimulant and vasodilator effects (Gupta et al. 1993). Of all *Piper* species only 10% (112 of 1,000+ known species world-wide) have been phytochemically investigated. These 112 species contain 667 different compounds distributed as follows: 190 alkaloids/amides, 49 lignans, 70 neolignans, 97 terpenes, 39 propenylphenols, 15 steroids, 18 kavapyrones, 17 chalcones/dihydrochalcones, 16 flavones, 6 flavanones, 4 piperolides (cinnamylidone butenolides) and 146 miscellaneous compounds (Dyer et al. 2004).

Ricinoleic acid, one active principle of castor oil, which was used as a leaf poultice for womb inflammation, possesses capsaicin-like dual pro-inflammatory and anti-inflammatory activity (Vieira et al. 2001).

When rats were given ruda (*Ruta graveolens*) for the first 4 to 10 d of pregnancy in two separate research studies it reduced their fertility and had a negative effect on embryo development; however, this effect was not seen in hamsters given ruda for the first week of pregnancy (Gutierrez-Pajares et al. 2003). De Freitas et al. (2005) did not find any negative effects when they gave mice the hydroalcoholic extract of *Ruta graveolens* aerial parts (1000 mg/kg per day) between the first and third Day of Pregnancy (DOP), between the fourth and sixth DOP or between the seventh and ninth DOP. However they, like other researchers quoted literature showing that deaths have resulted when ruda was used as an abortifacient by women.

Senna (*Cassia occidentalis*) is widely used as an expectorant and an anti-inflammatory agent. The anti-inflammatory activity is attributed to the constituents isolated from roots and stem. *Cassia occidentalis* treatment protected mice from cyclophosphamide-induced suppression of humoral immunity. The chemical constituents isolated from *Cassia occidentalis* include anthraquinones, fatty oils, flavonoids, gallactomannan, polysaccharides and tannins. Many of these constituents isolated from other plant sources have shown immunomodulatory activity (Bin-Hafeez et al. 2001). This study made the case that total plant extracts show more efficacy versus single constituents. The anthraquinone glycosides of the plant are reputed to be oxytocic and should not be given to pregnant women, unless it is to help delivery (Robineau 1991).

The methanolic extract of *Urena lobata* roots has broad spectrum antibacterial activity, antioxidant activity and inhibitory action against nitric oxide release from macrophages.

Isolated compounds from *Urena lobata* include mangiferin from the aerial parts, triglycerides and small quantities of quercetin (Morelli et al. 2006).

Wedelia paludosa and *Wedelia trilobata* contain a diterpene (kaurenoic acid), eudesmanolide lactones and luteolin (in leaves and stems) (Bohlmann et al. 1981, Block et al. 1998 a,b). Kaurenoic acid has antibacterial, larvicidal and trypanocidal activity; it is also a potent stimulator of uterine contractions. Luteolin exerts antitumoral, mutagenic and antioxidant effects, has depressant action on smooth muscles and a stimulant action on isolated guinea pig heart. Kaurenoic acid and luteolin in *Wedelia paludosa* showed pain relieving action more potent than the standard analgesic drugs (acetyl salicylic acid, acetaminophen, dipyrone and indomethacin). The root was the most potent part while the flower showed weak activity. Kaurenoic acid was 2–4 fold and luteolin was 8–16 fold more potent than acetyl salicylic acid, acetaminophen, dipyrone and indomethacin (some well-known analgesic drugs). N-hexane and ethyl-acetate extracts from *Wedelia trilobata* have antibacterial activity (Huang et al. 2006).

Conclusion

The plants used in Trinidad and Tobago for reproductive problems and childbirth need more detailed study because they are still widely used and if proved effective could play a role in modern reproductive care. Turmeric (*Curcuma longa*), annatto (*Bixa orellana*), vetivert (*Vetiveria zizanioides*) and hibiscus (*Hibiscus rosa-sinensis*) are usually recognized as safe, and are widely used in the food and beverage industries. *Ruta graveolens* has been used since antiquity but has serious safety problems. *Datura stramonium* may produce minor to serious side effects if ingested.

References

Achola, K.J. and R.W. Munenge. 1998. Bronchodilating and uterine activities of *Ageratum cozoides* extract. Pharmaceutical Biology 36: 93–96.

Acosta, S.L., L.V. Muro, A.L. Sacerio, A.R. Pena, and S.N. Okwei. 2003. Analgesic properties of *Capraria biflora* leaves aqueous extract. Fitoterapia 74: 686–8.

Almeida, F.C.G. and I.P. Lemonica. 2000. The toxic effects of *Coleus barbatus* B. on the different periods of pregnancy in rats. Journal of Ethnopharmacology 73: 53–60.

Baccelli, C., I. Navarro, S. Block, A. Abad, N. Morel, and J. Quetin-Leclercq. 2007. Vasorelaxant activity of diterpenes from *Croton zambesicus* and synthetic trachylobanes and their structure-activity relationships. Journal of Natural Products 70: 910–7.

Benie, T. and M-L. Thieulant. 2004. Mechanisms underlying antigonadotropic effects of some traditional plant extracts in pituitary cell culture. Phytomedicine 11: 157–164.

Bin-Hafeez, B., I. Ahmad, R. Haque, and S. Raisuddin. 2001. Protective effect of *Cassia occidentalis* L. on cyclophosphamide-induced suppression of humoral immunity in mice. Journal of Ethnopharmacology 75: 13–8.

Block, L.C., C. Scheidt, N.L. Quintao, A.R. Santos, and V. Cechinel-Filho. 1998a. Phytochemical and pharmacological analysis of different parts of *Wedelia paludosa* DC. (Compositae). Pharmazie 53: 716–718.

Block L.C., A.R. Santos, M.M. de Souza, C. Scheidt, R.A.Yunes, M.A. Santos, F.D. Monache, and V. Cechinel Filho. 1998b. Chemical and pharmacological examination of antinociceptive constituents of *Wedelia paludosa*. Journal of Ethnopharmacology 61: 85–89.

Boareto, A.C., J.C. Muller, A.C. Bufalo, G. Kasecker, S.L. de Araujo, M.A. Foglio, R.N. de Morais, and P.R. Dalsenter. 2008. Toxicity of Artemisinin [*Artemisia annua* L.] in two different periods of pregnancy in Wistar rats. Reproductive Toxicology 25: 239–46.

Bohlmann, F., J. Ziesche, R.M. King, and H. Robinson. 1981. Eudesmanolides and diterpenes from *Wedelia trilobata* and an ent-kaurenic acid derivative from *Aspilia parvifolia*. Phytochemistry 20: 751–756.

Bourdy, G. and A. Walter. 1992. Maternity and medicinal plants in Vanuatu. 1. The cycle of reproduction. Journal of Ethnopharmacology 37: 179–196.

Browner, C.H., B.R. Ortiz de Montellano, and A.J. Rubel. 1998. A methodology for cross-cultural ethnomedical research. Current Anthropology 29: 681–702.

de Freitas, T.G., P.M. Augusto, and T. Montanari, T. 2005. Effect of *Ruta graveolens* L. on pregnant mice. Contraception 71: 74–7.

Ding F., C. Qiong, and G. Xiaosong. 2008. The repair effects of *Achyranthes bidentata* extract on the crushed common peroneal nerve of rabbits. Fitoterapia 79: 161–7.

Dyer, L.A., C.D. Dodson, and J. Richards. 2004. Isolation, synthesis, and evolutionary ecology of *Piper amides*. Piper. In: L.A. Dyer and A.N. Palmer (eds.). A model genus for studies of evolution, chemical ecology, and trophic interactions. Kluwer Academic Publishers, Boston. pp. 117–139.

Ganguly, M., N. Devi, R. Mahant, and M.K. Borthakur. 2007. Effect of *Mimosa pudica* root extract on vaginal estrous and serum hormones for screening of antifertility activity in albino mice. Contraception 76: 482–485.

Gupta, M.P., D. Mireya, A. Correa, P.N. Solís, A. Jones, C. Galdames, and F. Guionneau-Sinclair. 1993. Medicinal plant inventory of Kuna Indians: Part 1. Journal of Ethnopharmacology 40: 77–109.

Gutiérrez-Pajares, J.L., L. Zúñiga, and J. Pino. 2003. *Ruta graveolens* aqueous extract retards mouse preimplantation embryo development. Reproductive Toxicology 17: 667–672.

Hernandez, T., M. Canales, B. Teran, O. Avila, A. Duran, A.M. Garcia, H. Hernandez, O. Angeles-Lopez, M. Fernandez-Araiza, and G. Avila. 2007. Antimicrobial activity of the essential oil and extracts of *Cordia curassavica* (Boraginaceae). Journal of Ethnopharmacology 111: 137–41.

Herskovits, M.J. and F.S. Herskovits. 1947 [1964]. Trinidad village. A. A. Knopf. Octagon Books Inc.New York.

Hiermann, A., H.W. Schramm, and S. Laufer. 1998. Anti-inflammatory activity of myricetin-3-O-beta-D-glucuronide and related compounds. Inflammation Research 47: 421–7.

Hodge, W.H. and D. Taylor. 1957. The ethnobotany of the island Caribs of Dominica. Webbia 12: 513–644.

Huang, X., S. Ou, Tang S., L. Fu, J. Wu. 2006. Simultaneous determination of trilobolide-6-O-isobutirates A and B in *Wedelia trilobata* by gas chromatography. Chinese Journal of Chromatography 24: 499–502.

Kasture, V.S., C.T. Chopde, and V.K. Deshmukh. 2000. Anticonvulsive activity of *Albizzia lebbeck*, *Hibiscus rosa-sinensis* and *Butea monosperma* in experimental animals. Journal of Ethnopharmacology 71: 65–75.

Kurzer, M.S. and X. Xu. 1997. Dietary phytoestogens. Annual Review of Nutrition 17: 353–381.

Lans, C. 2007. Ethnomedicines used in Trinidad and Tobago for reproductive problems. Journal of Ethnobiology and Ethnomedicine 15: 3–13.

Liu, I.M., T.F. Tzeng, S.S. Liou, and T.W. Lan. 2007. Improvement of insulin sensitivity in obese Zucker rats by myricetin extracted from *Abelmoschus moschatus*. Planta Medica 73: 1054–60.

Maia, J.G.S., Md., G.B. Zoghbi, E.H.A. Andrade, M.H.L. da Silva, A.I.R Luz, and J.D da Silva. 2002. Essential oils composition of *Eupatorium* species growing wild in the Amazon. Biochemical Systematics and Ecology 30: 1071–1077.

Medeiros, R.M.T., A.P.M. Figueiredo, T.M.A.De Venció, F.P.M. Dantas, and F. Riet-Correa. 2008. Teratogenicity of *Mimosa tenuiflora* seeds to pregnant rats. Toxicon 51: 316–9.

Moodie-Kublalsingh, S. 1994. The cocoa panyols of Trinidad: An oral record. British Academic Press, London.

Morelli, C.F., P. Cairoli, G. Speranza, M. Alamgir, and S. Rajia. 2006. Triglycerides from *Urena lobata*. Fitoterapia 77: 296–299.

Morton, J.F. 1981. Atlas of medicinal plants of middle America: Bahamas to Yucatan. Charles C. Thomas, Springfield, USA.

Mlynarcikova, A., M. Fickova, and S. Scsukova. 2005. Ovarian intrafollicular processes as a target for cigarette smoke components and selected environmental reproductive disruptors. Endocrine Regulations 39: 21–32.

Murthy, D.R., C.M. Reddy, and S.B. Patil. 1997. Effect of benzene extract of *Hibiscus rosa sinensis* on the estrous cycle and ovarian activity in albino mice. Biological and Pharmaceutical Bulletin 20: 756–8.

Nivsarkar, M., M. Patel, H. Padh, C. Bapu and N. Shrivastava. 2005. Blastocyst implantation failure in mice due to "nonreceptive endometrium": endometrial alterations by *Hibiscus rosa-sinensis* leaf extract. Contraception 71: 227–30.

Oliver-Bever, B. 1986. Medicinal Plants in Tropical West Africa. Cambridge University Press, New York.

Paumgartten, F.J., R.R. De-Carvalho, I.B. Araujo, F.M. Pinto, O.O. Borges, C.A. Souza, and S.N. Kuriyama. 2002. Evaluation of the developmental toxicity of annatto in the rat. Food and Chemical Toxicology 40: 595–601.

Prakash, A.O., A. Mathur, H. Mehta, and R. Mathur. 1990. Concentrations of Na^+ and K^+ in serum and uterine flushings of ovariectomized, pregnant and cyclic rats when treated with extracts of *Hibiscus rosa sinensis* flowers. Journal of Ethnopharmacology 28: 337–47.

Puel, C., J. Mathey, S. Kati-Coulibaly, M.J. Davicco, P. Lebecque, B. Chanteranne, M.N. Horcajada, and V. Coxam. 2005. Preventive effect of *Abelmoschus manihot* (L.) Medik. on bone loss in the ovariectomised rats. Journal of Ethnopharmacology 99: 55–60.

Robineau, L. (ed.). 1991. Towards a Caribbean ethnopharmacopoeia. TRAMIL 4 Workshop: Scientific Research and Popular Use of Medicinal plants in the Caribbean. UNAH, Enda-Caribe, Santo Domingo, DO.

Sampson, J.H., J.D. Phillipson, N.G. Bowery, M.J. O'Neill, J.G. Houston, and J.A. Lewis. 2000. Ethnomedicinally selected plants as sources of potential analgesic compounds: Indication of *in vitro* biological activity in receptor binding assays. Phytotherapy Research 14: 24–9.

Sewram, V., M.W. Raynor, D.A. Mulholland, and D.M. Raidoo. 2000. The uterotonic activity of compounds isolated from the supercritical fluid extract of *Ekebergia capensis*. Journal of Pharmaceutical and Biomedical Analysis 24: 133–45.

Sewram, V., M.W. Raynor, D.A. Mulholland, and D.M. Raidoo. 2001. Supercritical fluid extraction and analysis of compounds from *Clivia miniata* for uterotonic activity. Planta Medica 67: 451–5.

Shibeshi, W., E. Makonnen, L. Zerlhun, and A. Debella. 2006. Effect of *Achyranthes aspera* L. on fetal abortion, uterine and pituitary weights, serum lipids and hormones. African Health Sciences 6: 108–12.

Silva, M.J., F.R. Capaz, and M.R. Vale. 2000. Effects of the water soluble fraction from leaves of *Ageratum conyzoides* on smooth muscle. Phytotherapy Research 14: 130–2.

Sobo, E.J. 1996. Abortion traditions in rural Jamaica. Social Science and Medicine 42: 495 –508.

Sodsai, A., P. Piyachaturawat, S. Sophasan, A. Suksamrarn, and M. Vongsakul. 2007. Suppression by *Curcuma comosa* Roxb. of pro-inflammatory cytokine secretion in phorbol-12-myristate-13-acetate stimulated human mononuclear cells. International Immunopharmacology 7: 524–31.

Sun, H.X. 2006. Adjuvant effect of *Achyranthes bidentata* saponins on specific antibody and cellular response to ovalbumin in mice. Vaccine 24: 3432–9.

Sylvestre, M., A. Pichette, A. Longtin, F. Nagau, and J. Legault. 2006. Essential oil analysis and anticancer activity of leaf essential oil of *Croton flavens* L. from Guadeloupe. Journal of Ethnopharmacology 103: 99–102.

Telefo, P.B., P.F. Moundipa, and F.M. Tchouangue. 2004. Inductive effect of the leaf mixture extract of *Aloe buettneri, Justicia insularis, Dicliptera verticillata* and *Hibiscus macranthus* on *in vitro* production of estradiol. Journal of Ethnopharmacology 91: 225–230.

Valsala, S. and P.R. Karpagaganapathy. 2002. Effect of *Mimosa pudica* root powder on oestrous cycle and ovulation in cycling female albino rat, *Rattus norvegicus*. Phytotherapy Research 16: 190–2.

Vasconcellos, M.C., D.P. Becerra, A.M. Fonseca, M.R. Pereira, T.L. Lemos, O.D. Pessoa, C. Pessoa, M.O. Moraes, A.P. Alves, and L.V. Costa-Lotufo. 2007. Antitumor activity of biflorin, an o-naphthoquinone isolated from *Capraria biflora*. Biological & Pharmaceutical Bulletin 30: 1416–21.

Vasudeva, N. and S.K. Sharma. 2006. Post-coital antifertility activity of *Achyranthes aspera*. Linn. root. Journal of Ethnopharmacology 107: 179–81.

Verma, S., R. Shrivastava, P.K. Prasad, and V.K. Shrivastava. 2007. *Parthenium hysterophorus* induced changes in neurotransmitter levels in mouse brain. Phytotherapy Research 21: 183–5.

Vieira, C., S. Fetzer, S.K. Sauer, S. Evangelista, B. Averbeck, M. Kress, P.W. Reeh, R. Cirillo, R., A. Lippi, C.A. Maggi, and S. Manzini. 2001. Pro- and anti-inflammatory actions of ricinoleic acid: similarities and differences with capsaicin. Naunyn-Schmiedeberg's Archives of Pharmacology 364: 87–95.

Vyas, A.V. and N.B. Mulchandani. 1986. Polyoxygenated flavones from *Ageratum conyzoides*. Phytochemistry 25: 2625–2627.

Wenqiang, G., L. Shufen, Y. Ruixiang, and H. Yanfeng. 2006. Comparison of composition and antifungal activity of *Artemisia argyi* Lévl. Et Vant inflorescence essential oil extracted by hydrodistillation and supercritical carbon dioxide. Natural Products Research 20: 992–8.

Wong, W. 1976. Some folk medicinal plants from Trinidad. Economic Botany 30: 103–142.

Yamamoto, L.A., J.C. soldera, J.A. Emim, R.O. Godinho, C. Souccar, A.J. Lapa. 1991. Pharmacological screening of Ageratum conyzoides L. (Mentrasto). Memorias del Instituto Oswaldo Cruz 86 (suppl. 2): 145–147.

Chapter 5

Medicinal Value of Polyunsaturated and Other Fatty Acids in Ethnobotany

Sabreen F. Fostok, Antonios N. Wehbe, and *Rabih S. Talhouk**

Introduction

To date, numerous drugs have been developed for treating inflammatory and other disorders. However, dietary interventions, as preventive therapy, are gaining much interest these days due to better safety and tolerance. These include adoption of healthy nutritional habits as well as consumption of nutritional supplements, fish oil, which is rich in n-3 polyunsaturated fatty acids (PUFAs), being the most common.

PUFAs are those fatty acids (FAs) containing two or more double bonds and including the biologically important n-3 PUFAs, n-6 PUFAs and the conjugated linoleic acid (CLA) isomeric mixture. Many studies have demonstrated beneficial effects for these FAs in various disorders, including inflammatory diseases. Interestingly, several reports have shown that depending on the dietary PUFA composition, the onset of inflammatory diseases can be prevented, delayed or even promoted. For instance, western diets, characterized by low n-3 PUFAs and high vegetable oil, animal fat and meat content, which represent an important source of n-6 PUFAs, have been reported to augment the pathogenesis of inflammatory and auto-immune diseases (reviewed by Simopoulos 2006).

Department of Biology, Faculty of Arts and Sciences and Nature Conservation Center for Sustainable Futures (IBSAR), American University of Beirut, Beirut, Lebanon.

Corresponding address: Dr. Rabih S. Talhouk, Department of Biology, Faculty of Arts and Sciences, American University of Beirut, P.O. Box: 11-0236, Beirut, Lebanon.
E-mail: *rtalhouk@aub.edu.lb*

On the other hand, diets rich in n-3 PUFAs have been shown to exert preventive and therapeutic effects on inflammatory diseases (reviewed by La Guardia et al. 2005). Interest in n-3 PUFAs began when Kromann and Green (1980) asserted that Greenland Eskimos, whose diet mainly constituted of fish, exhibited a reduced incidence of myocardial infarction, auto-immune and inflammatory diseases. Since then, studies addressing the anti-inflammatory effects of n-3 PUFAs have considerably increased, and this is evident in a PubMed search, which gave 179 vs. at least 650 hits for 'omega-3 polyunsaturated fatty acids and inflammation' during the periods of 1980–1999 and 2000–2009, respectively.

Another class of PUFAs, CLA, highly abundant in dairy products and ruminant meat, became a topic of interest after Pariza and Hargraves (1985) reported an anticarcinogenic activity for CLA from grilled ground beef. Other beneficial biological activities of CLA were later discovered. One of the early reports demonstrating anti-inflammatory activity for CLA was published by Liu and Belury (1998), where it was noted that CLA decreased the inflammatory mediators prostaglandin (PG)E$_2$ and arachidonic acid (AA) in 12-O-tetradecanoyl-phorbol-13-acetate (TPA)-induced murine keratinocytes. This publication was followed by more reports on the anti-inflammatory activities of CLA, as reviewed by Zulet et al. (2005). Furthermore, an *in vitro* study in our laboratory demonstrated anti-inflammatory bioactivities in a FA mix extracted from a folk medicinal plant of the genus *Ranunculus* and attributed these effects to the possible occurrence of CLA in the plant (Fostok et al. 2009).

Although the plant kingdom represents a vital source of FAs, research has taken little notice of the biological effects of FAs extracted from plants, as compared to animal-derived CLA and fish-derived n-3 PUFAs, the most studied FAs. Few studies investigating plant-derived FAs have shown a wide range of biological activities for these FAs, including anti-inflammatory, antibacterial, hypoglycemic, hypocholesterolemic and antitumor activities (Lopez Ledesma et al. 1996, McGaw et al. 2002, Kohno et al. 2004, Chuang et al. 2006, Hua et al. 2006).

The general characteristics of FAs are described in this chapter. In addition, different dietary sources of essential fatty acids (EFAs) are listed, emphasizing the importance of plants, both edible and inedible species, in the supply of such FAs. Furthermore, the biological activities, mainly the anti-inflammatory ones, of plant-derived FAs, n-3 PUFAs and CLA are extensively reviewed. This chapter also outlines possible mechanisms of action for n-3 PUFAs and CLA.

Fatty acids (FAs)

A. Structure and nomenclature

A FA molecule consists of a hydrocarbon chain, typically unbranched, bearing a carboxyl (COOH) group at one end and a methyl (CH_3) group at the other. FAs commonly possess an even number of carbon atoms ranging between four and twenty-eight, and can be classified into short-, medium- and long-chain FAs. However, the occurrence of double bonds represents a more important scheme for their classification. While saturated fatty acids (SFAs) lack double bonds, those including such bonds in the acyl chain are referred to as unsaturated fatty acids (UFAs); FAs with one double bond are called monounsaturated fatty acids (MUFAs), while those containing two or more double bonds are known as PUFAs (Fig. 5.1A).

FAs are annotated according to the number of carbon atoms, the number of double bonds, if any, in the acyl chain and the position of the last double bond counted from the omega (designated as w or n) carbon, the carbon of the methyl group. Thus, a FA represented as 18:2n-6 is an 18-carbon PUFA having two double bonds, with its last double bond being located six carbons away from the methyl group (Fig. 5.1B). Note that UFAs are further subdivided into n-3, n-5, n-6, n-7 and n-9 classes.

B. Biosynthesis and metabolism

Although mammals possess the ability to synthesize FAs *de novo*, they lack the necessary enzymes required for the synthesis of the n-3 and n-6 series. Palmitic acid (16:0), the first FA released from FA synthase, undergoes desaturation (insertion of new double bonds) and/or elongation (expansion of the acyl chain) to produce other saturated, monounsaturated and polyunsaturated FA series. However, the desaturases responsible for the production of the biologically significant n-3 and n-6 PUFAs are absent in mammals. Thus, α-linolenic acid (ALA; 18:3n-3) and linoleic acid (LA; 18:2n-6), the parents of n-3 and n-6 PUFA families, respectively, are termed EFAs and must be obtained from the diet. Although these EFAs cannot be synthesized endogenously, they can be metabolized upon dietary supplementation through a series of chain elongation and desaturation reactions, a pathway that takes place mainly in the liver and gives rise to the long-chain PUFAs: AA (20:4n-6) and eicosapentaenoic acid (EPA; 20:5n-3) from LA and ALA, respectively. EPA can be further elongated and desaturated to give docosahexaenoic acid (DHA; 22:6n-3) (Fig. 5.2).

The metabolism of LA and ALA is catalyzed by common desaturase and elongase enzymes, indicating a competition between n-3 and n-6 PUFAs for metabolism. Accordingly, metabolism of the PUFA that is relatively more abundant in the diet predominates. However, two additional factors

A

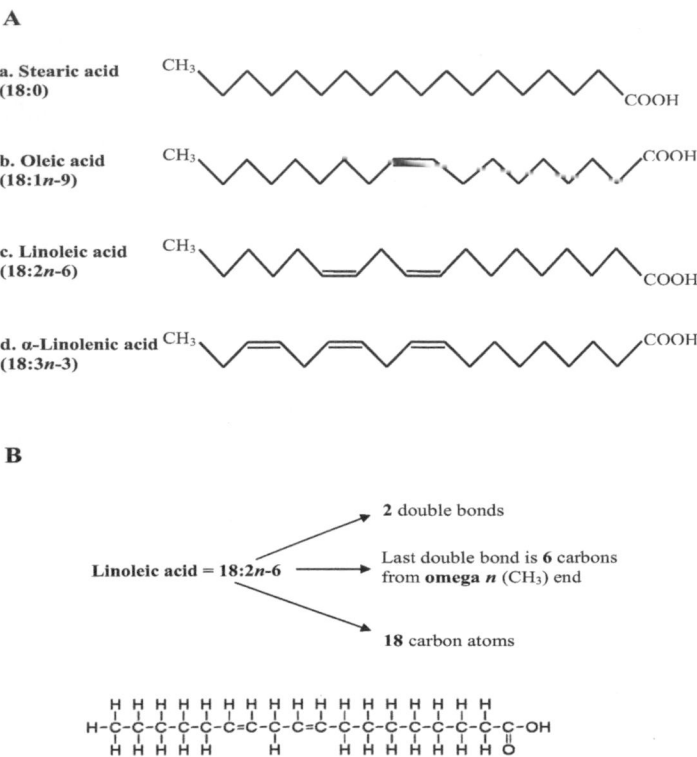

B

Fig. 5.1. Structure (A) and nomenclature (B) of some 18-carbon fatty acids **(A)** Saturated fatty acids (a) have no double bonds, while unsaturated fatty acids (b, c and d) are those having double bonds in the acyl chain. Fatty acids with one double bond are termed monounsaturated fatty acids (b), while those with two or more double bonds are referred to as polyunsaturated fatty acids (c and d). **(B)** Fatty acids are notated by the number of carbon atoms in the acyl chain, the number of double bonds, if any, and the position of the last double bond counted from the omega (CH$_3$) end. The example of linoleic acid is illustrated.

determine the rate of PUFA metabolism, regardless of their relative amounts: the enzymatic affinity for PUFAs and the efficiency of conversion reactions. Compared to n-6 PUFAs, n-3 ones represent a preferred substrate for Δ^6-desaturase, the rate-limiting enzyme in the metabolic pathway. Thus, theoretically, a diet containing equivalent amounts of LA and ALA is in favor of ALA metabolism. Taking into account the competition between these two PUFAs, increasing ALA and/or decreasing LA intake might enhance metabolism of the former; however, the conversion of ALA into EPA and DHA is inefficient in humans (reviewed by Gerster 1998, Davis and Kris-Etherton 2003 and Burdge and Calder 2005). Therefore, foods rich in EPA and DHA, rather than those limited in their n-3 content to ALA, are of an appreciable nutritional value. Given that 15–33% of ALA

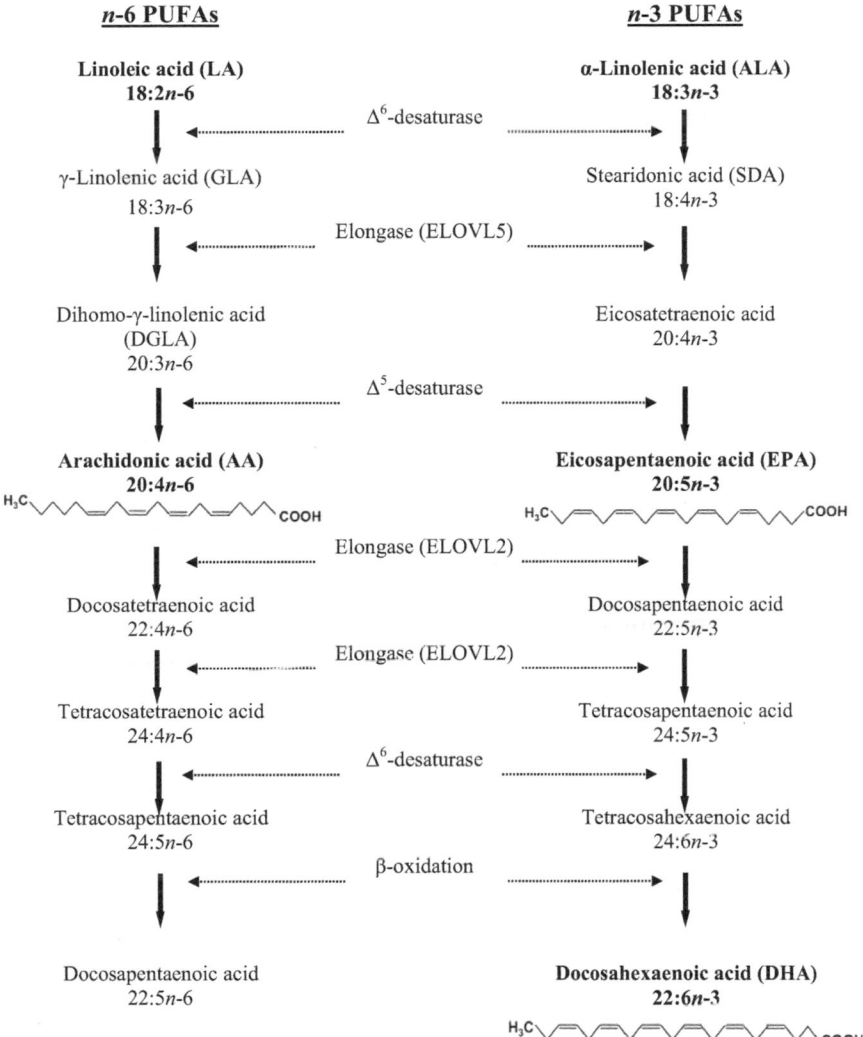

Fig. 5.2. The metabolic pathways of the *n*-6 and *n*-3 polyunsaturated fatty acids
PUFA: polyunsaturated fatty acid.

was partitioned towards β-oxidation in men, stable isotope tracer studies have shown that the conversion of ALA into EPA and DHA is greater in women than in men, perhaps due to an estrogen-dependent regulatory mechanism (reviewed by Burdge and Calder 2005). Consequently, these gender-related metabolic variations may be important in establishing a proper diet that meets the needs of both men and women.

C. Dietary sources and intake

Essential *n*-3 and *n*-6 PUFAs can be obtained from dietary sources, including plant and animal products, primarily meat and fish. They can be supplied in the form of parent PUFAs (LA and ALA) that undergo metabolism in the body, or end metabolites (AA, EPA and DHA).

Meat primarily supplies LA and AA, while fish, especially fatty fish (salmon, herring, tuna and mackerel), and their oils are characterized by their high EPA and DHA content. Essential PUFAs have been also reported to occur in algae and fungi, including yeast and basidiomycetes (Nerud and Musilek 1975, Ng and Laneelle 1977, Choi et al. 1987, Matsukawa et al. 1999, Guil-Guerrero et al. 2000, Otles and Pire 2001).

Plant seed oils provide an important source for EFAs. For instance, corn, sunflower, safflower and soybean oils are rich in LA. Besides soybean oil, other vegetable oils, such as perilla, rapeseed and flaxseed oils, contain ALA, with the latter being the richest source. In addition to seeds, other plant parts also contribute to LA and ALA supply. EFAs are commonly known to occur in a wide range of edible plants. Yet, certain medicinal plants with low edibility scores are rich in their essential PUFA content. A list of selected edible and inedible plant sources of these PUFAs is presented in Table 5.1, with emphasis on their medicinal ratings. As mentioned earlier, plant seeds represent a primary source for EFAs, mainly LA. Thus, most of the studies tended to analyze contents of seed oils. All studies investigating FA composition of seeds reported the presence of LA. In some of these studies, LA was the only EFA reported, as in the case of *Allium tuberosum* (garlic chives) seeds, where LA exceeded 55% of total FAs (Hu et al. 2005). Several other studies reported the occurrence of other EFAs, along with LA, such as linolenic acid, an 18:3 FA, in seeds. For instance, *Rosa canina* (dog rose) and *Juglans regia* (walnut) seeds contained significant amounts of linolenic acid, knowing that LA represented at least 50% of total FAs in these seeds (Ozcan 2002, Venkatachalam and Sathe 2006). In contrast, linolenic acid was detected in trace amounts in each of *Hibiscus sabdariffa* (roselle) and *Psophocarpus tetragonolobus* (winged bean) seeds (0.5% and 1%, respectively) (Homma et al. 1983, Rao 1996), and ALA was approximated at 2% in *Salicornia bigelovii* (dwarf glasswort) seeds (Anwar et al. 2002). Alternatively, ALA, followed by LA, was the most abundant FA at levels exceeding 55% in seeds of *Origanum onites* (pot marjoram) and *Origanum vulgare* (oregano) (Azcan et al. 2004). Abundant ALA was also reported in *Camelina sativa* (gold of pleasure) (Hrastar et al. 2009). In addition to LA and ALA, other EFAs were reported in seeds, such as γ-linolenic acid (GLA; 18:3*n*-6) and stearidonic acid (SDA; 18:4*n*-3), isolated from *Ribes nigrum* (blackcurrant) seed oil, with LA being the dominant among all FAs (Traitler et al. 1984). *Ribes nigrum* seed oil was shown to be the richest source of GLA amongst other *Ribes* species (Goffman and Galletti 2001).

Table 5.1. Edibility and medicinal rating of selected plants rich in essential PUFAs

Scientific Name	Common Name	Plant Part	PUFA	Edibility Rating*	Medicinal Rating*	Reference
Allium tuberosum	Garlic chives	Seed	LA	5	2	Hu et al. 2005
Rosa canina	Dog rose		Linolenic and LA	3	3	Ozcan 2002
Juglans regia	Walnut			4	3	Venkata-chalam and Sathe 2006
Hibiscus sabdariffa	Roselle			3	3	Rao 1996
Psophocarpus tetragonolobus	Winged bean			3	0	Homma et al. 1983
Salicornia bigelovii	Dwarf glasswort		LA and ALA	3	0	Anwar et al. 2002
Origanum onites	Pot marjoram			3	1	Azcan et al. 2004
O. vulgare	Oregano			4	3	
Camelina sativa	Gold of pleasure			3	0	Hrastar et al. 2009
Ribes nigrum	Blackcurrant		LA, ALA, GLA and SDA	5	3	Traitler et al. 1984, Goffman and Galletti 2001
Guizotia abyssinica	Niger seed		LA and EPA	3	1	Ramadan and Morsel 2003
Ribes nigrum	Blackcurrant	Leaf	ALA, SDA and GLA	5	3	Dobson 2000
Pinus sylvestris	Scot's pine	Stem	LA, DGLA and ALA	2	3	Piispanen and Saranpaa 2002
Asparagus officinalis	Asparagus	Aerial parts	LA	4	3	Jang et al. 2004

PUFA: polyunsaturated fatty acid.
LA: linoleic acid.
GLA: γ-linolenic acid (an *n*-6 PUFA obtained by desaturation of LA).
DGLA: dihomo-γ-linolenic acid (an *n*-6 PUFA obtained by elongation of GLA).
ALA: α-linolenic acid.
SDA: stearidonic acid (an *n*-3 PUFA obtained by desaturation of ALA).
EPA: eicosapentaenoic acid.
*Edibility and medicinal ratings have a range of 0-5, with 5 referring to highest edibility or medicinal value (www.pfaf.org).

Furthermore, EPA, uncommon in plants, was reported to occur in *Guizotia abyssinica* (Niger seed) seeds, where it represented less than 2% of total FAs, more than 60% of which was LA (Ramadan and Morsel 2003).

Besides seeds, EFAs were also reported to occur in leaves and stems. For instance, a study addressing the glycerolipid composition of *Ribes nigrum* leaves reported ALA accounting for more than 55% of total FAs, SDA and GLA were present at minor concentrations (4% and 1%, respectively) (Dobson 2000). Moreover, Piispanen and Saranpaa (2002) reported different lipid classes [triacylglycerols (TAGs), phospholipids and free FAs] in the wood of *Pinus sylvestris* (scot's pine) and determined their concentrations relevant to different stem heights and diameters. It was inferred from this study that LA and dihomo-γ-linolenic acid (DGLA; $20:3n$-6) were among the major FAs, LA being the most abundant of all FAs, in addition to the presence of minor ALA amounts. In another study, extracts of aerial parts of *Asparagus officinalis* (asparagus) were evaluated for their inhibitory effects on cyclooxygenase (COX). LA showed the most activity (Jang et al. 2004). Similarly, studies in our laboratory on a Ranunculaceous folk medicinal plant have shown that FA extracts from its aerial parts inhibit inflammatory mediators, as interleukin (IL)-6 and COX-2 (Fostok et al. 2009).

Human nutrition studies have shown that traditional diets of several old world cultures have had a balanced intake of n-3 and n-6 PUFAs. However, modern nutritional strategies have led to the disruption of this physiological balance resulting in a dramatic shift from 1/1 to 15–16.7/1 n-6/n-3 PUFA ratio (reviewed by Simopoulos 2002a, 2006 and 2008). This elevated ratio, typically found in western diets, is due to the increased consumption of LA- and AA-rich foods, including vegetable oils (such as corn, sunflower and safflower oils) and animal fats and meat (reviewed by Simopoulos 2002a, Covington 2004 and Stehr and Heller 2006). On the other hand, the Mediterranean diet is primarily built on the high intake of wild greens and dry fruits, both sources of ALA, moderate intake of fish, a rich source of EPA and DHA, and low intake of n-6 PUFAs (reviewed by de Lorgeril and Salen 2006, Galli and Marangoni 2006 and Manios et al. 2006). Vegetable salads and soups, foods of plant leaf and legume origin as well as nuts have made Mediterranean cuisine unique. Sardine, one of the largely consumed fish in Mediterranean countries, is characterized by its high EPA/DHA content among other fish varieties (Puglia et al. 2005). These facts recommend the Mediterranean diet as a nutritional model. This is also true for traditional diets from Far East cultures amongst others (Yamagishi et al. 2008, Elmadfa and Kornsteiner 2009).

D. Medicinal value

FAs serve a variety of physiological and pathophysiological functions. They constitute important structural components of membrane phospholipids, act as signaling molecules and provide substrates for the synthesis of eicosanoids(pro-inflammatorymediators),lipoxins(LXs;anti-inflammatory mediators), endocannabinoids (mood regulators) and hormones. They are also crucial for the proper development and performance of the nervous system. However, compelling evidence suggests the importance of dietary FA composition in determining the outcome of many diseases and, consequently, the health state of an individual. Various reports demonstrated the efficacy of certain FA classes, most notably n-3 PUFAs, in inhibiting carcinogenesis, treating major depression, hyperlipidemia, hypertension, diabetes, auto-immune and chronic inflammatory diseases and reducing the risk of cardiovascular diseases (CVDs) and mortality (reviewed by Harbige 1998, Simopoulos 2002b and Covington 2004). Among the early proofs validating the health benefits of n-3 PUFAs was an epidemiological study that reported a reduced incidence of myocardial infarction, auto-immune and inflammatory diseases in whaling and sealing Greenlanders (Kromann and Green 1980). Subsequent studies addressing the effects of dietary PUFAs have shown that frequent consumption of n-3 classes alleviates the symptoms of several diseases, such as atherosclerosis, inflammatory bowel disease (IBD), rheumatoid arthritis (RA) and IgA nephropathy (Kromhout et al. 1985, Donadio et al. 1994, Belluzzi et al. 1996, Volker et al. 2000), while high intakes of their n-6 counterparts promote inflammatory and cardiovascular diseases (Tappia and Grimble 1994, Bjorkkjaer et al. 2004; reviewed by Covington 2004). Thus, n-6-rich diets tend to possess pro-inflammatory activities, while those containing PUFAs of the n-3 family are anti-inflammatory (reviewed by Harbige 1998 and Covington 2004). In view of these findings, and as stated before, it has been established that manipulating the n-6/n-3 PUFA ratio in favor of n-3 PUFAs provides a preventive means and reduces the susceptibility to inflammation.

Plant-based fatty acids (FAs)

Many biological activities are attributed to the different classes of plant FAs, including SFAs, MUFAs and PUFAs, in addition to their derivatives (Table 5.2). These FAs are known to exert hypoglycemic, antitumor, antibacterial, hypocholesterolemic and anti-inflammatory activities. For instance, the fruit extract of *Momordica charantia* (bitter gourd), a plant traditionally used in Asia and Africa to treat diabetes and tumors among others, activated peroxisome proliferator-activated receptor (PPAR)-α and PPAR-γ, ligand-activated transcription factors belonging to the steroid hormone family of receptors and controlling the physiological homeostasis of glucose and lipids (Chao and Huang 2003). The bioactive component of this extract was later shown to be cis-9, trans-11, trans-13 conjugated linolenic acid (CLN) (Chuang et al. 2006). Taking into account the development of a novel class of type 2 diabetes medicines known as

Table 5.2. Biological effects of some plant-derived FAs

Scientific Name	Common Name	Plant Part	FA	Biological Effect	Reference
Momordica charantia	Bitter gourd	Fruit	cis-9, trans-11, trans-13 CLN	Hypoglycemic	Chao and Huang 2003, Chuang et al. 2006
		Seed		Apoptotic	Yasui et al. 2005
Punica granatum	Pomegranate	Seed	cis-9, trans-11, cis-13 CLN	Anticarcinogenic	Kohno et al. 2004
Biota orientalis	Biota	Seed	Juniperonic acid	Antiproliferative	Morishige et al. 2008
Coix lachryma-jobi	Job's tears	Seed	Palmitic, Oleic, Stearic and LA	Antitumor	Numata et al. 1994
Cola greenwayi	Hairy cola	Twig			Reid et al. 2005
Dombeya burgessiae	Tropical snowball				
Dombeya rotundifolia	Wild pear	Leaf	Plamitic acid		
Hermannia depressa	Rooi-opslang			Antibacterial	
Pentanisia prunelloides	Wild verbena	Root	Palmitic acid		Yff et al. 2002
Kigelia africana	Sausage tree	Fruit	Palmitic acid		Grace et al. 2002
Schotia brachypetala	Weeping schotia	Leaf	Linolenic acid and Methyl-5, 11, 14, 17-eicosatetraenoate		McGaw et al. 2002
Helichrysum pedunculatum	Isicwe	Leaf	Oleic and LA		Dilika et al. 2000
Persea americana	Avocado	Fruit	Oleic acid	Hypocholesterolemic Hypotriglyceridemic	Lopez Ledesma et al. 1996
Helianthus annuus	Sunflower	Seed	Oleic acid		Allman-Farinelli et al. 2005

Scientific Name	Common Name	Plant Part	FA	Biological Effect	Reference
Zostera japonica	Dwarf eelgrass	n.i.	Palmitic acid, Palmitic acid methyl ester, LA, LA methyl ester and Oleic acid methyl ester	Anti-inflammatory	Hua et al. 2006
Ehretia dicksonii	Cu kang shu	Leaf and Twig	9-hydroxy-trans-10, cis-12, cis-15-octadecatrienoic acid methyl ester		Dong et al. 2000
Brassica campestris	Wild turnip	Pollen	cis-9, cis-12, cis-15-octadecatrienoic acid sorbitol ester and (10, 11, 12)-trihydroxy-cis-7, cis-14-heptade-cadienoic acid*	Anticarcinogenic	Yang et al. 2009

FA: fatty acid.
LA: linoleic acid.
CLN: conjugated linolenic acid.
n.i.: not indicated.
*Four other known FA derivatives were isolated from this plant: N-(2-hydroxyethyl)-cis-9, cis-12, cis-15-octadecatrienamide, hexadecanoic acid sorbitol ester, (15, 16)-dihydroxy-cis-9, cis-12-octadecadienoic acid and cis-9, cis-12, cis-15-octadecatrienoic acid (2, 3)-dihydroxypropyl ester.

dual PPAR activators or PPAR-α/γ agonists, the presence of cis-9, trans-11, trans-13 CLN might be behind the reported hypoglycemic properties of the plant (Ojewole et al. 2006, Kumar et al. 2009). Moreover, bitter gourd seed oil, which is also rich in the CLN isomer, has been shown to upregulate GADD45, PPAR-γ and p53 in human colon cancer Caco-2 cells, and consequently lead to apoptosis (Yasui et al. 2005). Seed oil of *Punica granatum* (pomegranate), rich in cis-9, trans-11, cis-13 CLN, was reported to reduce colon carcinogenesis induced by azoxymethane (AOM), possibly due to the increase in CLA content of the liver and colon and/or the upregulation of PPAR-γ in colon mucosa (Kohno et al. 2004). The reported effect may be attributed to CLN, constituting more than 70% of the seed oil. These anticarcinogenic effects confirm the traditional value of pomegranate fruit, known for its anti-oxidant content. On the other hand, the pharmaceutical action of seeds of *Biota orientalis* (biota), a traditional Chinese medicinal plant used for its psycho-activity, may be attributed to juniperonic acid, a polymethylene-interrupted (PMI) FA, which exhibited an antiproliferative effect on bombesin-induced Swiss 3T3 cells. It was deduced that this activity is due to the juniperonic acid's *n*-3 double bond and not to its PMI structure since the *n*-3 EPA had comparable antiproliferative potency, while sciadonic acid, an *n*-6 analogue of juniperonic acid, did not display such an activity (Morishige et al. 2008). Another study of antitumor activities of FAs was reported by Numata et al. (1994) who fractionated the seed extract of *Coix lachryma-jobi* (Job's tears), a gramineous traditional plant used in China to cure tumors. The fractions were assayed for antitumor activity; the active fraction was acidic and contained the FAs: palmitic, oleic, stearic and LA.

The medicinal usage of some plants is due to the antibacterial effects of FAs they contain. For example, bioassay-guided fractionation of twig or leaf extracts from four Sterculiaceae species demonstrated that palmitic acid was a common antibacterial compound in all of these extracts. Other SFAs with antibacterial activity, such as myristic and stearic acid, were present in some of the extracts. All these FAs might be the reason behind using Sterculiaceae plants for treating coughs, stomach ache and diarrhea (Reid et al. 2005). Other plants used to combat bacterial infections are *Pentanisia prunelloides* (wild verbena) and *Kigelia africana* (sausage tree), both used in African folk medicine. Using bioautographic assaying of the former (Yff et al. 2002) and bioassay-guided fractionation of the latter (Grace et al. 2002), the active fractions of the root and fruit extracts of the plants, respectively, were isolated and shown to contain palmitic acid as the major compound. Furthermore, experiments demonstrated that palmitic acid was effective against both Gram-positive and Gram-negative bacteria. The antibacterial activity is not restricted to SFAs but is also displayed by UFAs. Indeed, bioactivity-guided fractionation of *Schotia brachypetala* (weeping schotia) leaf extract, a South African plant used for treating diarrhea and dysentery, lead to the isolation of linolenic acid and methyl-5, 11, 14, 17-eicosatetraenoate as the active compounds, which were more active against Gram-positive than Gram-negative bacteria (McGaw et al. 2002). Similarly, oleic acid and LA were isolated from the leaves of *Helichrysum pedunculatum* (isicwe), which are used in male circumcision rites in South Africa to prevent bacterial infections. These two UFAs were also active against gram-positive but not gram-negative bacterial species. Their activity against *Staphylococcus aureus* and their synergistic effect possibly justify the plant's usage in circumcision rites (Dilika et al. 2000).

Hypocholesterolemic activity is another effect mediated by plant-derived FA classes, namely MUFAs. This is evident from an experiment in which healthy and hypercholesterolemic volunteers, whether or not with hypertriglyceridemia, were fed *Persea americana* (avocado) fruit-rich diet. This fruit is a good source of MUFAs, mainly oleic acid, and is widely used to prepare traditional remedies in different regions of the world. The lipid levels in the serum of volunteers, measured before and after consuming this diet for 7 d, showed an increase in high-density lipoprotein (HDL) cholesterol, the so called 'good cholesterol', and a decrease in total serum cholesterol, low-density lipoprotein (LDL) cholesterol, the so called 'bad cholesterol', and TAG levels. Thus, the consumption of the MUFA-rich avocado fruit enhanced the lipid profile of both healthy and hypercholesterolemic individuals, but showed no major change in cholesterol or TAG levels of individuals on a control diet (Lopez Ledesma et al. 1996). The beneficial effects of oleic acid were confirmed by another study conducted by Allman-Farinelli et al. (2005). In this study, subjects consumed the oleic acid-containing seed oil of *Helianthus annuus* (sunflower), a folk medicinal plant, or a SFA-rich diet. As a result, oleic acid consumption reduced factor VII coagulant activity (FVIIc), LDL cholesterol and TAG levels, as compared to SFAs.

Additionally, anti-inflammatory activities are also associated with FA-containing plant extracts. For example, extracts from the seagrass *Zostera japonica* (dwarf eelgrass) possessed an anti-inflammatory activity, mainly due to palmitic acid, palmitic acid methyl ester, LA, LA methyl ester and oleic acid methyl ester. This extract inhibited lipopolysaccharide (LPS)-induced tumor necrosis factor (TNF)-α, IL-6 and IL-1β in murine macrophages dose-dependently (Hua et al. 2006).

As discussed above, plant-derived FAs possess a variety of biological activities; however, plants with FA derivatives can also be medicinally valuable. The leaves and twigs of the deciduous tree *Ehretia dicksonii* (cu kang shu) were discovered as a source of the derivative 9-hydroxy-trans-10, cis-12, cis-15-octadecatrienoic acid methyl ester. This derivative was shown to be anti-inflammatory as it reduced the TPA-induced mouse ear inflammation and suppressed soybean lipoxygenase (LOX) activity at doses of 500 μg and 10 μg/ml, respectively (Dong et al. 2000). The pollen of *Brassica campestris* (wild turnip) was also used to isolate FA derivatives, two of which were novel [cis-9, cis-12, cis-15-octadecatrienoic acid sorbitol ester and (10,11,12)-trihydroxy-cis-7, cis-14-heptadecadienoic acid] and four were previously known [N-(2-hydroxyethyl)-cis-9, cis-12, cis-15-octadecatrienamide, hexadecanoic acid sorbitol ester, (15, 16)-dihydroxy-cis-9, cis-12-octadecadienoic acid and cis-9, cis-12, cis-15-octadecatrienoic acid (2,3)-dihydroxypropyl ester]. Some of these FA derivatives strongly inhibited aromatase, an enzyme involved in the biosynthesis of estrogen, which promotes the growth of some breast cancers (Yang et al. 2009). Knowing that aromatase is upregulated in breast tumors, targeting this enzyme might render *Brassica campestris* useful for preventing breast cancer, especially since aromatase inhibitors are commonly used to treat such cancer.

Although the medicinal value of plant FAs has been reported in several studies, and plants remain a rich resource of such FAs, the integration of this knowledge into our dietary and health practices remains limited. On the other hand, CLA and fish-derived *n*-3 PUFAs are a subject of interest, and this is evident from the plentiful literature which reports their mechanisms of action and resulting health benefits, mainly the anti-inflammatory ones. Inferences from this literature re-emphasize the importance of n-3 PUFAs and CLA and the need to rethink of how best to harvest what remains to be a relatively untapped resource of plant-derived FAs into our dietary intake.

Omega (n)-3 polyunsaturated fatty acids (PUFAs)

A. Inflammatory diseases

Inflammation is a basic element of the body's innate response to a local injury, irritation or infection. It is characterized by a complex cascade of rapid reactions that aid in the elimination of pathogens or toxins causing the lesion while also repairing the damaged tissues. Although inflammation is normally a protective defense mechanism, it can become detrimental to the host tissues if it proceeds in an uncontrolled and unregulated manner. Such an uncontrolled response, whereby pro-inflammatory mediators are overly expressed, is a characteristic of acute and chronic inflammatory diseases.

In the past few years, numerous studies demonstrated anti-inflammatory bioactivities of long-chain n-3 PUFAs using both *in-vitro* and *in-vivo* models of inflammation (Tappia and Grimble 1994, Shoda et al. 1995, Lo et al. 1999, Li et al. 2005, Puglia et al. 2005). These findings were simultaneously supported by plentiful clinical data showing similar activities for these PUFAs in patients with chronic inflammatory diseases, mainly in IBD and RA patients (Bjorkkjaer et al. 2004, Bjorkkjaer et al. 2006, Goldberg and Katz 2007).

The healing effects of n-3 PUFAs have been best demonstrated clinically in Crohn's disease (CD) and ulcerative colitis (UC), collectively known as IBD and referring to a chronic inflammation of the gastrointestinal tract. Arslan et al. (2002) and later on Bjorkkjaer et al. (2004 and 2006) studied the effects of short-term seal oil administration on IBD. The n-6/n-3 PUFA ratio in rectal mucosa of IBD patients was significantly higher, compared to that in normal controls. Duodenal administration of the n-3-rich seal oil over a 10-d period, three times a day normalized both tissue and serum n-6/n-3 PUFA ratio, which was coupled to a reduction in IBD-associated joint pain. Unlike the n-6-rich soy oil, seal oil reduced the number of tender joints, duration of morning stiffness, pain intensity and disease activity. Furthermore, a follow up record of seal oil-treated patients for 6 mon post-treatment demonstrated a lasting effect for seal oil on joint pain. In support of these findings, a recent study by Goldberg and Katz (2007) established similar activities for n-3 PUFAs in patients with RA or IBD-associated joint pain. In addition, the expenditure of non-steroidal anti-inflammatory drugs (NSAIDs) by these patients was significantly reduced.

n-3 PUFAs have been reported, in addition, to reduce the frequency of relapses that characterize chronic inflammatory diseases. In a study by Belluzzi et al. (1996), fish oil was administered to CD patients with high chances of a relapse. At a dose of 9 capsules a day, more than 50% of these patients stayed in remission after 1-yr treatment.

The anti-inflammatory activities of *n*-3 PUFAs have been described in skin inflammation as well. Being rich in EPA and DHA, which were efficiently absorbed by human skin *in-vitro*, sardine oil extract proved effective in treating human UVB-induced erythema (Puglia et al. 2005).

The efficacy of *n*-3 PUFAs has also been well demonstrated in experimentally-induced inflammatory diseases. In a dextran sodium sulfate (DSS)-induced model of pig colitis, dietary supplementation with 1.33% fish oil for 42 d accelerated colonic regeneration (Bassaganya-Riera and Hontecillas 2006). In a similar model, rats fed on a diet of 2% perilla oil showed significant reduction in the ulcer index upon trinitro-benzene sulfonic acid (TNBS) challenge (Shoda et al. 1995). A different rat model of acute lung injury was used to manifest the anti-inflammatory effects of *n*-3 PUFAs. In this model, rats fed 1000 mg/kg/d of EPA, along with their standard diet, for 2 wk displayed significantly lower pulmonary edema than their control group following endotoxin (ET) injection (Sane et al. 2000). Other studies have shown that *n*-3 PUFAs inhibit the onset of inflammatory diseases. An *n*-3 PUFA-rich lipid extract obtained from green-lipped mussel powder exhibited such an activity when administered to rats per os (p.o.). Treated rats failed to develop arthritis in response to adjuvant or collagen II. The observed effect was at doses lower than those of NSAIDs and with no reported side effects. In addition, clinical data have demonstrated an anti-inflammatory effect for the same extract in RA and osteoarthritis (OA) patients (reviewed by Halpern 2000).

Interestingly, the anti-inflammatory effects of *n*-3 PUFAs have also been documented in transgenic animal models. For example, transgenic fat-1 mice expressing a *Caenorhabditis elegans* desaturase were used in a study by Schmocker et al. (2007). The ability of these mice to convert *n*-6 into *n*-3 PUFAs endogenously resulted in a balanced *n*-6/*n*-3 PUFA ratio and reduced the severity of injury and damage that accompany chemically-induced acute hepatitis.

B. Pro-inflammatory gene expression

The major pro-inflammatory signaling pathway is primarily initiated by the activation of the transcription factor nuclear factor (NF)-κB. This transcription factor is involved in the expression of genes coding for pro-inflammatory mediators, varying from chemokines and cytokines to nitric oxide (NO) and eicosanoids. Several studies have demonstrated anti-inflammatory effects for PUFAs, most commonly the *n*-3 members, which are mediated by blocking the activation of NF-κB and the expression of its downstream mediators (Fig. 5.3).

Li et al. (2005) showed that human kidney (HK)-2 cells incubated with EPA and DHA exhibit a reduced activation of NF-κB, compared to control HK-2 cells, following LPS stimulation. This reduction was coupled to an increase in PPAR-γ activity. Note that others have reported that PPARs' activation leads to the inhibition of inflammation in other cell types, possibly through a mechanism of inhibiting NF-κB activation, as suggested by Li et al. (2005).

Altering cytokine production is one mode of action by which *n*-3 PUFAs exert their anti-inflammatory effects downstream of NF-κB activation. For example, they can hinder the production of the pleiotropic cytokine IL-6. Khalfoun et al. (1997) showed that human endothelial cells (HECs) cultured in an *n*-3 PUFA-containing medium, produce less IL-6 in response to TNF-α, IL-4 or LPS. Similarly, unstimulated cells exhibited a reduction in their baseline IL-6 levels when supplemented with *n*-3 PUFAs. The effect of these FAs was dose-dependent, with EPA having stronger inhibitory action than DHA. *n*-3 PUFAs also mediate their actions by inhibiting the synthesis of other cytokines, such as IL-1 and TNF-α (Caughey et al. 1996,

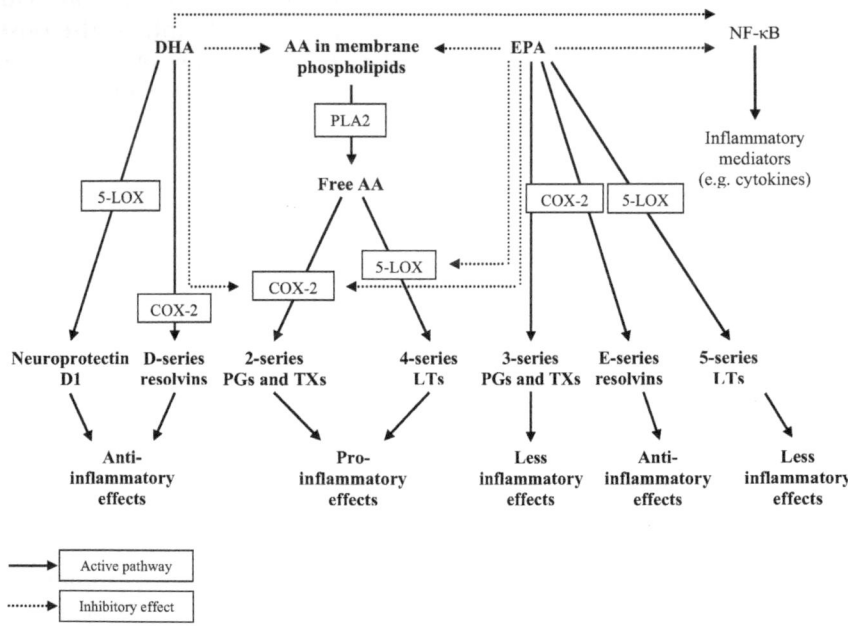

Fig. 5.3. The mechanism of action of *n*-3 polyunsaturated fatty acids
EPA: eicosapentaenoic acid; *DHA*: docosahexaenoic acid; *AA*: arachidonic acid; *NF-κB*: nuclear factor-κB; *PLA2*: phospholipase A2; *COX-2*: cyclooxygenase-2; *5-LOX*: 5-lipoxygenase; *PG*: prostaglandin; *TX*: thromboxane; *LT*: leukotriene.

Camuesco et al. 2005, Ferrucci et al. 2006, Rasic-Milutinovic et al. 2007, Schmocker et al. 2007).

The benefits of n-3 PUFAs have been further illustrated by their effects on endogenous anti-inflammatory mediators. The plasma concentration of these FAs was positively correlated with that of anti-inflammatory cytokines, such as transforming growth factor (TGF)-β and IL-10 (Ferrucci et al. 2006).

n-3 PUFAs are also known to act at the level of cytokine receptors. For instance, while AA amplified the expression of tumor necrosis factor receptors (TNFRs) by neutrophils, namely TNFR1 and TNFR2, EPA and DHA acted in an opposing manner (Moghaddami et al. 2007).

Another pro-inflammatory mediator whose synthesis is suppressed by n-3 PUFAs is NO. A study has reported that LPS-stimulated murine macrophage cell line RAW 264.7 produces fewer quantities of NO upon treatment with the n-3 PUFAs ALA, EPA and DHA. This effect varied with the administered dose of these FAs, while no effect was noted for stearic acid (SFA), oleic acid (MUFA) or LA (n-6 PUFA) (Ohata et al. 1997). In a similar study, Wohlers et al. (2005) tested the effect of fish oil-rich diets on inflammation in rats stimulated with carrageenan. This n-3 PUFA-rich oil reduced the production of NO, along with other inflammatory mediators, by rat macrophages.

Moreover, n-3 PUFAs appear to have an inhibitory effect on leukocyte chemotaxis, an event commonly associated with the inflammatory response. Schmidt et al. (1989) supplemented 12 healthy males with cod liver oil, which is rich in n-3 PUFAs, for a period of 6 wk. Using the under agarose technique, it was noted that monocyte and neutrophil chemotaxis towards chemoattractants decreases after the supplementation. Subsequent studies proved that n-3 PUFAs reduce chemotaxis by affecting the production of chemokines, a class of cytokines that act as chemoattractants. Li et al. (2005) showed that control HK-2 cells stimulated with LPS, express monocyte chemoattractant protein (MCP)-1 to a greater extent than cells pre-incubated with n-3 PUFAs. A similar study was conducted on macrophages, which displayed a reduced expression and secretion of MCP-1 when treated with EPA or DHA, compared to those treated with LA or SFAs (Wang et al. 2009).

Finally, n-3 PUFAs can attenuate the adhesion of leukocytes to endothelial surfaces, which accompanies chemotaxis. For instance, in an adhesion assay, EPA and DHA inhibited the adhesion of peripheral blood lymphocytes (PBLs) to HECs whether or not the latter had been treated with TNF-α, IL-4 or LPS, capable of upregulating adhesion molecules on cell surface. Similar inhibition was observed when PBLs or HECs were pretreated with the PUFAs. In addition, these PUFAs decreased the expression of vascular cell adhesion molecule (VCAM)-1 on stimulated

HECs and that of the adhesion molecule L-selectin on the surface of PBLs (Khalfoun et al. 1996). In another study, fish oil was shown to reduce the expression of E-selectin, intercellular adhesion molecule (ICAM)-1 and VCAM-1 in TNF-α-induced endothelial cells (Nohe et al. 2002). Knowing that leukocytes play a major role in tissue damage during inflammation, inhibiting their recruitment and adhesion provides an explanation for the healing and protective effects of n-3 PUFAs.

C. Eicosanoid synthesis

Eicosanoids belong to a family of pro-inflammatory mediators produced from 20-carbon n-6 PUFAs, mainly AA. They encompass thromboxanes (TXs), PGs and leukotrienes (LTs). AA is at first mobilized from membrane phospholipids by the enzyme phospholipase (PL)A$_2$, which is activated in response to pro-inflammatory stimuli. Subsequently, AA is metabolized by the NF-κB-inducible enzymes COX-2 and 5-LOX to produce the pro-inflammatory 2-series PGs and TXs and 4-series LTs, respectively (Fig. 5.3). However, a minor pathway involves 15-LOX, another isoform of LOX, that metabolizes AA to form anti-inflammatory compounds termed LXs (reviewed by Calder 2005).

Eicosanoids are responsible for the different signs that accompany inflammation, including fever, redness, edema and pain, and are thus key players in the inflammatory response. Therefore, inhibiting their synthesis provides important means for resolving inflammation and healing.

n-3 PUFAs have been shown by several studies to decrease the production of pro-inflammatory eicosanoids using various models of inflammation. In some of these studies, eicosanoid levels were measured in subjects fed n-3 PUFA-rich products. For instance, IBD patients exhibited a reduction in plasma LTB$_4$ following oral administration of seal oil or cod liver oil for 2 wk (Brunborg et al. 2008). In addition, using rat colitis models, Shoda et al. (1995) and Camuesco et al. (2005) demonstrated a reduction in both plasma and colonic LTB$_4$ by n-3 PUFA-supplemented diets. Moreover, eicosanoid reduction by n-3 PUFAs was demonstrated *ex-vivo*. For example, after consumption of a FA mix rich in n-3 PUFAs for 2 wk, LPS-stimulated whole blood of healthy volunteers produced less LTB$_4$ (Schubert et al. 2007). Similarly, fish oil diet resulted in a significant decrease in the production of PGE$_2$ and LTB$_4$ by rat peritoneal macrophages treated with calcium ionophore A23187 (Brouard and Pascaud 1990). Likewise, culturing cells in a medium enriched with n-3 PUFAs reduced eicosanoid production by these cells (Ishihara et al. 1998, Bousserouel et al. 2003).

n-3 PUFAs might block the production of eicosanoids by affecting their metabolic pathways or by their own conversion into less potent pro-inflammatory or even anti-inflammatory eicosanoids, thus reducing inflammation. This observed anti-inflammatory effect of n-3 PUFAs can be due to several reasons. First, when these FAs are present in certain amounts, they are incorporated into membrane phospholipids, partially displacing the n-6 AA (Lo et al. 1999). Consequently, less AA is available for eicosanoid synthesis. Second, n-3 PUFAs can compete with AA for metabolism and thus reduce the access of COX-2 and 5-LOX to AA resulting in diminished eicosanoid production (Lee et al. 1985, Ishihara et al. 1998, Bousserouel et al. 2003). Another reason might be the modification of gene expression. Through their ability to reduce NF-κB activation, n-3 PUFAs can indirectly alter eicosanoid levels by downregulating the expression of genes involved in their metabolic pathway. For example, rat smooth muscle cells (SMCs) stimulated with IL-1β expressed PLA_2 to a lesser extent when incubated with EPA or DHA in advance. In addition, PGE_2 production by these cells was reduced due to the inhibition of COX-2 expression (Bousserouel et al. 2003).

Furthermore, it has been reported that the metabolism of n-3 PUFAs generates eicosanoids with less potent pro-inflammatory activities than those derived from AA. These are the 3-series PGs and TXs and the 5-series LTs, produced by COX-2 and 5-LOX, respectively (Fig. 5.3). Compared to PGE_2, PGE_3 showed a 4-fold lower efficiency in upregulating COX-2 mRNA in fibroblasts and was less effective in inducing IL-6 synthesis in macrophages. Such effects suggest that n-3 PUFAs attenuate the cellular response to inflammatory stimuli (Bagga et al. 2003). Moreover, according to Wallace and McKnight (1990), LTC_5 and LTD_5 were five times less effective than their AA-derived 4-series analogues in promoting gastric mucosal damage and reducing gastric blood flow. In addition, LTB_5 showed 10- to 30-fold less potency than LTB_4 regarding their neutrophil chemotactic property (Goldman et al. 1983).

Recent reports suggest the generation of a novel class of anti-inflammatory eicosanoids, termed resolvins (resolution phase interaction products), from n-3 PUFAs. This occurs via conditioned COX-2-mediated conversion of EPA and DHA into E-series and D-series resolvins, respectively (Fig. 5.3). At doses of nanogram range, resolvins were reported to inhibit polymorphonuclear (PMN) leukocyte infiltration and to reduce leukocyte exudates by 40–80% *in vivo* (reviewed by Serhan 2005). In addition, a separate metabolic pathway involving several reactions catalyzed by 5-LOX converts DHA into a conjugated triene called neuroprotectin D1 (Fig. 5.3), which also possesses anti-inflammatory effects (Mukherjee et al. 2004).

Conjugated linoleic acid (CLA)

Besides the n-3 PUFAs, CLA is reported to have several physiological effects. CLA encompasses a group of positional and geometrical dienoic isomers derived from LA (cis-9, cis-12-octadecadienoic acid) and possessing conjugated rather than separated double bonds (Fig. 5.4). There are about

28 possible isomers, which can have conjugated double bonds in the 7 positions ranging from 6, 8 to 12, 14, where each double bond has either cis or trans configuration. The most commonly occurring isomeric form of CLA in nature is an *n*-7 cis-9, trans-11 CLA (reviewed by Zulet et al. 2005). However, all reported biological activities of CLA were due to the cis-9, trans-11 and trans-10, cis-12 (*n*-6) isomers (Fig. 5.4), with the former being the most biologically active possibly due to its unique incorporation into membrane phospholipids of animal tissues (Ha et al. 1990).

CLA is synthesized from LA, ALA and GLA in the guts of ruminants through microbial biohydrogenation, which involves an unusual enzymatic conversion mechanism. The amounts of CLA and the proportions of its isomers produced in the rumen and subsequently absorbed are influenced by the diet. Also, CLA is synthesized endogenously by Δ^9-desaturase

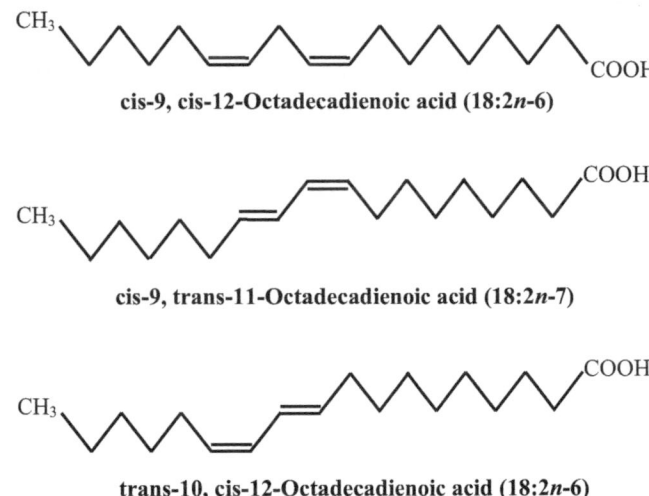

cis-9, cis-12-Octadecadienoic acid (18:2*n*-6)

cis-9, trans-11-Octadecadienoic acid (18:2*n*-7)

trans-10, cis-12-Octadecadienoic acid (18:2*n*-6)

Fig. 5.4. Structure of the two most biologically active isomers of conjugated linoleic acid The cis-9, trans-11 (middle) and the trans-10, cis-12 (bottom) isomer, both derived from linoleic acid (i.e. cis-9, cis-12-octadecadienoic acid; top).

enzyme, with its substrate being the MUFA vaccenic acid (trans-11, 18:1*n*-7), a product of rumen biohydrogenation pathways of LA, ALA and GLA (reviewed by Pariza et al. 2001). In addition, preliminary data in our laboratory on *Ranunculus* suggest the presence of one or more isomers of 18:2 FA with IL-6 inhibitory activity. Whether this/these isomer(s) is/are plant-derived CLA(s) remains to be determined (Fostok et al. 2009). CLA can also be produced by bacteria living in the large intestine of monogastric animals, but this CLA cannot be absorbed. Lastly, CLA can be synthesized in the laboratory by chemical methods that involve alkali isomerization of

LA or by total chemical synthesis (reviewed by Pariza et al. 2001, Bretillon et al. 2003 and Tanaka 2005).

Chin et al. (1992) developed a method for quantifying CLA in different foods, including meat, poultry, seafood, dairy products, plant oils and infant and processed foods. This study showed that animal products represent the main dietary source of CLA. Moreover, ruminant meat contained significantly greater amounts of CLA than that of non-ruminants, excluding turkey, with lamb meat having the highest CLA content. In contrast to seafood and plant oils that contained small amounts of CLA, dairy products possessed significant amounts of CLA. Additionally, the relative amounts of CLA isomers in foods were also examined. The cis-9, trans-11 CLA isomer constituted 75% and 90% of total CLA in animal and dairy products, respectively. Zlatanos et al. (2008) reported that the minimum CLA content of human plasma that would enable a person to benefit from its health effects is approximately 0.1% of the total FAs. This level can be maintained by a normal intake of dairy products (300–400 g cheese/wk) or by taking daily CLA supplements, such as CLA-containing capsules.

Interest in CLA first began when Pariza and Hargraves (1985) reported the anticarcinogenic activity of an extract from grilled ground beef. This led to the isolation of the active compounds in this extract, which were later referred to as CLA (Ha et al. 1987). Numerous studies reviewed the potential bioactivities of CLA, as evident from both *in vitro* and *in vivo* models (reviewed by Pariza et al. 2001, O'Shea et al. 2004 and Zulet et al. 2005). These include: anticarcinogenic, anti-atherosclerotic, antihypertensive, antidiabetic and anti-inflammatory effects. The anti-inflammatory effects are similar to those exerted by long-chain n-3 PUFAs. For instance, Hontecillas et al. (2002) reported that CLA can decrease tissue damage associated with bacterially-induced colitis. In this experiment, pigs fed CLA-rich diet showed a reduced growth inhibition and colonic mucosal damage that accompanied colitis. CLA is also protective against RA, as reported by Butz et al. (2007) who used the collagen antibody-induced arthritis model. This study involved feeding a group of mice with a mixture of CLA isomers (cis-9, trans-11 and trans-10, cis-12 CLA) and another control group with corn oil. Arthritis was subsequently induced in both groups through an iv. injection of monoclonal antibodies directed against type II collagen. As a result, redness and swelling were observed in the mice, for which an arthritis score was given according to the met criteria of inflammation. It appeared that CLA-fed mice had an arthritis score equivalent of 60% to that of corn oil-fed mice. Furthermore, CLA inhibits the activation of transcription factors (NF-κB) and downstream production of several pro-inflammatory mediators, including eicosanoids (LTB_4 and PGE_2), COX-2, pro-inflammatory cytokines [TNF-α, IL-6, IL-1β and interferon (IFN)-γ], chemokines (IL-8), adhesion molecules (VCAM-1) and inducible nitric oxide synthase (iNOS) (Fig. 5.5), as shown by different *in vitro* and *in vivo* studies (reviewed by O'Shea et al. 2004 and Zulet et al. 2005, Storey et al. 2007, Goua et al. 2008). CLA was also reported to inhibit leukocyte-endothelial adhesion, thus reducing inflammation (Sneddon et al. 2006). Alternatively, CLA promotes the production of anti-inflammatory cytokines (IL-10) and anti-inflammatory CLA-derived eicosanoids, as do *n*-3 PUFAs (reviewed by Zulet et al. 2005 and Bhattacharya et al. 2006).

CLA exerts its anti-inflammatory effects by several modes of action in a similar fashion to *n*-3 PUFAs. These were extensively reviewed by Zulet et al. (2005), and are summarized by the following: Since LA metabolism leads to the production of AA, the precursor of pro-inflammatory eicosanoids, then, interrupting this pathway would ameliorate inflammation. This is achieved by competition with CLA for desaturation and elongation in the

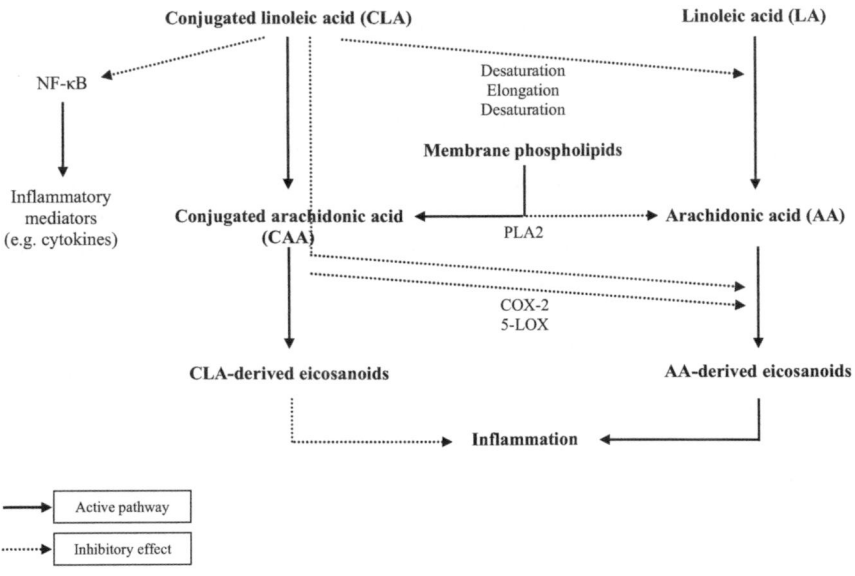

Fig. 5.5. The mechanism of action of conjugated linoleic acid
NF-κB: nuclear factor-κB; *PLA2*: phospholipase A2; *COX-2*: cyclooxygenase-2.
5-LOX: 5-lipoxygenase.

liver. Consequently, less AA is produced and, simultaneously, conjugated arachidonic acid (CAA) is generated. CLA and/or CAA are integrated into membrane phospholipids resulting in a decrease in the AA available for eicosanoid synthesis. In addition, CLA and CAA act as substrates for COX-2 and 5-LOX, thus reducing the enzymatic pool present for the metabolism of AA due to competing with this latter, and, as a result, less AA-derived eicosanoids are synthesized. At the same time, the CLA- and CAA-derived eicosanoids that are produced possess anti-inflammatory activities. Finally, in addition to acting directly via altering the metabolic pathway of pro-inflammatory eicosanoid synthesis, CLA can modulate pro-inflammatory gene expression through activation of PPAR and subsequent loss of NF-κB activation (Fig. 5.5).

Conclusion

EFAs are important for the normal physiological functioning of the body, yet their relative proportions (*n*-6/*n*-3 PUFAs) and absolute amounts in the diet can strongly influence the body's response to diseases and, consequently, the health of an individual. LA (*n*-6 PUFA) and ALA (*n*-3 PUFA), obtained from dietary sources, undergo metabolism to produce the long-chain *n*-6 PUFA, AA, and *n*-3 PUFAs, EPA and DHA, respectively.

It has been established that *n*-3 PUFAs, in contrast to their *n*-6 counterparts, are potential anti-inflammatory agents that can alleviate the symptoms and prevent the onset of inflammatory diseases, primarily IBD and RA, and this notion has been supported by many studies in the literature. Therefore, consuming foods rich in *n*-3 PUFAs (fish, plant leaves and certain seeds) while limiting the consumption of those rich in *n*-6 ones (meat and certain plant seeds) represents a strategy to correct the biased *n*-6/*n*-3 PUFA ratio, typical of western diets, and serves as a protective factor. CLA, another PUFA, has been reported to exert similar effects to those of *n*-3 PUFAs.

In view of the reported beneficial effects of fish oil and CLA, numerous marketable formulas of these products have been developed. However, the integration of plant-derived FAs into our diets, and recognition of their ethnobotanical value is yet to be realized. This chapter summarized our current knowledge of medicinal plants rich in PUFAs and other FAs and outlined their medicinal value. A better understanding of the biological effects of plant FAs and their role in ethnomedicine practices might provide means to adopt new herbal supplements into our diets to enhance disease resistance and improve our health status.

List of abbreviations

AA, Arachidonic acid
ALA, α-Linolenic acid
AOM, Azoxymethane
CAA, Conjugated arachidonic acid
CD, Crohn's disease
CLA, Conjugated linoleic acid
CLN, Conjugated linolenic acid
COX, Cyclooxygenase
CVDs, Cardiovascular diseases

DGLA, Dihomo-γ-linolenic acid
DHA, Docosahexaenoic acid
DSS, Dextran sodium sulfate
EFAs, Essential fatty acids
EPA, Eicosapentaenoic acid
ET, Endotoxin
FAs, Fatty acids
FVIIc, Factor VII coagulant activity
GLA, γ-Linolenic acid
HDL, High-density lipoprotein
HECs, Human endothelial cells
HK, Human kidney
IBD, Inflammatory bowel disease
ICAM, Intercellular adhesion molecule
IFN, Interferon
IL, Interleukin
iNOS, Inducible nitric oxide synthase
LA, Linoleic acid
LDL, Low-density lipoprotein
LOX, Lipoxygenase
LPS, Lipopolysaccharide
LTs, Leukotrienes
LXs, Lipoxins
MCP, Monocyte chemoattractant protein
MUFAs, Monounsaturated fatty acids
NF, Nuclear factor
NO, Nitric oxide
NSAIDs, Non-steroidal anti-inflammatory drugs
OA, Osteoarthritis
p.o., Per os
PBLs, Peripheral blood lymphocytes
PG, Prostaglandin
PL, Phospholipase
PMI, Polymethylene-interrupted
PMN, Polymorphonuclear
PPAR, Peroxisome proliferator-activated receptor
PUFAs, Polyunsaturated fatty acids
RA, Rheumatoid arthritis
SDA, Stearidonic acid
SFAs, Saturated fatty acids
SMCs, Smooth muscle cells
TAGs, Triacylglycerol
TGF, Transforming growth factor
TNBS, Trinitro-benzene sulfonic acid
TNF, Tumor necrosis factor
TNFRs, Tumor necrosis factor receptors
TPA, 12-O-tetradecanoyl-phorbol-13 acetate
TXs, Thromboxanes

UC, Ulcerative colitis
UFAs, Unsaturated fatty acids
VCAM, Vascular cell adhesion molecule

Acknowledgements

The authors are grateful to Mohamed-Bilal Fares for critical reading of the manuscript. This work was supported by grants from Lebanese National Council for Scientific Research (LNCSR), University Research Board (URB) at the American University of Beirut, Lebanon and HITECH, FZE, Dubai, UAE.

References

Allman-Farinelli, M.A., K. Gomes, E.J. Favaloro, and P. Petocz. 2005. A diet rich in high-oleic-acid sunflower oil favorably alters low-density lipoprotein cholesterol, triglycerides, and factor VII coagulant activity. Journal of the American Dietetic Association 105(7): 1071–1079.

Anwar, F., M.I. Bhanger, M.K. Nasir, and S. Ismail. 2002. Analytical characterization of Salicornia bigelovii seed oil cultivated in Pakistan. Journal of Agricultural and Food Chemistry 50(15): 4210–4214.

Arslan, G., L.A. Brunborg, L. Froyland, J.G. Brun, M. Valen, and A. Berstad. 2002. Effects of duodenal seal oil administration in patients with inflammatory bowel disease. Lipids 37(10): 935–940.

Azcan, N., M. Kara, B. Demirci, and K.H. Baser. 2004. Fatty acids of the seeds of *Origanum onites* L. and O. vulgare L. Lipids 39(5): 487–489.

Bagga, D., L. Wang, R. Farias-Eisner, J.A. Glaspy, and S.T. Reddy. 2003. Differential effects of prostaglandin derived from omega-6 and omega-3 polyunsaturated fatty acids on COX-2 expression and IL-6 secretion. Proceedings of the National Academy of Sciences of the United States of America 100(4): 1751–1756.

Bassaganya-Riera, J. and R. Hontecillas. 2006. CLA and n-3 PUFA differentially modulate clinical activity and colonic PPAR-responsive gene expression in a pig model of experimental IBD. Clinical Nutrition. 25(3): 454–465.

Bhattacharya, A., J. Banu, M. Rahman, J. Causey, and G. Fernandes. 2006. Biological effects of conjugated linoleic acids in health and disease. The Journal of Nutritional Biochemistry 17(12): 789–810.

Belluzzi, A., C. Brignola, M. Campieri, A. Pera, S. Boschi, and M. Miglioli. 1996. Effect of an enteric-coated fish-oil preparation on relapses in Crohn's disease. The New England Journal of Medicine 334(24): 1557–1560.

Bjorkkjaer, T., L.A. Brunborg, G. Arslan, R.A. Lind, J.G. Brun, M. Valen, B. Klementsen, A. Berstad, and L. Froyland. 2004. Reduced joint pain after short-term duodenal administration of seal oil in patients with inflammatory bowel disease: comparison with soy oil. Scandinavian Journal of Gastroenterology 39(11): 1088–1094.

Bjorkkjaer, T., J.G. Brun, M. Valen, G. Arslan, R. Lind, L.A. Brunborg, A. Berstad, and L. Froyland. 2006. Short-term duodenal seal oil administration normalised n-6 to n-3 fatty acid ratio in rectal mucosa and ameliorated bodily pain in patients with inflammatory bowel disease. Lipids in Health and Disease 5: 6.

Bousserouel, S., A. Brouillet, G. Bereziat, M. Raymondjean, and M. Andreani. 2003. Different effects of n-6 and n-3 polyunsaturated fatty acids on the activation of rat smooth muscle cells by interleukin-1 beta. Journal of Lipid Research 44(3): 601–611.

Bretillon, L., J.L. Sebedio, and J.M. Chardigny. 2003. Might analysis, synthesis and metabolism of CLA contribute to explain the biological effects of CLA? European Journal of Medical Research 8(8): 363–369.

Brouard, C., and M. Pascaud. 1990. Effects of moderate dietary supplementations with n-3 fatty acids on macrophage and lymphocyte phospholipids and macrophage eicosanoid synthesis in the rat. Biochimica et Biophysica Acta 1047(1): 19–28.

Brunborg, L.A., T.M. Madland, R.A. Lind, G. Arslan, A. Berstad, and L. Froyland. 2008. Effects of short-term oral administration of dietary marine oils in patients with inflammatory bowel disease and joint pain: a pilot study comparing seal oil and cod liver oil. Clinical Nutrition. 27(4): 614–622.

Burdge, G.C. and P.C. Calder. 2005. Conversion of alpha-linolenic acid to longer-chain polyunsaturated fatty acids in human adults. Reproduction, Nutrition, Development 45(5): 581–597.

Butz, D.E., G. Li, S.M. Huebner, and M.E. Cook. 2007. A mechanistic approach to understanding conjugated linoleic acid's role in inflammation using murine models of rheumatoid arthritis. American Journal of Physiology. Regulatory, Integrative and Comparative Physiology 293(2): R669–676.

Calder, P.C. 2005. Polyunsaturated fatty acids and inflammation. Biochemical Society Transactions 33(Pt 2): 423–427.

Calder, P.C. 2006. n-3 Polyunsaturated fatty acids, inflammation, and inflammatory diseases. The American Journal of Clinical Nutrition 83(6 Suppl): 1505S–1519S.

Camuesco, D., J. Galvez, A. Nieto, M. Comalada, M.E. Rodriguez-Cabezas, A. Concha, J. Xaus, and A. Zarzuelo. 2005. Dietary olive oil supplemented with fish oil, rich in EPA and DHA (n-3) polyunsaturated fatty acids, attenuates colonic inflammation in rats with DSS-induced colitis. The Journal of Nutrition 135(4): 687–694.

Caughey, G.E., E. Mantzioris, R.A. Gibson, L.G. Cleland, and M.J. James. 1996. The effect on human tumor necrosis factor alpha and interleukin 1 beta production of diets enriched in n-3 fatty acids from vegetable oil or fish oil. The American Journal of Clinical Nutrition 63(1): 116–122.

Chao, C.Y. and C.J. Huang. 2003. Bitter gourd (Momordica charantia) extract activates peroxisome proliferator-activated receptors and upregulates the expression of the acyl CoA oxidase gene in H4IIEC3 hepatoma cells. Journal of Biomedical Science 10(6 Pt 2): 782–791.

Chin, S.F., W. Liu, J.M. Storkson, Y.L. Ha, and M.W. Pariza. 1992. Dietary sources of conjugated dienoic isomers of linoleic acid, a newly recognized class of anticarcinogens. Journal of Food Composition and Analysis 5(3): 185–197.

Choi, K.J., Z. Nakhost, V.J. Krukonis, and M. Karel. 1987. Supercritical fluid extraction and characterization of lipids from algae Scenedesmus obliquus. Food Biotechnology 1(2): 263–281.

Chuang, C.Y., C. Hsu, C.Y. Chao, Y.S. Wein, Y.H. Kuo, and C.J. Huang. 2006. Fractionation and identification of 9c, 11t, 13t-conjugated linolenic acid as an activator of PPARα in bitter gourd (*Momordica charantia* L.). Journal of Biomedical Science 13(6): 763–772.

Covington, M.B. 2004. Omega-3 fatty acids. American Family Physician 70(1): 133–140.

Davis, B.C. and P.M. Kris-Etherton. 2003. Achieving optimal essential fatty acid status in vegetarians: current knowledge and practical implications. The American Journal of Clinical Nutrition 78(3 Suppl): 640S–646S.

de Lorgeril, M., and P. Salen. 2006. The Mediterranean-style diet for the prevention of cardiovascular diseases. Public Health Nutrition 9(1A): 118–123.

Dilika, F., P.D. Bremner, and J.J. Meyer. 2000. Antibacterial activity of linoleic and oleic acids isolated from Helichrysum pedunculatum: a plant used during circumcision rites. Fitoterapia 71(4): 450–452.

Dobson, G. 2000. Leaf lipids of Ribes nigrum: a plant containing 16:3, alpha-18:3, gamma-18:3 and 18:4 fatty acids. Biochemical Society Transactions 28(6): 583–586.

Donadio, J.V., Jr, E.J. Bergstralh, K.P. Offord, D.C. Spencer, and K.E. Holley. 1994. A controlled trial of fish oil in IgA nephropathy. Mayo Nephrology Collaborative Group. The New England Journal of Medicine 331(18): 1194–1199.

Dong, M., Y. Oda, and M. Hirota. 2000. (10E,12Z,15Z)-9-hydroxy-10,12,15-octadecatrienoic acid methyl ester as an anti-inflammatory compound from Ehretia dicksonii. Bioscience, Biotechnology, and Biochemistry 64(4): 882–886.

Elmadfa, I. and M. Kornsteiner. 2009. Dietary fat intake—a global perspective. Annals of Nutrition and Metabolism 54(Suppl 1): 8–14.

Ferrucci, L., A. Cherubini, S. Bandinelli, B. Bartali, A. Corsi, F. Lauretani, A. Martin, C. Andres-Lacueva, U. Senin, and J.M. Guralnik. 2006. Relationship of plasma polyunsaturated fatty acids to circulating inflammatory markers. The Journal of Clinical Endocrinology and Metabolism 91(2): 439–446.

Fostok, S.F., R.A. Ezzeddine, F.R. Homaidan, J.A. Al-Saghir, R.G. Salloum, N.A. Saliba, and R.S. Talhouk. 2009. Interleukin-6 and cyclooxygenase-2 downregulation by fatty-acid fractions of Ranunculus constantinopolitanus. BMC Complementary and Alternative Medicine 9: 44.

Galli, C. and F. Marangoni. 2006. N-3 fatty acids in the Mediterranean diet. Prostaglandins, Leukotrienes, and Essential Fatty Acids 75(3): 129–133.

Gerster, H. 1998. Can adults adequately convert α-linolenic acid (18:3n-3) to eicosapentaenoic acid (20:5n-3) and docosahexaenoic acid (22:6n-3)? International Journal for Vitamin and Nutrition Research 68(3): 159–173.

Goffman, F.D. and S. Galletti. 2001. γ-linolenic acid and tocopherol contents in the seed oil of 47 accessions from several *Ribes* species. Journal of Agricultural and Food Chemistry 49(1): 349–354.

Goldberg, R.J., and J. Katz. 2007. A meta-analysis of the analgesic effects of omega-3 polyunsaturated fatty acid supplementation for inflammatory joint pain. Pain 129(1-2): 210–223.

Goldman, D.W., W.C. Pickett, and E.J. Goetzl. 1983. Human neutrophil chemotactic and degranulating activities of leukotriene B_5 (LTB$_5$) derived from eicosapentaenoic acid. Biochemical and Biophysical Research Communications 117(1): 282–288.

Goua, M., S. Mulgrew, J. Frank, D. Rees, A.A. Sneddon, and K.W. Wahle. 2008. Regulation of adhesion molecule expression in human endothelial and smooth muscle cells by omega-3 fatty acids and conjugated linoleic acids: involvement of the transcription factor NF-kappaB? Prostaglandins, Leukotrienes, and Essential Fatty Acids 78(1): 33–43.

Grace, O.M., M.E. Light, K.L. Lindsey, D.A. Mulholland, J. Van Staden, and A.K. Jager. 2002. Antibacterial activity and isolation of active compounds from fruit of the traditional African medicinal tree Kigelia africana. South African Journal of Botany 68(2): 220–222.

Guil-Guerrero, J.L., E.H. Belarbi, and M.M. Rebolloso-Fuentes. 2000. Eicosapentaenoic and arachidonic acids purification from the red microalga Porphyridium cruentum. Bioseparation 9(5): 299–306.

Ha, Y.L., N.K. Grimm, and M.W. Pariza. 1987. Anticarcinogens from fried ground beef: heat-altered derivatives of linoleic acid. Carcinogenesis 8(12): 1881–1887.

Ha, Y.L., J. Storkson, and M.W. Pariza. 1990. Inhibition of benzo(a)pyrene-induced mouse forestomach neoplasia by conjugated dienoic derivatives of linoleic acid. Cancer Research 50(4): 1097–1101.

Halpern, G.M. 2000. Anti-inflammatory effects of a stabilized lipid extract of *Perna canaliculus* (Lyprinol). Allergie et Immunologie 32(7): 272–278.

Harbige, L.S. 1998. Dietary n-6 and n-3 fatty acids in immunity and autoimmune disease. The Proceedings of the Nutrition Society 57(4): 555–562.

Homma, S., M. Omachi, A. Tamura, E. Ishak, and M. Fujimaki. 1983. Lipid composition of winged bean (*Psophocarpus tetragonolobus*). Journal of Nutritional Science and Vitaminology 29(3): 375–380.

Hontecillas, R., M.J. Wannemeulher, D.R. Zimmerman, D.L. Hutto, J.H. Wilson, D.U. Ahn, and J. Bassaganya-Riera. 2002. Nutritional regulation of porcine bacterial-induced colitis by conjugated linoleic acid. The Journal of Nutrition 132(7): 2019–2027.

Hrastar, R., M.G. Petrisic, N. Ogrinc, and I.J. Kosir. 2009. Fatty acid and stable carbon isotope characterization of Camelina sativa oil: implications for authentication. Journal of Agricultural and Food Chemistry 57(2): 579–585.

Hu, G., Y. Lu and D. Wei. 2005. Fatty acid composition of the seed oil of *Allium tuberosum*. Bioresource Technology 96(14): 1630–1632.

Hua, K.F., H.Y. Hsu, Y.C. Su, I.F. Lin, S.S. Yang, Y.M. Chen, and L.K. Chao. 2006. Study on the antiinflammatory activity of methanol extract from seagrass Zostera japonica. Journal of Agricultural and Food Chemistry 54(2): 306–311.

Ishihara, K., M. Murata, M. Kaneniwa, H. Saito, K. Shinohara, and M. Maeda-Yamamoto. 1998. Inhibition of icosanoid production in MC/9 mouse mast cells by n-3 polyunsaturated fatty acids isoated from edible marine algae. Bioscience, Biotechnology, and Biochemistry 62(7): 1412–1415.

Jang, D.S., M. Cuendet, H.H. Fong, J.M. Pezzuto, and A.D. Kinghorn. 2004. Constituents of *Asparagus officinalis* evaluated for inhibitory activity against cyclooxygenase-2. Journal of Agricultural and Food Chemistry 52(8): 2218–2222.

Khalfoun, B., G. Thibault, P. Bardos, and Y. Lebranchu. 1996. Docosahexaenoic and eicosapentaenoic acids inhibit in vitro human lymphocyte-endothelial cell adhesion. Transplantation 62(11): 1649–1657.

Khalfoun, B., F. Thibault, H. Watier, P. Bardos, and Y. Lebranchu. 1997. Docosahexaenoic and eicosapentaenoic acids inhibit in vitro human endothelial cell production of interleukin-6. Advances in Experimental Medicine and Biology 400B: 589–597.

Kohno, H., R. Suzuki, Y. Yasui, M. Hosokawa, K. Miyashita, and T. Tanaka. 2004. Pomegranate seed oil rich in conjugated linolenic acid suppresses chemically induced colon carcinogenesis in rats. Cancer Science 95(6): 481–486.

Kromann, N. and A. Green. 1980. Epidemiological studies in the Upernavik district, Greenland. Incidence of some chronic diseases 1950–1974. Acta Medica Scandinavica 208(5): 401–406.

Kromhout, D., E.B. Bosschieter, and C. de Lezenne Coulander. 1985. The inverse relation between fish consumption and 20-year mortality from coronary heart disease. The New England Journal of Medicine 312(19): 1205–1209.

Kumar, R., S. Balaji, T.S. Uma, and P.K. Sehgal. 2009. Fruit extracts of *Momordica charantia* potentiate glucose uptake and up-regulate Glut-4, PPARgamma and PI3K. Journal of Ethnopharmacology. 126(3): 533–537.

La Guardia, M., S. Giammanco, D. Di Majo, G. Tabacchi, E. Tripoli, and M. Giammanco. 2005. Omega 3 fatty acids: biological activity and effects on human health. Panminerva Medica 47(4): 245–257.

Lee, T.H., R.L. Hoover, J.D. Williams, R.I. Sperling, J. Ravalese 3rd, B.W. Spur, D.R. Robinson, E.J. Corey, R.A. Lewis, and K.F. Austen. 1985. Effect of dietary enrichment with eicosapentaenoic and docosahexaenoic acids on in vitro neutrophil and monocyte leukotriene generation and neutrophil function. The New England Journal of Medicine 312(19): 1217–1224.

Li, H., X.Z. Ruan, S.H. Powis, R. Fernando, W.Y. Mon, D.C. Wheeler, J.F. Moorhead, and Z. Varghese. 2005. EPA and DHA reduce LPS-induced inflammation responses in HK-2 cells: evidence for a PPAR-gamma-dependent mechanism. Kidney International 67(3): 867–874.

Liu, K.L. and M.A. Belury. 1998. Conjugated linoleic acid reduces arachidonic acid content and PGE$_2$ synthesis in murine keratinocytes. Cancer Letters 127(1–2): 15–22.

Lo, C.J., K.C. Chiu, M. Fu, R. Lo, and S. Helton. 1999. Fish oil augments macrophage cyclooxygenase II (COX-2) gene expression induced by endotoxin. The Journal of Surgical Research 86(1): 103–107.

Lopez Ledesma, R., A.C. Frati Munari, B.C. Hernandez Dominquez, S. Cervantes Montalvo, M.H. Hernandez Luna, C. Juarez, and S. Moran Lira. 1996. Monounsaturated fatty acid (avocado) rich diet for mild hypercholesterolemia. Archives of Medical Research 27(4): 519–523.

Manios, Y., V. Detopoulou, F. Visioli, and C. Galli. 2006. Mediterranean diet as a nutrition education and dietary guide: misconceptions and the neglected role of locally consumed foods and wild green plants. Forum of Nutrition 59: 154–170.

Matsukawa, R., K. Hatakeda, S. Ito, Y. Numata, H. Nakamachi, Y. Hasebe, S. Uchiyama, M. Notoya, Z. Dubinsky, and I. Karube. 1999. Eicosapentaenoic acid release from the red alga Pachymeniopsis lanceolata by enzymatic degradation. Applied Biochemistry and Biotechnology 80(2): 141–150.

McGaw, L.J., A.K. Jager, and J. van Staden. 2002. Isolation of antibacterial fatty acids from Schotia brachypetala. Fitoterapia 73(5): 431–433.

Moghaddami, N., J. Irvine, X. Gao, P.K. Grover, M. Costabile, C.S. Hii, and A. Ferrante. 2007. Novel action of n-3 polyunsaturated fatty acids: inhibition of arachidonic acid-induced increase in tumor necrosis factor receptor expression on neutrophils and a role for proteases. Arthritis and Rheumatism 56(3): 799–808.

Morishige, J., N. Amano, K. Hirano, H. Nishio, T. Tanaka, and K. Satouchi. 2008. Inhibitory effect of juniperonic acid (Delta-5c,11c,14c,17c-20:4, omega-3) on bombesin-induced proliferation of Swiss 3T3 cells. Biological and Pharmaceutical Bulletin 31(9): 1786–1789.

Mukherjee, P.K., V.L. Marcheselli, C.N. Serhan, and N.G. Bazan. 2004. Neuroprotectin D1: a docosahexaenoic acid-derived docosatriene protects human retinal pigment epithelial cells from oxidative stress. Proceedings of the National Academy of Sciences of the United States of America 101(22): 8491–8496.

Nerud, F. and V. Musilek. 1975. Lipids in fruiting bodies of the basidiomycete Oudemansiella mucida. Folia Microbiologica 20(1): 24–28.

Ng, K.H. and M.A. Laneelle. 1977. Lipids of the yeast Hansenula anomala. Biochimie 59(1): 97–104.

Nohe, B., H. Ruoff, T. Johannes, C. Zanke, K. Unertl, and H.J. Dieterich. 2002. A fish oil emulsion used for parenteral nutrition attenuates monocyte-endothelial interactions under flow. Shock. Augusta, Ga. 18(3): 217–222.

Numata, M., A. Yamamoto, A. Moribayashi, and H. Yamada. 1994. Antitumor components isolated from the Chinese herbal medicine Coix lachryma-jobi. Planta Medica 60(4): 356–359.

O'Shea, M., J. Bassaganya-Riera, and I.C. Mohede. 2004. Immunomodulatory properties of conjugated linoleic acid. The American Journal of Clinical Nutrition 79(6 Suppl): 1199S–1206S.

Ohata, T., K. Fukuda, M. Takahashi, T. Sugimura, and K. Wakabayashi. 1997. Suppression of nitric oxide production in lipopolysaccharide-stimulated macrophage cells by omega 3 polyunsaturated fatty acids. Japanese Journal of Cancer Research: Gann 88(3): 234–237.

Ojewole, J.A., S.O. Adewole, and G. Olayiwola. 2006. Hypoglycaemic and hypotensive effects of Momordica charantia Linn (Cucurbitaceae) whole-plant aqueous extract in rats. Cardiovascular Journal of South Africa 17(5): 227–232.

Otles, S. and R. Pire. 2001. Fatty acid composition of Chlorella and Spirulina microalgae species. Journal of AOAC International 84(6): 1708–1714.

Ozcan, M. 2002. Nutrient composition of rose (Rosa canina L.) seed and oils. Journal of Medicinal Food 5(3): 137–140.

Pariza, M.W. and W.A. Hargraves. 1985. A beef-derived mutagenesis modulator inhibits initiation of mouse epidermal tumors by 7,12-dimethylbenz[a]anthracene. Carcinogenesis 6(4): 591–593.

Pariza, M.W., Y. Park, and M.E. Cook. 2001. The biologically active isomers of conjugated linoleic acid. Progress in Lipid Research 40(4): 283–298.

Piispanen, R. and P. Saranpaa. 2002. Neutral lipids and phospholipids in Scots pine (*Pinus sylvestris*) sapwood and heartwood. Tree Physiology 22(9): 661–666.

Puglia, C., S. Tropea, L. Rizza, N.A. Santagati, and F. Bonina. 2005. In vitro percutaneous absorption studies and in vivo evaluation of anti-inflammatory activity of essential fatty acids (EFA) from fish oil extracts. International Journal of Pharmaceutics 299(1-2): 41–48.

Ramadan, M.F. and J.T. Morsel. 2003. Determination of the lipid classes and fatty acid profile of Niger (*Guizotia abyssinica* Cass.) seed oil. Phytochemical Analysis: PCA 14(6): 366–370.

Rao, P.U. 1996. Nutrient composition and biological evaluation of mesta (*Hibiscus sabdariffa*) seeds. Plant Foods for Human Nutrition. Dordrecht, Netherlands 49(1): 27–34.

Rasic-Milutinovic, Z., G. Perunicic, S. Pljesa, Z. Gluvic, S. Sobajic, I. Djuric, and D. Ristic. 2007. Effects of N-3 PUFAs supplementation on insulin resistance and inflammatory biomarkers in hemodialysis patients. Renal Failure 29(3): 321–329.

Reid, K.A., A.K. Jager, M.E. Light, D.A. Mulholland, and J. Van Staden. 2005. Phytochemical and pharmacological screening of Sterculiaceae species and isolation of antibacterial compounds. Journal of Ethnopharmacology 97(2): 285–291.

Sane, S., M. Baba, C. Kusano, K. Shirao, T. Andoh, T. Kamada, and T. Aikou. 2000. Eicosapentaenoic acid reduces pulmonary edema in endotoxemic rats. The Journal of Surgical Research 93(1): 21–27.

Schmidt, E.B., J.O. Pedersen, S. Ekelund, N. Grunnet, C. Jersild, and J. Dyerberg. 1989. Cod liver oil inhibits neutrophil and monocyte chemotaxis in healthy males. Atherosclerosis 77(1): 53–57.

Schmocker, C., K.H. Weylandt, L. Kahlke, J. Wang, H. Lobeck, G. Tiegs, T. Berg, and J.X. Kang. 2007. Omega-3 fatty acids alleviate chemically induced acute hepatitis by suppression of cytokines. Hepatology. Baltimore, Md. 45(4): 864–869.

Schubert, R., R. Kitz, C. Beermann, M.A. Rose, P.C. Baer, S. Zielen, and H. Boehles. 2007. Influence of low-dose polyunsaturated fatty acids supplementation on the inflammatory response of healthy adults. Nutrition 23(10): 724–730.

Serhan, C.N. 2005. Novel omega—3-derived local mediators in anti-inflammation and resolution. Pharmacology and Therapeutics 105(1): 7–21.

Shoda, R., K. Matsueda, S. Yamato, and N. Umeda. 1995. Therapeutic efficacy of N-3 polyunsaturated fatty acid in experimental Crohn's disease. Journal of Gastroenterology 30 Suppl 8: 98–101.

Simopoulos, A.P. 2002a. The importance of the ratio of omega-6/omega-3 essential fatty acids. Biomedicine and Pharmacotherapy 56(8): 365–379.

Simopoulos, A.P. 2002b. Omega-3 fatty acids in inflammation and autoimmune diseases. Journal of the American College of Nutrition 21(6): 495–505.

Simopoulos, A.P. 2006. Evolutionary aspects of diet, the omega-6/omega-3 ratio and genetic variation: nutritional implications for chronic diseases. Biomedicine and Pharmacotherapy 60(9): 502–507.

Simopoulos, A.P. 2008. The importance of the omega-6/omega-3 fatty acid ratio in cardiovascular disease and other chronic diseases. Experimental Biology and Medicine 233(6): 674–688.

Sneddon, A.A., E. McLeod, K.W. Wahle, and J.R. Arthur. 2006. Cytokine-induced monocyte adhesion to endothelial cells involves platelet-activating factor: suppression by conjugated linoleic acid. Biochimica et Biophysica Acta 1761(7): 793–801.

Stehr, S.N. and A.R. Heller. 2006. Omega-3 fatty acid effects on biochemical indices following cancer surgery. Clinica Chimica Acta; International Journal of Clinical Chemistry 373(1–2): 1–8.

Storey, A., J.S. Rogers, F. McArdle, M.J. Jackson, and L.E. Rhodes. 2007. Conjugated linoleic acids modulate UVR-induced IL-8 and PGE_2 in human skin cells: potential of CLA isomers in nutritional photoprotection. Carcinogenesis 28(6): 1329–1333.

Tanaka, K. 2005. Occurrence of conjugated linoleic acid in ruminant products and its physiological functions. Animal Science Journal 76(4): 291–303.

Tappia, P.S. and R.F. Grimble. 1994. Complex modulation of cytokine induction by endotoxin and tumour necrosis factor from peritoneal macrophages of rats by diets containing fats of different saturated, monounsaturated and polyunsaturated fatty acid composition. Clinical Science. 87(2): 173–178.

Traitler, H., H. Winter, U. Richli, and Y. Ingenbleek. 1984. Characterization of gamma-linolenic acid in Ribes seed. Lipids 19(12): 923–928.

Venkatachalam, M. and S.K. Sathe. 2006. Chemical composition of selected edible nut seeds. Journal of Agricultural and Food Chemistry 54(13): 4705–4714.

Volker, D., P. Fitzgerald, G. Major, and M. Garg. 2000. Efficacy of fish oil concentrate in the treatment of rheumatoid arthritis. The Journal of Rheumatology 27(10): 2343–2346.

Wallace, J.L. and G.W. McKnight. 1990. Comparison of the damage-promoting effects of leukotrienes derived from eicosapentaenoic acid and arachidonic acid on the rat stomach. The Journal of Experimental Medicine 171(5): 1827–1832.

Wang, S., D. Wu, S. Lamon-Fava, N.R. Matthan, K.L. Honda, and A.H. Lichtenstein. 2009. In vitro fatty acid enrichment of macrophages alters inflammatory response and net cholesterol accumulation. The British Journal of Nutrition 102(4): 497–501.

Wohlers, M., R.A. Xavier, L.M. Oyama, E.B. Ribeiro, C.M. do Nascimento, D.E. Casarini, and V.L. Silveira. 2005. Effect of fish or soybean oil-rich diets on bradykinin, kallikrein, nitric oxide, leptin, corticosterone and macrophages in carrageenan stimulated rats. Inflammation 29(2–3): 81–89.

www.pfaf.org.

Yamagishi, K., H. Iso, C. Date, M. Fukui, K. Wakai, S. Kikuchi, Y. Inaba, N. Tanabe, and A. Tamakoshi. 2008. Fish, omega-3 polyunsaturated fatty acids, and mortality from cardiovascular diseases in a nationwide community-based cohort of Japanese men and women the JACC (Japan Collaborative Cohort Study for Evaluation of Cancer Risk) Study. Journal of the American College of Cardiology 52(12): 988–996.

Yang, N.Y., K. Li, Y.F. Yang, and Y.H. Li. 2009. Aromatase inhibitory fatty acid derivatives from the pollen of *Brassica campestris* L. var. *oleifera* DC. Journal of Asian Natural Products Research 11(2): 132–137.

Yasui, Y., M. Hosokawa, T. Sahara, R. Suzuki, S. Ohgiya, H. Kohno, T. Tanaka, and K. Miyashita. 2005. Bitter gourd seed fatty acid rich in 9c,11t,13t-conjugated linolenic acid induces apoptosis and up-regulates the GADD45, p 53 and PPARgamma in human colon cancer Caco 2 cells. Prostaglandins, Leukotrienes, and Essential Fatty Acids 73(2): 113–119.

Yff, B.T., K.L. Lindsey, M.B. Taylor, D.G. Erasmus, and A.K. Jager. 2002. The pharmacological screening of *Pentanisia prunelloides* and the isolation of the antibacterial compound palmitic acid. Journal of Ethnopharmacology 79(1): 101–107.

Zlatanos, S.N., K. Laskaridis, and A. Sagredos. 2008. Conjugated linoleic acid content of human plasma. Lipids in Health and Disease 7: 34.

Zulet, M.A., A. Marti, M.D. Parra, and J.A. Martinez. 2005. Inflammation and conjugated linoleic acid: mechanisms of action and implications for human health. Journal of Physiology and Biochemistry 61(3): 483–494.

Chapter 6

Smoke of Ethnobotanical Plants used in Healing Ceremonies in Brazilian Culture

Raquel de Luna Antonio,[1] Nayara Scalco,[1] Tamiris Andrade Medeiros,[1] Julino A. R. Soares Neto,[2] and Eliana Rodrigues[2]*

Introduction

It is believed that human beings have always interacted with plants. Surely the first men to inhabit our planet, even their four footed walking ancestors, used the potential of plants for feeding, medicines, clothing, shelter, dyeing, transportation and ritualistic articles. Simpson and Ogorzaly (2001) have indicated that the relationship between man and plants was set before the human condition, owing to the fact that plants were used by our ancestors.

The primordial man, who depended on nature to survive, discovered his natural medicines consciously through observation and empiric experimentation (Balick and Cox 1996, Di Stasi 1996, Heinrich et al. 2004). Balick and Cox (1996) affirm that once we can extract multiple resources from vegetables plants, more so than animals, can be the material basis for cultural development in the majority of the Earth's population.

The oldest archaeological data uncovered supporting the theory on the use of medicinal plants was found at Shanidar IV—a flower burial in northern Iraq where several species of pollen were discovered. At this

[1]Department of Psychobiology, Universidade Federal de São Paulo. Rua Botucatu, 862-1º Andar Edifício de Ciências Biomédicas, 04023-062-São Paulo, SP, Brazil.
[2]Department of Preventive Medicine, Universidade Federal de São Paulo. Rua Borges Lagoa, 1341-1º Andar, 04038-034, São Paulo-SP, Brazil.
[*]E-mail: *elirodri@psicobio.epm.br*

archaeological site, which goes back to the Neanderthal man dating more than 50,000 years ago, seven vegetable species were found that are still used as medicines in Iraq today (Solecki 1975). While it cannot be assumed that the use of these plants were medicinal, there is evidence pointing towards this (Heinrich et al. 2004). A curious detail, which reinforces the idea of the medicinal use found from these species, is that there was as much dry material as ashes suggesting that the plants were not only there to be burnt to produce fire (Leroi-Gourhan 1975). This may be the first ancient historical record of burning plants for a purpose other than fire production.

The first written document of common ground registering the medicinal use of plants is *Pen Tsao* canon, by the Chinese emperor Shen Nung. This document dates back to the 3rd millennium of the ancient era and its contents embody the use of 250 plants (Balick and Cox 1996, Simpson and Ogorzaly 2001). Once again in the East, the Indian book of *Vedas—Atharva Veda*, containing information on over 1000 medicinal plants, was written about 1,000 B.C.E. (Zimmer 2005). A later article including both medical and pharmaceutical knowledge relating to plants, Ebers Papyrus, is the main document from ancient Egypt, dating B.C.E. The document opens with the sentence *"Here begins the book on the preparation of medicine for all parts of the body"* (Haas 1999, Heinrich et al. 2004).

In 4th century B.C.E. Greece, Aristotle's disciple, Theophrastus, who was devoted to the study of medicinal plants, left writings on this subject (Balick and Cox 1996). During the first century of the common era, the Greek Pedanius Dioscorides produced the work *De Materia Medica*, which influenced medicine for the following thousand years, and described more than 600 types of plants (Balick and Cox 1996, Schulz et al. 2002, Heinrich et al. 2004). Nor can one forget Avicena (980–1037 C.E.) who, after the fall of the Roman Empire during the Middle Ages, emerged with the first Arabic knowledge in his treatise *Al Qanoon Fil Tib*, which mentions the therapeutic use of smoke (Mohagheghzadeh et al. 2006).

Nowadays, few studies focus on the medicinal use of smoke, as Mohagheghzadeh's et al. (2006) review, while burning aromatic plants, inhalation and simply standing close to smoke are practices observed in several human cultures throughout history, which ascend to religious and healing practices (Lewington 2003). It is important to note that in human cultural development we can often observe a link between cosmogonic myths and myths of medicinal origin (Eliade 2008), setting the relationship between religion and cure. In those cases, the purifying power of smoke is quite clear. Nevertheless, it was through the Egyptians that the act of burning aromatic herbs became acclaimed—with famous Egyptians perfumes, oils and incenses. The word 'perfume' comes from *per* (through) *fumum* (smoke), indicating that this use begun with the burning of aromatic plants (Lewington 2003). In Egypt, "aromatic woods and resins were burned to produce scented smoke that would carry offerings or entreaties to the gods" (Simpson and Ogorzaly 2001).

The natural occurrence of fire could be seen by humans, which later led to the burning of aromatic plants for a purpose. The ritualistic use of smoke can be understood through the symbolism of this subtle matter—smoke always rises to the heavens, the abode of the gods. Eliade (1991) describes the 'ascension symbolism' found in the various human cultures (ritualistically represented by stairs, trees, ropes, etc.) as "self-elevation of the human soul and union with God", expressing the notion of transcendence. Many sacred houses, spaces where religious cults are held, have a hole to output smoke (chimney), representing the connection axis between the profane and the sacred worlds. The author claims that a "mystical experience always involves a heavenly ascent", something that more than justifies the use of smoke in both religious and healing rituals, situations in which we seek transcendence of the ordinary.

> *"The smoke coming from the incense, along with the prayers of the saints, ascended before God from the angel's hand."*
> —(Christian Bible, Revelation 8:4)

In Brazil, smoke is used ethnopharmacologically in healing rituals. As an example, Mercante (2006) observed in Barquinha, a Brazilian ayahuasca religious system, mediums smoking their clients with tobacco, along with imposition of hands, to perform a cleanup. The use of smoke in two Brazilian cultural practices: Shamanism and Umbanda will be discussed later.

Olfactory physiology

Smell is the sense responsible for the perception of odors. In association with other sense organs that are connected to the central nervous system, smell helps us with multiple functions, among them the ability to recognize the environment and its dangers (Nef 1998).

While humans are capable of discriminating about 10,000 odors, animals have a great ability to distinguish the different odors of volatile chemicals. Data suggest that humans and mice have a similar basic structure of the olfactory system and olfactory receptors, which facilitates comparative studies. Research provides evidence that the animal's olfactory system uses a combination of different olfactory receptors to discriminate and code smells. Thus, an olfactory receptor can recognize multiple odors and odor can be recognized by multiple receptors, working together to code the identity of various odors and to allow discrimination of a multitude of odoriferous substances (Malnic et al. 1999, Menini et al. 2004). However, after our receivers are desensitized they process about 50% after the first second (Guyton and Hall 2006), while after a minute we cannot perceive the smell in a conscious way (Nef 1998).

With the discovery of approximately 1,000 genes present in the olfactory receptors of mammals (humans have between 300 and 400 functional genes), the combined approach of molecular genetics, imaging, and electrophysiological have provided a major breakthrough in understanding smell and the structures involved with this system (Menini et al. 2004, Mombaerts et al. 1996).

Smell is closely linked to taste, being one of the specialized organs of the nervous system that are among the oldest structures in the brain. The olfactory cells are responsible for the olfactory sensations through the olfactory chemoreceptors, located in the upper nasal cavity, which respond to chemical-specific substances dissolved in nasal mucus (Van De Graaff 2003, Guyton and Hall 2006).

After connecting the scent molecules to olfactory chemoreceptors, nerve impulses are generated. The transmission of these olfactory signals to the central nervous system occurs through the relay to the olfactory bulb neurons, an expansion of brain tissue at the base of the skull consisting of mitral cells and olfactory nerve fibers which are related to specialized areas of the cortex. The olfactory pathways branch out into different structures with different functions: the medial olfactory pathway, a very old olfactory system associated with the hypothalamus and other regions that control behavior; and lateral olfactory pathway where the signals are directed to less primitive limbic structures, such as hippocampus (Guyton and Hall 2006).

Thousands of odors can be distinguished by the sense of smell, but only the volatile substances have odor and can be noticeable when airborne and inhaled (Nef 1998). Regardless of the basic mechanisms of chemical stimulation of the olfactory cells, the stimulating substances must be at least partially hydrosoluble to get through the mucus and partially liposoluble so that might not be repelled by the lipid constituents of the cell membrane (Guyton and Hall 2006).

The upper airways can also be used to administer drugs topically or to provide systemic action. The nasal mucosa presents a typical absorbing mechanism, which hydrosoluble drugs intranasally enter by passive diffusion in aqueous channels and have a quick increase in plasma concentrations peaks. The lung's permeability, at alveolar epithelium and capillary endothelium, is high for water, most gases, and lipophilic substances. However, there is an effective barrier to large particles, many substances of hydrophilic nature, and molecular ionic species (Washington et al. 2001). In addition, Gilman et al. (2003) points out that drugs in aerosol form, when inhaled, have almost instant absorption into the blood, preventing losses from the hepatic first pass.

Shamanism practices

According to Langdon (1996), shamanism is seen as a collective representation expressed and renovated not only by the shaman's actions and the rituals he performs, but also in the way other people think and deal both with their everyday life and the disease. This view allows us to understand the practice of shamanism as a manifestation of a social and cosmological institution present in many societies (Pérez-Gil 2001).

By having a wide geographical distribution throughout history, shamanism has essential elements for understanding humanity and clarifies many points about the process of healing and understanding of one's health (Money 1997).

Shamanism can be interpreted as techniques that enable shamans to access information that is not 'ordinarily accessible'. The term *shaman* corresponds to the social construction that refers to the individual responsible for serving the psychological and spiritual needs of the community (Krippner 2000). Thus, he not only has the role of a healer but also the intermediary between men and gods or spirits—defending their community from evil spirits, as well as suggesting the best places to hunt and fish, controlling atmospheric phenomena, facilitating deliveries, and revealing future events (Eliade 1998).

In Brazil, shamanism is practiced by indigenous peoples, bringing the number of nations currently using this practice to a total of 220 (ISA 2009). It is stated by Krippner (2007) that although there are reports of shamanic practices have existed for 500 years in Brazil, the essence of this ritual still remains a mystery. The author also reports that traditional indigenous shamanism is a very threatened practice.

In South America, as well as in other continents where it occurs, there are variations in the purpose of shamanism. There are groups where this ritual can heal, but it may also have malicious purposes. In other groups, shamanic techniques and witchcraft techniques are distinct (Barcelos Neto 2006). Whether shamanism is used for beneficial or malevolent purposes, it remains a real practice existing in various human groups around the world (Storrie 2006).

A lot of literature devoted to this practice in Brazil, supports the idea that shamanism is practiced primarily by men, the shamans. However, there is several ethnographic evidence of the existence of women in the role of shamans in certain Amazonic societies. These women harmonize the roles of mother and shaman, have social functions credited only to the men and reach the later stages of shamanic power (Colpron 2005). Among Krahô Indians the presence of one shaman woman among 58 men was observed.

As elsewhere, in Brazil the main function of the shaman is to cure diseases whose origin may be natural or supernatural. Regardless of the origin of the disease, the 'ecstatic journey' of the shaman is essential. The trance state is part of the treatment, being the means by which the shaman finds the exact cause of the disease and its effective treatment (Eliade 1998). To achieve this ecstasy state, various elements—dependent on the indigenous ethnic group—such as music, circular and repeated movements, plants, and other psychoactive substances, can be used by the shaman during the ritual.

Diseases of supernatural origin are marked by the escape of the soul or introduction of 'magical objects' in the patient's body (Ackerknecht 1985). Shamanic healing in these cases involves the use (by the shaman) of cigarettes made with plants which provide contact with their spiritual guides, whose smoke enables both the diagnosis and cure, often by extracting the 'magical objects'. The smoking process, which is the focus of this text, involves the use of different plant species. The most common species of plant used among the natives of South America are those of the *Nicotiana* species.

Although there is a striking similarity in the function and procedures of the shamans among the various indigenous groups in Brazil, this similarity does not occur in the process of becoming a shaman. The explanations are heredity, spirituality, vocational involvement, God's choice, and their own will followed by learning, among others. Laraia (2005) describes that the pathway to the Tupi-Guarani involves a gift to be discovered and developed by learning. The author adds that among the Assurinis of the Tocantins River, "there is a ritual called *opetimo* that aims to identify, among young people, those who have the potential to become a shaman. Between songs and dances, the candidate smokes a big cigar, swallowing the smoke. Those who get sick, feeling nausea or vomiting are discarded. Those who faint are chosen, so the shaman responsible for the ritual clamors '*Oman*', 'he died'. It is in 'dying' that one can travel to the other world, enabling contact with the ancestors".

Some studies conducted among Brazilian indigenous groups point to the plants used by shamans. A review of plants indicated by 26 ethnic groups for disorders of the central nervous system, performed by Rodrigues et al. (2006), shows 25 plants indicated to the hallucinogen category, many of which are utilized for shamanic practices. These plants work by supposedly altering the perception of the shaman in order to facilitate contact with the spiritual world, they are also used for the ritual of cure and for their therapeutic function. Among them are: *Anadenanthera peregrina* (L.) Speg., Ayahuasca (*Banisteripsis caapi* (Spruce ex Griseb.) and *Psychotria viridis* Ruiz and Pav.), *Mimosa hostilis* (Mart.) Bent., *Virola elongata* (Benth.) Warb., *Nicotiana tabacum* L. and *Theobroma subincanum* Martius. Kerr (1970 *apud*

Camargo 2005/2006) reports the use of a long pipe by Kayapo Indians. The substance used with the Kayapo pipe is a combination of tobacco (*Nicotiana tabacum* L.), peanut leaves (*Arachys hypogaea* L.), ginger (*Zingiber officinale* Rox.), pigeon pea (*Cajanus cajan* (L.) Mill.) and to improve the odor and reduce the toxicity they add mashed cumaru leaves (*Dipterix odorata* Aub.). The Wauja have three shamanic experts: *yakapá, pukaiwekeho e yatamá*. The last one has the function of relieving pain by using the tobacco smoke (Barcelos Neto 2001, 2006). For the Kaingang, 'cure' is a ritual practice, where through herbal baths, smokes, and ashes one can obtain powers of nature (Silva 2002).

Epidemics by inter-ethnic contact is also seen to be "caused by the smoke produced by burning of things belonging to non-indigenous". According to Smiljanic (1999 *apud* Vidille 2006), "through this smoke, invisible by ordinary people, come many cannibal spirits (*xawararibë*) that devour the vital principle of people". Vidille (2006) explains that in the view of Ianomäe Indians, for whom the smoke, besides having a curative role, is linked to the disease process, what these people feel is due to reversals in the balance of the cosmos (this *cosmos* consists of paths and forests through which superhuman beings transit).

Tobacco

"Shamanism, whether used for beneficial or malevolent, is a very real presence in the world. It does not exist as part of a field of extraordinary experience and is not seen as belonging to an isolated sphere of practice. Everyone has a shamanic aspect, "ho", and all have direct access to the shamanic ambient through their dreams, the raw material of shamanism. At night, before going to bed, all adults compress a strong mix of prepared tobacco between the gum and lower lip, so that the dreams are more intense".

—(Storrie 2006).

Plant species of the Solanaceae family from the genus *Nicotiana* are, without doubt, the most widespread and commonly used plants in shamanic rituals of South American Indians. There are several forms of tobacco use by participants of the rituals such as: chewing, drinking, licking, enema, snuffing and smoking (Cooper 1987, Wilbert 1987).

In addition to being an essential element of magic and religion amongst the Indians of South America, tobacco goes through similar procedures among shamans such as: rolling it, lighting it, smoking it, and having a smoke to the heavens and spirits (Camargo 2005/2006).

The practice of smoking tobacco is performed by both sexes of various ages in rituals, board reunions, singing, and births, amongst other situations. This is easily observed when looking at the table on methods of use of tobacco in Wilbert's book (1987), where, of the 289 indigenous peoples of South America, only 55 do not use tobacco in the smoked form. This is, for sure, the most 'sacred' plant of the continent. Eliade (2008) reports that "the man of archaic societies tends to live as possible in the sacred or very close to sacred items." He also goes on to say "this is understandable; for the 'primitive' the sacred is equivalent to power and, ultimately, the reality par excellence".

The act of smoking tobacco by the Indians along the Brazilian coast is documented by André Thevet in 1557 when he contacted the Tupinambas, an ethnic group that inhabited the coastal area of southeastern Brazil and used smoke in recreation rituals. In 1578, Jean De Léry (1951) confirmed the observations of Thevet saying that the Tupinambás spent three to four days without eating anything, just smoking and using tobacco; the women never used tobacco.

In a review conducted by Rodrigues et al. (2006), tobacco (*Nicotiana tabacum* L.) was included under the category hallucinogen since it was considered psychoactive. Schultes (1984) explains that tobacco is definitely psychoactive in any method of use; the enigma remains as to how, under which conditions and in various methods of use, *Nicotiana* can have strong psychoactive effects in aboriginal societies—providing contact with the supernatural—since this effect is not observed outside of these contexts. A possible explanation could be due not only to the pharmacological action of this plant, but also to other substances present in the cigarette, in a synergistic action. Some communities in the Brazilian Amazon use the inner bark of *tauari* (*Couratari* sp.) to wrap the tobacco in a cigarette form. Maybe the substances present in this and other plants utilized to wrap the cigarettes act synergistically with tobacco. Another hypothesis is that the tobacco species used in the past by American Indians was *Nicotiana rustica* L., which possesses greater presence of nicotine and other alkaloids, and could increase its effect (Ritz and Orth 2000). In Brazil, as reported by Wilbert (1987), the most commonly used species is *Nicotiana petunioides* (Griseb.), Millan, although *N. rustica* L. and *N. tabacum* L. are also generally used in South America.

Presence of smoke in shamanic rituals of two Brazilian indigenous ethnic groups

Fieldwork carried out in Brazil by two of the authors, between the ethnic groups Guarani (N. Scalco) and Krahô (E. Rodrigues), observed the use of smoke in rituals and its role in the process of shamanic healing.

Guarani indians

The Guarani people occupy southern South America in the boundaries of Brazil, Paraguay, Uruguay and Argentina. Because they lived near the coast in the Atlantic, they became some of the first ethnic groups to be influenced by the Europeans who arrived in that region in the early sixteenth century. Currently inhabiting the mid-west, southeast, and southern parts of Brazil in addition to those countries previously mentioned, they preserve their language Tupi-Guarani, for its cosmological and spiritual importance (Clastres 1978).

Since the first contact of the Europeans with the Guarani people, the use of tobacco smoke was observed, drawing the attention of naturalists and travelers who described this practice in their work (Ferri 1980).

Considered a sacred plant for these Indians, tobacco is the only specimen smoked from the elderly to children. Currently, it is used in the form of a 'smoking rope', where the leaves of *Nicotiana* sp. are collected and distributed in piles to dehydrate and when droughts are braided to form a rope. This is cut into very thin and small pieces that are rubbed between the fingers to be separated. This is then placed in *petynguá*, a kind of wooden pipe with a coal on top and then smoked.

The Guarani people see this act as a way to raise their spirit to *Nhanderu*, the creator of the world in the Guarani cosmology, and therefore to establish an approach to the spiritual world. This feature makes *Nicotiana* the main genus plant used in healing rituals of this community. This ritual is conducted by the shaman (known as *pajé*) and occurs primarily within the *Opy* (house of prayer). The sick person is brought to the center of the *Opy*, which is a bamboo and feather structure symbolizing the presence of elements of the spiritual world, opposite the 'altar', where he sits and takes off all the clothes from the upper body. The shaman blows tobacco smoke by the *petynguá* around the patient. The smoke is directed to a region, signaling the place where the disease is concentrated. Then, seeing this signaling, the shaman, sucks the 'disease' with his mouth and inside his body begins a struggle between his own spirit and the spirit of the disease. This fight makes the shaman loose his balance and he needs to be supported by other indigenous people – then he begins to tremble and drool. The battle ends with the materialization of the disease within the body of the shaman, taking form of a small stone that is thrown out and smoked with tobacco smoke.

This shamanic ritual occurs whenever necessary in Guarani villages, in the *Opy* and at night. It is interesting to note that even without the necessity of healing, the village meets every night in the *Opy*, where they sing, smoke *petyngua*, and drink tea of erva-mate (*Ilex paraguariensis* A. St.-Hil.). This ritual occurs as an acknowledgment, requesting protection and communication with the spiritual world, especially with *Nhanderu*.

Krahô indians

The Krahô people inhabit the central region of Brazil, an area of cerrado savannahs. Their history of contact with the non-indigenous people is about two centuries old and their language is Timbira.

Among the Krahô Indians, shamans are known as *wajacas*, who are recognized as the keepers of the knowledge of herbal remedies and healing processes, for which they receive instructions and help from their respective *pahis* (spiritual guides, generally represented by the spirits of animals, plants, minerals, objects or even the deceased). He may heal or kill another, acting as a *wajaca* or a sorcerer. Each *wajaca* is a specialist in one or more illnesses, such as fever, diarrhea, snakebites, those brought by the wind, or even spells cast by other *wajacas*. The healing process involves two parts: the first is a ceremony conducted by the *wajacas*, mainly at night; during this practice they smoke tobacco, marijuana, or some other native plants. The act of smoking could help in communicating with the *pahi* or in furnishing more power at the moment of the healing, according to the interviewees. The exhaled smoke is blown at the patient, spreading out the illness so it can 'be more clearly diagnosed' or, even, to 'collect' the illness which is spread throughout the body of the patient to a single point so that it can then be 'aspirated' by the shaman, 'removing' the illness from the patient's body. In the second part, after the ceremony, the *wajaca* chooses one or more plants to be utilized in the treatment and returns several times to the patient's home to follow up on the effects of the remedy administered (Rodrigues and Carlini 2005).

It was observed that 23 plants were used in these healing ceremonies; the preference for each of them depends on the *wajaca*, although there is a general preference for marijuana (*Cannabis sativa* L.), known among the Krahô as *iumhô*. Before the introduction of this plan, however, the *wajacas* used the other 22, including: tobacco, *togré hô*, *pênjarahihhô*, *cumxê*, *tingui*, *carājatxy* and several kinds of *ahkroré*, *ahkrô*, *māputréhô*, *pjejapac* e *caprānkohiréhô*. A special pipe called *cót*, made of buriti straw (*Mauritia flexuosa* L. f) is used for smoking these plants, however, some prefer to use cigarettes made with paper.

The Krahô explain their preference for marijuana because of its safety compared to the other native plants of the cerrado savannahs. They say that marijuana can be consumed in high doses, unlike the other ones where there is not much knowledge of their effects in high doses or chronic use. For the Kraho, the 13 native plants are dangerous because 'it takes long time to exit the body', so they need to be careful with their consumption. If the doses are frequent or exaggerated they could have hallucinations. Furthermore, marijuana, known as the 'queen', can be consumed freely by most of the Krahô, as they feel 'it is eliminated quickly from the body.'

All these plants can also be used as cigarettes, in social contexts beyond shamanism. However, in a few cases, there were reports that marijuana could not be consumed by some individuals of the Krahô who were considered 'especially sick', as the misuse of alcohol associated with marijuana could trigger 'madness' in these people, as reported.

Umbanda religion

The slave trade of Africans has profoundly marked Brazil's history. More than four million men, women, and children were estimated to have landed here between the sixteenth and mid-nineteenth centuries. Despite the repression, their contribution to music, religion, food, dance, and language is outstanding in Brazilian culture (IBGE 2000).

This trade subjugated human beings, who were transported by ship from the African continent to America in unsanitary conditions, forced to change their living habits and perform hard labor for theirs 'masters' (landowners who bought them as slaves in Brazil). These factors are largely responsible for the poor health of slaves, eventually leading to diseases as there was no access to medical care (Pôrto 2006).

Umbanda, a Brazilian religion widely practiced in this area, was founded by the Africans who arrived here as slaves. Historically, the factors that contributed to the construction of Umbanda in Brazil are religions of African origin (Voodoo and Candomblé); Catholicism, brought to Brazil by the Europeans; indigenous elements and Allan Kardec Spiritism (Camargo 1961, Di Paolo 1979).

According to the 2000 census, it was estimated that there are over 397 thousand followers of Umbanda living in urban and rural areas of Brazil (IBGE 2009). But these numbers are underestimated. Many followers of African-Brazilian religions do not declare their religion as historically Catholicism was the only tolerated religion in Brazil, thus making it a factor for social inclusion. Even with the political changes in the nineteenth century, when there was more tolerance to religions other than Catholicism, they continued to proclaim themselves as Catholics (Prandi 2004).

The differentiation of all the elements is clear within the umbandistic syncretism. For example, Catholicism, the official and compulsory religion of colonization, was imposed on Africans and Indians by the Jesuit catechism, and according to Bastide (1971) "–could not stand without reconciling more or less with Christianity". Catholicism contributed with its saints and prayers for the formation of Umbanda (Silva 2000).

From the indigenous cults, Umbanda inherited interpersonal relations which remained as in the past and formed a harmonious group "like a large tent" in devotion to the spirits of 'caboclo'—spirits of the Indians ancestors who after death revolutionize Umbanda religion (Ortiz 1978, Di Paolo 1979).

European Spiritism brought the need for standardization, coding, and reinterpretation to Umbanda. The intellectuals of Spiritism, which established the form to Umbanda, were adding explanations of the spirit world, ideas about reincarnation, mediumship theories about the 'semantic extension of the master' instead of 'healer', and even the name of the spirit manifested in the medium (Camargo 1961, Ortiz 1978, Monteiro 1985).

Africans of diverse ethnic and tribal groups, who met each other in Brazil, away from their homelands and cultures, aggregated all these elements present in the rituals. To survive, these groups hold on to cultural background, relying on their beliefs and values, but being always permeable to changes from contact with other groups (Cancone 1987).

The coexistence and interdependence amongst the physical, spiritual and psychological dimensions of health is genuinely addressed with knowledge. Each African-Brazilian religious tradition uses its various features or combinations of treatment. The use of herbs, baths, diets, advice, sets of shells, as well as initiation rites observed as treatments in Umbanda may be associated with conventional therapies (Da Silva 2007, Alves and Seminotti 2009).

Western doctors and other health professionals who work especially in sub-Saharan Africa face different cultural realities that limit their medical intervention. To these western professionals, generally, anthropocentric view predominates over the elements of nature. African societies are aware of the physiological and natural causes of phenomena, but understand the factors of life and death, health and disease, as originated by man and his culture, this being their first and fundamental explanation. The imbalance of these factors may be negatively affected by: the services of a sorcerer, failure or moral transgression, punishment of the gods (or associated with these), (Table 6.1) and the ancestors. The prosperity and good health are seen as increasing the *aché*[1] (Munanga 2007).

Table 6.1. Relationship between gods and some health problems according to Ketu Candomblé's nation (Da Silva 2007)

Symptoms, disorders and diseases	Gods/Goddesses
Epidemic diseases (smallpox, AIDS) and skin diseases	Obaluaiê
Abortion, female infertility, menstrual problems, etc.	Iemanjá and Oxum
Impotence and male infertility	Xangô and Exu
Eyesight problems	Oxum
Asthma, shortness of breath and respiratory problems	Iansã
Emotional disorders	Oxossi and Ossaín
Disorders of liver, gallbladder and stomach ulcers	Oxossi and Logun-Edé
Obesity	Iemanjá, Oxum and Xangô

---

[1]According to umbandists aché is "life force", "good energy".

Umbanda is a religion characterized by a state of trance, manifested according to Cancone (1987) through possession. A trance can be defined as a state of 'altered consciousness' that allows behavior 'manifestations' coming from the deeper strata of the 'personality' (Bello 1960), but it does not explain its ritualistic presence.

> *"It is clear, therefore, that the possession trance is part of a ritual that can only be interpreted over its own and entire universe."*
>
> —(Cancone 1987)

In Umbanda, the elements of nature are decoded in a harmonious connection—the elements of water, earth, fire, and air are represented in the rituals. Water plays a magnet role for negative vibes; the earth is represented by various ritual uses, being 'taken as the space limit'; the fire is represented by heat, flames and light, symbolizing the warmth of the divine and cosmic energy; finally, air is one of the most important elements, being represented by the smoke of a *'defumador'*—an ambient smoker—(Fig. 6.1), pipes and cigars, and even the burning of gunpowder. These smoking elements are used to "produce smoke, that has the magical power to clear the psychic world, removing negative entities[2], attracting the positive ones, protecting against the *mau-olhado*[3] and opening the

Fig. 6.1. *Pai-de-santo* smoking with a *defumador* at Centro Caboclo Ubirajara e Exú Ventania, August 2008.

Color image of this figure appears in the color plate section at the end of the book.

[2]Entity, to Umbanda, is a spirit incorporated by mediums.
[3]Mau-olhado, literally "bad eye", is a folk belief that someone's envy can cause misfortunes or bad luck.

ways" (Lima 1979). Each element has its function and symbolism within the rituals of Umbanda, especially in healing rituals, where there are herbs, poultices, teas, baths, and *defumadores*.

These *defumadores* are very diverse, from the industrialized ones, which can be purchased at specialized Umbanda stores, to the 'fresh' ones, prepared from dried herbs burned with charcoal. The herbs most commonly used in this context are: rosemary (*Rosmarinus officinalis* L. Lamiaceae); basil (*Ocimum basilicum* L. Lamiaceae); rue (*Ruta graveolens* L. Rutaceae); lavander (*Lavandula officinalis* Chaix Lamiaceae); *jurema* leaves (*Mimosa hostilis* (Mart.) Benth. Fabaceae); fennel (*Pimpinella anisum* L. Apiaceae); eucalyptus leaves (*Eucalyptus* sp. Myrtaceae); Indian clove (*Syzygium aromaticum* (L.) Merr. and L.M. Perry Myrtaceae) (Lima 1979).

It is important to emphasize the sacral role of plants in Umbanda rituals, since *in* them or *through* them devotees "will seek solutions to the problems that afflict them". It can still be said that the importance given to the plants in Umbanda is such that it would not be possible to admit the existence of this religion without them (Camargo 2000, 2005/2006).

All Umbanda rituals begin by smoking the ambient with a *defumador*. The smoking is always done by the 'boss' of the yard, the *Pai* or *Mãe-de-santo* (literally 'saint's mother' or 'saint's father'), who smokes the place from rear towards the front door—never crossing the diagonal. After the smoking of space, the smoking of those presents begins. This is done with respect to the religion's hierarchy, whereby the senior mediums are the first to be smoked, followed by the newer mediums, and lastly the 'assistance', as the participants are known. Therefore, without the presence of 'smoke' by the *defumador*, the ritual doesn't begin.

Other common objects used are *pitos* (cigarettes), pipes, and cheroots, all having the task of eliminating harmful environmental fluids that are understood to be negative energies. In Umbanda, cigars are used by entities known as *caboclo*, who smoke a lot during the religious sessions, while the pipe is primarily used by *preto-velhos*—the spirits of former slaves (Ortiz 1978, Lima 1979).

> *"Why smoke has so Aché? The great strength of the magic smoke relates with the fact that it represents releasing and immediate dissolution in the surrounding breath—the air element. This allows you to act as an intermediary between the world of form and formless world, engaging and invisible, and becomes a magical agent, able to transform into effective result the intention of the released smoke".*
>
> —(Lima 1979)

At the umbandistic literature *defumar* (pass the smoke on the environment and on the participants) is defined as 'a ritual of high magic', where its main function is to ward off evil spirits, which are represented by the smoke itself. It is believed that the burning of herbs is designed as an act of great magic, being directly related to the forces of nature and the sacredness of plants. The smoke produced by burning plant "brings, by evocations, a high vibration sense, because through that smoke magical powers and high irradiation of spiritual flow are manifested". Still, according to Umbandists, those seeking peace and tranquility "should constantly be smoking their homes" (Pinto 1975).

As Umbanda is a syncretic religion, we can find the ritual use of *defumadores*, tobacco, and incense in the different religions that were connected to form it.

Silva (2000) shows that at the time of colonization in Brazil, Catholicism allowed the presence of the aromatic smoke of incense, cleansing the altar consecrated with relics of bones and of the clothing of saints "as if there was an open and privileged access to the supernatural world". For the Indigenous, the smoke is derived from burning tobacco and assumes an important ritual role, as noted above.

This legacy can also be seen in the rituals of *Catimbó*. In this northeastern Brazilian religion, that mixes African and Indigenous elements, tobacco "is a sacred plant and its smoke is the cure to diseases". In *Catimbó* there is a unique way of smoking the pipe, it is used in an inverted position, the illuminated part is placed in the mouth and the smoke comes out the tube of the pipe. This way of smoking is found today in Umbanda, which is not surprising seeing that the contact with *Catimbó* brought forth a movement which pioneered the way for Umbanda. This syncretism inherited the cure by smoking the environment (Bastide 2001).

But the power of smoke goes beyond all that. In Umbanda it is believed that smoke magic is 'an immediate dissolution of the surrounding breath'. It acts as a medium between 'the form world and formless world', a connection between natural and supernatural worlds. This is an extremely important idea in religions with the presence of trance, such as Umbanda and other African-Brazilian religions.

In general, the smoke in Umbanda healing rituals is used for communication with the supernatural. It seeks a cure for the ills of the soul and body, thus seeking to dispel the 'negative energies' (the supernatural and negative force that are acting on the individual). For this 'magic cure' treatment to occurs, one must be familiar with the concepts and interpretations of disease within this religious system.

"The disease becomes a significant factor only when associated with the idea of a general negativity, with the concept of a disorder that goes beyond the individual body to encompass social relations and the organization of the supernatural world. With this broad negativity that magical thinking seeks to understand and neutralize..."

—(Monteiro 1985)

And it is acting against the 'negativity' that, according to the Umbandists, smoke acts in the healing process. Below is a music sung in the rituals of Umbanda during the time of smoking by the *defumador*:

> *"Smokes with jurema herbs*
> *Smokes with rue and guiné...*
> *Rosemary, Beizoim and lavender;*
> *Lets smoke faithful sons!*
> *Smokes..."*

Another quite common element is gunpowder, burning when necessary, forms a dense curtain of gray smoke, with the same function of cleaning the environment and the body.

However, it is observed that with the constant revision of Umbanda, some elements and rituals are being simplified, such as smoking ritual by the *defumador*. With the presence of Spiritism, smoke is seen as a kind of primitivism, thus being eliminated little by little (Camargo 1961, Lima 1979, Silva 2000).

The smoking occupies an essential place within the Umbanda religious framework. As observed in field trials, smoking has the function of calming people who participate in the ritual.

Rodrigues et al. (2008) observed the use of a cigarette in a healing context known as *tira-capeta* (removing the devil) in an Umbanda practicing quilombola[4] community named Sesmaria Mata-Cavalos, in the state of Mato Grosso do Sul. Mr. Cezário, the former spiritual and political leader of this community, reported that the cigarette was smoked when people with headaches and malaise sought blessings. The healer blew the smoke up the patient's body instead of blessing them, helping the person to calm down. The cigarette can also be smoked by the patients to 'fortify the head', prevent flu, sinusitis, and to sleep. The cigarette is composed of nine plants: 'guiné'—*Petiveria alliacea* L. (Phytolaccaceae), 'eucalipto'—*Eucalyptus globulus* Labill. (Myrtaceae), 'alecrim-do-norte'—*Anemopaegma arvense* (Vell.) Stellfeld ex de Souza (Bignoniaceae), 'negramina'—*Siparuna guianensis* Aubl. (Monimiaceae), 'arruda'—*Ruta graveolens* L. (Rutaceae) and 'hortelã-da-várzea'—*Hyptis cana* Pohl ex Benth (Lamiaeae), rhizomes of 'caiá-piá'—*Dorstenia asaroides* Hook. (Moraceae), the flowers of 'cravo-da-Índia'—*Syzygium aromaticum* (L.) Merr. and L.M. Perry (Myrtaceae) and finally, the skin of one bulb of garlic—*Allium sativum* L. (Liliaceae).

Some plants used in Umbanda

The following plants listed below are utilized as smokes during Umbanda therapeutic rituals. Although there are presented some data concerning

---

[4]Quilombolas are descendant of Afro-Brazilian runaway slaves living in hideouts up-country.

its pharmacological activities, most of them were not observed by the inhalation route.

Rue

Scientific name: *Ruta graveolens* L.

Family: Rutaceae

Origin: Originally from Europe / Mediterranean.

Geographic distribution: Over Brazil.

Curiosities: Since the ancient times, magical values are assigned to Rue. These values can be found even today and it is not uncommon to see a branch of rue behind the ear of even the most skeptical and gullible of people. According to Pio Corrêa (1926) rue is closely linked to Greeks and Romans history and is considered 'an all around cure' from diseases to 'getting rid of bad business'. According to the same author, rue can also be found to be associated with spells and superstitions from the African slaves in Brazil, and is still today considered for the people of this origin to be effective against *mau-olhado*.

Active principle: Rutin.

Ethnopharmacology: Childbirth and infertility (Lans 2007), memory (Adsersen et al. 2006), mental disorders (Stafford et al. 2008), abortive and emmenagogue (Pollio et al. 2008), contraceptive (Harat et al. 2007), anaphrodisiac, cough, fever, hysteria and otalgia (Duke 2009).

Pharmacology: Anti-inflammatory (Conway and Slocumb 1979, Atta and Alkofahi 1998, Ciganda and Laborde 2003, Raghav et al. 2006); anti-fertility (Kong et al. 1989, Gandhi et al. 1991); antifungal (Oliva et al. 2003); cytotoxic (Ivanova et al. 2005); anti-spasmodic, diuretic, sedative, anti-rheumatic and analgesic (Khouri and El-Akawi 2005); antitumor (Preethi et al. 2006); palpitations and heart protection (Seak and Lin 2007) and anti-arrhythmia (Khori et al. 2008).

Guiné

Scientific name: *Petiveria alliacea* L.

Family: Phytolaccacea

Origin: Native (Amazon region, Africa and Tropical America).

Geographic distribution: Over Brazil.

Curiosities: Guiné is popularly known as 'taming sir', making a link with slave use, where it was a weapon against their masters. Slaves prepared 'magic drinks' that they administered to their masters in small doses over a long period of time to 'weaken the brain', so that their masters would enter starvation and die slowly (Camargo 2007). Other studies have also focused on the use of this plant in focusing the mind, as Rodrigues and Carlini (2004), who in a survey with *quilombolas* umbandists in the state of Mato Grosso, recorded this plant, which was included under the category 'mess with your head'. Reinforcing this idea, a quote from Pio Corrêa (1952) says "it is considered toxic and will lead to imbecility, aphasia, and even death to those who consume it".

Ethnopharmacology: 'Wound Healing' (Schmidt et al. 2009); 'Kidney problems' (Lans 2006); Used for snake bites, scorpion stings and success in hunting (Lans et al. 2001); Protozoan infections (Cáceres et al. 1998); abortive, antiseptic, aphrodisiac, snakebites, odont cavity, catarrh, depurative, diuretic, dysmenorrhea, emmenagogue, expectorant, fever, hysteria, insecticide, nerves, paralysis, birth, vermifuge, repellent (bats), repellent (insect), uterus, sedative, rheumatism (Duke 2009).

Pharmacology: Potential depressant and anticonvulsant (Gomes 2008); antinociceptive (De Lima et al. 1991, Gomes et al. 2005); anti-inflammatory effect and analgesic (Lopes-Martins et al. 2002); antimicrobial activity (Von Szczepanski et al. 1972); antimycotic (Malpezzi et al. 1994); antibacterial and antifungal (Benevides et al. 2001, Kim et al. 2005); anti-oxidant activity (Okada et al. 2008).

Alfazema

Scientific name: *Lavandula* cf. *dentata* L.

Family: Lamiaceae

Origin: Exotic/European origin.

Geographic distribution: Over Brazil.

Curiosities: It is an extremely aromatic plant, and according to Pio Corrêa (1926) is "stimulating to the nervous system and the brain".

Ethnopharmacology: Phlegm and sore (Duke 2009).

Pharmacology: No data found.

Colônia

Scientific name: *Alpinia* cf. *zerumbet* (Pers.) Burtt and Smith

Family: Zingiberaceae

Origin: Exotic.

Geographic distribution: Over Brazil.

Ethnopharmacology: Sedative (Berg 1984).

Pharmacology: 'Behavioral changes in rats' (Murakami 2009);

antihypertensive (Lahlou et al. 2003, De Moura et al. 2005); antinociceptive (De Araújo et al. 2005); Cardiovascular effect (Lahlou et al. 2003).

Melissa

Scientific name: *Melissa officinalis* L.

Family: Lamiaceae

Origin: Mediterranean region (Europe / Africa).

Geographic distribution: Almost over Brazil.

Curiosities: According to Pio Corrêa (1974) is a medicinal plant 'common and universal', used against nerve diseases, and is also anti-spasmodic and anti-neuralgic.

Ethnopharmacology: Antiseptic, soothing, carminative, cosmetic, digestive, emmenagogue, fever, perfume, sclerosis, sedative, stomachic, stimulant, diaphoretic, tonic and spasmolytic (Duke 2009); obesity control (Lee et al. 2008); Alzheimer (Perry et al. 1998, Perry et al. 1999, Dos Santos Neto et al. 2006, Akhondzadeh and Abassi 2006).

Pharmacology: Antinociceptive (Guginski 2009); antioxidant (De Sousa et al. 2004, Marongiu et al. 2004, Mimica-Dukic et al. 2004, Pereira et al 2008); decreases shaking (Ballard et al. 2002, Abuhamdah et al. 2008) anxiolytic (Kennedy and Scholey 2006, Kennedy et al. 2006); decreases the effects of stress (Kennedy et al. 2004); antitumoral (De Sousa et al. 2004) Alzheimer (Kennedy et al. 2003, Akhondzadeh et al. 2003); relaxing (Sadraei et al. 2003).

Manjericão

Scientific name: *Ocimum basilicum* L.

Family: Lamiaceae

Origin: Exotic/Africa and Asia.

Geographic distribution: Almost over Brazil.

Curiosities: Tradition says that the basil was found growing around Christ's tomb after the resurrection. Consequently, some Orthodox churches began to use it to prepare a kind of holy water, as well as sometimes planting it in pots beneath the altars of churches. In India, there is a belief that basil was imbued with a divine essence (Missouri Botanical Gardens 2009).

Ethnopharmacology: Cosmetics, fever, sore throat, emmenagogue, fungicide and coughs (Duke 2009); repellent (Seyoum et al. 2002, Erler et al. 2006).

Pharmacology: Bactericidal (Nguefack et al. 2004); mosquito repellent (Erler et al. 2006); antimicrobial activity (Viyoch et al. 2006); anti giardiasis

(De Almeida et al. 2007); anti-inflammatory (Singh 1999a, 1999b, Benedec et al. 2007); antioxidant (Trevisan et al. 2006; Gülçin et al. 2007); anthelmintic (Asha et al. 2001); anti-ulcer (Singh 1999a).

Conclusion

The use of plant smoke in healing ceremonies is not well described in scientific literature. In Brazil, the practice carried out mainly by indigenous groups in shamanism and practitioners of Umbanda points to different plants used in these contexts; exotic ones are predominate among Umbanda, since it is a religion created primarily by Africans, while in shamanism it is the Brazilian native ones that are predominate. It is interesting that in both of these groups who smoke the plant-made cigarette it is the curator who enters a state of trance and then acts beneficially on behalf the patient; all the while inhaling the smoke. The exception is the *tira-capeta*, used by a *quilombola* community, in which besides being used as mentioned above, patients also smoke the plant in search of therapeutic benefit. All these uses indicate activity in the central nervous system of the plants in question, emphasizing the importance of more research in this area.

Acknowledgements

We appreciate the help of AFIP (Associação de Incentivo à Psicofarmacologia), for providing finnancial support, and also we thank Stephanie Antonio for helping with the English translation.

References

Abuhamdah, S., L. Huang, M.S. Elliott, M.J. Howes, C. Ballard, C. Holmes, A. Burns, E.K. Perry, P.T. Francis, G. Lees, and P.L. Chazot. 2008. Pharmacological profile of an essential oil derived from *Melissa officinalis* with anti-agitation properties: focus on ligand-gated channels. Journal of Pharmacy and Pharmacology 60 (3): 377–384.

Ackerknecht, E.H. 1985. Medicina y antropologia social: estúdios varios. Ediciones Akal. Madrid.

Adsersen, A., B. Gauguin, L. Gudiksen, and A.K. Jager. 2006. Screening of plants used in Danish folk medicine to treat memory dysfunction for acetylcholinesterase inhibitory activity. Journal of Ethnopharmacology 104(3): 418–422.

Akhondzadeh, S. and S.H. Abbasi. 2006. Herbal medicine in the treatment of Alzheimer's disease. American Journal of Alzheimer's Disease and Other Dementias 21(2): 113–118.

Akhondzadeh, S., M. Noroozian, M. Mohammadi, S. Ohadinia, A.H. Jamshidi, and M. Khani. 2003. *Melissa officinalis* extract in the treatment of patients with mild to moderate Alzheimer's disease: a double blind, randomized, placebo controlled trial. Journal of Neurology and Neurosurgery Psychiatry 74(7): 863–866.

Alves, M.C. and N. Seminotti. 2009. Atenção à saúde em uma comunidade tradicional de terreiro. Revista Saúde Pública 43(1): 85–91.

Asha, M.K., D. Prashanth, B. Murali, R. Padmaja, and A. Amit. 2001. Anthelmintic activity of essential oil of *Ocimum sanctum* and eugenol. Fitoterapia 72(6): 669–670.

Atta, A.H. and A. Alkofahi. 1998. Antinociceptive and anti-inflammatory effects of some Jordanian medicinal plant extracts. Journal of Ethnopharmacology 60(2): 117–124.

Balick, M.J. and P.A. Cox. 1996. Plants, people and culture: the science of ethnobotany. Scientific American Library. New York.

Ballard, C.G., J.T. O'brien, K. Reichelt, and E.K. Perry. 2002. Aromatherapy as a safe and effective treatment for the management of agitation in severe dementia: the results of a double-blind, placebo-controlled trial with melissa. Journal of Clinical Psychiatry 63(7): 553–558.

Barcelos Neto, A. 2001. O universo visual dos xamãs wauja (Alto Xingu). Journal de la Société des Americanistes 87: 137–161.

Barcelos Neto, A. 2006. De divinações xamânicas e acusações de feitiçaria: imagens wauja da agência letal. Mana 12(2): 285–313.

Bastide, R. 1971. Religiões africanas no Brasil. Pioneira. São Paulo.

Bastide, R. 2001. Imagens do nordeste místico em branco e preto. O Cruzeiro: Rio de Janeiro, 1945. In: R. Prandi. (org.). Encantaria brasileira: o livre dos mestres e cablocos e encantados. Pallas. Rio de Janeiro.

Bello, J. 1960. Trance in Bali. Columbia University Press. New York.

Benedec, D., A.E. Parvu, I. Oniga, A. Toiu, and B. Tiperciuc. 2007. Effects of *Ocimum basilicum* L. extract on experimental acute inflammation. Revista Medico-Chirurgicală a Societății de Medici si Naturaliști Din Lași 111(4): 1065–1069.

Benevides, P.J., M.C. Young, A.M. Giesbrecht, N.F. Roque, and V.S. Bolzani. 2001. Antifungal polysulphides from *Petiveria alliacea* L. Phytochemistry 57(5): 743–747.

Berg, M. and Van Den. Elisabeth. 1984. Ver o peso: the ethnobotany of Amazonian market. In: G.T. Prance and. J.A. Kallunki (eds.). Society for Economic Botany (U.S.). New York Botanical Garden. Ethnobotany in the Neotropics. New York Botanical Garden. New York.

Cáceres, A., B. López, S. González, I. Berger, I. Tada, and J. Maki. 1998. Plants used in Guatemala for the treatment of protozoal infections. I. screening of activity to bacteria, fungi and *American trypanosomes* of 13 native plants. Journal of Ethnopharmacology 62(3): 195–202.

Camargo, C.P.F. 1961. Kardecismo e Umbanda: uma interpretação sociológica. Livraria Pioneira Editora. São Paulo.

Camargo, M.T.L.A. 2000. A etnobotânica e as plantas rituais Afro-Brasileiros In: C. Martins R. Lody (org.). Faraimará, o caçador traz alegria: mãe Stella, 60 anos de iniciação. Pallas. Rio de Janeiro.

Camargo, M.T.L.A. 2005/2006. Os poderes das plantas sagradas numa abordagem etnofarmacobotânica. Revista do Museu de Arqueologia e Etnologia da Universidade de São Paulo—MAE 15/16: 1–22.

Camargo, M.T.L.A. 2007. Contribuição etnofarmacobotânica ao estudo de *Petiveria alliacea* L.—Phytolacaceae—("amansa-senhor") e a atividade hipoglicemiante relacionada a transtornos mentais. Dominguezia 23(1): 21–28.

Cancone, M.H.V.B. 1987. Umbanda: uma religião brasileira. FFLCH/CER. São Paulo.

Carvalho, V.C. 1983 Umbanda: os seres superiores e os orixás/santo. Edições Loyola. São Paulo.

Ciganda, C. and A.J. Laborde. 2003. Herbal infusions used for induced abortion. Journal of Toxicology—Clinical Toxicology 41(3): 235–239.

Clastres, H. 1978. Terra sem mal—o profetismo tupi-guarani. Brasiliense. São Paulo.

Colpron, A.M. 2005. Monopólio masculino do xamanismo amazônico: o contra exemplo das mulheres xamãs Shipibo-Conibo. Mana 11(1): 95–128.

Conway, G.A., and J.C. Slocumb. 1979. Plants used as abortifacients and emmenagogues by Spanish new Mexicans. Journal of Ethnopharmacology 1(3): 241–261.

Cooper, J.M. 1987. Estimulantes e narcóticos. In: D. Ribeiro (org.). Handbook of South American Indians. Finep. Petrópolis.

Da Silva, J.M. 2007. Religiões e saúde: a experiência da rede nacional de religiões Afro-Brasileiras e saúde. Revista Saúde e Sociedade 16(2): 171–177.

De Almeida, I., D.S. Alviano, D.P. Vieira, P.B. Alves, A.F. Blank, A.H. Lopes, C.S. Alviano, and M.S. Rosa. 2007. Antigiardial activity of *Ocimum basilicum* essential oil. Parasitology Research 101(2): 443–452.

De Araújo, P.F., A.N. Coelho-De-Souza, S.M. Morais, S.C. Ferreira, and J.H. Leal-Cardoso. 2005. Antinociceptive effects of the essential oil of *Alpinia zerumbet* on mice. Phytomedicine 12(6-7): 482–486.

De Lima, T.C., G.S. Morato, and R.N. Takahashi. 1991. Evaluation of antinociceptive effect of *Petiveria alliacea* (guiné) in animals. Memória Instintuto Oswaldo Cruz 86(Suppl. 2): 153–158.

De Moura, R.S., A.F. Emiliano, L.C. De Carvalho, M.A. Souza, D.C. Guedes, T. Tano, and A.C.C. Resende. 2005. Antihypertensive and endothelium-dependent vasodilator effects of *Alpinia zerumbet*, a medicinal plant. Journal of Cardiovascular Pharmacology 46(3): 288–294.

De Sousa, A.C., D.S. Alviano, A.F. Blank, P.B. Alves, C.S. Alviano, and C.R. Gattass. 2004. *Melissa officinalis* L. essential oil: antitumoral and antioxidant activities. Journal of Pharmacy and Pharmacology 56(5): 677–681.

Di Paolo, P. 1979. Umbanda e integração social. Editora Boitempo Ltda. Pará.

Di Stasi, L.C. 1996. Plantas medicinais: arte e ciência: um guia de estudo interdisciplinar. Editora da Universidade Estadual Paulista. São Paulo.

Dos Santos Neto, L.L., T.M.A. De Vilhena, S.P. Medeiros, and G.A. De Souza. 2006. The use of herbal medicine in Alzheimer's disease—a systematic review. Alternative Medicine 3(4): 441–445.

Duke's phytochemical and ethobotanical database. Accessed in September 2009. Available at http://www.ars-grin.gov/duke/

Eliade, M. 1991. Imagens e símbolos: ensaio sobre o simbolismo mágico-religioso. Martins Fontes. São Paulo.

Eliade, M. 1998. O xamanismo e as técnicas arcaicas do êxtase. Ed. Martins Fontes. São Paulo.

Eliade, M. 2008. O sagrado e o profano: a essência das religiões. Martins Fontes. 2nd ed. São Paulo.

Erler, F., I. Ulug, and B. Yalcinkaya. 2006. Repellent activity of five essential oils against *Culex pipiens*. Fitoterapia 77(7-8): 491–494.

Ferri, M.G. 1980. História Da Botânica No Brasil. In: M.G. Ferri, and .S. Motoyama (eds.). História das ciências no brasil. Epu—Ed. Universidade de São Paulo. São Paulo.

Gandhi, M. and R. Lal, A. Sankaranarayanan, and P.L. Sharma. 1991. Post-coital antifertility action of *Ruta graveolens* in female rats and hamsters. Journal of Ethnopharmacology 34(1): 49–59.

Gilman, A.G., T.W. Rall, A.S. Nies and, P. Taylor. 2003. Goodman and Gilman: as bases farmacológicas da terapêutica. 10th ed. Editora Guanabara Koogan. Rio de Janeiro.

Gomes, P.B. 2008. Central effects of isolated fractions from the root of *Petiveria alliacea* L. (tipi) in mice. Journal of Ethnopharmacology 120(2): 209–214.

Gomes, P.B., M.M. Oliveira, C.R. Nogueira, E.C. Noronha, L.M. Carneiro, J.N. Bezerra, M.A. Neto, S.M.M. Vasconcelos, M.M.F. Fonteles, G.S.B. Viana, and F.C.F. Sousa. 2005. Study of antinociceptive effect of isolated fractions from *Petiveria alliacea* L. Biological and Pharmaceutical Bulletin 28(1): 42–46.

Guginski, G., A.P. Luiz, M.D. Silva, M. Massaro, D.F. Martins, J. Chaves, R.W. Mattos, D. Silveira, V.M. Ferreira, J.B. Calixto, and A.R. Santos. 2009. Mechanisms involved in the antinociception caused by ethanolic extract obtained from the leaves of *Melissa officinalis* (lemon balm) in mice. Pharmacology, Biochemistry, and Behavior 93(1): 10–16.

Gülçin, I., M. Elmastaş, and H.Y. Aboul-Enein. 2007. Determination of antioxidant and radical scavenging activity of basil (*Ocimum basilicum* L. family Lamiaceae) assayed by different methodologies. Phytotherapy Research 21(4): 354–361.

Guyton, A.C. and J.E. Hall. 2006. Fundamentos de Guyton: tratado de fisiologia médica. 10th ed. Guanabara Koogan. Rio de Janeiro.

Haas, L.F. 1999. Papyrus of Ebers and Smith. Journal of Neurology Neurosurgery and Psychiatry 67(5): 578.

Harat, Z.N., M.R. Sadeghi, H.R. Sadeghipour, M. Kamalinejad, and M.R. Eshraghian. 2007. Immobilization effect of *Ruta graveolens* L. on human sperm: a new hope for male contraception. Journal of Ethnopharmacology 115(1): 36–41.

Heinrich, M., J. Barnes, S. Gibbons, and E.M. Williamson. 2004. Fundamentals of pharmacognosy and phytotherapy. Churchill Livingstone. London.

IBGE—Instituto Brasileiro de Geografia e Estatística. 2000. Brasil: 500 anos de povoamento. IBGE. Rio de Janeiro.

IBGE—Instituto Brasileiro de Geografia e Estatística. Censo demográfico de 2000. Accessed in September 2009. Available at http://www.ibge.gov.br

Instituto Sociambiental - ISA. Accessed in August 2009. Available at http://www.socioambiental.org.

Ivanova, A., B. Mikhova, H. Najdenski, I. Tsvetkova, and I. Kostova. 2005. Antimicrobial and cytotoxic activity of *Ruta graveolens*. Fitoterapia 76(3-4): 344–347.

Kennedy, D.O. and A.B. Scholey. 2006. The psychopharmacology of European herbs with cognition-enhancing properties. Current Pharmaceutical Design 12(35): 4613–4623

Kennedy, D.O., G. Wake, S. Savelev, N.T. Tildesley, E.K. Perry, K.A. Wesnes, and A.B. Scholey. 2003. Modulation of mood and cognitive performance following acute administration of single doses of *Melissa officinalis* (lemon balm) with human CNS nicotinic and muscarinic receptor-binding properties. Neuropsychopharmacology 28(10): 1871–1881.

Kennedy, D.O., W. Little, and A.B. Scholey. 2004. Attenuation of laboratory-induced stress in humans after acute administration of *Melissa officinalis* (lemon balm). Psychosomatic Medicine 66(4): 607–613.

Kennedy, D.O., W. Little, C.F. Haskell, and A.B. Scholey. 2006. Anxiolytic effects of a combination of *Melissa officinalis* and *Valeriana officinalis* during laboratory induced stress. Phytotherapy Research 20(2): 96–102.

Khouri, N.A. and Z. El-Akawi. 2005. Antiandrogenic activity of *Ruta graveolens* L in male albino rats with emphasis on sexual and aggressive behavior. Neuro Endocrinology Letters 26(6): 823–829.

Khouri, V., M. Nayebpour, S. Semnani, M.J. Golalipour, and A. Marjani. 2008. Prolongation of AV nodal refractoriness by *Ruta graveolens* in isolated rat hearts. Potential role as an anti-arrhythmic agent. Saudi Medical Journal 29(3): 357–363.

Kim, S., R. Kubec, and R.A. Musah. 2005. Antibacterial and antifungal activity of sulfur-containing compounds from *Petiveria alliacea* L. Journal of Ethnopharmacology 104(1-2): 188–192.

Kong, Y.C., C.P. Lau, K.H. Wat, K.H. Ng, P.P. But, K.F. Cheng, and P.G. Waterman. 1989. Antifertility principle of *Ruta graveolens*. Planta Medica 55(2): 176–178.

Krippner, S. 2000. The epistemology and technologies of shamanic states of consciousness. Journal of Consciousness Studies 7: 93–118.

Krippner, S. 2007. Os primeiros curadores da humanidade: abordagens psicológicas e psiquiátricas sobre os xamãs e o xamanismo. Revista de Psiquiatria Clínica 34(1): 17–24.

Lahlou, S., L.F. Interaminense, J.H. Leal-Cardoso, and G.P. Duarte. 2003. Antihypertensive effects of the essential oil of *Alpinia zerumbet* and its main constituent, terpine-4-ol, in doca-salt hypertensive conscious rats. Fundamental and Clinical Pharmacology 17: 323–330.

Langdon, J.M. 1996. Xamanismo no Brasil: novas perspectivas. Florianópolis: Ed. Ufsc.

Lans, C. 2006. Ethnomedicines used in Trinidad and Tobago for urinary problems and diabetes mellitus. Journal Ethnobiology Ethnomedicine 2(45) S/P.

Lans, C. 2007. Ethnomedicines used in Trinidad and Tobago for reproductive problems. Journal of Ethnobiology and Ethnomedicine 3(13) S/P.

Lans, C., T. Harper, K. Georges, and E. Bridgewater. 2001. Medicinal and ethnoveterinary remedies of hunters in Trinidad. BMC Complementary Alternative Medicine 1(10) S/P.

Laraia, R.B. 2005. As religiões indígenas: o caso tupi-guarani. Revista Usp 67: 6–13.

Lee, J., K. Chae, J. Ha, B.Y. Park, H.S. Lee, S. Jeong, M.Y. Kim, and M. Yoon. 2008. Regulation of obesity and lipid disorders by herbal extracts from *Morus alba*, *Melissa officinalis* and *Artemisia capillaris* in high-fat diet-induced obese mice. Journal of Ethnopharmacology 116(3): 576.

Lent, R. 2004. Cem bilhões de neurônios conceitos fundamentais de neurociências. Editora Atheneu. Belo Horizonte.

Leroi-Gourhan, A. 1975. The flowers found with Shanidar IV, a Neanderthal burial in Iraq. Science 190: 562–564.

Léry, J. 1951. Viagem à terra do brasil. 2nd ed. Livraria Martins. São Paulo.

Lewington, A. 2003. Plants for people. Eden Project Books. London.

Lima, D.B.F. 1979. Malungo, decodificação do Umbanda: contribuição à história das religiões. Civilização Brasileira. Rio de Janeiro.

Lopes-Martins, R.A., D.H. Pegoraro, R. Woisky, S.C. Penna, and J.A. Sertié. 2002. The anti-inflammatory and analgesic effects of a crude extract of *Petiveria alliacea* L. (Phytolaccaceae). Phytomedicine 9(3): 245–248.

Malnic, B., J. Hirono, T. Sato, and L.B. Buck. 1999. Combinatorial receptor codes for odors. Cell 96: 713–723.

Malpezzi, E.L., S.C. Davino, L.V. Costa, J.C. Freitas, A.M. Giesbrecht, and N.F. Roque. 1994. Antimitotic action of extracts of *Petiveria alliacea* on sea urchin egg development. Brazilian Journal of Medical and Biological Research 27(3): 749–754.

Marongiu, B., S. Porcedda, A. Piras, A. Rosa, M. Deiana, and M.A. Dessi. 2004. Antioxidant activity of supercritical extract of *Melissa officinalis* subsp. *officinalis* and *Melissa officinalis* subsp. *inodora*. Phytotherapy Research 18(10): 789–792.

Menini, A., L. Lagostena, and A. Boccaccio. 2004. Olfaction: drom odorant molecules to the olfactory cortex. News Physiology Science 19: 101–104.

Mercadante, M.S. 2006. Images of healing: spontaneous mental imagery and healing process of the Barquinha, a Brazilian ayahuasca religious system. PhD Thesis, Saybrook Graduate School and Research Center. San Francisco.

Mimica-Dukic, N., B. Bozin, M. Sokovic, and N. Simin. 2004. Antimicrobial and antioxidant activities of *Melissa officinalis* L. (Lamiaceae) essential oil. Journal of Agricultural and Food Chemistry 52(9): 2485–2489.

Missouri Botanical Gardens. Accessed in September 2009. Available at Http://Mobot.Mobot. Org/Pick/Search/Pick.Html.

Mohagheghzadeh, A., P. Faridi, M. Shams Ardakani, and Y. Ghasemi. 2006. Medicinal smokes. Journal of Ethnopharmacology 108: 161–184.

Mombaerts, P., F. Wang, C. Dulac, S.K. Chao, A. Nemes, M. Mendelsohn, J. Edmondson, and R. Axel. 1996. Visualizing an olfactory sensory map. Cell 87: 675–686.

Money, M. 1997. Shamanism and complementary therapy. Complementary Therapies in Nursing and Midwifery 3: 131–135.

Monteiro, P. 1985. Da doença a desordem: a magia na Umbanda. Edições Graal. Rio de Janeiro.

Munanga, K. 2007. Saúde e diversidade. Revista Saúde e Sociedade 16(2): 13–18.

Murakami, S., M. Matsuura, T. Satou, S. Hayashi, and K. Koike. 2009. Effects of the essential oil from leaves of *Alpinia zerumbet* on behavioral alterations in mice. Natural Product Communications 4(1): 129–132.

Nef, P. 1998. How we smell: the molecular and cellular bases of olfaction. News Physiology Science 1(13): 1–5.

Nguefack, J., B.B. Budde, and M. Jakobsen. 2004. Five essential oils from aromatic plants of Cameroon: their antibacterial activity and ability to permeabilize the cytoplasmic membrane of *Listeria innocua* examined by flow cytometry. Letters in Applied Microbiology 39(5): 395–400.

Okada, Y., K. Tanaka, E. Sato, and H. Okajima. 2008. Antioxidant activity of the new thiosulfinate derivative, s-benzyl phenylmethanethiosulfinate, from *Petiveria alliacea* l.org. Biomolecular Chemistry 21(6): 1097–1102.

Oliva, A., K.M. Meepagala, D.E. Wedge, D. Harries, A.L. Hale, G. Aliotta, and S.O. Duke. 2003. Natural fungicides from *Ruta graveolens* l. leaves, including a new quinolone alkaloid. Journal of Agricultural and Food Chemistry 51(4): 890–896.

Ortiz, R. 1978. A morte branca do feiticeiro negro: Umbanda e sociedade brasileira. Editora Vozes, Petrópolis.

Pereira, R.P., R. Fachinetto, P.A. De Souza, R.L. Puntel, G.N.S. Da Silva, B.M. Heinzmann, T.K. Boschetti, M.L. Athayde, M.E. Bürger, A.F. Morel, V.M. Morsch, and J.B. Rocha. 2008. Antioxidant effects of different extracts from *Melissa officinalis*, *Matricaria recutita* and *Cymbopogon citratus*. Neurochemical Research S/P.

Pérez-Gil, L. 2001. O sistema médico yawanáwa e seus especialistas: cura, poder e iniciação xamânica. Caderno de Saúde Pública 17(2): 333–344.

Perry, E.K., A.T. Pickering, W.W. Wang, P. Houghton, and N.S. Perry. 1998. Medicinal plants and Alzheimer's disease: integrating ethnobotanical and contemporary scientific evidence. Journal of Alternative and Complementary Medicine 4(4): 419–428.

Perry, E.K., A.T. Pickering, W.W. Wang, P.J. Houghton, and N.S. Perry. 1999. Medicinal plants and Alzheimer's disease: from ethnobotany to phytotherapy. Journal of Pharmacy and Pharmacology 51(5): 527–534.

Pinto, A. 1975. Dicionário da Umbanda. Editora Eco. Rio de Janeiro.

Pio Corrêa, M. 1926; 1952; 1974. Dicionário das plantas úteis do brasil e das exóticas cultivadas. Imprensa Nacional. Rio de Janeiro.

Pollio, A., A. De Natale, E. Appetiti, G. Aliotta, and A. Touwaide. 2008. Continuity and change in the Mediterranean medical tradition: *Ruta* spp. (Rutaceae) in hippocratic medicine and present practices. Journal of Ethnopharmacology 116(3): 469–482.

Porto, A. 2006. O sistema de saúde do escravo no Brasil do século XIX: doenças, instituições e práticas terapêuticas. História, Ciência, Saúde—Manguinhos 13(4): 1019–1027.

Prandi, R. 2004. O Brasil com axé: candomblé e umbanda no mercado religioso. Estudos Avançados 18(52): 223–238.

Preethi, K.C., G. Kuttan, and R. Kuttan. 2006. Anti-tumor activity of *Ruta graveolens* extract. Asian Pacific Journal of Cancer Prevention. 7(3): 439–443.

Raghav, S.K., B. Gupta, C. Agrawal, K. Goswami, and H.R. Das. 2006. Anti-inflammatory effect of *Ruta graveolens* L. in murine macrophage cells. Journal of Ethnopharmacology 104(1-2): 234–239.

Raven, P.H., R.F. Evert, and S.E. Eichhorn. 2007. Biologia Vegetal. 6th ed. Guanabara Koogan. Rio de Janeiro.

Ritz, E., and S.R. Orth. 2000. The cultural history of smoking. Contributions to Nephrology 130: 1–10.

Rodrigues, E. 2001. Usos rituais de plantas que indicam ações sobre sistema nervoso central pelos índios krahô, com ênfase nas psicoativas. PhD Thesis. Universidade Federal de São Paulo. São Paulo.

Rodrigues, E. and E.A. Carlini. 2004. Plants used by a quilombola group in Brazil with potential central nervous system effects. Phytotherapy (9): 748–753.

Rodrigues, E. and E.A. Carlini. 2005. Ritual use of plants with possible action on the central nervous system by the Krahô Indians, Brazil. Phytotherapy Research 19(2): 129–135

Rodrigues, E., F.R. Mendes, and G. Negri. 2006. Plants indicated by Brazilian Indians to central nervous system disturbances: a bibliographical approach. Current Medicinal Chemistry—Central Nervous System Agents 6: 211–244.

Rodrigues, E., B. Gianfratti, R. Tabach, G. Negri, and F.R. Mendes. 2008. Preliminary investigation of the central nervous system effects of 'tira-capeta' (removing the devil), a cigarette used by some quilombolas living in pantanal wetlands of Brazil. Phytotherapy Research 22: 1248–1255.

Sadraei, H., A. Ghannadi, and K. Malekshahi. 2003. Relaxant effect of essential oil of *Melissa officinalis* and citral on rat ileum contractions. Fitoterapia 74(5): 445–452.

Salah, S.M. and A.K. Jager. 2005. Screening of traditionally used Lebanese herbs for neurological activies. Journal of Ethnopharmacology 97(1): 145–149.

Schmidt, C., M. Fronza, M. Goettert, F. Geller, S. Luik, E.M. Flores, C.F. Bittencourt, G.D. Zanetti, B.M. Heinzmann, S. Laufer, and I. Merfort. 2009. Biological studies on Brazilian plants used in wound healing. Journal of Ethnopharmacology 122(3): 523–532.

Schultes, R.E. 1984. Fifteen years of study of psychoactive snuffs of South America: 1967–1982—a review. Journal of Ethnopharmacology 11: 17–32.

Schulz, V., R. Hänsel, and V.E. Tyler. 2002. Rational phytotherapy—a physicians' guide to herbal medicine. Springer-Verlag. New York.

Seak, C.J., and C.C. Lin. 2007. *Ruta graveolens* intoxication. Clinical Toxicology 45(2): 173–175.

Seyoum, A., K. Palsson, S. Kung'a, E.W. Kabiru, W. Lwande, G.F. Killeen, A. Hassanali, and B.G. Knols. 2002. Traditional use of mosquito-repellent plants in western Kenya and their evaluation in semi-field experimental huts against *Anopheles gambiae*: ethnobotanical studies and application by thermal expulsion and direct burning. Transactions of the Royal Society of Tropical Medicine and Hygiene 96(3): 225–231.

Silva, S.B. 2002. Dualismo e cosmologia kaingang: o xamã e o domínio da floresta. Horizontes Antropológicos 18: 189–209.

Silva, V.G. 2000. Candomblé e umbanda: caminhos da devoção brasileira. Ática. São Paulo.

Simpson, B.B. and M.C. Ogorzaly. 2001. Economic botany: plants in our world. 3rd ed. McGraw-Hill. New York.

Singh, S. 1999a. Evaluation of gastric anti-ulcer activity of fixed oil of *Ocimum Basilicum* Linn. and its possible mechanism of action. Indian Journal of Experimental Biology 37(3): 253–257.

Singh, S. 1999b. Mechanism of action of antiinflammatory effect of fixed oil of *Ocimum basilicum* Linn. Indian Journal of Experimental Biology 37(3): 248–252.

Solecki, R.S. 1975. Shanidar IV, a Neanderthal flower burial in Northern Iraq. Science 190: 880–881.

Stafford, G.I., M.E. Pedersen, J. Van Staden, and A.K. Jager. 2008. Review on plants with CNS-effects used in traditional South African medicine against mental diseases. Journal of Ethnopharmacology 119(3): 513–537.

Storrie, R. 2006. A política do xamanismo e os limites do medo. Revista de Antropologia 49(1): 357–391.

Thevet, A. 1557. Les singvlaritez da la france antartique, avtrement nommée amerique: de plusieurs terres isles decouuertes de nostre temps. Chez Les Heritiers De Maurice De La Porte. Paris.

Trevisan, M.T., M.G.S. Vasconcelos, B. Pfundstein, B. Spiegelhalder, and R.W. Owen. 2006. Characterization of the volatile pattern and antioxidant capacity of essential oils from different species of the genus *Ocimum*. Journal of Agricultural Food Chemistry 54(12): 4378–4382.

Van De Graaff, K.M. 2003. Anatomia humana. 6th ed. Manole. Barueri.

Vidille, W. 2006. Xamãs e os espíritos ancestrais. Psychê 19: 47–64.

Viyoch, J., N. Pisutthanan, A. Faikreua, K. Nupangta, K. Wangtorpol, and J. Ngokkuen. 2006. Evaluation of in vitro antimicrobial activity of Thai basil oils and their micro-emulsion formulas against propionibacterium acnes. International Journal of Cosmetic Science 28(2): 125–133.

Von Szczepanski, C., P. Zgorzelak, and G.A. Hoyer. 1972. Isolation, structural analysis and synthesis of an antimicrobial substance from *Petiveria alliacea* L. Arzneimittelforschung 22(11): 1975–1976.

Washignton, N., C. Washington, and C.G. Wilson. 2001. Physiological pharmaceutics: barriers to drug absorption. 2nd ed, Taylor and Francis. London.

Wilbert, J. 1987. Tobacco and shamanism in South America. Yale University Press. London.

Zimmer, H. 2005. Filosofias da Índia. 3rd ed. Palas Athena. São Paulo.

Chapter 7

Traditional Medicines as the Source of Immuno-modulators and Stimulators and their Safety Issues

Mahmud Tareq Hassan Khan

Introduction

The practice of medicinal herbs for treating diseases is well known from ancient times (Wasik 1999). More than 80% of the world's population is dependent on indigenous or traditional or complementary and alternative medicines (CAMs) (Zhang 2002, Geethangili and Tzeng 2009) and a large number of CAM therapies that have been shown to affect the immune system (Mainardi et al. 2009).

Natural products have long been regarded as excellent sources for drug discovery, given their structural diversity and a wide variety of biological activities (Fu et al. 2008). A number of synthetic medicines have been derived from medicinal herbs. Many modern magic molecules, like digoxin, salicin, reserpine, ephedrine, quinine, vincristine, vinblastine, paclitaxel, artemisinin, hypericin and silymarin, etc., have been used to develop more potent drug molecules. These are regarded as active constituents of the herbs and are present in standardized form in the herbal extracts, if not isolated as a single entity (Singh 2006).

Present address: GenØk—Center for Biosafety, FellesLab, IMB, MHB, University of Tromsø, 9037 Tromsø, Norway.
E-mail: *mahmud.khan@uit.no, mthkhan2002@yahoo.com*

Plant products have traditionally provided 30–40% of modern anti-microbial and anti-cancer drugs to the pharmaceutical industry, but we still have barely touched the tip of the iceberg in our efforts to design more highly active drug molecules from plants resources (Khan et al. 2005, Chattopadhyay and Khan 2008, Chattopadhyay et al. 2009). In recent years scientific reports revealed several potential biochemical mechanisms involving the immunomodulatory pathway of many supplemental vitamins (A, D, and E) that appear to affect the differentiation of CD4⁺ cell Th1 and Th2 subsets (Mainardi et al. 2009). Some reports suggested that plant derived molecules, like resveratrol, quercetin, and magnolol, etc., may affect transcription factors such as nuclear factor-κB (NFκB) and the signal transducer and activator of transcription (STAT)/Janus kinase pathways with resultant changes in cytokines and inflammatory mediators (Mainardi et al. 2009). There have been hundreds of clinical trials looking at the effect of CAM on asthma, allergic rhinitis, and atopic dermatitis involving immunochemical pathways (Mainardi et al. 2009).

The term 'Immuno-modulation' represents change of the indicators of cellular and humoral immunity and nonspecific defense issues. The essence of immunomodulation is that a pharmacological agent acting under various doses and time regimens displays an immunomodulating effect (Flemming 1985, Kowalozyk-Bronisz and Paegelow 1986). The immunomodulating action is irreversible and requires maintaining the dose of a preparation. The extreme manifestations of immunomodulating action of biologically active substances are immunosuppression (known as 'immunosuppressant', depression of the immune response) and immunostimulation (immunopotentiation or strengthening of the immune reactions) (Smirnov and Suskova 1989). 'Immunostimulants' augment the general immunity of the host, and present a non-specific immune response against the microbial pathogens. They also enhance humoral and cellular immune responses, either by increasing the cytokine secretion, or by the direct stimulation of B- or T-lymphocytes (Tan and Vanitha 2004).

In this chapter some important examples are discussed which have been utilized in the drug discovery process of potential candidate(s) for immunomodulators or -stimulators .

Traditional medicines for immunomodulators or –stimulators: examples

Traditional medical systems of different countries play a major role in that particular area or country or locality, like Ayurveda, Unani, traditional Chinese medicine (TCM), Sidha, Chakma (Khan et al. 2005a, Chattopadhyay and Khan 2008, Chattopadhyay et al. 2009), etc. Several reports suggested that traditional medicines or herbal drugs or natural products as well as isolated pure molecules from them could be a good source of newer medications for different disease conditions (Jabbar et al. 2001, Khan et al. 2001, Khan et al. 2002, Shaphiullah et al. 2003, Ahmad et al. 2004, Jabbar et al. 2004, Costa-Lotufo et al. 2005, Khan et al. 2005, Lambertini et al. 2005, Lampronti et al. 2005, Shaheen et al. 2005, Sultankhodzhaev et al. 2005, Afroz et al. 2006, Azhar Ul et al. 2006, Khan et al. 2006a, Khan et al. 2006b, Devkota et al. 2007, Previati et al. 2007, Ullah et al. 2007, Lampronti et al. 2008, Penolazzi et al. 2008, Khan et al. 2009). They are a very rich source of immunomodulators or –stimulators (Svoboda 1998, Prasher et al. 2008). Traditional herbal pharmacotherapy is famous for combining plant species that results in complex phytochemical mixtures in the attempts to ameliorate pathophysiological processes. Although research is necessary on isolated constituents and single herbal extracts to provide information about the molecular modes of activity, such studies have limited relevance to the practical use of herbs due to the traditional custom of dispensing herbal medicine in formulas (Walker 2006).

Ayurveda is an ancient system of personalized medicine documented and practiced in greater India since 1500 B.C. According to this system an individual's basic constitution to a large extent determines predisposition and prognosis to diseases as well as therapy and life-style regime. Ayurveda describes seven broad constitution types (Prakritis) each with a varying degree of predisposition to different diseases. Amongst these, the three most contrasting types, Vata, Pitta, Kapha, are the most vulnerable to diseases (Prasher et al. 2008). Many of the traditional medicines are prepared from combinations of medicinal plants which may influence numerous molecular pathways. These effects may differ from the sum of effects from the individual plants and therefore, research demonstrating the effects of the formula is crucial for insights into the effects of traditional remedies (Burns et al. 2009). The most common studies of cytokines were interleukin (IL)-4, IL-6, IL-10, tumor necrosis factor (TNF) and interferon (IFN)γ and the majority of the formulas researched derived from TCM. The following formulas had activity on at least three cytokines; Chizukit N, CKBM, Daeganghwal-tang, Food Allergy Formula, Gamcho-Sasim-Tang, Hachimi-jio-gan, Herbkines, Hochuekki, Immune System Formula, Jeo-Dang-Tang, Juzen-taiho-to, Kakkon-to, Kan jang, Mao-Bushi-Saishin-to, MSSM-002, Ninjin-youei-to, PG201, Protec, Qing-huo-bai-du-yin, Qingfu Guanjieshu, Sambucol Active Defense, Seng-fu-tang, Shin-Xiao-Xiang, Tien Hsien, Thuja formula, Unkei-to, Vigconic, Wheeze-relief-formula, Xia-Bai-San, Yangyuk-Sanhwa-Tang, Yi-fey Ruenn-hou, and Yuldahansotang (Burns et al. 2009). Numerous formulas demonstrated activity on both gene and protein expression. Therapeutic success using these formulas may be partially due to their effects on cytokines. Further study of phytotherapy on cytokine related diseases or syndromes is necessary (Burns et al. 2009).

In a very recent report mentioning the effect of a tea fortified with five herbs, such as *Withania somnifera, Glycyrrhiza glabra, Zingiber officinale, Ocimum sanctum* and *Elettaria cardamomum*, selected from Ayurveda for their putative immunoenhancing effect on innate immunity. *Ex vivo* natural killer (NK) cell activity was assessed after consumption of fortified tea compared with regular tea in two independent double-blind intervention studies. This study indicated that regular consumption of the tea fortified with Ayurvedic herbs enhanced NK cell activity, which is an important aspect of the (early) innate immune response to infections (Bhat et al. 2009).

In another study, 178 ethanolic plant extracts from the pharmacopoeia of Tacana, an ethnic group from Bolivia, have been screened for immunomodulatory activity using complement cascade inhibition and ADP-induced platelet aggregation inhibition assays (Deharo et al. 2004). Six impaired both complement pathways (classical and alternative): stem bark from the roots of *Astronium urundeuvea* (Anacardiaceae), *Cochlospermum vitifolium* (Cochlospermaceae), *Terminalia amazonica* (Combretaceae), *Triplaris americana* (Polygonaceae), *Uncaria tomentosa* (Rubiaceae) and *Euterpe precatoria* (Arecaceae). Inhibition of complement cascade was independent of essential ion complexation, and was not due to direct hemolytic activity on target red blood cells (Deharo et al. 2004). For *A. urundeuvea, C. vitifolium*, and *T. amazonica*, anti-inflammatory activity relied on cyclooxygenase inhibition. Four of these species (*A. urundeuva, T. americana, U. tomentosa* and *E. precatoria*) are used traditionally to treat inflammatory processes (Deharo et al. 2004).

Saireito, a traditional herbal medicine, is widely used for treating ulcerative colitis in Japan. Watanabe et al. recently, in 2009, analyzed the immunological characteristics of an oxazolone-induced colitis model and examined the effects of Saireito on this model (Watanabe et al. 2009). The transcription levels of Th2 cytokines were significantly upregulated in the spleen and middle colon of the induced colitis mice, whereas those of the Th1 cytokine IFN-γ decreased in the spleen and increased in the middle colon. Saireito significantly ameliorated induced colitis (Watanabe et al. 2009). In the middle colon of the saireito-treated mice, enhanced expression of Th2 cytokine mRNAs was markedly downregulated, while that of IFN-γ mRNA was further upregulated. In contrast, in the spleen, saireito had no effect on the transcription of either type of cytokine. After global transcriptome analysis, real-time (RT) polymerase chain reaction (PCR) analysis revealed that Saireito greatly downregulated the enhanced expression of the suppressor of cytokine signaling (SOCS)-3 mRNA in the middle colon of induced colitis mice (Watanabe et al. 2009). Finally in this report, the authors concluded that Saireito exhibits inhibitory effects on induced colitis by the induction of Th1-polarized immune responses in the mucosal immune system of the colon (Watanabe et al. 2009).

CGX is a modification of the herbal drug that originated from a traditional Korean formula composed of 13 herbs and other items used for 'liver cleaning' which is used to treat various chronic liver disorders in oriental clinics. In a study Wang et al. (2009) investigated the antifibrotic effects and associated mechanisms of CGX (Wang et al. 2009). Their results demonstrated that the antifibrotic effects of CGX and the corresponding mechanisms associated with sustaining the antioxidative system and inhibiting hepatic stellate cell activation via the downregulation of fibrogenic cytokines (Wang et al. 2009).

Molecules from the plant kingdom that impact the immune system

Numerous outstanding research has been undertaken by Chinese and Japanese scientists on their traditional remedies revealing great importance to immunology. However, the same attention is rarely given to traditional European or North American herbs. This is primarily because of research grants, not because of an inherent lack of value in the plants. Perhaps if nettles (*Urtica dioica*) were given the same quality of attention that *Astragalus* has garnered for itself, we might have the immunological 'proof' of its profound effects (Hoffmann 2009).

Some of the examples of plant originated compounds shown to be immunomodulators or stimulators are given below (Hoffmann 2009).

Small molecules

- **Alkaloids and other nitrogen containing compounds.** Large number of alkaloids and nitrogenous compounds shown to have immunological activities, for example, aristolochic acid, from Serpentary (*Aristolochia clematis*); cepharanthine from *Stephania cepharantha*; tylophorine from *Tylophora indicea* (Mohla et al. 1989, Garbe 1993, Danishefsky et al. 2000, Lai 2002, Dinkova-Kostova 2008, Pla et al. 2008, Hoffmann 2009).

- **Phenylethanoid glycosides.** Phenylethanoid glycosides are naturally occurring compounds of plant origin and are structurally characterized with a hydroxyphenylethyl moiety to which a glucopyranose is linked through a glycosidic bond. Until now several hundred such types of compounds have been reported from plant sources and further pharmacological studies *in vitro* or *in vivo* have shown that these compounds possess a broad array of biological activities including antibacterial, antitumor, antiviral, anti-inflammatory, neuroprotective, antioxidant, hepatoprotective, immunomodulatory, and tyrosinase inhibitory actions (Fu et al. 2008).

- **Terpenes**. Important sesquiterpene immunostimulants include helenalin, tenulin, eupahyssopin. These groups of compounds abound in much of our materia medica. Research on *Arnica montana* suggests that this sesquiterpenes component is pivotal to many of its actions. Many saponins, the more complex triterpenes, are considered to be 'immuno-adjuvants' and are widely used as such in Japan, either in the form of the plant source or as the extracted chemical (Heitzman et al. 2005, Kuipers and van den Elsen 2007, Ajikumar et al. 2008, Hoffmann 2009).

- **Phenols.** A number of aromatic (in the chemical sense) molecules have demonstrated immunological effects. Aromatic acids are ubiquitous amongst flowering plants, fruits and vegetables. Laboratory research has found that ferulic acid, named after Asafoetida (*Ferula foetida*), increases phagocytosis in mice; anethol, found in aniseed oil, increase the leucocyte count in blood; the widely found pseudotannin catechol stimulates granulocytes. The more complex lignans are proving to be important, with a range of effects including stimulation of phagocytic activity (Hirata et al. 1984, Umehara et al. 1996, Visnjic et al. 1997, Seo et al. 2004a, Seo et al. 2004b) in polymorphonuclear granulocytes, cytotoxicity and induction of IFN (Hussaini et al. 2000, Liu et al. 2002, Katiyar 2003, Hermann et al. 2005, Hoffmann 2009, Kim et al. 2009, Kokoska and Janovska 2009).

Large molecules

These appear to be the most important immunostimulants from natural resources, especially the lectins (also known as glycoproteins) and polysaccharides. It has been suggested that their impact is based upon some interaction with the membranes of 'immunocompetent' cells (Hoffmann 2009).

- **Lectins.** Sugar binding, carbohydrate specific proteins that agglutinate some cells, such as the erythrocytes of certain blood types, and precipitate some molecules. Originally found in plants they were called phytohemagglutinins, however they have also been found in bacteria, fungi, invertebrates, and most vertebrates. Some well known plant toxicity problems are due to lectins, e.g. ricin from Castor bean. A range of effects have been shown in the laboratory, including stimulation of mitosis in lymphocytes, inhibition of protein synthesis in bacteria, agglutination of malignant cells (tumor-specific lectins). Some of the lectins containing plants are *Phytolacca decandra*, *Viscum album* (Gabius et al. 1994, Beuth 1997, Hajto et al. 1999, Gabius 2001, Rabinovich et al. 2002, Gibbs 2005, Ilarregui et al. 2005, Elluru et al. 2006, Hoffmann 2009, Johnston et al. 2009).

- **Polysaccharides.** It has been suggested that several polysaccharides are being used as herbal immunomodulators (Barta 1986, Domer et al. 1988, Baker and Hraba 1994, Wong et al. 1994, Jacob 1995, Ovodov Iu 1998, Reynolds and Dweck 1999, Ooi and Liu 2000, Chang 2002, Caroff and Karibian 2003, Monro 2003, Ramesh and Tharanathan 2003, Tzianabos et al. 2003, Ellerbroek et al. 2004, Lin and Zhang 2004, Thomas and Harn 2004, Zjawiony 2004, Ng and Wang 2005, Mazmanian and Kasper 2006, Prakash and Martoni 2006, Schepetkin and Quinn 2006, Castro et al. 2007, Hokke et al. 2007, Moradali et al. 2007, Hamman 2008, Comstock 2009, Hoffmann 2009, Johnston et al. 2009).

Some examples

In this section some plants have been described for their role in herbal as well as modern medical systems.

Panax ginseng

Ginseng, e.g. *Panax ginseng*, is the root of the Araliaceous plant. *P. ginseng*, the most popular species, is a well-known medicinal herb native to China and Korea, and has been used as a herbal remedy in eastern Asia for thousands of years. However, there is different evidence of ginseng efficacy between TCM, modern pharmacological experiments and clinical trials. In TCM, ginseng is a highly valued herb and has been applied to a variety of pathological conditions and illnesses such as hypodynamia, anorexia, shortness of breath, palpitation, insomnia, impotence, hemorrhage and diabetes. Ginsenosides (structure shown in Fig. 7.1) are a special group of triterpenoid saponins that can be classified into two groups by the skeleton of their aglycones, namely dammarane- and oleanane-type. Ginsenosides are found nearly exclusively in *Panax* species (ginseng) and up to now more than 150 naturally occurring ginsenosides have been isolated from different parts of the plant, like roots, leaves, stems, fruits, flower heads (Christensen 2009). They have been the target of many research works as they are believed to be the main active principles behind the claims of ginseng's efficacy (Christensen 2009). Modern pharmacological experiments have proved that ginseng possesses several highly pharmacologically active molecules like, ginsenosides, polysaccharides, peptides, polyacetylenic alcohols, etc. and their pharmacological actions e.g. central nervous system effects, neuroprotective effect, immunomodulation, anticancer, etc. are also reported (Xiang et al. 2008, Christensen 2009).

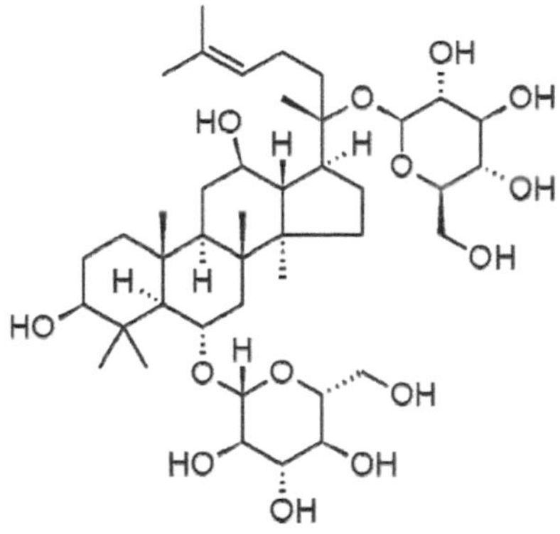

Fig.7.1. Molecular structure of ginsenoside.

Ginsenosides as the active ingredients, especially have antioxidant, anti-inflammatory, anti-apoptotic and immunostimulant properties. Recently, ginseng has been studied in a number of randomized controlled trials investigating its effect mainly on physical and psychomotor performance, cognitive functions, immunomodulation (Yang and Ma 1988, Kenarova et al. 1990, Joo et al. 2005, Lee et al. 2005, Yu et al. 2005a, Yu et al. 2005b, Rhule et al. 2006, Yang et al. 2008, Christensen 2009, Lee et al. 2009), diabetes mellitus, cardiovascular risk factors, quality of life, as well as adverse effects. Equivocal results have been demonstrated for many of these indications. Because of the poor quality of most clinical trials on ginseng, reliable clinical data in humans are still lacking. Therefore, a broader understanding of medical knowledge and reasoning on ginseng is necessary (Xiang et al. 2008).

Curcuma longa

Turmeric (*Curcuma longa*) is known for its multiple health restoring properties, and has been used in treating several diseases including several respiratory disorders. This is a common spice used in culinary preparations in South and East Asian countries. The active component of turmeric is curcumin (the structure shown in Fig. 7.2), a polyphenolic phytochemical, with anti-inflammatory, anti-amyloid, antiseptic, antitumor, and anti-oxidative properties. Curcumin was reported to have anti-allergic properties with inhibitory effect on histamine release from mast cells. The results have been verified by further investigation using the murine model of allergy. Results indicated marked inhibition of allergic response in animals treated with curcumin suggesting a major role for the molecule in reducing the allergic response. Findings needed further evaluation, extrapolation, and confirmation before using curcumin for controlling allergy and asthma in humans.

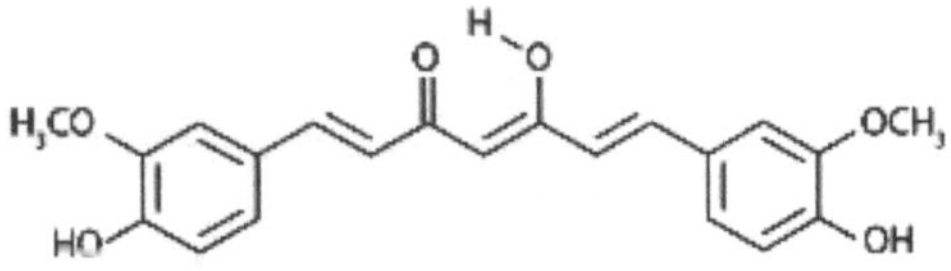

Fig. 7.2. Molecular structure of curcumin (diferuloylmethane).

A large number of reports claiming immunomodulatory or related activities and their mechanism of actions of curcumin have been published in recent years (Antony et al. 1999, Churchill et al. 2000, Rabinovich et al. 2000, Cole et al. 2004, Gao et al. 2004, Cao et al. 2005, Yadav et al. 2005, Barta et al. 2006, Cao et al. 2006, Gautam et al. 2007, Jagetia and Aggarwal 2007, Kurup and Barrios 2008, Pathak and Khandelwal 2008, Shirley et al. 2008, Strimpakos and Sharma 2008, Tong et al. 2008, Varalakshmi et al. 2008, Villegas et al. 2008, Allam 2009, Jeong et al. 2009, Liang et al. 2009, Yang et al. 2009). The anti-inflammatory properties of turmeric have been documented, as also its ability to retard some of the progression of acquired immune deficiency syndrome (AIDS) (Bentley and Trimen 1880).

Aloe vera

A number of pharmacological activities associated with *Aloe vera* have been attributed to polysaccharides contained in the gel of the leaves. These biological activities include promotion of wound healing, antifungal activity, hypoglycemic or antidiabetic effects, anti-inflammatory, anticancer, immunomodulatory (Hart et al. 1988, t'Hart et al. 1989, Karaca et al. 1995, Robinson 1998, Reynolds and Dweck 1999, Lee et al. 2001, Tan and Vanitha 2004, Im et al. 2005, Iljazovic et al. 2006, Zhang et al. 2006, Akev et al. 2007, Akev et al. 2007, Hamman 2008) and gastroprotective properties (Hamman 2008).

Glycyrrhiza (licorice)

The roots and rhizomes of the licorice (*Glycyrrhiza*) species have long been used worldwide as a herbal medicine and natural sweetener (Asl and Hosseinzadeh 2008). Licorice species are perennial herbs native to the Mediterranean region, central to southern Russia, and Asia Minor to Iran, now widely cultivated throughout Europe, the Middle East and Asia (Blumenthal et al. 2000). The genus *Glycyrrhiza* (Fabaceae) consists of about 30 species including *G. uralensis*, *G. inflata*, *G. aspera*, *G. korshinskyi* and *G. eurycarpa*, *G. glabra* and also includes three varieties: Persian and Turkish licorices are assigned to *G. glabra* var. *violacea*, Russian licorice is *G. glabra* var. *gladulifera*, and Spanish and Italian licorices are *G. glabra* var. *typical* (Nomura et al. 2002).

Licorice roots contain triterpenoid saponins (4–20%), mostly glycyrrhizin, a mixture of potassium and calcium salts of glycyrrhizic acid (also known as glycyrrhizic or glycyrrhizinic acid, and a glycoside of glycyrrhetinic acid) which is 50 times as sweet as sugar (Blumenthal et al. 2000). The molecular structures of glycyrrhizin or glycyrrhizic acid and glycyrrhetinic acid are shown in Fig. 7.3.

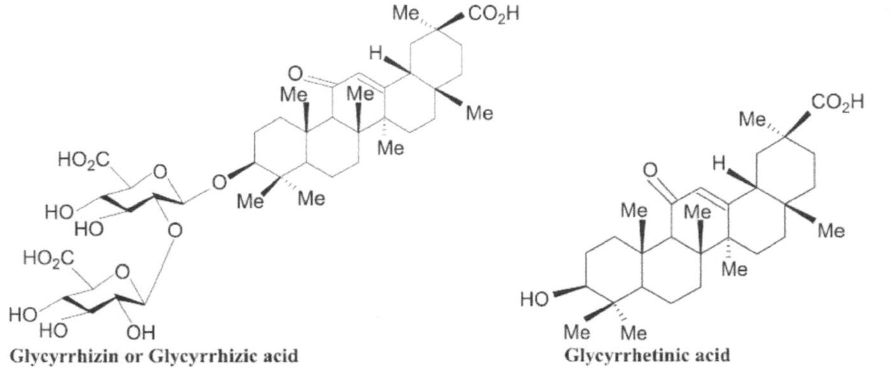

Glycyrrhizin or Glycyrrhizic acid   Glycyrrhetinic acid

Fig. 7.3. The molecular structures of glycyrrhizin or glycyrrhizic acid and glycyrrhetinic acid.

In past years several immunomodulatory activities have been recognized with glycyrrhizin (Zhang et al. 1993, Kroes et al. 1997, Xu et al. 1997) and glycyrrhetinic acid (Chavali et al. 1987, Zhang et al. 1993, Kroes et al. 1997, Shaneyfelt et al. 2006, Asl and Hosseinzadeh 2008). Glycyrrhizin selectively activated extrathymic T cells in the liver and in human T cell lines and glycyrrhizic acid enhanced Fas-mediated apoptosis without alteration of caspase-3-like activity (Kimura et al. 1992, Ishiwata et al. 1999).

Safety considerations

Large amounts of licorice may result in severe hypertension, hypokalemia and other signs of mineralocorticoid overload (Asl and Hosseinzadeh 2008). The hypertension is mainly caused by decreased 11β-hydroxysteroid dehydrogenase activity, which is responsible for the renal conversion of cortisol to cortisone. Therefore, licorice leads to the activation of renal mineralocorticoid receptors by cortisol, consequential apparent mineralocorticoid excess and suppression of the rennin angiotensin system (Conn et al. 1968, Koster and David 1968, Lefebvre and Marc-Aurele 1968, Anonymous 1990, Latif et al. 1990, Izzo et al. 2005, van Uum 2005, Gerritsen et al. 2009, Johns 2009, Templin et al. 2009). In another report the toxicity of licorice extract was shown in the liver of the fish

Black molly (Radhakrishnan et al. 2005). In a review Asl and Hosseinzadeh (2008) reported several hazard issues of the excessive uses of licorice and its different side effects.

Conclusion

Even though there are considerable advances in synthetic as well as medicinal chemistry, molecules from natural resources correspond more completely in our view to the requirements of contemporary medicine (Bakuridze et al. 1993). The action of natural substances, not foreign to the organism, participating as natural agents in the metabolism of substances, practically excludes the generation of allergic reactions on treatment with medicinal plants (Kovaleva 1971, Loss and Khisamutdinova 1980, Gammerman et al. 1983, Brekhman et al. 1984).

The use of plant products as immunostimulants has a traditional history. However, the isolation of the active principles involved did not gain momentum till the 19th century (Phillipson 2001). The successful derivation of pure bioactive compounds from *Ganoderma lucidum, Panax ginseng* and *Zingiber officinale* supports the traditional practice of using these plants to stimulate the immune system. As many modern drugs are often patterned after phytochemicals, studying the influence of each compound on immune cells as well as microbes can provide useful insights to the development of potentially useful new pharmacological agents (Tan and Vanitha 2004).

Experimental data published in recent years on the persuasion of herbal formulas targeting cytokine activities is limited by the paucity of human studies. Despite the fact that the majority of the research is *in vitro* or animal models, there does appear to be substantial historical, empirical, and increasing scientific evidence that herbal formulas may provide therapeutic application in the modulation of cytokines or other systems. In addition, pharmacodynamic and pharmacokinetic potentiating of active constituents, as well as buffering of toxic constituents, appear to be common strategies of traditional formulas (Burns et al. 2009). It has also been observed that the majority of research work has been studied and published directly on traditional medicine or plant extracts, which probably have been matured enough to go further for the drug development considering active-single molecule(s) from the potential natural resources. The active-single molecule(s) can be synthesized or partially synthesized or make different analogs, if possible, which is necessary to study the structure-activity relationships of those active molecules to further design and develop such immuno–modulators or –stimulators.

It is anticipated that the information presented here will serve to stimulate further research not only on this topic but also related issues.

References

Afroz, S., M. Alamgir, M.T. Khan, S. Jabbar, N. Nahar, and M.S. Choudhuri. 2006. Antidiarrhoeal activity of the ethanol extract of *Paederia foetida* Linn. (Rubiaceae). J. Ethnopharmacol. 105(1-2): 125–30.

Ahmad, V.U., F. Ullah, J. Hussain, U. Farooq, M. Zubair, M.T. Khan, and M.I. Choudhary 2004. Tyrosinase inhibitors from *Rhododendron collettianum* and their structure-activity relationship (SAR) studies. Chem. Pharm. Bull. (Tokyo) 52(12): 1458–61.

Ajikumar, P.K., K. Tyo, S. Carlsen, O. Mucha, T.H. Phon, and G. Stephanopoulos 2008. Terpenoids: opportunities for biosynthesis of natural product drugs using engineered microorganisms. Mol. Pharm. 5(2): 167–90.

Akev, N., G. Turkay, A. Can, A. Gurel, F. Yildiz, H. Yardibi, E.E. Ekiz, and H. Uzun. 2007. Effect of Aloe vera leaf pulp extract on Ehrlich ascites tumours in mice. Eur. J. Cancer Prev. 16(2): 151–7.

Akev, N., G. Turkay, A. Can, A. Gurel, F. Yildiz, H. Yardibi, E.E. Ekiz, and H. Uzun 2007. Tumour preventive effect of Aloe vera leaf pulp lectin (Aloctin I) on Ehrlich ascites tumours in mice. Phytother. Res. 21(11): 1070–5.

Allam, G. 2009. Immunomodulatory effects of curcumin treatment on murine schistosomiasis mansoni. Immunobiology 214(8): 712–27.

Anonymous. 1990. Licorice, tobacco chewing, and hypertension. N. Engl. J. Med. 322(12): 849–50.

Antony, S., R. Kuttan, and G. Kuttan 1999. Immunomodulatory activity of curcumin. Immunol Invest 28(5-6): 291–303.

Asl, M.N. and H. Hosseinzadeh. 2008. Review of pharmacological effects of *Glycyrrhiza* sp. and its bioactive compounds. Phytother. Res. 22(6): 709–24.

Azhar, U.L., H.A. Malik, M.T. Khan, U.L.H. Anwar, S.B. Khan, A. Ahmad, and M.I. Choudhary. 2006. Tyrosinase inhibitory lignans from the methanol extract of the roots of *Vitex negundo* Linn. and their structure-activity relationship. Phytomedicine 13(4): 255–60.

Baker, P.J., and T. Hraba. 1994. T-cell mediated immunosuppression and its implications for the development of protective immunity. Folia Biol. (Praha) 40(6): 349–58.

Bakuridze, A.D., M.S. Kurtsikidze, V.M. Pisarev, R.V. Makharadze, and D.T. Berashvili. 1993. Medicinal plants: Immunomodulators of plant origin (review). Khimiko-farmatsevticheskii Zhurnal 27(8): 43–47.

Barta, I., P. Smerak, Z. Polivkova, H. Sestakova, M. Langova, B. Turek, and J. Bartova 2006. Current trends and perspectives in nutrition and cancer prevention. Neoplasma 53(1): 19–25.

Barta, O. 1986. Immunomodulatory effect of serum on lymphocytes—consequences for the study of immunostimulants *in vitro* and in vivo. Comp. Immunol. Microbiol. Infect. Dis. 9(2-3): 193–203.

Bentley, R. and H. Trimen. 1880. Medicinal Plants. Churchill. London.

Beuth, J. 1997. Clinical relevance of immunoactive mistletoe lectin-I. Anticancer Drugs 8 Suppl. 1: S53–5.

Bhat, J., A. Damle, P.P. Vaishnav, R. Albers, M. Joshi, and G. Banerjee. 2009. In vivo enhancement of natural killer cell activity through tea fortified with Ayurvedic herbs. Phytother Res.

Blumenthal, M., A. Goldberg, and J. Brinckmann. 2000. Herbal Medicine: Expanded Commission E Monographs. American Botanical Council. Newton.

Brekhman, I.I., N.R. Deryana, M.A. Grinevich, and A. S. Saratikov 1984. Immunomodulators of plant origin (review). Rast. Res. 19(4): 438–444.

Burns, J.J., L. Zhao, E.W. Taylor, and K. Spelman. 2009. The influence of traditional herbal formulas on cytokine activity. Toxicology.

Cao, W.G., M. Morin, C. Metz, R. Maheux, and A. Akoum. 2005. Stimulation of macrophage migration inhibitory factor expression in endometrial stromal cells by interleukin 1β involving the nuclear transcription factor NFκB. Biol. Reprod. 73(3): 565–70.

Cao, W.G., M. Morin, V. Sengers, C. Metz, T. Roger, R. Maheux, and A. Akoum. 2006. Tumour necrosis factor-alpha up-regulates macrophage migration inhibitory factor expression in endometrial stromal cells via the nuclear transcription factor NF-κB. Hum. Reprod. 21(2): 421–8.

Caroff, M. and D. Karibian 2003. Structure of bacterial lipopolysaccharides. Carbohydr. Res. 338(23): 2431–47.

Castro, G.R., B. Panilaitis, E. Bora, and D.L. Kaplan. 2007. Controlled release biopolymers for enhancing the immune response. Mol. Pharm. 4(1): 33–46.

Chang, R. 2002. Bioactive polysaccharides from traditional Chinese medicine herbs as anticancer adjuvants. J. Altern. Complement. Med. 8(5): 559–65.

Chattopadhyay, D., and M.T.H. Khan. 2008. Ethnomedicines and ethnomedicinal phytophores against herpes viruses. Biotechnol. Annu. Rev. 14: 297–348.

Chattopadhyay, D., M.C. Sarkar, T. Chatterjee, R. Sharma Dey, P. Bag, S. Chakraborti, and M.T.H. Khan. 2009. Recent advancements for the evaluation of anti-viral activities of natural products. N. Biotechnol. 25(5): 347–68.

Chavali, S.R., T. Francis, and J.B. Campbell 1987. An *in vitro* study of immunomodulatory effects of some saponins. Int. J. Immunopharmacol 9(6): 675–83.

Christensen, L.P. 2009. Ginsenosides chemistry, biosynthesis, analysis, and potential health effects. Adv. Food Nutr. Res. 55: 1–99.

Churchill, M., A. Chadburn, R.T. Bilinski, and M.M. Bertagnolli 2000. Inhibition of intestinal tumors by curcumin is associated with changes in the intestinal immune cell profile. J. Surg. Res. 89(2): 169–75.

Cole, G. M., T. Morihara, G.P. Lim, F. Yang, A. Begum, and S.A. Frautschy 2004. NSAID and antioxidant prevention of Alzheimer's disease: lessons from *in vitro* and animal models. Ann. N Y Acad. Sci. 1035: 68–84.

Comstock, L.E. 2009. Importance of glycans to the host-bacteroides mutualism in the mammalian intestine. Cell Host Microbe 5(6): 522–6.

Conn, J.W., D.R. Rovner, and E.L. Cohen. 1968. Licorice-induced pseudoaldosteronism. Hypertension, hypokalemia, aldosteronopenia, and suppressed plasma renin activity. JAMA 205(7): 492–6.

Costa-Lotufo, L.V., M.T. Khan, A. Ather, D.V. Wilke, P.C. Jimenez, C. Pessoa, M.E. de Moraes, and M.O. de Moraes. 2005. Studies of the anticancer potential of plants used in Bangladeshi folk medicine. J. Ethnopharmacol. 99(1): 21–30.

Danishefsky, S.J., M. Inoue, and D. Trauner. 2000. Synthesis of immunomodulatory marine natural products. Ernst Schering Res. Found Workshop.(32): 1–24.

Deharo, E., R. Baelmans, A. Gimenez, C. Quenevo, and G. Bourdy. 2004. *In vitro* immunomodulatory activity of plants used by the Tacana ethnic group in Bolivia. Phytomedicine 11(6): 516–22.

Devkota, K.P., M.T. Khan, R.Ranjit, A.M. Lannang, Samreen, and M.I. Choudhary. 2007. Tyrosinase inhibitory and antileishmanial constituents from the rhizomes of Nat. Prod. Res. 21(4): 321–7.

Dinkova-Kostova, A.T. 2008. Phytochemicals as protectors against ultraviolet radiation: versatility of effects and mechanisms. Planta Med. 74(13): 1548–59.

Domer, J., K. Elkins, D. Ennist, and P. Baker. 1988. Modulation of immune responses by surface polysaccharides of *Candida albicans*. Rev. Infect. Dis. 10 Suppl. 2: S419–22.

Ellerbroek, P.M., A.M. Walenkamp, A.I. Hoepelman, and F.E. Coenjaerts. 2004. Effects of the capsular polysaccharides of Cryptococcus neoformans on phagocyte migration and inflammatory mediators. Curr. Med. Chem. 11(2): 253–66.

Elluru, S., J.P. Van Huyen, S. Delignat, F. Prost, J. Bayry, M.D. Kazatchkine, and S.V. Kaveri. 2006. Molecular mechanisms underlying the immunomodulatory effects of mistletoe (*Viscum album* L.) extracts Iscador. Arzneimittelforschung 56(6A): 461–6.

Flemming, K.B. 1985. Reticuloendothelial System: A Comprehensive Treatise. Plenum Press, New York.

Fu, G., H. Pang, and Y.H. Wong. 2008. Naturally occurring phenylethanoid glycosides: potential leads for new therapeutics. Curr. Med. Chem. 15(25): 2592–613.

Gabius, H.J. 2001. Probing the cons and pros of lectin-induced immunomodulation: case studies for the mistletoe lectin and galectin-1. Biochimie. 83(7): 659–66.

Gabius, H.J., S. Gabius, S.S. Joshi, B. Koch, M. Schroeder, W.M. Manzke, and M. Westerhausen 1994. From ill-defined extracts to the immunomodulatory lectin: will there be a reason for oncological application of mistletoe? Planta Med. 60(1): 2–7.

Gammerman, A.F., G.I. Kadaev, and A.A. Yashenko-Khmelevskii. 1983. Medicinal Plants (Plant Healing Agents). Moscow.

Gao, X., J. Kuo, H. Jiang, D. Deeb, Y. Liu, G. Divine, R.A. Chapman, S.A. Dulchavsky, and S.C. Gautam. 2004. Immunomodulatory activity of curcumin: suppression of lymphocyte proliferation, development of cell-mediated cytotoxicity, and cytokine production *in vitro*. Biochem. Pharmacol. 68(1): 51–61.

Garbe, C. 1993. Chemotherapy and chemoimmunotherapy in disseminated malignant melanoma. Melanoma. Res. 3(4): 291–9.

Gautam, S.C., X. Gao, and S. Dulchavsky. 2007. Immunomodulation by curcumin. Adv. Exp. Med. Biol. 595: 321–41.

Geethangili, M. and Y.-M. Tzeng. 2009. Review of pharmacological effects of *Antrodia Camphorata* and its bioactive compounds. eCAM: nep108.

Gerritsen, K.G., J.Meulenbelt, W. Spiering, I.P. Kema, A. Demir, and V.J. van Driel. 2009. An unusual cause of ventricular fibrillation. Lancet 373(9669): 1144.

Gibbs, B.F. 2005. Human basophils as effectors and immunomodulators of allergic inflammation and innate immunity. Clin. Exp. Med. 5(2): 43–9.

Hajto, T., K. Hostanska, and R. Saller. 1999. Mistletoe therapy from the pharmacologic perspective. Forsch Komplementarmed 6(4): 186–94.

Hamman, J.H. 2008. Composition and applications of *Aloe vera* leaf gel. Molecules 13(8): 1599–616.

Hart, L.A., P.H. van Enckevort, H. van Dijk, R. Zaat, K.T. de Silva, and R.P. Labadie. 1988. Two functionally and chemically distinct immunomodulatory compounds in the gel of *Aloe vera*. J. Ethnopharmacol. 23(1): 61–71.

Heitzman, M.E., C.C. Neto, E. Winiarz, A.J. Vaisberg, and G.B. Hammond 2005. Ethnobotany, phytochemistry and pharmacology of *Uncaria* (Rubiaceae). Phytochemistry 66(1): 5–29.

Hermann, F., F. Ruschitzka, L. Spieker, I. Sudano, G. Noll, and R. Corti. 2005. The sweet secret of dark chocolate. Ther. Umsch. 62(9): 635–7.

Hirata, M., T. Inamitsu, T. Hashimoto, and T. Koga 1984. An inhibitor of lipoxygenase, nordihydroguaiaretic acid, shortens actin filaments. J Biochem 95(3): 891–4.

Hoffmann, D.L. 2009. Immuno-stimulation, Immuno-modulation or what? Retrieved 24/09/09, 2009, from http://www.healthy.net/scr/article.asp?ID=1803.

Hokke, C.H., J.M. Fitzpatrick, and K.F. Hoffmann. 2007. Integrating transcriptome, proteome and glycome analyses of Schistosoma biology. Trends. Parasitol. 23(4): 165–74.

Hussaini, I.M., Y.H. Zhang, J.J. Lysiak, and T.Y. Shen. 2000. Dithiolane analogs of lignans inhibit interferon-gamma and lipopolysaccharide-induced nitric oxide production in macrophages. Acta. Pharmacol. Sin. 21(10): 897–904.

Ilarregui, J.M., G.A. Bianco, M.A. Toscano, and G.A. Rabinovich. 2005. The coming of age of galectins as immunomodulatory agents: impact of these carbohydrate binding proteins in T cell physiology and chronic inflammatory disorders. Ann. Rheum. Dis. 64 Suppl. 4: iv96–103.

Iljazovic, E., D. Ljuca, A. Sahimpasic, and S. Avdic. 2006. Efficacy in treatment of cervical HRHPV infection by combination of beta interferon, and herbal therapy in women with different cervical lesions. Bosn. J. Basic. Med. Sci. 6(4): 79–84.

Im, S.A., S.T. Oh, S. Song, M.R. Kim, D.S. Kim, S.S. Woo, T.H. Jo, Y.I. Park, and C. K. Lee. 2005. Identification of optimal molecular size of modified Aloe polysaccharides with maximum immunomodulatory activity. Int. Immunopharmacol. 5(2): 271–9.

Ishiwata, S., K. Nakashita, Y. Ozawa, M. Niizeki, S. Nozaki, Y. Tomioka, and M. Mizugaki. 1999. Fas-mediated apoptosis is enhanced by glycyrrhizin without alteration of caspase-3-like activity. Biol. Pharm. Bull. 22(11); 1163–6.

Izzo, A.A., G. Di Carlo, F. Borrelli, and E. Ernst. 2005. Cardiovascular pharmacotherapy and herbal medicines: the risk of drug interaction. Int. J. Cardiol. 98(1): 1–14.

Jabbar, S., M.T. Khan, and M.S. Choudhuri. 2001. The effects of aqueous extracts of *Desmodium gangeticum* DC. (Leguminosae) on the central nervous system. Pharmazie 56(6): 506–8.

Jabbar, S., M.T. Khan, M.S. Choudhuri, and B.K. Sil. 2004. Bioactivity studies of the individual ingredients of the Dashamularishta. Pak. J. Pharm. Sci. 17(1): 9–17.

Jacob, G.S. 1995. Glycosylation inhibitors in biology and medicine. Curr. Opin. Struct. Biol. 5(5): 605–11.

Jagetia, G.C. and B.B. Aggarwal. 2007. "Spicing up" of the immune system by curcumin. J. Clin. Immunol. 27(1): 19–35.

Jeong, Y.I., S.W. Kim, I.D. Jung, J.S. Lee, J.H. Chang, C.M. Lee, S.H. Chun, M.S. Yoon, G.T. Kim, S.W. Ryu, J.S. Kim, Y.K. Shin, W.S. Lee, H.K. Shin, J.D. Lee, and Y. M. Park. 2009. Curcumin suppresses the induction of indoleamine 2,3-dioxygenase by blocking the Janus-activated kinase-protein kinase Cdelta-STAT1 signaling pathway in interferon-gamma-stimulated murine dendritic cells. J. Biol. Chem. 284(6): 3700–8.

Johns, C. 2009. Glycyrrhizic acid toxicity caused by consumption of licorice candy cigars. CJEM 11(1): 94–6.

Johnston, M.J., J.A. MacDonald, and D.M. McKay. 2009. Parasitic helminths: a pharmacopeia of anti-inflammatory molecules. Parasitology 136(2): 125–47.

Joo, S.S., T.J. Won, and D.I. Lee. 2005. Reciprocal activity of ginsenosides in the production of proinflammatory repertoire, and their potential roles in neuroprotection in vivo. Planta Med. 71(5): 476–81.

Karaca, K., J.M. Sharma, and R. Nordgren. 1995. Nitric oxide production by chicken macrophages activated by Acemannan, a complex carbohydrate extracted from *Aloe vera*. Int. J. Immunopharmacol. 17(3): 183–8.

Katiyar, S.K. 2003. Skin photoprotection by green tea: antioxidant and immunomodulatory effects. Curr. Drug. Targets. Immune. Endocr. Metabol. Disord. 3(3): 234–42.

Kenarova, B., H. Neychev, C. Hadjiivanova, and V.D. Petkov. 1990. Immunomodulating activity of ginsenoside Rg1 from *Panax ginseng*. Jpn. J. Pharmacol. 54(4): 447–54.

Khan, M.T., T. Matsui, Y. Matsumoto, and S. Jabbar. 2001. *In vitro* ACE inhibitory effects of some Bangladeshi plant extracts. Pharmazie 56(11): 902–3.

Khan, M.T., I. Lampronti, D. Martello, N. Bianchi, S. Jabbar, M.S. Choudhuri, B.K. Datta, and R. Gambari. 2002. Identification of pyrogallol as an antiproliferative compound present in extracts from the medicinal plant *Emblica officinalis*: effects on *in vitro* cell growth of human tumor cell lines. Int. J. Oncol. 21(1): 187–92.

Khan, M.T., M.I. Choudhary, R. Atta ur, R.P. Mamedova, M.A. Agzamova, M.N. Sultankhodzhaev, and M.I. Isaev. 2006a. Tyrosinase inhibition studies of cycloartane and cucurbitane glycosides and their structure-activity relationships. Bioorg. Med. Chem. 14(17): 6085–8.

Khan, M.T., S.B. Khan, and A. Ather. 2006b. Tyrosinase inhibitory cycloartane type triterpenoids from the methanol extract of the whole plant of *Amberboa ramosa* Jafri and their structure-activity relationship. Bioorg. Med. Chem. 14(4): 938–43.

Khan, M.T., I. Orhan, F.S. Senol, M. Kartal, B. Sener, M. Dvorska, K. Smejkal, and T. Slapetova. 2009. Cholinesterase inhibitory activities of some flavonoid derivatives and chosen xanthone and their molecular docking studies. Chem. Biol. Interact 181(3): 383–9.

Khan, M.T.H., A. Ather, K.D. Thompson, and R. Gambari. 2005a. Extracts and molecules from medicinal plants against herpes simplex viruses. Antiviral Res. 67(2): 107–19.

Khan, S.B., U.L.H. Azhar, N. Afza, A. Malik, M.T. Khan, M.R. Shah, . and M.I. Choudhary. 2005b. Tyrosinase-inhibitory long-chain esters from *Amberboa ramosa*. Chem. Pharm. Bull. (Tokyo) 53(1): 86–9.

Kim, J.Y., J.S. Kang, H.M. Kim, Y.K., Kim, H.K. Lee, S. Song, J.T. Hong, Y. Kim, and S. B. Han. 2009. Inhibition of phenotypic and functional maturation of dendritic cells by manassantin A. J. Pharmacol. Sci. 109(4): 583–92.

Kimura, M., H. Watanabeh, and T. Abo. 1992. Selective activation of extrathymic T cells in the liver by glycyrrhizin. Biotherapy 5: 167–176.

Kokoska, L. and D. Janovska. 2009. Chemistry and pharmacology of *Rhaponticum carthamoides*: a review. Phytochemistry 70(7): 842–55.

Koster, M. and G.K. David. 1968. Reversible severe hypertension due to licorice ingestion. N. Engl. J. Med. 278(25): 1381–3.

Kovaleva, N.G. 1971. Treatment with Plants—Outlines of Phytotherapy. Moscow.

Kowalozyk-Bronisz, S.H., and I. Paegelow. 1986. The influence of immunomodulators on lymphokine secretion of radiation-damaged lymphocytes. Allerg. Immunol. 32: 57–64.

Kroes, B.H., C.J. Beukelman, A.J. van den Berg, G.J. Wolbink, H. van Dijk, and R.P. Labadie. 1997. Inhibition of human complement by β-glycyrrhetinic acid. Immunology 90(1): 115–20.

Kuipers, H.F. and P.J. van den Elsen. 2007. Immunomodulation by statins: inhibition of cholesterol vs. isoprenoid biosynthesis. Biomed Pharmacother 61(7): 400–7.

Kurup, V.P. and C.S. Barrios. 2008. Immunomodulatory effects of curcumin in allergy. Mol. Nutr. Food Res. 52(9): 1031–9.

Lai, J.H. 2002. Immunomodulatory effects and mechanisms of plant alkaloid tetrandrine in autoimmune diseases. Acta Pharmacol Sin 23(12): 1093–101.

Lambertini, E., I. Lampronti, L. Penolazzi, M.T. Khan, A. Ather, G. Giorgi, R. Gambari, . and R. Piva 2005. Expression of estrogen receptor alpha gene in breast cancer cells treated with transcription factor decoy is modulated by Bangladeshi natural plant extracts. Oncol. Res. 15(2): 69–79.

Lampronti, I., M.T. Khan, N. Bianchi, A. Ather, M. Borgatti, L. Vizziello, E. Fabbri, and R. Gambari. 2005. Bangladeshi medicinal plant extracts inhibiting molecular interactions between nuclear factors and target DNA sequences mimicking NF-κB binding sites. Med. Chem. 1(4): 327–33.

Lampronti, I., M.T. Khan, M. Borgatti, N. Bianchi, and R. Gambari. 2008. Inhibitory Effects of Bangladeshi Medicinal Plant Extracts on Interactions between Transcription Factors and Target DNA Sequences. Evid Based Complement Alternat Med. 5(3): 303–312.

Latif, S.A., T.J. Conca, and D.J. Morris. 1990. The effects of the licorice derivative, glycyrrhetinic acid, on hepatic 3 α- and 3 β-hydroxysteroid dehydrogenases and 5 α- and 5 β-reductase pathways of metabolism of aldosterone in male rats. Steroids 55(2): 52–8.

Lee, D.C., C.L. Yang, S.C. Chik, J.C. Li, J.H. Rong, G.C. Chan, and A. S. Lau. 2009. Bioactivity-guided identification and cell signaling technology to delineate the immunomodulatory effects of *Panax ginseng* on human promonocytic U937 cells. J. Transl. Med. 7: 34.

Lee, J.K., M.K. Lee, Y.P. Yun, Y. Kim, J.S. Kim, Y.S. Kim, K. Kim,S.S. Han, and C. K. Lee. 2001. Acemannan purified from *Aloe vera* induces phenotypic and functional maturation of immature dendritic cells. Int. Immunopharmacol. 1(7): 1275–84.

Lee, T.K., R.M. Johnke, R.R. Allison, K.F. O'Brien, and L. J. Dobbs Jr, 2005. Radioprotective potential of ginseng. Mutagenesis 20(4): 237–43.

Lefebvre, R F. and J. Marc-Aurele. 1968. Licorice and hypertension. Can Med. Assoc. J. 99(5): 230–1.

Liang, G., H. Zhou, ,Y. Wang, E.C. Gurley, B. Feng, L. Chen, J.Xiao, S.Yang, and X. Li. 2009. Inhibition of LPS-induced production of inflammatory factors in the macrophages by mono-carbonyl analogues of curcumin. J. Cell. Mol. Med. 13(9b): 3370–9.

Lin, Z.B. and H.N. Zhang. 2004. Anti-tumor and immunoregulatory activities of *Ganoderma lucidum* and its possible mechanisms. Acta Pharmacol. Sin. 25(11): 1387–95.

Liu, Z.L., S. Tanaka, H. Horigome, T., Hirano, and K. Oka. 2002. Induction of apoptosis in human lung fibroblasts and peripheral lymphocytes *in vitro* by Shosaiko-to derived phenolic metabolites. Biol. Pharm. Bull. 25(1): 37–41.

Loss, S.M. and T.F. Khisamutdinova. 1080 First Ukrainian Republic Conference on Medicinal Botany, Kiev.

Mainardi, T., S. Kapoor, and L. Bielory. 2009. Complementary and alternative medicine: herbs, phytochemicals and vitamins and their immunologic effects. J. Allergy Clin. Immunol. 123(2): 283–94; quiz 295–6.

Mazmanian, S.K. and D.L. Kasper. 2006. The love-hate relationship between bacterial polysaccharides and the host immune system. Nat. Rev. Immunol. 6(11): 849–58.

Mohla, S., M.J. Humphries, S.L. White, K. Matsumoto, S.A. Newton, C.C. Sampson, D. Bowen, and K. Olden. 1989. Swainsonine: a new antineoplastic immunomodulator. J. Natl. Med. Assoc. 81(10): 1049–56.

Monro, J.A. 2003. Treatment of cancer with mushroom products. Arch. Environ. Health 58(8): 533–7.

Moradali, M.F., H. Mostafavi, S. Ghods, and G.A. Hedjaroude. 2007. Immunomodulating and anticancer agents in the realm of macromycetes fungi (macrofungi). Int. Immunopharmacol. 7(6): 701–24.

Ng, T.B. and H.X. Wang. 2005. Pharmacological actions of Cordyceps, a prized folk medicine. J. Pharm. Pharmacol. 57(12): 1509–19.

Nomura, T., T. Fukai, and T. Akiyama. 2002. Chemistry of phenolic compounds of licorice (*Glycyrrhiza* species) and their estrogenic and cytotoxic activities. Pure Appl. Chem. 74: 1199–1206.

Ooi, V.E. and F. Liu. 2000. Immunomodulation and anti-cancer activity of polysaccharide-protein complexes. Curr. Med. Chem. 7(7): 715–29.

Ovodov Iu, S. 1998. Polysaccharides of flower plants: structure and physiological activity. Bioorg Khim 24(7): 483–501.

Pathak, N. and S. Khandelwal. 2008. Comparative efficacy of piperine, curcumin and picroliv against Cd immunotoxicity in mice. Biometals 21(6): 649–61.

Penolazzi, L., I. Lampronti, M. Borgatti, M.T. Khan, M. Zennaro, R. Piva, and R. Gambari. 2008. Induction of apoptosis of human primary osteoclasts treated with extracts from the medicinal plant *Emblica officinalis*. BMC Complement Alternat. Med. 8: 59.

Phillipson, J.D. 2001. Phytochemistry and medicinal plants. Phytochemistry 56(3): 237–43.

Pla, D., F.Albericio, and M. Alvarez. 2008. Recent advances in lamellarin alkaloids: isolation, synthesis and activity. Anticancer Agents Med. Chem. 8(7): 746–60.

Prakash, S., and C. Martoni. 2006. Toward a new generation of therapeutics: artificial cell targeted delivery of live cells for therapy. Appl. Biochem. Biotechnol. 128(1): 1–22.

Prasher, B., S. Negi, S. Aggarwal, A.K. Mandal, T.P. Sethi, S.R. Deshmukh, S.G. Purohit, S. Sengupta, S. Khanna, F. Mohammad, G. Garg, S.K. Brahmachari, and M. Mukerji. 2008. Whole genome expression and biochemical correlates of extreme constitutional types defined in Ayurveda. J. Transl. Med. 6: 48.

Previati, M., E. Corbacella, L. Astolfi, M. Catozzi, M.T. Khan, I. Lampronti, R. Gambari, S.Capitani, and A. Martini. 2007. Ethanolic extract from *Hemidesmus indicus* (Linn) displays otoprotectant activities on organotypic cultures without interfering on gentamicin uptake. J. Chem. Neuroanat 34(3-4): 128–33.

Rabinovich, G.A., C.R. Alonso, C.E. Sotomayor, S. Durand, J.L. Bocco, and C.M. Riera. 2000. Molecular mechanisms implicated in galectin-1-induced apoptosis: activation of the AP-1 transcription factor and downregulation of Bcl-2. Cell Death Differ 7(8): 747–53.

Rabinovich, G.A., N. Rubinstein, and M.A. Toscano. 2002. Role of galectins in inflammatory and immunomodulatory processes. Biochim. Biophys. Acta 1572(2-3): 274–84.

Radhakrishnan, N., A. Gnanamani, and S. Sadulla. 2005. Effect of licorice (*Glycyhrriza glabra* Linn.), a skin-whitening agent on Black molly (*Poecilia latipinnaa*). J. Appl. Cosm. 23: 149–158.

Ramesh, H.P. and R.N. Tharanathan. 2003. Carbohydrates—the renewable raw materials of high biotechnological value. Crit. Rev. Biotechnol. 23(2): 149–73.

Reynolds, T. and A.C. Dweck. 1999. *Aloe vera* leaf gel: a review update. J. Ethnopharmacol 68(1-3): 3–37.

Rhule, A., S. Navarro, J.R. Smith, and D.M. Shepherd. 2006. *Panax notoginseng* attenuates LPS-induced pro-inflammatory mediators in RAW264.7 cells. J. Ethnopharmacol 106(1): 121–8.

Robinson, M. 1998. Medical therapy of inflammatory bowel disease for the 21st century. Eur. J. Surg Suppl(582): 90–8.

Schepetkin, I.A. and M.T. Quinn. 2006. Botanical polysaccharides: macrophage immunomodulation and therapeutic potential. Int. Immunopharmacol. 6(3): 317–33.

Seo, B.R., K.W. Lee, J. Ha, H.J. Park, J.W. Choi, and K.T. Lee. 2004a. Saucernetin-7 isolated from *Saururus chinensis* inhibits proliferation of human promyelocytic HL-60 leukemia cells via G0/G1 phase arrest and induction of differentiation. Carcinogenesis 25(8): 1387–94.

Seo, B.R., C.B. Yoo, H.J. Park, J.W. Choi, K. Seo, S.K., Choi, and K.T. Lee. 2004b. Saucernetin-8 isolated from *Saururus chinensis* induced the differentiation of human acute promyelocytic leukemia HL-60 cells. Biol. Pharm Bull. 27(10): 1594–8.

Shaheen, F., M. Ahmad, M.T. Khan, S. Jalil, A. Ejaz, M.N. Sultankhodjaev, M. Arfan, M.I. Choudhary, and Atta-ur-Rahman. 2005. Alkaloids of Aconitum leaves and their anti-inflammatory antioxidant and tyrosinase inhibition activities. Phytochemistry 66(8): 935–40.

Shaneyfelt, M.E., A.D. Burke, J.W. Graff, M.A. Jutila, and M.E. Hardy. 2006. Natural products that reduce rotavirus infectivity identified by a cell-based moderate-throughput screening assay. Virol. J. 3: 68.

Shaphiullah, M., S.C. Bachar, J.K. Kundu, F. Begum, M.A. Uddin, S.C. Roy, and M.T. Khan. 2003. Antidiarrheal activity of the methanol extract of *Ludwigia hyssopifolia* Linn. Pak. J. Pharm. Sci. 16(1): 7–11.

Shirley, S.A., A.J. Montpetit, R.F. Lockey, and S.S. Mohapatra. 2008. Curcumin prevents human dendritic cell response to immune stimulants. Biochem. Biophys. Res. Commun. 374(3): 431–6.

Singh, A.P. 2006. Distribution of steroid like compounds in Plant Flora. Pharmacognosy Magazine 2(6): 87–89.

Smirnov, L.D. and V.S. Suskova. 1989. Trufakin, cellular factors in the regulation of immunogenesis. Khim.-farm. Zh. 7: 773–784.

Strimpakos, A.S. and R.A. Sharma. 2008. Curcumin: preventive and therapeutic properties in laboratory studies and clinical trials. Antioxid Redox Signal 10(3): 511–45.

Sultankhodzhaev, M.N., M.T. Khan, M. Moin, M.I. Choudhary, and R. Atta ur. 2005. Tyrosinase inhibition studies of diterpenoid alkaloids and their derivatives: structure-activity relationships. Nat. Prod. Res. 19(5): 517–22.

Svoboda, R.E. 1998. Ayurveda's role in preventing disease. Indian J. Med. Sci. 52(2): 70–7.

Tan, B.K. and J. Vanitha. 2004. Immunomodulatory and antimicrobial effects of some traditional chinese medicinal herbs: a review. Curr Med Chem 11(11): 1423–30.

Templin, C., M. Westhoff-Bleck, and J.R. Ghadri. 2009. Hypokalemic paralysis with rhabdomyolysis and arterial hypertension caused by liquorice ingestion. Clin Res Cardiol 98(2): 130–2.

t'Hart, L.A., A.J. van den Berg, L. Kuis, H. van Dijk, and R.P. Labadie. 1989. An anti-complementary polysaccharide with immunological adjuvant activity from the leaf parenchyma gel of *Aloe vera*. Planta Med. 55(6): 509–12.

Thomas, P.G. and D.A. Harn Jr. 2004. Immune biasing by helminth glycans. Cell Microbiol. 6(1): 13–22.

Tong, K.M., D.C. Shieh, C.P. Chen, C.Y. Tzeng, S.P. Wang, K.C. Huang, Y.C. Chiu, Y.C. Fong, and C. H. Tang. 2008. Leptin induces IL-8 expression via leptin receptor, IRS-1, PI3K, Akt cascade and promotion of NF-κB/p300 binding in human synovial fibroblasts. Cell Signal 20(8): 1478–88.

Tzianabos, A , J.Y. Wang, and D.L. Kasper. 2003. Biological chemistry of immunomodulation by zwitterionic polysacchai ides. Carbohydr Res. 338(23): 2531–8.

Ullah, F., H. Hussain, J. Hussain, I.A. Bukhari, M.I. Khan, M.I. Choudhary, A.H. Gilani, and V.U. Ahmad. 2007. Tyrosinase inhibitory pentacyclic triterpenes and analgesic and spasmolytic activities of methanol extracts of *Rhododendron collettianum*. Phytother Res. 21(11): 1076–81.

Umehara, K., M. Nakamura, T. Miyase, M. Kuroyanagi, and A. Ueno. 1996. Studies on differentiation inducers. VI. Lignan derivatives from Arctium fructus. (2). Chem. Pharm. Bull. (Tokyo) 44(12): 2300–4.

van Uum, S.H. 2005. Liquorice and hypertension. Neth. J. Med. 63(4): 119–20.

Varalakshmi, C., A.M. Ali, B.V. Pardhasaradhi, R.M. Srivastava, S. Singh, S. and A. Khar. 2008. Immunomodulatory effects of curcumin: *in-vivo*. Int. Immunopharmacol. 8(5): 688–700.

Villegas, I., S. Sanchez-Fidalgo, and C. Alarcon de la Lastra. 2008. New mechanisms and therapeutic potential of curcumin for colorectal cancer. Mol. Nutr. Food Res. 52(9): 1040–61.

Visnjic, D., D. Batinic, and H. Banfic. 1997. Arachidonic acid mediates interferon-gamma-induced sphingomyelin hydrolysis and monocytic marker expression in HL-60 cell line. Blood 89(1): 81–91.

Walker, A.F. 2006. Herbal medicine: the science of the art. Proc. Nutr. Soc. 65(2): 145–52.

Wang, J.H., J. W.Shin, J.Y. Son, J.H. Cho, and C.G. Son. 2010. Antifibrotic effects of CGX, a traditional herbal formula, and its mechanisms in rats. J. Ethnopharmacol. 127(2): 534–42

Wasik, J. 1999. The truth about herbal supplements. Consumer's Digest. (July/August 1999) pp. 75–76, 78–79.

Watanabe, T., T. Yamamoto, M. Yoshida, K. Fujiwara, N. Kageyama-Yahara, H. Kuramoto, Y. Shimada, and M. Kadowaki. 2009. The traditional herbal medicine saireito exerts its inhibitory effect on murine oxazolone-induced colitis via the induction of Th1-polarized immune responses in the mucosal immune system of the colon. Int. Arch. Allergy Immunol. 151(2): 98–106.

Wong, C.K., K.N. Leung, K.P. Fung, and Y. M. Choy. 1994. Immunomodulatory and anti-tumour polysaccharides from medicinal plants. J. Int. Med. Res. 22(6): 299–312.

Xiang, Y. Z., H.C. Shang, X.M. Gao, and B.L. Zhang. 2008. A comparison of the ancient use of ginseng in traditional Chinese medicine with modern pharmacological experiments and clinical trials. Phytother Res. 22(7): 851–8.

Xu, Q., J. Lu, R. Wang, F. Wu, J. Cao, and X. Chen. 1997. Liver injury model induced in mice by a cellular immunologic mechanism—study for use in immunopharmacological evaluations. Pharmacol Res. 35(4): 273–8.

Yadav, V. S., K. P.Mishra, D.P. Singh, S. Mehrotra, and V.K. Singh. 2005. Immunomodulatory effects of curcumin. Immunopharmacol. Immunotoxicol. 27(3): 485–97.

Yang, C.S., S.R. Ko, B.G. Cho, D.M. Shin, J.M. Yuk, S. Li, J.M. Kim, R.M. Evans, J.S., Jung, D.K. Song, and E. K. Jo. 2008. The ginsenoside metabolite compound K, a novel agonist of glucocorticoid receptor, induces tolerance to endotoxin-induced lethal shock. J. Cell Mol. Med. 12(5A): 1739–53.

Yang, G.Z. and T.H. Ma. 1988. Immunomodulatory effect of ginsenoside on cell-mediated immunity with operative stress in the mouse. Zhong Xi Yi Jie He Za Zhi 8(8): 479–80, 454.

Yang, H., J.A. Zonder, and Q.P. Dou. 2009. Clinical development of novel proteasome inhibitors for cancer treatment. Expert Opin. Investig. Drugs 18(7): 957–71.

Yu, J.L., D.Q. Dou, X.H. Chen, H.Z. Yang, N. Guo, and G. F. Cheng. 2005a. Protopanaxatriol-type ginsenosides differentially modulate type 1 and type 2 cytokines production from murine splenocytes. Planta Med. 71(3): 202–7.

Yu, J.L., D.Q. Dou, X.H. Chen, H.Z.Yang, X.Y., Hu, and G. F. Cheng. 2005b. Ginsenoside-Ro enhances cell proliferation and modulates Th1/Th2 cytokines production in murine splenocytes. Yao Xue Xue Bao 40(4): 332–6.

Zhang, X. 2002. WHO Traditional Medicine Strategy 2002–2005. WHO, Geneva.

Zhang, X.F., H.M. Wang, Y.L. Song, L.H. Nie, L.F. Wang, B. Liu, P.P. Shen, and Y. Liu. 2006. Isolation, structure elucidation, antioxidative and immunomodulatory properties of two novel dihydrocoumarins from *Aloe vera*. Bioorg Med. Chem. Lett. 16(4): 949–53.

Zhang, Y.H., K. Isobe, F. Nagase, T. Lwin, M. Kato, M. Hamaguchi, T. Yokochi, and I. Nakashima. 1993. Glycyrrhizin as a promoter of the late signal transduction for interleukin-2 production by splenic lymphocytes. Immunology 79(4): 528–34.

Zjawiony, J.K. 2004. Biologically active compounds from Aphyllophorales (polypore) fungi. J. Nat. Prod. 67(2): 300–10.

Chapter 8

Traditional Knowledge about Indian Antimicrobial Herbs: Retrospects and Prospects

Deepak Acharya[1] and *Mahendra Rai[2]*

Introduction

Ever since the birth of humanity, a terrible battle has been raging between man and his various ailments. The modern age has seen the development of different methods for the diagnosis, prevention and cure for many of these diseases. But some dreadful diseases like cancer and AIDS still remain elusive and continue to plague mankind. Recently, an awakening has generated among people regarding herbs and natural medicines, which has more to offer in the battle against diseases like cancer. Plants had been used for medicinal purposes long before recorded history. Recently, the World Health Organization estimated that 80% of the people worldwide rely on herbal medicines for some aspect of their primary healthcare. For most herbs, the specific ingredient that causes the therapeutic effect is not known. Whole herbs contain many ingredients, and it is likely that they work together to produce the desired medicinal effect. Herbalists treat many conditions such as asthma, eczema, premenstrual syndrome, rheumatoid arthritis, migraine, menopausal symptoms, chronic fatigue, irritable bowel syndrome etc. New medicines have been discovered with traditional, empirical and molecular approaches.

[1]Abhumka Herbal Private Limited, 502, 5th Floor, Shreeji Chambers, B/h Ford Cargo CG Road, Ahmedabad- 380006, Gujarat.
E-mail: *deep_acharya@rediffmail.com*
[2]Department of Biotechnology, SGB Amravati University, Amravati-444 602; Maharashtra, India.
E-mail: *mkrai123 @rediffmail.com*

Traditional medicine is a priceless heritage which was created in the historical course of prevention and treatment of diseases over a long period. Today, traditional systems of medicine, which utilize mostly plant-derived prescriptions, remain the source of primary health care for more than 3/4ths of the Third World population. It is estimated that a third of all world pharmaceuticals are of plant origin, or if algae, fungi and bacteria are included, then two thirds of all pharmaceuticals are plant based. Therefore, traditional medicine and medicinal plants are indispensable in practice. The rich traditional ethnopharmacopoeia of the Third World's tropical flora is, indeed, indicative of the high utility of indigenous medicinal plants.

Plants are indispensable to man for his livelihood. Traditional medicine (TM) occupies a central place among rural communities of developing countries for the provision of health care in the absence of an efficient primary health care system (World Health Organisation 1995; Sheldon et al. 1997; Teh 1998; Tabuti et al. 2003). The existence of TM depends on plant species diversity and related knowledge of their use as herbal medicines (Sheldon et al. 1997; Svarstad and Dhillion 2000 and Laird 2002). In addition, both plant species and traditional knowledge (TK) are important to the herbal medicine trade and the pharmaceutical industry, whereby plants provide raw materials, and TK the prerequisite information (Johns et al. 1990; Sheldon et al. 1997; Dhillion and Ampornpan 2000; Carlson et al. 2001; Dhillion et al. 2002 and Laird 2002). Unfortunately both plant species and TK are threatened in various ways. Medicinal plants species or their populations are threatened by habitat modification and unsustainable rates of exploitation (World Bank 1992; Sheldon et al. 1997 and Dhillion and Ampornpan 2000), while TK is threatened by loss of plant diversity (Farooque and Saxena, 1996 and Tabuti et al. 2003), urbanisation, modernisation and low income of traditional medicine practitioners (Tsey 1997; Ugent 2000 and Tabuti et al. 2003). Against this background, it is important that immediate steps are taken to protect both plant species diversity and associated TK.

It is been more than 50 years since the first clinical trial of penicillin took place in early 1941. Around 10000 natural antibiotics have been isolated and identified and at least 100000 analogues have seen synthesized till now. The last 20 years have witnessed a new era of science with the growth of some inter-disciplinary branches of botany—mainly termed as ethnobotany, ecological biochemistry, ethnopharmacology, ethnopharmacognosy. Interestingly the last 20 years were crucial for both public and private participatory sectors as both of them were deeply engaged in searching natural products to cure a range of health ailments. Plants were the major concern of that research. Modern science has now started accepting the role of herbs in curing various health ailments. Many plant origin products are being marketed as functional food, a food that serves as medicine. Looking at the global aspect herbal medicines and the public concern about it, modern science has been trying to invent certain tools that can help in isolating the marker chemical compounds of plant species.

Plants are the integral part of rural and tribal life style (Chitme et al. 2003). Almost 25% of prescribed drugs in the world are basically plant derived. There are as many as 119 secondary metabolites isolated or extracted from 90 different species of higher plants which are being used as prescribed drugs, whereas, more than 250000 species of higher plants are yet to be studied for obtaining any compound that can be used as a drug. It takes billions of dollars and more than 15 years for a new drug development programme. Industries are more reluctant to put in their investment in such a long lasting or never ending approach of finding the appropriate molecule or chemical compound as the results of many plants or their compounds are negative at the end of the drug development research. Traditional knowledge about herbs can serve a short route in such sort of programmes. Potential herbal formulations retrieved or documented from local herbal healers can be taken into consideration and proper validation should be proposed with a scientific explanation. Literally thousands of phytochemicals with inhibitory effects on microorganisms have been found to be active *in vitro*. One may argue that these compounds have not been tested *in vivo* and therefore activity can not be claimed but one must take in to consideration that many of these plants have been used for centuries by various cultures in the treatment of disease.

Tribals in various remote regions of India still rely on their traditional herbal knowledge to combat microbial infections. Various researchers from modern science have already proved the antimicrobial potential of many herbs used by the local tribesmen in India. Different herbs which they apply against microbial infections are well examined and researched for their microbicidal activities (Upadhyay et al. 1998; Rai et al. 1999; Shoba and Thomas 2001; Ramesh et al. 2004; Jeevan Ram et al. 2004; Greeshma et al. 2006; Parekh and Chanda 2007a; Pritima and Pandian 2007; Swain et al. 2008 and Chitravadivu et al. 2009).

History and status of antimicrobial study of traditional Indian herbs

There are a number of investigators who contributed a lot on antimicrobial activity of medicinal plants (Jain and Agrawal 1976; Misra and Dixit 1980; Tripathi and Dixit 1981; Deshmukh et al. 1986; Rai 1987, 1988, 1989, Rai and Vasanth 1995; Nakhare and Garg 1996; Rai et al. 1999; Jeevan Ram et al. 2004; Greeshma et al. 2006; Parekh and Chanda 2007b and Pritima and Pandian 2007). Probably, Acharya and Chaterjee (1974) were the first in India who studied the activity of chrysophanic acid-9 anthron against *Trichophyton rubum, T. mentagrophytes, M. canis, M. gypseum* and *Geotrichum candidum.* Ray and Majumdar (1975, 1976a and 1976b) isolated active principles from rhizome of *Alpina officinarum* and *Saussurea lappa,* which exhibited strong antifungal action against dermatophytes. Eugenol acetate, geranyl acetate and methyl heptanone (the major component of *Ocimum* and essential oil of *archangelica*) showed inhibitory nature to *Microsporum gypseum* complex (Goutam et al. 1980 and Jain et al. 1980). Screening of Indian plants for a wide range of activity (antimalarial antiprotozoic, antiviral, antiheminthic, anticancer, antifungal, etc.) have been carried out by various investigators (Dhar et al. 1968; Bhakuni et al. 1969,1971; Dhar et al. 1973, 1974; Dhawan et al. 1977, 1980 and Rai and Acharya 2000). However, Bhakuni et al. (1971) reported that fungitoxic activity was observed only in 3-extracts out of 300 plants which prove that a more thorough investigation is required for the search of antifungal activity.

The investigators of Central Institute of Medicinal and Aromatic Plants, Lucknow, have screened more than 3500 plants for their biological activity (Dhar et al. 1968, 1973, 1974; Atal et al. 1978; Aswal et al. 1984a, 1984b, 1996; Abraham et al. 1986; Bhakuni et al. 1988, 1990 and Goel et al. 2002) and have reported many important findings.

Current status

Universities and the private sector are now deeply engaged in isolating marker compounds from herbs which have been used by the local folk for treatment of microbial infections. Many commercially proven drugs used in modern medicine were initially used in a crude form in traditional or folk healing practices, or for other purposes that suggested potentially useful biological activity. These herbal medicines are relatively cheap, safe and eco-friendly and offer more profound therapeutic benefits alongwith the affordable price. There are many potential herbs which have shown remarkable potential against microorganisms.

In India, many researchers have studied herbs and herbal products for the cure of infectious diseases (Mukherjee et al. 1998; Bagchi et al. 1999; Kumar et al. 2001; Mazumder et al. 2001; Shoba and Thomas 2001; Gnanamani et al. 2003; Shirwaikar et al. 2003; Chamundeeswari et al. 2004; Mazumder et al. 2004; Singh et al. 2005a; Nair and Chanda 2005; Duraipandiyan et al. 2006; Parekh and Chanda 2007a, 2007b; Dabur et al. 2007; Kumaraswamy et al. 2008; Srikrishna et al. 2008 and Ayyanar and Ignacimuthu 2009). Ahmed et al. (1998) reported that out of 82 Indian medicinal plants, 56 demonstrated antibacterial activity against one or more of the tested pathogens. Further, Ahmed and Beg (2001) studied antimicrobial activity of those plants against multi-drug resistant human pathogens. Perumal Samy et al. (1998) reported 34 plant species belonging to 18 different families for antibacterial activity against *Escherichia coli, Klebsiella aerogenes, Proteus vulgaris,* and *Pseudomonas aerogenes* (Gram-negative bacteria) and they reported 16 plants showing significant antibacterial activity against the tested bacteria. Importantly, they had selected the plants on the basis of folklore medicinal reports practised by the tribal people of the Western Ghats, India. Aqueous extracts of 50 Indian medicinal plants were tested for their antimicrobial activity against 14 microbial pathogens by Srinivasan et al. (2001). Among 50 plants 72% showed antimicrobial activity. Only nine plant extracts were found to have an antifungal activity. The authors concluded that plants demonstrating a broad spectrum of activity, may help to discover a new chemical class of antibiotics that could serve as selective agents for maintenance of animal or human health and provide biochemical tools for the study of infectious diseases (Srinivasan et al. 2001). Valsaraj and co-workers (1997) studied 78 Indian traditional medicinal plants based on their use in the treatment of infections. They had screened ethanol extracts of plants against four bacteria and two fungi to test their antimicrobial activity. They concluded that more plant extracts were active against bacteria than against fungi and the antimicrobial activity against Gram-positive bacteria was more pronounced than against Gram-negative bacteria.

Ethnomedical knowledge of plants used by the Kunabi Tribe of Karnataka in India was studied by Harsha et al. (2002). Medical ethnobotany of the tribals of Sonaghati of Sonbhadra district, Uttar Pradesh, India was studied by Singh et al. (2002). With a view to develop protective clothing from inherent microbial activity Singh et al. (2005a) studied the antimicrobial activity of some plant derived dyes. They tested four natural dyes against common pathogenic bacteria and found that textile materials impregnated with these natural dyes showed 10–25% reduction in bacterial growth (Singh et al. 2005a). Antimicrobial activity of butanolic extracts of six terrestrial herbs and seaweeds was tested against shrimp pathogen *Vibrio parahaemolyticus* by Immanuel et al. (2004). Antibacterial and antifungal activities of *Trewia polycarpa* roots were studied by Chamundeeswari et al. (2004). Ramesh et al. (2002) investigated antimicrobial activity of phytochemicals of various extracts of the leaves of *Begonia malabarica* and isolated and identified six known compounds. The aqueous and organic solvent extracts of this plant were tested against ten human pathogenic bacteria and four fungi. They concluded that all extracts of leaves were devoid of antifungal activity against the tested fungi. Some of the extracts were found effective against pathogenic bacteria causing respiratory infections, diarrhoea and skin lesions. According to Biswas and Mukherjee (2003) and Kumar et al. (2007), as many as 163 plant species belonging to various families have been used as wound healing agents in various systems of medicines in India. Many plant species such as *Aloe vera, Azadirachta indica, Berberis aristata, Carica papaya, Celosia argentea, Centella asiatica, Cinnamomum zeylanicum, Curcuma longa, Cynodon dactylon, Euphorbia nerifolia, Ficus bengalensis, Ficus racemosa, Glycyrrhiza glabra, Nelumbo nucifera, Ocimum sanctum, Phyllanthus emblica, Plumbago zeylanica, Pterocarpus santalinus, Rubia cordifolia, Symplocos racemosa, Terminalia arjuna* and *Terminalia chebula* are commonly used by Indian tribes for wound healing purposes (Biswas and Mukherjee 2003 and Kumar et al. 2007).

Therapeutic role of antimicrobials

Around one-half of all deaths in tropical countries are results of infectious disorders (Iwu et al. 1999). A great deal of research has been carried out worldwide to change the scenario of such devastating death data. Scientists are looking forward to seek out result oriented products from plants, as plants have been a wonderful source of antibiotics.

The majority of Indian plants have shown enormous activity against various microbes. These mainly include: *Aloe vera, Malus sylvestris, Aegle marmelos, Ocimum basilicum, Piper nigrum, Carum carvi, Matricaria chamomilla, Capsicum annuum, Allium sativum, Camellia sinensis, Allium cepa, Citrus sinensis, Carica papaya, Artemisia dracunculus, Curcuma longa*, etc. Continued and further exploration of plant antimicrobials has to be carried out. On one side, a number of plants has shown potential results against many diseases while, on the other hand, many have proven effective in mitigating side effects often associated with synthetic antimicrobials. It shows that phytomedicines may have multiple effects on the body (Iwu et al. 1999). Their actions often act beyond the symptomatic treatment of disease. Likewise, *Hydrastis canadensis* is not only an antimicrobial plant, but also increases blood supply to the spleen promoting optimal activity of the spleen to release mediating compounds (Murray 1995).

Important Indian ethnomedicinal herbs with antimicrobial potential

Based on the indigenous knowledge of Indian tribesmen, there are as many as 50 medicinal plants most commonly used by them for treatment of different microbial infections (Acharya and Shrivastava 2008). The authors herewith provide brief information about these plants and also a short profile along with the vernacular names. Plant names are arranged alphabetically.

1. *Allium sativum* L. (Family: Liliaceae)

Vernacular Names: Lashan, Lashun, Rasun (Bengali); Garlic (English); Lasan (Gujarati); Lahsan, Lahsun (Hindi); Belluli (Kannada); Velluli, Velutha Ulli (Malayalam); Lasunas (Marathi); Lashuna (Sanskrit); Vallai-pundu (Tamil); Velluli Tella-gadda (Telugu).

Plant Profile and Distribution: Small, perennial herbs; bulbs of many united cloves, with strong odour, white to yellowish, covered with white, thin, peeling off flakes; leaves long, linear, flat, acute, fistular, sheathing the stem; scapes slender, smooth; flowers white, enclosed in long, beaked spathes. The plant is commonly cultivated for the edible bulbs throughout in India.

Antimicrobial studies: Abdou et al. 1972; Misra and Dixit 1977; Dankert et al. 1979; Davis et al. 1990; Rees et al. 1993; Pai and Platt 1995; Khan and Katiyar 2000; Samuel et al. 2000; Harris et al. 2001; Benkeblia 2004 and Satya et al. 2005.

2. *Anogeissus latifolia* (Roxb. ex DC.) Wall. ex Guill. & Perr. (Family: Combretaceae)

Vernacular Names: Dhavdo (Gujarati); Dhaura, Dhawa (Hindi); Bejjalu, Dinduga (Kannada); Marukanjiram (Malayalam); Dabria, Dandua, Dhaura, Dhavda (Marathi); Dohu (Oriya); Vekay Naga, Vekkali, Vella Nava (Tamil); Chiri-manu, Tiruman, Yella Maddi (Telugu).

Plant Profile and Distribution: Moderate-sized deciduous trees, with smooth, white-grey bark; leaves opposite, obtuse, coriaceous, silky tomentose when young; flowers pinkish-yellow, in racemose heads; bracts leafy; fruits orbicular, glabrous, with entire wings. The tree is very common in the deciduous forests of western and central India, particularly in mixed dry deciduous forests.

Antimicrobial studies: Chaturvedi et al. 1987; Raghavan et al. 2004; Govindarajan et al. 2006; Begum et al. 2007; Mohammad et al. 2007 and Mann et al. 2009.

3. *Argemone mexicana* L. (Family: Papaveraceae)

Vernacular Names: Shialkanta (Bengali); Mexican Poppy, Prickly Poppy (English); Darudi (Gujarati); Bharbhand, Brahmadundi, Satyanashi, Peeli Kateri (Hindi); Datturigidda (Kannada); Bhrahmadanti (Malayalam); Pilva Dhotra (Marathi); Kantakusham (Oriya); Bhatkateya (Punjabi); Bhrahmadandi (Sanskrit); Kurukkum (Tamil); Brahmadandi (Telugu).

Plant Profile and Distribution: Erect, prickly, divaricately branched annual herbs, with yellow latex; leaves sessile, pinnatified; segments spiny along margins and veins, glaucous-green; flowers solitary, terminal or axillary, bright yellow, caducous, sparsely prickly outside; capsules prickly; seeds numerous, pitted, blackish-brown, occurs as a weed in cultivated fields and open wastelands.

Antimicrobial studies: Chang et al. 2003; Sangameswaran et al. 2004; Bhattacharjee et al. 2006; Uma Reddy et al. 2008 and Mashiar et al. 2009.

4. *Argyreia nervosa* (Burm. f.) Bojer (Family: Convolvulaceae)

Vernacular Names: Bichtarak (Bengali); Elephant Creeper, Woolly Morning-glory (English); Samudrasoka (Gujarati, Marathi); Samundar-ka-pat (Hindi); Chandhrapada (Kannada, Telugu); Bryddhotareko (Oriya); Samudrapalaka (Sanskrit); Samunddirapacchai (Tamil).

Plant Profile and Distribution: Large, white-tomentose, twining shrubs, with milky sap; leaves orbicular to ovate-cordate, glabrous above, silvery-white tomentose beneath; flowers rose-purple or white, in subcapitate cymes. Occasionally found in the forests, often planted for ornamental purposes in homes, gardens and pilgrimage sites as wall and gate climbers.

Antimicrobial studies: Batra and Mehta 1986; Shukla et al. 1999 and Habhu et al. 2009.

5. *Artocarpus heterophyllus* Lam. (Family: Moraceae)

Vernacular Names: Kathal (Assamese); Kanthal (Bengali); Jack Fruit (English); Phanas (Gujarati, Marathi); Kanthal, Kathal, Panasa (Hindi); Halasu, Hebhalasu (Kannada); Chakka, Pilavu (Malayalam); Ichodopholo, Kantokalo, Ponoso (Oriya); Ashaya, Atibrihatphala, Panasa, Phanasa (Sanskrit); Murasabalam, Pala, Pila, Pila Palam (Tamil); Panasa, Verupanasa (Telugu).

Plant Profile and Distribution: Large, evergreen trees, with exuding sticky milky latex; bark rough, greenish-black; leaves broadly obovate-elliptic, decurrent, glabrous, glossy, leathery; flowers monoecious, borne on trunk; syncarps oblong, greenish-yellow, covered with numerous hard knobs; flesh custard yellow, fragrant. It is cultivated for edible fruits throughout India.

Antimicrobial studies: Siddiqui et al. 1992; Sato et al. 1996 and Khan et al. 2003.

6. *Azadirachta indica* A. Juss. (Family: Meliaceae)

Vernacular Names: Nim (Bengali); Indian Lilac, Margosa Tree, Neem Tree (English); Limbado (Gujarati); Nim, Nimb (Hindi); Bevinamara (Kannada); Veppa (Malayalam); Limba (Marathi); Nimba (Oriya); Arishta, Nimba (Sanskrit); Vembu, Veppam (Tamil); Veepachettu, Yapachettu (Telugu).

Plant Profile and Distribution: Large trees with peeling-off bark; leaves imparipinnate compound; leaflets sub-opposite or alternate, oblique at base, coarsely crenate-serrate; flowers white, fragrant, in panicles; drupes ovoid-oblong, yellow after ripening, 1-seeded. Very common tree species of social forestry, commonly planted along road-sides and gardens as a shade tree; also grown near sacred places and habitations.

Antimicrobial studies: Aderounmu et al. 2003; Alzoreky and Nakahara 2003; Chea et al. 2007; Nautiyal et al. 2007; Pyo et al. 2007 and Ramos et al. 2007.

7. *Blumea mollis* (D. Don) Merrill (Family: Asteraceae)

Vernacular Names: Kukursunga (Bengali); Blumea (English); Kolhar, Pilo Kapurio (Gujarati); Jangli Muli, Kakronda (Hindi); Gabbusoppu (Kannada); Bhamurda, Burando (Marathi); Kukundara, Mridu chhada (Sanskrit); Kattumullangi, Narakkarandai (Tamil); Kukka Pogaaku (Telugu).

Plant Profile and Distribution: Erect, aromatic, densely viscid, annual herbs or undershrubs, leaves ovate-oblong, serrate; heads purple to lilac, globose, in compact panicles; involucral bracts reflexed; achenes oblong, pubescent, brown, with white pappus. Commonly found as a weed in wastelands and agricultural fields, more frequent in the Gangetic plain and peninsular India.

Antimicrobial studies: Rai and Acharya 1999, 2000, 2002; Rai et al. 2002, 2003 and Gautam et al. 2003.

8. *Bryonopsis laciniosa* (L.) Naud. (Family: Asteraceae)

Vernacular Names: Lollipop Climber (English); Bilanja, Garunaru, Shivlingi (Hindi); Kavdoli (Marathi); Baja, Citraphalah, Lingini, Sivalingi (Sanskrit); Aiveli, Aiviral Kovai, Aivirali, Iviralikovai, Ivoralikovai (Tamil).

Plant Profile and Distribution: More or less scabrous, climbing annuals; leaves deeply palmately 5-lobed, denticulate to sub-crenulate; tendrils 2-fid; male flowers greenish-yellow, pedunculate; female ones fasciculate; fruits spherical, yellowish-green, narrowly six-striped; seeds belted, acute, with raised projections on both faces, grey. Occasionally found climbing on bushes and shrubs on the outskirts of the tropical forests.

Antimicrobial studies: Cowan 1999; Mosaddik et al. 2003; Sivakumar et al. 2004; Khan et al. 2008 and Bonyadi et al. 2009.

9. *Butea monosperma* (Lam.) Taub. (Family: Fabaceae)

Vernacular Names: Palas (Bengali, Malayalam, Marathi); Bastard Teak, Bengal Kino Tree, Flame of the Forest (English); Khakharo (Gujarati); Dhak, Palas, Palash (Hindi); Muttuga (Kannada); Porasu (Oriya); Chichra, Dhak, Palas (Punjabi); Kesuda, Khamkra (Rajasthani); Palasa (Sanskrit); Parasa, Pilasu (Tamil); Mooduga, Palasamu (Telugu).

Plant Profile and Distribution: Medium-sized, deciduous trees, with longitudinally fissured bark. Leaves trifoliolate, densely tomentose beneath; flowers orange-red; pods flat, tomentose, 1-seeded. Very common tree of tropical and subtropical forests almost forms a pure stand or different associations with other dominant tree species.

Antimicrobial studies: Mehta et al. 1983a; Porwal et al. 1988; Bandara et al. 1989; Zafar et al. 1989 and Gurav et al. 2008.

10. *Calendula officinalis* L. (Family: Asteraceae)

Vernacular Names: Calendula, English Garden Marigold, Pot-marigold (English); Zergul (Hindi); Akbelulmulk, Saladbargh (Punjabi); Thulukka Saamanthi (Tamil).

Plant Profile and Distribution: Erect, aromatic, glandular hairy, annual herbs, with angular stems; upper leaves lanceolate; flowers of variable colours, yellow to orange, in terminal heads; achenes boat-shaped, faintly ribbed. The plant is commonly grown in gardens and homes as an ornamental plant throughout India. Occasionally it is found in wastelands.

Antimicrobial studies: Araki and Abe 1980; Ognean et al. 1994; Kalvatchev et al. 1997; Kasiram et al. 2000; Aghili et al. 2001; Mulabadi et al. 2005; Ketzis and Laux 2006; Kulataeva et al. 2006 and Oliveira et al. 2007.

11. *Calotropis procera* (Ait.) Ait. f. (Family: Asclepiadaceae)

Vernacular Names: Dead Sea Apple, Milkweed, Swallow-wort (English); Akado, Nano Akado (Gujarati); Ak, Akada (Hindi); Orka (Marathi, Oriya); Ak, Madar (Punjabi); Aak, Akaro (Rajasthani); Alkarka (Sanskrit); Vellerukku (Tamil); Chinna Jilleedu (Telugu).

Plant Profile and Distribution: Erect, pubescent shrubs, with milky latex; leaves opposite, decussate; flowers in axillary and terminal, corymbose cymes, purple; follicles in pairs, recurved, many-seeded; seeds flat, pale brown, with long, silky coma. The plant is a very common weed of wastelands, road-side fallow lands and scrub forests of arid and semi-arid regions.

Antimicrobial studies: Adoum et al. 1997; Kishore et al. 1997; Larhsini et al. 1997; 1999; Ali et al. 2001 and Tahir and Chi 2002.

12. *Cardiospermum helicacabum* L. (Family: Sapindaceae)

Vernacular Names: Kopalphuta (Assamese); Lataphatkari, Nayaphatki, Sibjhul (Bengali); Balloon Vine, Blister Creeper, Winter Cherry (English); Kagdoliyo, Karolio (Gujarati); Kanphuti (Hindi); Agniballe, Bekkinabuddigida, Kakaralata (Kannada); Jyotishmati, Katabhi Paluruvan, Uzhinja (Malayalam); Kanphuti, Kapala-phodi (Marathi); Fatkari (Punjabi); Kapal-phori Latafatkari (Rajasthani); Jyotishmati, Karnasphota, Paravatanghi, Sakralata (Sanskrit); Moedakottan, Samuttiram (Tamil); Buddakaakaraeega, Tapaakaayateega (Telugu).

Plant Profile and Distribution: Annual climbing herbs; leaves biternate, alternate; flowers white, in extra-axillary, umbellate peduncles, with two circinate tendrils at the apex and the third branch subtending the cyme; capsules shortly stalked, winged; seeds globose, black, smooth, arillate. Frequenly grows in peripheral zones of tropical forests and hedges of cultivated fields.

Antimicrobial studies: Ali et al. 1995; Parekh et al. 2005 and Banso 2007.

13. *Carthamus tinctorius* L. (Family: Asteraceae)

Vernacular Names: Kusukphal, Kusum (Bengali); Bastard Saffron, False Saffron, Safflower (English); Kasumbo (Gujarati); Karrah, Kusum (Hindi); Kusumba, Kusume (Kannada); Kardai, Kurdi (Marathi); Kusumba (Punjabi); Kusumbha (Sanskrit); Kusumba, Sethurangam (Tamil); Kusumbalu (Telugu).

Plant Profile and Distribution: Much-branched, annual herbs; leaves lanceolate, entire or spinulose-serrate; flowers variable in colour, orange-red to yellow or white, in globular heads; achenes 4-angled; pappus absent. The plant is commonly cultivated as an oilseed crop throughout India, also planted in homes and gardens for ornamental purposes.

Antimicrobial studies: Ray and Majumdar 1975; Sedigheh et al. 2000.

14. *Cassia angustifolia* Vahl (Family: Caesalpiniaceae)

Vernacular Names: Sannamakki, Sonpat (Bengali); Alexandrian Senna, Tinnevelly Senna (English); Middiawal, Senamakki (Gujarati); Bhuikhakhasa, Hindisana (Hindi); Nelavarike, Soonamukhi (Kannada); Nilavaka, Sunnamukhi (Malayalam); Bhuitarvada, Shonamakhi (Marathi); Shonamukhi (Oriya); Senna (Rajasthani); Bhumiari, Pitapushpi, Swarnamukhi, Swarnapatrika (Sanskrit); Nattunelavarai, Nelavagai, Sooratnilla Avarai (Tamil); Neelaponna, Neelatangeedu (Telugu).

Plant Profile and Distribution: Erect, much-branched shrubs; leaves imparipinnate; leaflets bluish green; flowers bright yellow, in terminal racemes; pods thin, flat, pubescent, dark brown to black after maturity; seeds dark brown. An introduced species, mostly cultivated and also naturalized in some parts of India.

Antimicrobial studies: Müller et al. 1989; Sydiskis et al. 1996; Goswami. and Reddi 2004 and Srivastava et al. 2006.

15. *Cassia fistula* L. (Family: Caesalpiniaceae)

Vernacular Names: Honalu, Sonaru (Assamese); Amultas, Bandarlati, Sundali (Bengali); Golden Shower, Purging Cassia (English); Garmalo (Gujarati); Amaltas, Bandarlauri, Jhagda (Hindi); Kakkemara, Rajataru (Kannada); Kanikonna, Kritamalam, Svarnaviram (Malayalam); Bahava, Boya, Chimkani (Marathi); Soturongulo, Sunari (Oriya); Alash, Ali, Kaniar, Karangal (Punjabi); Aragwadha, Kritamala, Svannavriksha (Sanskrit); Arakkuvadam, Konnei, Sarakkondai (Tamil); Aragvadhamu, Kolaponna, Rellachettu (Telugu).

Plant Profile and Distribution: Medium-sized, deciduous trees with rough and dark brown bark; leaflets 3–8 pairs; flowers yellow, in axillary, pendulous, racemes; pods cylindric, become dark brownish-black after ripening; seeds flat, glossy brown. Very common element of tropical forests, sometimes form pure patches on hill-slopes and in valleys, also planted in gardens and along roadsides.

Antimicrobial studies: Samy et al. 1998; Sharma 1998; Kavitha et al. 2001; Kumar et al. 2006; Mahida and Mohan 2006; Balasubramanian et al. 2007

16. *Cassia tora* L. (Family: Caesalpiniaceae)

Vernacular Names: Chakunda, Panevar (Bengali); Foetid Cassia, Sickle Senna, Wild Senna (English); Kawario, Konariya (Gujarati); Chakavat, Chakunda, Panevar, Chirota, Titi (Hindi); Gandutogache, Taragasi (Kannada); Chakramandarakam, Takara (Malayalam); Takala, Tankli, Tarota (Marathi); Chakunda (Oriya); Chakunda, Panwar, Pawas (Punjabi); Chakuada, Panwar, Pumaria (Rajasthani); Chakramarda, Dadamari, Prishnaparni (Sanskrit); Senavu, Tagarai, Vindu (Tamil); Chinnakasinda, Tellakasinda (Telugu).

Plant Profile and Distribution: Annual herbs or undershrubs; leaflets 3 pairs, with linear-cylindric glands; flowers yellow, axillary, solitary or in 1 to 3-flowered racemes; with 7 fertile stamens; pods linear-cylindric, beaked, 15 to 25-seeded; seeds rhomboidal, glossy, dark coloured. Common weed of roadsides and wastelands, also found in degraded forest areas forming ground vegetation.

Antimicrobial studies: Kim et al. 2004 and Sinha et al. 2004.

17. *Cichorium intybus* L. (Family: Asteraceae)

Vernacular Names: Chicory, Wild Succory (English); Kasni (Hindi); Cikkari (Malayalam); Kachani (Marathi); Hinduba, Kasani (Sanskrit); Cikkari Kacinivittu Panaviyakkirai (Tamil); Kasini, Kasini-vittulu (Telugu); Nim Kofta, Tukhm Kasni.

Plant Profile and Distribution: Deep-rooted, annual or perennial herbs; stems terete, with milky sap; radical leaves short petioled, lyrate-pinnatifid; upper ones sessile, entire or dentate; heads blue, homogamous, sessile, clustered in axillary and terminal panicles; achenes turbinate, 5-angular, with 2 to 3-seriate pappus. Commonly found as weeds in agricultural fields and wastelands, often cultivated as a fodder crop.

Antimicrobial studies: Jain et al. 1987; Hatano et al. 1999; Kavitha et al. 2001; Kim et al. 2004 and Sinha et al. 2004.

18. *Coleus forskohlii* Briq. (Family: Lamiaceae)

Vernacular Names: Patherchur (Bengali); Coleus, Country Borage, Indian Borage (English); Garmalu (Gujarati); Patharchur (Hindi); Makandiberu (Kannada); Pathurchur (Marathi); Pashan Bhedi (Sanskrit); Karpuravalli (Tamil).

Plant Profile and Distribution: Perennial, aromatic herbs, with thick root-stalk, distributed in subtropical Himalayas of Kumaon and Nepal ascending to 8, 000 ft. and in the Deccan peninsula, Gujarat and Bihar. It is common on dry, barren hills, and is cultivated in Bombay and other regions for the roots which are pickled and eaten. This species is considered to be the wild ancestor of all the tuber varieties known as Kaffir Potatoes.

Antimicrobial studies: Collier and van de Pijl 1949; Gupta et al. 1993; Sasidharan 1997 and Nilani et al. 2006.

19. *Commiphora wightii* (Arn.) Bhandari (Family: Burseraceae)

Vernacular Names: Guggul (Bengali, Gujarati, Hindi); Indian Bdellium Tree (English); Guggule (Kannada, Marathi); Guggulu, Koushikaha, Devadhupa (Sanskrit); Maishakshi Gukkal (Tamil); Guggul (Telugu).

Plant Profile and Distribution: Much-branched, deciduous shrubs, with silvery, papery bark; leaves 1 to 3-foliolate; flowers red or pinkish-white, sessile, solitary or in 2 to 3-flowered fascicles; drupes ovoid, shortly beaked, red when ripe. Rarely found in gravelly and rocky habitats, forming thickets, often cultivated for gum used in medicine industry.

Antimicrobial studies: Zhu et al. 2001 and Celine et al. 2007.

20. *Costus speciosus* (Koenig) Sm. (Family: Zingiberaceae)

Vernacular Names: Tara (Assamese); Kew, Kust (Bengali); Wild or Spiral Ginger, Canereed (English); Parkarmula (Gujarati); Kenaa, Keu, Keokand (Hindi); Changalakoshta, Chikke, Karikutu, Pushkaramoola (Kannada); Anakuva, Channakuva, Koettam (Malayalam); Penva, Pushkaramoola (Marathi); Chittorokudha, Ghauapophora, Kuduo (Oriya); Kemuka (Sanskrit); Koeshtam, Kottam, Kuruvam (Tamil); Bogakhikadumpalu, Bommakachika, Changalvakoshtu (Telugu).

Plant Profile and Distribution: Perennial, erect, succulent herbs; rhizomes clothed with sheaths in lower parts, leafy upwards; leaves spirally arranged, thick, silky beneath; flowers large, white, in cone-like terminal spikes; bracts bright-red; capsules 3-gonous, red, with black seeds. It is commonly cultivated medicinal crop, grows naturally in moist shady places of tropical forests, also planted as ornamental plants.

Antimicrobial studies: Bandara et al. 1988; Singh et al. 1992; Ali et al. 1996; Malabadi 2005 and Malabadi and Kumar 2005.

21. *Cuminum cyminum* L. (Family: Apiaceae)

Vernacular Names: Jira (Bengali); Cumin (English); Jira, Jeera, Zeera (Hindi); Jeerige (Kannada); Jorekam (Malayalam); Jiregire (Marathi); Zira-sufed (Punjabi); Jiraka, Jira (Sanskrit); Siragam (Tamil); Jilakara, Jiraka (Telugu).

Plant Profile and Distribution: Small, much-branched, annual herbs, with striated stem; leaves 2 to 3-partite, linear, bluish-green; flowers white or purplish, in compound umbels; fruits tapering towards both ends, laterally compressed, greyish. Cultivated extensively as spice crop almost throughout India, particularly in northern and western India.

Antimicrobial studies: Iacobellis et al. 2005; Erturk 2006; Parihar and Bohra 2006; Proestos et al. 2006 and Purohit 2006.

22. *Curcuma amada* Roxb. (Family: Zingiberaceae)

Vernacular Names: Amada (Bengali); Mango-Adrak, Amada (English); Am Haldi, Ama Haldi, Amba Haldi (Hindi); Amba Haladi (Marathi); Karpura-haridra (Sanskrit); Mangai Inji (Tamil); Mamidiallam (Telugu).

Plant Profile and Distribution: Small, rhizomatous, annual herbs, with thick, ovoid rootstock and cylindric tubers having characteristic odour of raw mangoes; leaves oblong-lanceolate; flowers pale yellow, in spikes. The plant is commonly cultivated as a vegetable crop, also planted in gardens, occasionally found naturally in hilly tracts of the Western Ghats and southern India.

Antimicrobial studies: Gupta and Banerjee 1972; Venkitraman 1978 and Ghosh et al. 1980.

23. *Curcuma longa* L. (Family: Zingiberaceae)

Vernacular Names: Haldi, Halada (Bengali, Gujarati, Hindi, Marathi); Turmeric (English); (Kannada) Arishina; Haridra (Sanskrit); Manjal (Tamil); Pasupu (Telugu).

Plant Profile and Distribution: Small, rhizomtous, perennial herbs, with short stem; rhizomes short, thick, yellow; leaves large, oblong-lanceolate, acuminate, distinctly nerved, tufted; flowers yellow, funnel-shaped, in long spikes. It is cultivated extensively on a large scale as a vegetable and spice crop thoughout India, particularly in the southern states.

Antimicrobial studies: Chen et al. 2008; Babu et al. 2007; Balasubramanian et al. 2007; Cho et al. 2006 and Jagannath and Radhika 2006.

24. *Elephantopus scaber* L. (Family: Asteraceae)

Vernacular Names: Gojialata, Shamdulum (Bengali); Prickly Leaves Elephant's Foot (English); Bhopathari (Gujarati); Gobhi, Samudulan, Gojivha, Jangli Tambakhu (Hindi); Hakkarike (Kannada); Anashovadi (Malayalam, Tamil); Hastipata, Pathari (Marathi); Gojihva, Karipadam (Sanskrit); Hastikasaka (Telugu).

Plant Profile and Distribution: Rigid, scabrid, perennial herbs; leaves radical, forming a rosette, large, oblong-obovate or elliptic-lanceolate; flowers in terminal, compound, homogamous, violet heads; achenes oblong, truncate, 10-ribbed, pubescent, with 1-seriate, white pappus. Commonly grows as a weed in pasture lands and roadside wastelands throughout the hotter regions of India.

Antimicrobial studies: Chen et al. 1989; Ali et al. 1996; Mohamed et al. 1996; Avani and Neeta 2005 and Vrushabendra et al. 2006.

25. *Embelia ribes* Burm. f. (Family: Myrsinaceae)

Vernacular Names: Biranga, Baibirang (Bengali); False Black Pepper (English); Vyvirang, Vavading (Gujarati); Baberang, Wawrung (Hindi); Vayuvilanga (Kannada, Tamil, Telugu); Vizhal (Malayalam); Kakannie, Vavdinga, Vaivaranag (Marathi); Babrung (Punjabi); Vidanga, Vrishanasana (Sanskrit).

Plant Profile and Distribution: Large, scandent shrubs, with slender branches; leaves elliptic-lanceolate, gland-dotted; fruits globose, wrinkled or warty, of various colours dull red to black, aromatic; seed one, covered with a membrane, embedded in brittle pericarp. Commonly found almost in all parts of India.

Antimicrobial studies: Narang et al. 1961; Rao et al. 1987 and Chitra et al. 2003.

26. *Impatiens balsamina* L. (Family: Balsaminaceae)

Vernacular Names: Dupati (Bengali); Garden Balsam (English); Raktendi, Pantambol (Gujarati); Raktendhi, Tivdi (Hindi); Mecchingom (Malayalam); Terada (Marathi); Haragaura (Oriya); Bantil, Trual, Halu, Talura, Tilphar, Juk (Punjabi); Dushpatrijati (Sanskrit); Kasillumbai (Tamil).

Plant Profile and Distribution: Erect, annual herbs; leaves alternate, serrate on margins; flowers of various colours; spur incurved, filiform; wings deeply notched; capsules dehiscent, tomentose. Commonly planted in gardens and homes for the pretty flowers; also grows in forests as undergrowth and in open grasslands and cultivated fields.

Antimicrobial studies: Lee et al. 2000; Oku and Ishiguro 2000; Thevissen et al. 2005; Nair et al. 2006 and Pandey et al. 2007.

27. *Lawsonia inermis* L. (Family: Lythraceae)

Vernacular Names: Mehedi, Mendi (Bengali), Henna, Egyptian Privet (English); Medi, Mendi (Gujarati); Mehndi (Hindi); Mayilanchi, Gorante (Kannada); Mailanchi, Pontlasi (Malayalam); Mendhi (Marathi); Benjati (Oriya); Mehndi (Punjabi); Mendika, Raktgarbha, Ragangi (Sanskrit); Marithondi, Maruthani (Tamil); Goranti (Telugu).

Plant Profile and Distribution: Large, glabrous shrubs, with terete, often thorny branches; leaves opposite, subsessile, lanceolate or oblanceolate; flowers creamy, fragrant, in terminal, panicled cymes; fruits depressed-globose, reddish-brown, tipped with persistent style; seeds numerous, minute, angular, brown. Commonly planted as hedges close to agricultural fields.

Antimicrobial studies: Ahmed and Beg 2001; Ali et al. 2001; Mouhajir et al. 2001; Sule and Olusanys 2001; Metwali 2003 and Habbal et al. 2005.

28. *Lycopersicon esculentum* Mill. (Family: Solanaceae)

Vernacular Names: Tamatar, Vilayithi Baingan (Bengali, Hindi); Tomato (English); Vilayithi Vengan (Gujarati); Tamataa, Vel Vangi (Marathi); Tamaatar (Punjabi); Vrintakam (Sanskrit); Takkali (Tamil).

Plant Profile and Distribution: Perennial, much-branched, sprawling, pubescent, aromatic herbs; leaves alternate, petiolate, irregularly pinnate; leaflets elliptic or ovate, entire to sinuate, greyish-green, curled, hairy; flowers yellow; berries globose, fleshy, green villous when young, shining red after ripening; seeds numerous, flat, kidney-shaped. It is cultivated extensively as a vegetable crop throughout India.

Antimicrobial studies: Nidiry 1999 and Kravchenko et al. 2003.

29. *Momordica dioica* Roxb. ex Willd. (Family: Cucurbitaceae)

Vernacular Names: Bhatkarela (Assamese); Ban-karela (Bengali); Kaksa, Golkandra, Katwal (Hindi); Karlikai (Kannada); Kartoli (Marathi); Kakaura, Kirara, Dhar Karela (Punjabi); Tholoopavai, Paluppakai (Tamil); Agakara (Telugu).

Plant Profile and Distribution: Perennial, dioecious climbing herbs; leaves entire or 3 to 5-lobed; tendrils simple, filiform; male flowers axillary, solitary, with sessile bracts, yellow; female flowers ebracteate; fruits elliptic-ovate, beaked, densely echinate with soft spines; seeds pyriform, slightly corrugate, pale yellow. Commonly found growing on large herbs and undershrubs in the forests and on hedges of cultivated fields.

Antimicrobial studies: Misra et al. 1991 and Bumrela et al. 2009.

30. *Murraya koenigii* (L.) Spreng. (Family: Rutaceae)

Vernacular Names: Narasingha, Bishahari (Assamese); Barsanga, Kariaphulli (Bengali); Curry Leaf Tree (English); Goranimb, Kadhilimbdo (Gujarati); Kathnim, Mitha Neem (*Azadirachta indica*), Kurry Patta, Gandhela, Barsanga, Meethi Neema (Hindi); Karibevu (Kannada); Kariveppilei (Malayalam); Karhinimb, Poospala, Gandla, Jhirang (Marathi); Barsan, Basango, Bhursunga (Oriya); Karivempu, Karuveppilei, Kattuveppilei (Tamil); Karepaku (Telugu).

Plant Profile and Distribution: Strongly scented, large shrubs or small trees, with brownish-black bark; leaves imparipinnate; flowers creamish-white, in terminal, paniculate cymes; fruits ovoid, immature green, black when ripe; seeds embedded in mucilage. Commonly planted near habitations and gardens; sometimes also found as escape in open forests.

Antimicrobial studies: Goutam and Purohit 1974; Pandey and Dubey 1994; Nutan et al. 1998 and Rahman and Gray 2005.

31. *Musa paradisiaca* L. (Family: Musaceae)

Vernacular Names: Edible Banana, Plantain (English); Kela (Hindi); Bale (Kannada); Vazha (Malayalam); Kadali, Rambha (Sanskrit); Vazhai (Tamil); Arati, Anati (Telugu).

Plant Profile and Distribution: Stoloniferous perennial herbs, with hollow stem; leaves spirally arranged; flowers in terminal, drooping spikes, yellow; male flowers in the upper part; female ones in the lower part; bracts spathaceous; berries yellowish-green. Commonly cultivated throughout India for its edible fruits; also planted in homes and near pilgrimage sites.

Antimicrobial studies: Sharma et al. 1989; Ono et al. 1998 and Agarwal et al. 2009.

32. *Nerium indicum* Mill. (Family: Apocynaceae)

Vernacular Names: Karabi (Bengali); Indian Oleander, Sweet-Scented Oleander (English); Kagaer (Gujarati); Kaner, Karber, Kuruvira (Hindi); Kanagalu (Kannada); Areli (Malayalam); Kanher, Kaneri (Marathi); Konero, Korobiro (Oriya); Arali (Tamil); Ganneru, Kastoori Pattelu (Telugu).

Plant Profile and Distribution: Large, evergreen shrubs, with milky sap; leaves linear-lanceolate, entire, coriaceous; flowers red, fragrant, in terminal, corymbose cymes; follicles cylindrical; seeds villous, with caduceus, light brown coma. It is commonly planted near pilgrimage sites, habitations, gardens and road-dividers for its pretty flowers.

Antimicrobial studies: Ahmed et al. 1993; Huq et al. 1999; Khafagi 1999; Erturk et al. 2000 and Amaresh and Nargund 2003.

33. *Ocimum basilicum* L. (Family: Lamiaceae)

Vernacular Names: Sweet Basil, Common Basil (English); Damaro, Nasabo, Sabza (Gujarati); Babui Tulsi, Gulal Tulsi, Kali Tulsi, Marua (Hindi); Kama Kasturi, Sajjagida (Kannada); Marva, Sabza (Marathi); Dhala Tulasi, Kapur Kanti (Oriya); Furrunj-mushk, Baburi, Niyazbo, Panr (Punjabi); Munjariki, Surasa, Varvara (Sanskrit); Tirnirupachai, Karpura Tulasi (Tamil); Bhutulasi, Rudrajada, Vepudupachha (Telugu).

Plant Profile and Distribution: Erect, much-branched, fragrant herbs, with white, eglandular hairs on stem; leaves gland-dotted on both surfaces; flowers in compact whorls, white to purplish; nutlets punctate, brownish black. Commonly found in open plateaus and in moist and deciduous forests; also grows around habitations and wastelands as a weed.

Antimicrobial studies: Adiguzel et al. 2005; Chiang et al. 2005; Abd El-Zaher et al. 2006; Demirci et al. 2007 and El-Mougy and Abdel-Kader 2007.

34. *Ocimum sanctum* L. (Family: Lamiaceae)

Vernacular Names: Tulsi (Bengali, Gujarati); Sacred Basil, Holy Basil (English); Tulsi, Baranda (Hindi); Vishnu Tulasi, Kari Tulasi, Sri Tulasi (Kannada); Trittavu (Malayalam); Tulasa, Tulasi Chajadha (Marathi); Ajaka, Brinda, Manjari (Sanskrit); Thulasi (Tamil); Tulasi, Brynda, Gaggera, Krishna Tulasi, Nalla Tulasi (Telugu).

Plant Profile and Distribution: Erect, much-branched, aromatic annual undershrubs; branches quadrangular, softly hairy; leaves elliptic-oblong, entire or sub-serrate, hairy on both surfaces, minutely gland-dotted; flowers purplish, in whorls on long, simple or branched racemes; nutlets ellipsoid, dark brown. The plant is commonly grown in homes and temples throughout India.

Antimicrobial studies: Dabur et al. 2004; Sinha et al. 2004; Harikrishnan and Balasundaram 2005; Singh et al. 2005b; Mukherjee 2006 and Viyoch et al. 2006.

35. *Pennisetum americanum* (L.) Leeke (Family: Poaceae)

Vernacular Names: Bajra, Lahra (Bengali, Hindi); Pearl Millet, Bulrush Millet, Spiked Millet (English); Bajri (Gujarati, Marathi); Sajje (Kannada); Kambu (Tamil); Sajja, Ganti (Telugu).

Plant Profile and Distribution: Tall, robust, annuals, with slender or stout culms; leaves long, lanceolate; panicles long, terminal, compact, cylindrical, greenish-yellow, slightly pinkish tinged, with wooly rachis, densely packed with spikelets and bristles; caryopsis pale yellow or white. It is commonly cultivated as minor millet and fodder crop in plains of dry regions, particularly in western India and the Gangetic Plains.

Antimicrobial studies: Sunitha and Chandrashekar 1994.

36. *Picrorhiza kurroa* Royle ex Benth. (Family: Scrophulariaceae)

Vernacular Names: Kuru, Kutki (Bengali, Hindi); Kadu (Gujarati); Katukarogani, Kadugurohini (Malayalam, Tamil, Telugu); Kutaki (Marathi); Karru (Punjabi); Katuka, Katurohini (Sanskrit).

Plant Profile and Distribution: Small, somewhat hairy, perennial, creeping herbs, with stout rhizome; leaves mostly radical, whorled, spathulate, sharply crenate-serrate; flowers white or purple, borne in a dense terminal spicate raceme; capsules ovoid, with numerous minute seeds. Frequently cultivated in the temperate regions from the north-west to eastern Himalayas, also grows in the wild in open areas.

Antimicrobial studies: Vohora et al. 1972 and Rani and Khullar 2004.

37. *Plumbago zeylanica* L. (Family: Plumbaginaceae)

Vernacular Names: Chita, Chitarak, Chitra (Bengali, Hindi); Leadwort, Ceylon Lead Root (English); Chitaro, Chitrak (Gujarati); Chitramula, Vahni (Kannada); Tumba Koduveli, Vellakoduvel (Malayalam); Chitramula, Chitraka (Marathi); Chitamulo, Chitapru, Krisanu, Ongi (Oriya); Cithiramulam (Tamil); Agnimata, Chitramoolam (Telugu).

Plant Profile and Distribution: Erect or straggling, glandular hairy, perennial undershrubs; leaves ovate-lanceolate, undulate-crispy; flowers in axillary and terminal, leafy, panicled spikes, white, with glandular hairy peduncles and calyx; capsules longitudinally 5-grooved. Common in tropical moist and deciduous forests of India ; also planted in homes and temples.

Antimicrobial studies: Goud et al. 2002, 2005; Ram et al. 2004; Wang and Huang 2005; Gebre-Mariam et al. 2006 and Nair et al. 2006.

38. *Pongamia pinnata* (L.) Pierre (Family: Fabaceae)

Vernacular Names: Karchaw (Assamese); Karanj, Karanja (Bengali, Gujarati, Hindi, Marathi); Pongam Oil Tree, Indian Beech (English); Honge (Kannada); Pungu, Punnu (Malayalam); Koranjo (Oriya); Sukhchein, Karanj, Paphri (Punjabi); Ponga, Pongam (Tamil); Gaanuga, Pungu (Telugu).

Plant Profile and Distribution: Medium-sized, glabrous trees, with greyish barks; leaves imparipinnate; leaflets elliptic or ovate-oblong; flowers in axillary racemes, pinkish-white; pods compressed, woody, indehiscent, glabrous; seeds reniform, dirty white with brown streaks. Very common avenue tree, planted along roadsides, in the forests and near habitations.

Antimicrobial studies: Baswa et al. 2001; Simonsen et al. 2001 and Kumar and Khanam 2004.

39. *Pterocarpus marsupium* Roxb. (Family: Fabaceae)

Vernacular Names: Pitshal (Bengali); Indian Kino Tree, Malabar Kino Tree (English); Biyo, Hiradakhan (Gujarati); Bijasal, Bija, Beejasar, Bijasar (Hindi); Honne, Bange (Kannada); Venga (Malayalam); Dhorbenla, Asan, Bibla (Marathi); Byasa (Oriya); Vengai (Tamil); Yegi, Peddagi (Telugu).

Plant Profile and Distribution: Large, deciduous trees, with crooked stem; bark rough, longitudinally fissured, grey, with blood-red gum-resin; leaves imparipinnate; flowers pale yellow, in paniculate racemes, fragrant; pods orbicular, flat, winged, 1 to 2-seeded. Commonly found in tropical deciduous forests in plains and hilly terrains throughout the Deccan Plateau, Gangetic Plains and Central India, also planted as shade trees.

Antimicrobial studies: Mathew et al. 1984.

40. *Rubia cordifolia* L. (Family: Rubiaceae)

Vernacular Names: Majathi (Assamese); Manjistha (Bengali); Indian Madder (English); Manjit, Majith, Majeth (Hindi); Siomalate, Siragatti, Manjushtha (Kannada); Poont, Manjetti (Malayalam); Manjestha (Marathi); Barheipani, Manjistha (Oriya); Kukarphali, Tiuru, Manjit, Sheni, Mitu Runang (Punjabi); Manjistha, Kala-meshika (Sanskrit); Shevelli, Manjitti (Tamil); Taamaravalli, Chiranji, Manjestateega (Telugu).

Plant Profile and Distribution: Extensive, creeping or climbing, sweet-scented, perennial shrubs, with 4-angled branches; roots flexuose with thin, red bark; leaves much variable, whorled; flowers small, white to greenish-yellow or red tinged, in terminal panicles of cymes; fruits globose, purple or black, fleshy, 2-seeded. Commonly found throughout India.

Antimicrobial studies: Jin et al. 1989; Qiao et al. 1990 and Basu et al. 2005.

41. *Senna auriculata* (L.) Roxb. (Family: Caesalpiniaceae)

Vernacular Names: Tanner's Cassia, Tanner's Senna (English); Aval (Gujarati); Awal, Tarval (Hindi); Avarike, Ollethangadi (Kannada); Avara, Aviram, Ponnaviram (Malayalam); Arsual, Taravada (Marathi); Anwala (Rajasthani); Avartaki, Hemapushpam, Mayahari (Sanskrit); Avaram, Semmalai (Tamil); Merakatangeedu, Tangeedu (Telugu).

Plant Profile and Distribution: Tall, spreading, evergreen shrubs, with reddish-brown bark and downy branches; leaves pinnate, with large, auricled stipules; flowers yellow, in axillary and terminal, corymbose racemes; pods linear, flat, papery, beaked, pale brown; seeds laterally compressed. Commonly planted in gardens and as a hedge plant, also grows gregariously in scrub forests, wastelands and along roads, railway tracks etc.

Antimicrobial studies: Kavitha et al. 2001 and Silva et al. 1997.

42. *Sesbania sesban* Merrill (Family: Fabaceae)

Vernacular Names: Jintri, Jayantri (Assamese); Jainti, Jayanti (Bengali); Common Sesban, Egyptlan Rattle Pod (English); Jayati, Raishingin (Gujarati); Jainti, Jait, Rawasan (Hindi); Arisina Jeenangi, Karijeenangimara (Kannada); Sempa, Nellithalai, Kedangu (Malayalam); Shewarie, Jayat, Jarjan (Marathi); Thaitimul, Baryajantis, Joyontri (Oriya); Jaint, Jait (Punjabi); Jayantika, Jayanti (Sanskrit); Chithagathi, Karunchembai, Champai (Tamil); Samintha, Suiminta (Telugu).

Plant Profile and Distribution: Tall, weak, annual herbs or undershrubs; armed with hooked prickles; leaves paripinnate, with prickly rachis; leaflets 15–40 pairs; flowers yellow, in axillary, drooping racemes; pods long, linear, slightly falcate, torulose, beaked, 15 to 20-seeded; seeds brown. Frequently cultivated as fodder crop and green manure; occasionally found near water courses in grasslands and wastelands.

Antimicrobial studies: Jayalaxmi et al. 1982.

43. *Solanum nigrum* L. (Family: Solanaceae)

Vernacular Names: Pichkati (Assamese); Gurkamai, Kakmachi, Tulidun (Bengali); Black Nightshade (English); Piludi (Gujarati); Makoi (Hindi); Kamuni, Ghati, Mako (Marathi); Mako, Kambei, Kachmach, Riaungi (Punjabi); Kakamachi (Sanskrit); Munatakali (Tamil); Kachchipundu, Kachi, Kamanchi, Gajju Chettu (Telugu).

Plant Profile and Distribution: Erect, much-branched, annual herbs; leaves ovate-oblong to oblong-lanceolate, entire to sinuate-dentate; flowers in umbelliform, extra-axillary cymes, white; anthers yellow; berries globose, yellowish-red. Commonly found in moist shady places in fallow lands, in forests as undergrowth and in agricultural fields as a weed.

Antimicrobial studies: Shamim et al. 2004; Al-Fatimi et al. 2006; Balasubramanian et al. 2007 and Jamil et al. 2007.

44. *Syzygium aromaticum* (L.) Merrill & Perry (Family: Myrtaceae)

Vernacular Names: Laung, Lavang (Bengali, Gujarati, Hindi, Marathi); Clove Tree (English); Lavanga (Kannada); Karayampu, Krambu (Malayalam); Lavang (Sanskrit); Kirambu (Tamil); Lavangamuchettu (Telugu).

Plant Profile and Distribution: Moderate-sized, conical, aromatic, evergreen trees, with smooth, grey bark; leaves in pairs, lanceolate, gland-dotted, strongly fragrant; flowers greenish, pink with age, aromatic, in small clusters at the ends of branches; drupes fleshy, dark pink. It is commonly grown as a spice crop in western coasts and southern states in India, particularly in Kerala and Tamilnadu.

Antimicrobial studies: Abu-Shanab et al. 2004; Lopez et al. 2005; Parihar and Bohra 2006; Betoni et al. 2006 and Fabio et al. 2007.

45. *Tagetes erecta* L. (Family: Asteraceae)

Vernacular Names: Genda (Bengali); Aztec or African Marigold (English); Guljharo, Makhanala (Gujarati); Genda, Gultera (Hindi); Seemeshamantige, Chandumallige (Kannada); Chendumalli (Malayalam); Rajia-cha-phul, Zendu (Marathi); Gendu (Oriya); Tangla, Mentok, Genda (Punjabi); Sthulapushpa, Sandu, Ganduga (Sanskrit); Tulukka-samandi (Tamil); Bantichettu (Telugu).

Plant Profile and Distribution: Stout, branched, annual herbs; leaves strongly aromatic, pinnately compound, finely dissected into oblong or lanceolate segments, serrate on margins; flower-heads bright yellow to orange, solitary; achenes with scaly pappus. Commercially cultivated for ornamental flowers, often planted in homes, gardens and sacred places throughout India.

Antimicrobial studies: Rai and Acharya 1999, 2000, 2002; Rai et al. 2002, 2003 and Swami and Mukadam 2006.

46. *Tamarindus indica* L. (Family: Caesalpiniaceae)

Vernacular Names: Tetuli (Assamese); Tentul, Anbli (Bengali); Tamarind Tree (English); Amli, Ambli (Gujarati); Imli, Amli, Anbli (Hindi); Huli, Amli (Kannada); Puli, Amlam (Malayalam); Chinch, Chicha (Marathi); Tentuli, Konya (Oriya); Imbli (Punjabi); Puli, Amilam (Tamil); Chintachettu, Sintachettu, Chintapandu (Telugu).

Plant Profile and Distribution: Large, evergreen trees, with pinnately compound leaves; flowers yellow with red blotches, in few-flowered, terminal racemes; pods sub-torulose, falcate, with compressed, shining, dark brown seeds. Commonly planted near habitations, temples and along roads; occasionally grows on the outskirts of forests.

Antimicrobial studies: Bautista-Banos et al. 2002; Metwali 2003; Al-Fatimi et al. 2006; Melendez and Capriles 2006 and Voravuthikunchai and Limsuwan 2006.

47. *Trigonella foenum-graecum* L. (Family: Fabaceae)

Vernacular Names: Methi, Methi-shak, Methuka, Hoemgreeb (Bengali); Fenugreek (English); Methi, Methini, Bhaji (Gujarati); Methi, Muthi (Hindi); Menthya, Mentesoppu, Menk-palle, Mente (Kannada); Uluva, Venthiam (Malayalam); Methi (Marathi); Methi, Methini, Methri, Methra (Punjabi); Methika, Chandrika, Asumodhagam (Sanskrit); Vendayam (Tamil); Mentikoora, Mentulu (Telugu).

Plant Profile and Distribution: Small, aromatic, annual herbs; leaves 3-foliolate; leaflets obscurely serrate-dentate; flowors pale yellow, in leaf-axils; pods straight, turgid, beaked, many-seeded; seeds brown, deeply grooved, aromatic. The plant is extensively cultivated as a vegetable and fodder crop in the plains, almost throughout India.

Antimicrobial studies: Purohit and Bohra 2002; Aqil and Ahmad 2003; Amalraj et al. 2005; Olli and Kirti 2006 and Parihar and Bohra 2006.

48. *Vernonia anthelmintica* (L.) Willd. (Family: Asteraceae)

Vernacular Names: Ironweed, Purple Feabane (English); Sumraji, Kadu Jeera (Hindi); Avalguja, Kalijiri, Vakuchi (Sanskrit).

Plant Profile and Distribution: Herb, annual or perennial, erect upto 60 cm high, stem terete, ribbed, greyish pubescent, glandular. Many leaves, obtuse or acute, repand serrate, undulate or almost entire upto 8.5–3.5 cm, glabrous, glandular beneath, petioled or sub-sessile. Heads in corymbose panicles, 4–5 mm across, 18–20 flowered. Involucral bracts 4- seriate, lanceolate, pointed at the tip or recurved with few glands, outer

1.5 mm long, inner 4.5 mm long. Achenes terete 1.5 mm long, faintly ribbed, sometimes dimorphic. Pappus hairs white or fulvous, biseriate. The herb is found abundantly in Central and western India.

Antimicrobial studies: Rai and Acharya 1999, 2000, 2002; Rai et al. 2002, 2003.

49. *Withania somnifera* (L.) Dunal (Family: Solanaceae)

Vernacular Names: Ashvaganda (Bengali); Asan, Asoda, Ghodakun (Gujarati); Asgandh, Ashwagandha (Hindi); Viremaddlinagadde, Pannaeru, Kiremallinagida (Kannada); Asgund, Askandha (Marathi); Asgand, Isgand (Punjabi); Chirpotan (Rajasthani); Ashwagandha, Turangi-gandha (Sanskrit); Amukkura, Amkulang, Amukkuram-kilangu (Tamil); Pulivendram, Panneru-gadda, Panneru (Telugu).

Plant Profile and Distribution: Erect, much-branched undershrubs, covered with stellate, greenish-white hairs; leaves ovate to ovate-oblong; flowers greenish-yellow, in subsessile, axillary, few-flowered, umbellate cymes; berries globose, orange-red and enclosed in inflated calyx. Commonly grows in moist, shady places along roads, often abundant near habitations.

Antimicrobial studies: Jamil et al. 2007; Girish et al. 2006; Saadabi 2006; Owais et al. 2005 and Arora et al. 2004.

50. *Xanthium indicum* Koen. ex Roxb. (Family: Asteraceae)

Vernacular Names: Agara (Assamese); Banokra, Chotadhatura (Bengali); Cocklebur, Burweed (English); Gadariun (Gujarati); Adhasisi, Banokra, Chota-gokhru, Gokhru (Hindi); Maruluummatti (Kannada); Shankeshvara, Dutundi (Marathi); Wangan Tsuru, Chiru (Punjabi); Arishta (Sanskrit); Maruloomatham (Tamil); Marulamathangi (Telugu).

Plant Profile and Distribution: Erect, unarmed, monoecious, annual herbs; leaves ovate-suborbicular, palmately 3-lobed, dentate, hispid; heads greenish-yellow, in terminal and axillary racemes; inner involucral bracts covered with hooked bristles, terminating into 2, strong, hooked, divergent beaks; pappus absent. Weeds are commonly found on roadsides, open wastelands and near habitations, naturalized almost throughout India.

Antimicrobial studies: Mehta et al. 1983b; Rai and Acharya 1999, 2000, 2002; Rai et al. 2002, 2003.

Validation and Value addition Efforts and Factors Affecting Botanical Preparations

There are many factors which are basically responsible for affecting botanical preparations. Climatic conditions, soil type, topography, seed material, dynamism, physical and chemical characters of soil, genetic influence are among the few built in factors. For example, accumulation of hypericin in narrow leafed *Hypericum perforatum* has greater concentration than the broader leafed variety (Southwell and Christopher 2001). A recent study on content of essential oil composition of *Ocimum basilicum* L. showed greater variations in wild than in the cultivated populations. The active chemical compound concentration may vary depending upon the prevailing conditions. Therefore, this becomes a major hurdle in validation of sampling material. According to herbal healers, the time of harvesting a certain herb or improper field collection can also hamper the activity of the herb. Therefore, it is necessary to document herbal practices with the correct protocol so that all this important information can be pooled properly. Contaminated herbs, pesticidal residues and other extrinsic factors may also affect the quality and efficacy of herbs. There should be a total control over all these issues.

Conclusion

Tribals in India rely upon herbal medication and they have been using plant resources for curing various microbial infections. The plants in tribal regions like Patalkot (a deep valley amidst the giant hills of Satpura in Central India), the Dang District in Gujarat and the Aravallis in Rajasthan should be screened for the search of antimicrobials. It appears that one of the best sources for finding plant species to test is still the healer's pouch, because such plants have often been tested by generations of indigenous people. Traditional herbal knowledge can be the major source of new antimicrobials. In the 20th century, however, advances in molecular biology and pharmacology led to a precipitous decline in the importance of ethnobotany in drug discovery programmes but the pendulum is slowly swinging back. It is possible to accomplish in a few minutes what once took months to analyze in the laboratory by making TMP (Traditional Medicinal Practices) a source of information. Indeed, it is an age old tried, tested and trusted practice.

References

Abd El-Zaher, F.H., M.A. Fayez, H.K. El-Maksoud, and I. Hosny. 2006. Antimicrobial activity of some Egyptian medicinal and aromatic plant waste extracts. Bulletin of the National Research Centre (Cairo) 31: 1–20.

Abdou, I.A., A.A. Abou-Zeid, M.R. El-Sherbeeny, and Z.H. Abou-El-Gheat. 1972. Antimicrobial activities of *Allium sativum, Allium cepa, Raphanus sativus, Capsicum frutescens, Eruca sativa, Allium kurrat* on bacteria. Plant Foods for Human Nutrition 22 (1): 29–35.

Abraham, Z., D.S. Bhakuni, H.S. Garg, A.K. Goel, B.N. Mehrotra, and G.K. Patnaik. 1986. Screening of Indian Plants for Biological Activity :Part XII. Indian Journal of Experimental Biology 24: 48–68.

Abu-Shanab, B., G. Adwan, D. Abu-Safiya, N. Jarrar, and K. Adwan. 2004. Antibacterial activities of some plant extracts utilized in popular medicine in Palestine. Turkish Journal of Biology 28: 99–102.

Acharya, D. and A. Shrivastava. 2008. Indigenous Herbal Medicines: Tribal Formulations and Traditional Herbal Practices, Aavishkar Publishers Distributor, Jaipur- India. ISBN 9788179102527.

Acharya, T.K. and I.B. Chatterjee. 1974. Isolation of Chrysophonic acid-5-anthrone: A fungicidal compound from *Cassia tora* Linn. Sci. and Cul. 40 (7): 316.

Aderounmu, A.O., A.K. Bolad, B.N. Thomas, A.F. Fagbenro-Beyioku, and K. Berzins. 2003. The antimalarial activity of four commonly used medicinal plants in Nigeria. American Journal of Tropical Medicine and Hygiene 69.

Adiguzel, A., M. Gulluce, M. Sengul, H. Ogutcu, F. Sahin, and I. Karaman. 2005. Antimicrobial effects of *Ocimum basilicum* (Labiatae) extract. Turkish Journal of Biology 29: 155–160.

Adoum, O.A., N.T. Dabo, and M.O. Fatope. 1997. Bioactivities of some savanna plants in the brine shrimp lethality test and *In vitro* antimicrobial assay. International Journal of Pharmacognosy 35: 334–337.

Agarwal, P.K., A. Singh, K. Gaurav, S. Goel, H.D. Khanna, and P.K. Goyal. 2009. Evaluation of wound healing activity of extracts of plantain banana (*Musa sapientum* var. *paradisiacal*). Indian Journal of Experimental Biology 47: 32–40.

Aghili, S.R., B. Shabankhani, and H. Asgari-Rad. 2001. Evaluation of therapeutic efficacy of *Calendula officinalis* extract in dermatophytose on guinea-pigs. International Journal of Antimicrobial Agents 17.

Ahmed, B.A., K.D. Sulayaman, A.A. Aziz, A.A. Abd, and L.J. Rashan. 1993. Antibacterial activity of the leaves of *Nerium oleander*. Fitoterapia 64: 273–274.

Ahmed, I. and A.Z. Beg. 2001. Antimicrobial and phytochemical studies on 45 Indian medicinal plants against multi-drug resistant human pathogens. Journal of Ethnopharmacology 74: 113–123.

Ahmed, I., Z. Mehmood, and F. Mohammad.1998. Screening of some Indian medicinal plants for their antimicrobial properties. J. Ethnopharmacol. 62: 183–193.

Al-Fatimi, M., M. Wurster, G. Schroder, and U. Lindequist. 2006. Antimicrobial, antioxidant and cytotoxic activities of selected medicinal plants from Yemen. Planta Medica 72: 1063.

Ali, A.M., S.H. El-Sharkawy, J.A. Hamid, N.H. Ismail, and N.H. Lajis. 1995. Antimicrobial activity of selected Malaysian plants. Pertanika Journal of Tropical Agricultural Science 18: 57–61.

Ali, A.M., M.M. Mackeen, S.H. El-Sharkawy, J.A. Hamid, N.H. Ismail, F.B.H. Ahmad, and N.H. Lajis. 1996. Antiviral and cytotoxic activities of some plants used in Malaysian indigenous medicine. Pertanika Journal of Tropical Agricultural Science 19: 129–136.

Ali, M.S., I. Azhar, Z. Amtul, V.U. Ahmed, and K. Usmanghani. 1999. Antimicrobial screening of some Caesalpiniaceae. Fitother. 70: 299–304.

Ali, N.A.A., W.D. Juelich, C. Kusnick, and U. Lindequist. 2001. Screening of Yemeni medicinal plants for antibacterial and cytotoxic activities. Journal of Ethnopharmacology 74: 173–179.

Alzoreky, N.S. and K. Nakahara. 2003. Antibacterial activity of extracts from some edible plants commonly consumed in Asia. Int. J. Food Microbiol. 80: 223–230.

Amalraj, A., A. Balasubramanian, E. Edwin, and E. Sheeja. 2005. Antimicrobial activity of Fenugreek seeds and leaves. Indian Journal of Natural Products 21: 35–36.

Amaresh, Y.S. and V.B. Nargund. 2003. Antifungal activity of some plant extracts on *Alternaria helianthi* (Hansf.) Tubaki and Nishihara causing leaf blight of sunflower. Indian Journal of Plant Protection 31: 129–130.

Aqil, F. and I. Ahmad. 2003. Broad-spectrum antibacterial and antifungal properties of certain traditionally used Indian medicinal plants. World Journal of Microbiology & Biotechnology 19: 653–657.

Araki, T. and O. Abe. 1980. A Sensitive Radial Diffusion Method for Detection of Esterolytic Activities of Proteases. Agricultural and Biological Chemistry 44: 1931–1932.

Arora, S., S. Dhillon, G. Rani, and A. Nagpal. 2004. The *In vitro* antibacterial/synergistic activities of *Withania somnifera* extracts. Fitoterapia 75: 385–388.

Aswal, B.S., D.S. Bhakuni, A.K. Goel, B.N. Mehrotra, and K.C. Mukherjee. 1984a. Screening of Indian Plants for Biological Activity: Part X. Indian Journal of Experimental Biology 22: 312–332.

Aswal, B.S., D.S. Bhakuni, A.K. Goel, K. Kar, and B.N. Mehrotra. 1984b. Screening of Indian Plants for Biological Activity-Part XI. Indian Journal of Experimental Biology 22: 487–504.

Aswal, B.S., A.K. Goel, D.K. Kulshrestha, B.N. Mehrotra, and G.K. Patnaik. 1996. Screening of Indian plants for biological activity: Part XV. Indian Journal of Experimental Biology 34: 444–467.

Atal, C.K., J.B. Srivastava, B.K. Wali, R.V. Chakravarty, B.N. Dhawan, and R.P. Rastogi. 1978. Screening of Indian Plants for Biological Activity: Part VIII. Indian Journal of Experimental Biology 16: 330–349.

Athar, M. and Z. Ahmed. 2004. Taxonomy, distribution and medicinal uses of legume trees of Pakistan. SIDA 21(2): 951–962.

Avani, K. and S. Neeta. 2005. A study of the antimicrobial activity of *Elephantopus scaber*. Indian Journal of Pharmacology 37: 126–127.

Ayyanar, M. and S. Ignacimuthu. 2009. Herbal medicines for wound healing among tribal people in Southern India: Ethnobotanical and Scientific evidences. International Journal of Applied Research in Natural Products 2(3): 29–42.

Babu, G.D.K., V. Shanmugam, S.D. Ravindranath, and V.P. Joshi. 2007. Comparison of chemical composition and antifungal activity of *Curcuma longa* L. leaf oils produced by different water distillation techniques. Flavour and Fragrance Journal 22: 191–196.

Bagchi, G.D., A. Singh, S.P.S. Khanuja, R.P. Bansal, S.C. Singh, and S. Kumar. 1999. Wide spectrum antibacterial and antifungal activities in the seeds of some coprophilous plants of north Indian plains. J. Ethnopharmacol. 64: 69–77.

Balasubramanian, G., M. Sarathi, S.R. Kumar, and A.S.S. Hameed. 2007. Screening the antiviral activity of Indian medicinal plants against white spot syndrome virus in shrimp. Aquaculture 263: 15–19.

Bandara, B.M.R., N.S. Kumar, and K.M.S. Samaranayake. 1989. An Antifungal Constituent from the Stem Bark of *Butea-Monosperma*, Journal of Ethnopharmacology 25: 73–76.

Bandara, B.M.R., C.M. Hewage, V. Karunaratne, and N.K.B. Adikaram. 1988. Methyl Ester of P Coumaric Acid Antifungal Principle of the Rhizome of *Costus-Speciosus*. Planta Medica 54: 477–478.

Banso, A. 2007. Comparative studies of antimicrobial properties of *Cardiospermum grandiflorum* and *Cardiospermum halicacabum*. Nigerian Journal of Health and Biomedical Sciences 0(1): 21–24.

Basu, S., A. Ghosh, and B. Hazra. 2005. Evaluation of the antibacterial activity of *Ventilago madraspatana* Gaertn., *Rubia cordifolia* Linn. and *Lantana camara* Linn. Isolation of emodin and physcion as active antibacterial agents. Phytotherapy Research 19: 888–894.

Baswa, M., C.C. Rath, S.K. Dash, and P.K. Mishra. 2001. Antibacterial activity of Karanj (*Pongamia pinnata*) and Neem (*Azadirachta indica*) seed oil: A preliminary report. Microbios 105: 183–189.

Batra, A. and B.K. Mehta. 1986. Chromatographic Analysis and Antibacterial Activity of the Seed Oil of *Argyreia speciosa*. Fitoterapia 56: 357–359.

Bautista-Banos, S., L.L. Barrera-Necha, L. Bravo-Luna, and K. Bermudez-Torres. 2002. Antifungal activity of leaf and stem extracts from various plant species on the incidence of *Colletotrichum gloeosporioides* of papaya and mango fruit after storage. Revista Mexicana de Fitopatologia 20: 8–12.

Begum, J., M. Yusuf, J.U. Chowdhury, S. Khan, and M. Nural Anwar. 2007. Antifungal Activity of Forty Higher Plants against Phytopathogenic Fungi. Bangladesh J. Microbiol. 24(1): 76–78.

Benkeblia, N. 2004. Antimicrobial activity of essential oil extracts of various onions (*Allium cepa*) and garlic (*Allium sativum*). Lebensmittel-Wissenschaft und Technologie 37(2): 263–268.

Betoni, J.E.C., R. Passarelli Mantovani, L. Nunes Barbosa, L.C. Di Stasi, and A. Fernandes Jr. 2006. Synergism between plant extract and antimicrobial drugs used on *Staphylococcus aureus* diseases. Memorias do Instituto Oswaldo Cruz 101: 387–390.

Bhakuni, D.S., M.L. Dhar, M.M. Dhar, B.N. Dhawan, and B.N. Mehrotra. 1969. Screening of Indian Plants for Biological Activity: Part II. Indian Journal of Experimental Biology 7: 250–262.

Bhakuni, D.S., M.L. Dhar, M.M. Dhar, B.N. Dhawan, B. Gupta, and R.C. Srimal. 1971. Screening of Indian Plants for Biological Activity: Part III. Indian Journal of Experimental Biology 9: 91–102.

Bhakuni, D.S., A.K. Goel, A.K. Goel, S. Jain, B.N. Mehrotra, and R.C. Srimal. 1990. Screening of Indian plants for biological activity : Part XIV. Indian Journal of Experimental Biology 28: 619–637.

Bhakuni, D.S., A.K. Goel, S. Jain, B.N. Mehrotra, G.K. Patnaik, and Ved Prakash. 1988. Screening of Indian Plants for Biological Activity: Part XIII. Indian Journal of Experimental Biology 26: 883–904.

Bhattacharjee, I., S.K. Chatterjee, S. Chatterjee, and G. Chandra. 2006. Antibacterial potentiality of *Argemone mexicana* solvent extracts against some pathogenic bacteria. Mem. Inst. Oswaldo Cruz 101(6): 645–648.

Biswas, T.K. and B. Mukherjee. 2003. Plant medicines of Indian origin for wound healing activity: A review. Lower Extr Wounds 2: 25–39.

Bonyadi, R.E., V. Awad, and B.N. Kunchiraman. 2009. Antimicrobial activity of the ethanolic extract of *Bryonopsis laciniosa* leaf, stem, fruit and seed. African Journal of Biotechnology 8(15): 3565–3567.

Bumrela, S., A. Samleti, P. Melisa, and M. Saxena. 2009. Evaluation of antimicrobial and antioxidant properties of *Momordica dioica* Roxb. (Ex Willd). Journal of Pharmacy Research 2(6): 1075–1078.

Camporese, A., M.J. Balick, R. Arvigo, R.G. Esposito, N. Morsellino, F.D. Simone, and A. Tubaro. 2003. Screening of antibacterial activity of medicinal plants from Belize (Central America). J. Ethnopharmacol. 87: 103–107.

Carlson, T.J., B.M. Foula, J.A. Chinnock, S.R. King, G. Abdourahmaue, B.M. Sannoussy, A. Bah, S.A. Cisse, M. Camara, and R.K. Richter. 2001. Case study on medicinal plant research in Guinea: Prior Informed Consent, Focused Benefit Sharing, and Compliance with the Convention on Biological Diversity. Economic Botany 55(4): 478–491.

Celine, C., A. Sindhu, and M.P. Muraleedharn. 2007. Microbial growth inhibition by aparajitha dhooma choornam. Ancient Science of Life 26(3–4): 4–8.

Chamundeeswari, D., J. Vasantha, S. Gopalakrishnan, and E. Sukumar. 2004. Antibacterial and antifungal activities of *Trewia polcarpa* roots. Fitotherapia 75: 85–88.

Chang, Y.C., P.W. Hsieh, F.R. Chang, R.R. Wu, C.C. Liaw, K.H. Lee, and Y.C. Wu. 2003. New protopine and the anti-HIV alkaloid 6-acetonyldihydrochelerythrine from Formosan *Argemone mexicana*. Planta Med. 69: 148–152.

Chaturvedi, S.K., V.K.S. Bhadouriya, and K.R. Naidu. 1987. Antifungal activity of the volatile oil of *Anogeissus latifolia* leaves. Indian Perfumer 31(3): 238–239.

Chea, A., M.C. Jonville, S.S. Bun, M. Laget, R. Elias, G. Dumenil, and G. Balansard. 2007. *In vitro* antimicrobial activity of plants used in Cambodian traditional medicine. American Journal of Chinese Medicine 35: 867–873.

Chen, C.P., C.C. Lin, and T. Namba. 1989. Screening of Taiwanese Crude Drugs for Antibacterial Activity against *Streptococcus mutans*. Journal of Ethnopharmacology 27: 285–296.

Chen, I.N., C.C. Chang, C.C. Ng, C.Y. Wang, Y.T. Shyu, and T.L. Chang. 2008. Antioxidant and antimicrobial activity of Zingiberaceae plants in Taiwan. Plant Foods for Human Nutrition (Dordrecht) 63: 15–20.

Chiang, L.C., L.T. Ng, P.W. Cheng, W. Chiang, and C.C. Lin. 2005. Antiviral activities of extracts and selected pure constituents of *Ocimum basilicum*. Clinical and Experimental Pharmacology and Physiology 32: 811–816.

Chitme, H.R., R. Chandra, and S. Kaushik. 2003. Studies on anti-diarrheal activity of *calotropis gigantea* R. Br. in experimental animals. J. Pharm. Pharmaceut. Sci. 7: 70–75.

Chitra, M., C.S.S. Devi, and E. Sukumar. 2003. Antibacterial activity of embelin. Fitoterapia 74: 401–403.

Chitravadivu, C., M. Bhoopathi, T. Elavazhagan, S. Jayakumar, and V. Balakrishnan. 2009. Screening of antimicrobial activity of medicinal plant oils prepared by herbal venders, South India. Middle-East Journal of Scientific Research 4(2): 115–117.

Cho, J.Y., G.J. Choi, S.W. Lee, K.S. Jang, H.K. Lim, C.H. Lim, S.O. Lee, K.Y. Cho, and J.C. Kim. 2006. Antifungal activity against *Colletotrichum* spp. of curcuminoids isolated from *Curcuma longa* L. rhizomes. Journal of Microbiology and Biotechnology 16: 280–285.

Coelho de Souza, G., A.P.S. Haas, G.L. Poser, E.E.S. Schapoval, and E. Elisabetsky. 2004. Ethnopharmacological studies of antimicrobial remedies in the south of Brazil. J. Ethnopharmacol. 90: 135–143.

Collier, W.A. and L. van de Pijl. 1949. The antibiotic action of plants, especially the higher plants, with results with Indonesian plants. Chron. Nat. 105: 8–15.

Cowan, M.M. 1999. Plant Products as Antimicrobical Agents. Clin. Microbiol. Rev.12: 564–582.

Dabur, R., H. Singh, A.K. Chhillar, M. Ali, and G.L. Sharma. 2004. Antifungal potential of Indian medicinal plants. Fitoterapia 75: 389–391.

Dabur, R., A. Gupta, T.K. Mandal, D.D. Singh, V. Bajpai, A.M. Gurav, and G.S. Lavekar. 2007. Antimicrobial activity of some Indian medicinal plants. Afr. J. Trad. Comp. Alt. Med, 4: 313–318.

Dankert, J., T.F. Tromp, H. de Vries, and H.J. Klasen. 1979. Antimicrobial activity of crude juices of *Allium ascalonicum*, *Allium cepa* and *Allium sativum*. Zentralblatt für Bakteriologie, Parasitenkunde, Infektionskrankheiten und Hygiene, Erste Abteilung, Reihe A: Medizinische Mikrobiologie und Parasitologie 245(1–2): 229–239.

Davis, L.E., J.K. Shen, and Y. Cai. 1990. Antifungal activity in human cerebrospinal fluid and plasma after ingestion of garlic (*Allium sativum*). Antimicrobial Agents and Chemotherapy 34(4): 651–653.

Demirci, F., G. Iscan, B. Demirci, and K.H.C. Baser. 2007. Anticandidal activity of essential oils from commercial herbal spices from Turkey. Planta Medica 73, 860.

Deshmukh SK, P.C.Jain, and S.C. Agrawal (1986) A note on mycotoxicity of some essential oils. Fitoterapia 67. 295–296.

Dhar, M.L., M.M. Dhar, B.N. Dhawan, B.N. Mehrotra, and C. Ray. 1968. Screening of Indian Plants for Biological Activity: Part IX. Indian Journal of Experimental Biology 6: 232–247.

Dhar, M.L., M.M. Dhar, B.N. Dhawan, B.N. Mehrotra, R.C. Srimal, and J.S. Tandon. 1973. Screening of Indian Plants for Biological Activity: Part IV. Indian Journal of Experimental Biology 11: 43–54.

Dhar, M.L., B.N. Dhawan, C.R. Prasad, R.P. Rastogi, K.K. Singh, and J.S. Tandon. 1974. Screening of Indian Plants for Biological Activity: Part V. Indian Journal of Experimental Biology 12: 512–523.

Dhawan, B.N., G.K. Patnaik, R.P. Rastogi, K.K. Singh, and J.S. Tandon. 1977. Screening of Indian Plants for Biological Activity: Part VI. Indian Journal of Experimental Biology 15: 208–219.

Dhawan, B.N., M.P. Dubey, B.N. Mehrotra, and J.S. Tandon. 1980. Screening of Indian Plants for Biological Activity: Part IX. Indian Journal of Experimental Biology 18: 594–602.

Dhillion, S.S. and L. Ampornpan. 2000. Bioprospecting and phytomedicines in Thailand: Conservation, benefit sharing and regulations. In: H. Svarstad and S.S. Dhillion (eds.). Responding to bioprospecting: from biodiversity in the South to medicines in the North. Spartacus Forlag As, Oslo. ISBN 82-430-0163-8

Dhillion, S.S., H. Svarstad, C. Amundsen, and H.C. Bugge. 2002. Bioprospecting: effects on environment and development. Ambio 31(6): 491–493.

Duraipandiyan, V., M. Ayyanar, and S. Ignacimuthu. 2006. Antimicrobial activity of some ethnomedicinal plants used by Paliyar tribe from Tamil Nadu, India. BMC Comp. Alt. Med. 6: 35–41.

El-Mougy, N.S. and M.M. Abdel-Kader. 2007. Antifungal effect of powdered spices and their extracts on growth and activity of some fungi in relation to damping-off disease control. Journal of Plant Protection Research 47: 267–278.

Erturk, O. 2006. Antibacterial and antifungal activity of ethanolic extracts from eleven spice plants. Biologia (Bratislava) 61: 275–278.

Erturk, O., Z. Demirbag, and A.O. Belduz. 2000. Antiviral activity of some plant extracts on the replication of *Autographa californica* nuclear polyhedrosis virus. Turkish Journal of Biology 24: 833– 844.

Fabio, A., C. Cermelli, G. Fabio, P. Nicoletti, and P. Quaglio. 2007. Screening of the antibacterial effects of a variety of essential oils on microorganisms responsible for respiratory infections. Phytotherapy Research 21: 374–377.

Farooque, N.A. and K.G. Saxena. 1996. Conservation and utilization of medicinal plants in high hills of the central Himalayas. Environmental Conservation 23(1): 75–80.

Feresin, G.E., A. Tapin, S.N. Lopez, and S.A. Zacchino. 2001. Antimicrobial activity of plants used in traditional medicine of San Juan province, Argentine. J. Ethnopharmacol. 78: 103–107.

Gautam, K., P.B. Rao, and S.V.S. Chauhan. 2003. Efficacy of some botanicals of the family Compositae against *Rhizoctonia solani* Kuhn. Journal of Mycology and Plant Pathology 33: 230–235.

Gebre-Mariam, T., R. Neubert, P.C. Schmidt, P. Wutzler, and M. Schmidtke. 2006. Antiviral activities of some Ethiopian medicinal plants used for the treatment of dermatological disorders. Journal of Ethnopharmacology 104: 182–187.

Gertsch, J., K.G. Niomave Roost, and O. Sticher. 2004. *Phyllanthus piscatorum*, ethnopharmacological studies on a women's medicinal plant of the Yanomami Amerindians. J. Ethnopharmacol. 91: 181–188.

Ghosh, S.B., S. Gupta, and A.K. Chandra. 1980. Anti Fungal Activity in Rhizomes of *Curcuma amada*. Indian Journal of Experimental Biology 18: 174–176.

Girish, K.S., K.D. Machiah, S. Ushanandini, K.H. Kumar, S. Nagaraju, M. Govindappa, A. Vedavathi, and K. Kemparaju. 2006. Antimicrobial properties of a non-toxic glycoprotein (WSG) from *Withania somnifera* (Ashwagandha). Journal of Basic Microbiology 46: 365–374.

Gnanamani, A., K.S. Priya, N. Radhakrishnan, and M. Babu. 2003. Antibacterial activity of two plant extracts on eigth burn pathogens. J. Ethnopharmacol. 86: 59–61.

Goel, A.K., D.K. Kulshrestha, M.P. Dubey, and S.M. Rajendran. 2002. Screening of Indian plants for biological activity: Part XVI. Indian Journal of Experimental Biology 40: 812–827.

Goswami, A. and A. Reddi. 2004. Antimicrobial activity of flavonoids of medicinally important plant *Cassia angustifolia in vivo*. Journal of Phytological Research 17(2): 179–181.

Goud, P.S.P., K.S.R. Murthy, T. Pullaiah, and G.V.A.K. Babu. 2005. Screening for antibacterial and antifungal activity of some medicinal plants of Nallamalais, Andhra Pradesh, India. Journal of Economic and Taxonomy Botany 29: 704–708.

Goud, P.S.P., K.S.R. Murthy, T. Pullaiah, and G.V.A.K. Babu. 2002. Screening for antibacterial and antifungal activity of some medicinal plants of Nallamalais, Andhra Pradesh, India. Journal of Economic and Taxonomy Botany 26: 677–684.

Goutam, M.P. and R.M. Purohit. 1974. Anti Microbial Activity of the Essential Oil of the Leaves of *Murraya koenigii* Indian Curry Leaf. Indian Journal of Pharmacy 36: 11–12.

Goutam, M.P., P.C. Jain, and K.V. Singh. 1980. Activity of some essential oils against dermatophytes. Indian Drugs 17: 269–270.

Govindarajan, R., M. Vijayakumar, M. Singh, C.H.V. Rao, A. Shirwaikar, A.K.S. Rawat, and P. Pushpangadan. 2006. Antiulcer and antimicrobial activity of *Anogeissus latifolia*. Journal of Ethnopharmacology 106(1): 57–61.

Greeshma, A.G., B. Srivastava, and K. Srivastava. 2006. Plants used as antimicrobials in the preparation of traditional starter cultures of fermentation by certain tribes of Arunachal Pradesh. Bulletin of Arunachal Forest Research 22(1-2): 52–57.

Gupta, S.K. and A.B. Banerjee. 1972. Screening of Selected West Bengal Plants for Anti Fungal Activity. Economic Botany 26: 255–259.

Gupta, S., J.N.S. Yadava, and J.S. Tandon. 1993. Antisecretory (antidiarrhoeal) activity of Indian medicinal plants against *Escherichia coli* enterotoxin-induced secretion in rabbit and guinea pig ileal loop models. Pharmaceutical Biology 31(3): 198–204.

Gurav Shailendra, S., D.N. Vijay, J. Gulkari Duragkar, and A.T. Patil. 2008. Antimicrobial activity of *Butea monosperma* Lam. Gum. Iranian Journal of Pharmacology and Therapeutics 7(1): 21–24.

Habbal, O.A., A.A. Ai-Jabri, A.H. El-Hag, Z.H. Al-Mahrooqi, and N.A. Al-Hashmi. 2005. *In-vitro* antimicrobial activity of *Lawsonia inermis* Linn (henna)—A pilot study on the Omani henna. Saudi Medical Journal 26: 69–72.

Habhu, P.V., K.M. Mahadevan, R.A. Shastry, and H. Manjunatha. 2009. Antimicrobial activity of flavanoid sulphates and other fractions of *Argyreia speciosa* (Burm.f) Boj. Indian J. Exp. Biol. 47(2): 121–128.

Harikrishnan, R. and C. Balasundaram. 2005. Antimicrobial activity of medicinal herbs *In vitro* against fish pathogen *Aeromonas hydrophila*. Fish Pathology 40: 187–189.

Harris, J.C., S. Cottrell, S. Plummer, and D. Lloyd. 2001. Antimicrobial properties of *Allium sativum* (garlic). Applied Microbiology and Biotechnology 57(3): 282–286.

Harsha, V.H., S.S. Hebbar, G.R. Hegde, and V. Shripathi. 2002. Ethnomedical knowledge of plants used by Kunabi Tribe of Karnataka in India. Fitoterapia 73: 281–287.

Hatano, T., H. Uebayashi, H. Ito, S. Shiota, T. Tsuchiya, and T. Yoshida. 1999. Phenolic constituents of Cassia seeds and antibacterial effect of some naphthalenes and anthraquinones on methicillin-resistant *Staphylococcus aureus*. Chemical and Pharmaceutical Bulletin (Tokyo) 47: 1121–1127.

Hernandez, N.E., M.L. Tereschuk, and L.R. Abdala. 2000. Antimicrobial activity of flavonoides in medicinal plants from Tafi del valle (Tucuman, Argentina). J. Ethnopharmacol. 73: 317–322.

Huq, M.M., A. Jabbar, M.A. Rashid, and C.M. Hasan. 1999. A novel antibacterial and cardiac steroid from the roots of *Nerium oleander*. Fitoterapia 70: 5–9.

Iacobellis, N.S., P. Lo Cantore, F. Capasso, and F. Senatore. 2005. Antibacterial activity of *Cuminum cyminum* L. and *Carum carvi* L. essential oils. Journal of Agricultural and Food Chemistry 53: 57–61.

244 *Ethnomedicinal Plants*

Immanuel, G., V.C. Vincybai, V. Sivaram, A. Palavesam, and M.P. Marian. 2004. Effect of butanolic extracts from terrestrial herbs ans seaweeds in the survival, growth and pathogen (*Vibrio parahaemolyticus*) load on shrimp *Penaeus indicus* juveniles. Aquaculture 236: 53–65.

Iwu, M.M., A.R. Duncan, and C.O. Okunji. 1999. New antimicrobials of plant origin. In: J. Janick (ed.). Perspectives on new crops and new uses. ASHS press, Alexandria. pp. 457–462.

Jagannath, J.H. and M. Radhika. 2006. Antimicrobial emulsion (coating) based on biopolymer containing neem (*Melia azardichta*) and turmeric (*Curcuma longa*) extract for wound covering. Bio-Medical Materials and Engineering 16: 329–336.

Jain, P.C. and S.C. Agrawal. 1976. Activity of plant extracts against some keratinophilic species of *Nannizzia*. Indian Drugs 13(12): 25–26.

Jain, P.C., C.K. Jain, and K. Jain. 1980. A note on the activity of odoriferous compound against dermatophytes. Indian Drugs 17(12): 397–98.

Jain, P.P., R.K. Suri, S.K. Deshmukh, and K.C. Mathur. 1987. Fatty Oils from Oilseeds of Forest Origin as Antibacterial Agents. Indian Forester 113: 297–299

Jamil, A., M. Shahid, M.M. Khan, and M. Ashraf. 2007. Screening of some medicinal plants for isolation of antifungal proteins and peptides. Pakistan Journal of Botany 39: 211–221.

Jayalaxmi, S., R. Shrivastava, and D.V. Kohli. 1982. Anti bacterial and pharmacological studies of unsaponifiable matter of seeds of *Sesbania sesban*. Indian Journal of Hospital Pharmacy 19: 67–69.

Jeevan Ram, A., L. Md. Bhakshu, and R.R. Venkata Raju. 2004. *In vitro* antimicrobial activity of certain medicinal plants from Eastern Ghats, India, used for skin diseases. Journal of Ethnopharmacology 90(2-3): 353–357.

Jin, Y.H., Y.K. Wang, and B.L. Gu. 1989. The Study of Antiviral Effect of Methanol Extract of *Rubia cordifolia* against Herpes Simplex Virus 2 *in-Vitro*. Virologica Sinica 4: 345–349.

Johns, T., J.O. Kokwaro, and E.K. Kimanani. 1990. Herbal remedies of the Luo of Siaya district, Kenya: Establishing quantitative criteria for consensus. Economic Botany 44(3): 369–381.

Kalvatchev, Z., R. Walder, and D. Garzaro. 1997. Anti-HIV activity of extracts from *Calendula officinalis* flowers. Biomedicine and Pharmacotherapy 51: 176–180.

Kasiram, K., P.R. Sakharkar, and A.T. Patil. 2000. Antifungal activity of *Calendula officinalis*. Indian Journal of Pharmaceutical Sciences 62: 464–466.

Kavitha, S., V.N. Devi, R. Sekar, and G.N. Subbiah. 2001. Antifungal activity in thirteen species of Cassia. Geobios (Jodhpur) 28: 261–262.

Ketzis, J.K. and M.T. Laux. 2006. *In vitro* anti-microbial activity of the essential oils and extracts in a plant-based skin cream. Planta Medica 72: 1043–1044.

Khafagi, I.K. 1999. Screening *In vitro* cultures of some Sinai medicinal plants for their antibiotic activity. Egyptian Journal of Microbiology 34: 613–627.

Khan, A., M. Rahman, and M.S. Islam. 2008. Antibacterial, antifungal and cytotoxic activities of amblyone isolated from *Amorphophallus campanulatus*. Indian J. Pharmacol. 40: 41–44.

Khan, M.R., A.D. Omoloso, and M. Kihara. 2003. Antibacterial activity of *Artocarpus heterophyllus*. Fitoterapia 74: 501–505.

Khan Zafar, K. and Katiyar Ratna. 2000. Potent antifungal activity of garlic (*Allium sativum*) against experimental murine disseminated cryptococcosis. Pharmaceutical Biology 38(2): 87–100.

Kim, Y.M., C.H. Lee, H.G. Kim, and H.S. Lee. 2004. Anthraquinones isolated from *Cassia tora* (Leguminosae) seed show an antifungal property against phytopathogenic fungi. Journal of Agricultural and Food Chemistry 52: 6096–6100.

Kishore, N., A.K. Chopra, and Q. Khan. 1997. Antimicrobial properties of *Calotropis procera* Ait. in different seasons, a study *in vitro*. Biological Memoirs 23: 53–57.

Kravchenko, L.V., T.S. Azarova, E.I. Leonova-Erko, A.I. Shaposhnikov, N.M. Makarova, and I.A. Tikhonovich. 2003. Root Exudates of Tomato Plants and Their Effect on the Growth and Antifungal Activity of Pseudomonas Strains, Microbiology 72(1): 37–41.

Kulataeva, A.K., R.N. Pak, B.A. Ermekbaeva, and S.M. Adekenov. 2006. Antimicrobial and regenerating properties of medicinal plants ethanol extracts and essential oils phytocomposition. Rastitel'nye Resursy 42: 102–109.

Kumar, B., M. Vijayakumar, R. Govindarajan, and P. Pushpangadan. 2007. Ethnopharmacological approaches to wound healing—Exploring medicinal plants of India. J. Ethnopharmacol 114: 103–113.

Kumar, G.S. and S. Khanam. 2004. Anti-acne activity of natural products. Indian Journal of Natural Products 20: 7–9.

Kumar, S., S. Dewan, H. Sangraula, and V.L. Kumar. 2001. Anti-diarrhoeal activity of the latex of *Calotropis procera*. J. Ethnopharmacol. 76: 115–118.

Kumar, V.P., N.S. Chauhan, H. Padh, and M. Rajani. 2006. Search for antibacterial and antifungal agents from selected Indian medicinal plants. Journal of Ethnopharmacology 107: 182–188.

Kumaraswamy, M.V., H.U. Kavitha, and S. Satish. 2008. Antibacterial Potential of Extracts of *Woodfordia fruticosa* Kurz on Human Pathogens. World Journal of Medical Sciences 3(2): 93–96.

Laird, S.A. 2002. Biodiversity and traditional knowledge: equitable partnerships in practice. Earthscan, London.

Larhsini, M., M. Bousaid, H.B. Lazrek, M. Jana, and H. Amarouch. 1997. Evaluation of antifungal and molluscicidal properties of extracts of *Calotropis procera*. Fitoterapia 68: 371–373.

Larhsini, M., L. Oumoulid, H.B. Lazrek, S. Wataleb, M. Bousaid, K. Bekkouche, M. Markouk, and M. Jana. 1999. Screening of antibacterial and antiparasitic activities of six Moroccan medicinal plants. Therapie (London) 54: 763–765.

Lee, D.G., D.H. Kim, M.Y. Seo, S.Y. Jang, and K.S. Hahm. 2000. Antifungal mechanism of Ib-AMP1 against *Candida albicans*. Biochemical Society Transactions 28.

Lopez, A., J.B. Hudson, and G.H.N. Towers. 2001. Antiviral and antimicrobial activities of Colombian medicinal plants. J. Ethnopharmacol. 77: 189–196.

Lopez, P., C. Sanchez, R. Batlle, and C. Nerin. 2005. Solid- and vapor-phase antimicrobial activities of six essential oils: Susceptibility of selected food borne bacterial and fungal strains. Journal of Agricultural and Food Chemistry 53: 6939–6946.

Mahida, Y. and J.S.S. Mohan. 2006. Screening of Indian plant extracts for antibacterial activity. Pharmaceutical Biology 44: 627–631.

Mahidol, C., S. Ruchirawat, H. Prawat, S. Pisutjaroenpong, S. Engprasert, P. Chumsri, T. Tengchaisri, S. Sirisinha, and P. Picha. 1998. Biodiversity and natural product drug discovery. Pure. Appli. Chem. 70(11): 2065–2072.

Malabadi, R.B. 2005. Antibacterial activity in the rhizome extracts of *Costus speciosus* (Koen.). Journal of Phytological Research 18: 83–85.

Malabadi, R.B. and S.V. Kumar. 2005. Assessment of anti-dermatophytic activity of some medicinal plants. Journal of Phytological Research 18: 103–106.

Mann, A., J.O. Amupitan, A.O. Oyewale, J.I. Okogun, and K. Ibrahim. 2009. Chemistry of secondary metabolites and their antimicrobial activity in the drug development process: A review of the genus *Anogeissus*. International Journal of Phytomedicines and Related Industries 1(2).

Mashiar Rahman, M., Jahangir Alam Md, S.S. Akhtar, M. Mizanur Rahman, A. Rahman, and M.F. Alam. 2009. *In vitro* Antibacterial Activity of *Argemone mexicana* L. (Papaveraceae). CMU. J. Nat. Sci. 8(1): 77.

Mathew, J., A.V.S. Rao, and S. Rambhav. 1984. Propterol an Antibacterial Agent from *Pterocarpus marsupium*. Current Science 53: 576–577.

Mazumder, R., S.G. Dastidar, S.P. Basu, A. Mazumder, and S.K. Singh. 2004. Antibacterial potentiality of *Mesua ferrea* Linn. flowers. Phytother. Res. 18: 824–826.

Mazumder, U.K., M. Gupta, L. Manikandan, and S. Bhattacharya. 2001. Antibacterial activity of *Urena lobata* root. Fitother. 72: 927–929.

Mehta, B.K., A. Dubey, M.M. Bokadia, and S.C. Mehta. 1983a. Isolation and *In-Vitro* Antimicrobial Efficiency of *Butea-Monosperma* Seed Oil on Human Pathogenic Bacteria and Phytopathogenic Fungi. Acta Microbiologica. Hungarica 30: 75–78.

Mehta, P., S. Chopra, and A. Mehta. 1983b. Antimicrobial properties of some plant extracts against bacteria. Folia Microbiol (Praha) 28(6): 467–469.

Melendez, P.A. and V.A. Capriles 2006. Antibacterial properties of tropical plants from Puerto Rico. Phytomedicine (Jena) 13: 272–276.

Metwali, M.R. 2003. Study of antimicrobial potencies of some Yemeni medicinal plants. Egyptian Journal of Microbiology 38: 105–114.

Misra, P., N.L. Pal, P.Y. Guru, J.C. Katiyarm, and J.S. Tandon. 1991. Antimalarial Activity of Traditional Plants against Erythrocytic Stages of *Plasmodium berghei*. International Journal of Pharmacognosy 29: 19–23.

Misra, S.B. and S.N. Dixit. 1977. Antifungal properties of *Allium sativum*. Science and Culture 43(11): 487–488.

Misra, S.S. and S.N. Dixit. 1980. Antifungal principle of *Ranunculus scleratus*. Econ. Bot. 34: 362–367.

Mohamed, S., S. Saka, S.H. El-Sharkawy, A.M. Ali, and S. Muid. 1996. Antimycotic screening of 58 Malaysian plants against plant pathogens. Pesticide Science 47: 259–264.

Mohammad, S.R., Z.R. Mohammad, A.B.M.R. Ahad Uddin, and A.R. Mohammad. 2007. Steriod and Triterpenoid from *Anogeissus latifolia*. Dhaka University Journal of Pharmaceutical Sciences 6(1): 47–50.

Mosaddik, A.M.H. and M. Ekramul. 2003. Cytotoxicity and antimicrobial activity of goniothalamin isolated from *Bryonopsis laciniosa*. Phytotherapy Research 17(10): 1155–1157.

Mosaddik, M.A., L. Banbury, P. Forster, R. Booth, J. Markham, D. Leach, and P.G. Waterman. 2004. Screening of some Australian Flacourtiaceae species for *In vitro* antioxidant, cytotoxic and antimicrobial activity. Phytomed. 11: 461–466.

Mouhajir, F., J.B. Hudson, M. Rejdali, and G.H.N. Towers. 2001. Multiple antiviral activities of endemic medicinal plants used by Berber peoples of Morocco. Pharmaceutical Biology 39: 364–374.

Mukherjee, P.K., K. Saha, T. Murugesan, S.C. Mandal, M. Pal, and B.P. Saha. 1998. Screening of anti-diarrhoeal profile of some plant extracts of a specific region of West Bengal, India. J. Ethnopharmacol. 60: 85–89.

Mukherjee, R. 2006. Antibacterial and therapeutic potential of Ocimum sanctum in bovine sub clinical mastitis. Indian Veterinary Journal 83: 522–524.

Müller, B.M., J. Kraus, and G. Franz. 1989. Chemical structure and biological activity of water-soluble polysaccharides from *Cassia angustifolia* leaves. Planta Medica 55(6): 536–539.

Murray, M. 1995. The healing power of herbs. Prima Publishing. Rocklin, CA. pp. 162–171.

Nair, R. and S. Chanda. 2005. Antibacterial activity of *Punica granatum* in different solvents. Ind. J. Pharm. Sci. 67: 239–243.

Nair, R., Y. Vaghasiya, and S. Chanda. 2006. Screening of some medicinal plants of Gujarat for potential antibacterial activity. Indian Journal of Natural Products 22: 17–19.

Nakhare, S. and S.C. Garg. 1996. Antimicrobial activity of essential oil of *Artemisia pallens* Wall. Indian Perfumer 40(4): 118–120.

Narang, G.D., L.C. Garg, and R.K. Mehta. 1961. Preliminary studies on antibacterial activity of *Embelia ribes* Myrsinaceae. Jour Vet and Animal Husbandry Res. 5: 73–77.

Nautiyal, C.S., P.S. Chauhan, and Y.L. Nene. 2007. Medicinal smoke reduces airborne bacteria. Journal of Ethnopharmacology 114: 446–451.

Nidiry, E.S.J. 1999. Antifungal activity of tomato seed extracts. Fitoterapia 2(1): 181–183.

Nilani, P., B. Duraisamy, P.S. Dhanabal, S. Khan, B. Suresh, V. Shankar, K.Y. Kavitha, and G. Syamala. 2006. Antifungal activity of some Coleus species growing in Nilgiris. Ancient Science of Life 26(1-2): 82–84.

Nostro, A., M.P. Germano, V. D'Angelo, A. Marino, and M.A. Cannatelli. 2000. Extraction methods and bioautography for evaluation of medicinal plant antimicrobial activity. Letter. Appl. Microbiol. 30: 379–384.

Nutan, M.T.H., A. Hasnat, and M.A. Rashid. 1998. Antibacterial and cytotoxic activities of *Murraya koenigii*. Fitoterapia 69: 173–175.

Ognean, L., C. Statov, V. Cozma, I. Oana, D. Neculoiu, and V. Miclaus. 1994. Assessment of the therapeutic value of some substances with antimycotic action in mastitis in cows. Buletinul Universitatii de Stiinte Agricole Cluj-Napoca Seria Zootehnie si Medicina Veterinara 48: 403–408.

Oku, H. and K. Ishiguro. 2000. Anti-pruritic and anti-dermatitic effect of extract and compounds of *Impatiens balsamina* L in atopic dermatitis model NC mice. Phytomedicine (Jena) 7.

Oliveira, D.F., A.C. Pereira, H.C.P. Figueiredo, D.A. Carvalho, G. Silva, A.S. Nunes, D.S. Alves, and H.W.P. Carvalho. 2007. Antibacterial activity of plant extracts from Brazilian southeast region. Fitoterapia 78: 142–145.

Olli, S. and P.B. Kirti. 2006. Cloning, characterization and antifungal activity of defensin Tfgd1 from *Trigonella foenum-graecum* L. Journal of Biochemistry and Molecular Biology 39: 278–283.

Ono, H., S. Tesaki, S. Tanabe, and M. Watanabe. 1998. 6-methylsulfinylhexyl isothiocyanate and its homologues as food-originated compounds with antibacterial activity against *Escherichia coli* and *Staphylococcus aureus*. Bioscience Biotechnology and Biochemistry 62: 363–365.

Otshudi, A.L., A. Vercruysse, and A. Foriers. 2000. Contribution to the ethnobotanical, phytochemical and pharmacological studies of traditionally used medicinal plants in the treatment of dysentery and diarrhoea in Lomela area, Democratic Republic of Congo (DRC). J. Ethnopharmacol. 71: 411–423.

Owais, M., K.S. Sharad, A. Shehbaz, and M. Saleemuddin. 2005. Antibacterial efficacy of *Withania somnifera* (ashwagandha) an indigenous medicinal plant against experimental murine salmonellosis. Phytomedicine (Jena) 12: 229–235.

Pai, S.T. and M.W. Platt. 1995. Antifungal effects of *Allium sativum* (garlic) extract against the *Aspergillus* species involved in otomycosis. Lett. Applied Microbiology 20(1): 14–18.

Pandey, M.K., B.K. Sarma, D.P. Singh, and U.P. Singh. 2007. Biochemical investigations of sclerotial exudates of *Sclerotium rolfsii* and their antifungal activity. Journal of Phytopathology (Berlin) 155: 84–89.

Pandey, V.N. and N.K. Dubey. 1994. Antifungal potential of leaves and essential oils from higher plants against soil phytopathogens. Soil Biology and Biochemistry 26: 1417–1421.

Parekh, J. and S. Chanda. 2007a. *In vitro* antimicrobial activity and phytochemical analysis of some Indian medicinal plants. Turk. J. Biol. 31: 53–58.

Parekh, J. and S. Chanda. 2007b. *In vitro* screening of antibacterial activity of aqueous and alcoholic extracts of various Indian plant species against selected pathogens from Enterobacteriaceae. African Journal of Microbiology Research 1(6): 092–099.

Parekh, J., D. Jadeja, and S. Chanda. 2005. Efficacy of aqueous and methanol extracts of some medicinal plants for potential antibacterial activity. Turkish Journal of Biology 29: 203–210.

Parihar, L. and A. Bohra. 2006. Antimicrobial activity of stem extracts of some spices plants. Advances in Plant Sciences 19: 391–395.

Pennacchio, M., A.S. Kemp, R.P. Taylor, K.M. Wickens, and L. Kienow. 2005. Interesting biological activities from plants traditionally used by Native Australians. J. Ethnopharmacol. 96: 597–601.

Perumal Samy, R., S. Ignacimuthu, and A. Sen. 1998. Screening of 34 Indian medicinal plants for antibacterial properties. Journal of Ethnopharmacology 62(2): 173–181.

Porwal, K.M., S. Sharma, and B.K. Metha. 1988. *In-Vitro* Antimicrobial Efficacy of *Butea-Monosperma* Outer Seed Coat Extracts. Fitoterapia 59: 134–135.

Pritima, R.A. and R S, Pandian. 2007. Activity of *Coleus aromaticus* (Benth) against microbes of reproductive tract infections among women. African Journal of Infectious Diseases 1(1): 18–24.

Proestos, C., I. S. Boziaris, G.J.E. Nychas, and M. Komaitis. 2006. Analysis of flavonoids and phenolic acids in Greek aromatic plants: Investigation of their antioxidant capacity and antimicrobial activity. Food Chemistry 95: 664–671.

Purohit, P.S. 2006. Spicy antifungal. Journal of Phytological Research 19: 139–141.

Purohit, P. and A. Bohra. 2002. Antifungal activity of various spice plants against phytopathogenic fungi. Advances in Plant Sciences 15: 615–617.

Pyo, D. and H.H. Oo. 2007. Supercritical fluid extraction of drug-like materials from selected Myanmar natural plants and their antimicrobial activity. Journal of Liquid Chromatography & Related Technologies 30: 377–392.

Qiao, Y.F., S.X. Wang, L.J. Wu, X. Li, and T.R. Zhu. 1990. Studies on Antibacterial Constituents from the Roots of *Rubia cordifolia* L. Yaoxue Xuebao 25: 834–839.

Raghavan, G., V. Madhavan, C.V. Rao, A. Shirwaikar, S. Mehrotra, and P. Pushpangadan. 2004. Healing potential of *Anogeissus latifolia* for dermal wounds in rats. Acta Pharm. 54: 331–338.

Rahman, M.M. and A.I. Gray. 2005. A benzoisofuranone derivative and carbazole alkaloids from *Murraya koenigii* and their antimicrobial activity. Phytochemistry (Amsterdam) 66: 1601–1606.

Rai, M.K. 1987. A first report of *Curvularia verruculosa*, an opportunistic human pathogen from scalp and *In vitro* sensitivity of this fungus to different plant extracts. Indian Drugs 24(11): 818–520.

Rai, M.K. 1988. *In vitro* sensitivity of *Microsporum nanum* to some plants extract. Indian Drugs 25: 521–523.

Rai, M.K. 1989. Mycosis in Man due to *Arthrinium phaeospermum* var. *indicum*. First case report. Mycoses 32(9): 472–475.

Rai, M.K. and S. Vasanth. 1995. Laboratory evaluation of sensitivity of three keratinophilic fungi to some vicolides. Hindustan Antibio. Bull. 37(1-4): 48–50.

Rai, M.K. and D. Acharya. 1999. Screening of some Asteraceous plants for antimycotic activity. Compositae Newsletter 34: 37–43.

Rai, M.K. and D. Acharya. 2000. Search for fungitoxic potential in essential oils of Asteraceous plants. Compositae Newsletter 35: 18–23.

Rai, M.K. and D. Acharya. 2002. *In vitro* susceptibility of *Trichophyton mentagrophytes* to different concentrations of three Asteraceous essential oils. Compositae Newsletter 39: 98–105.

Rai, M.K., S. Qureshi, and A.K. Pandey. 1999. *In vitro* susceptibility of opportunistic *Fusarium* spp. to essential oils. Mycoses 42: 97–101.

Rai, M.K., K.K. Soni, and D. Acharya. 2002. *In vitro* effect of five Asteraceous essential oils against *Saprolegnia ferax*, a pathogenic fungus isolated from fish. The Antiseptic 99(4): 136–137.

Rai, M.K., D. Acharya, and P. Wadegaonkar. 2003. Plant derived-antimycotics: Potential of Asteraceous plants, In: Plant-derived antimycotics:Current Trends and Future prospects, Haworth Press, N-York, Londin, Oxford. pp. 165–185.

Ram, A.J., L.M. Bhakshu, and R.R.V. Raju. 2004. *In vitro* antimicrobial activity of certain medicinal plants from Eastern Ghats, India, used for skin diseases. Journal of Ethnopharmacology 90: 353–357.

Ramesh, N., M.B. Viswanathan, A. Saraswathy, K. Balakrishna, P. Brinda, and P. Lakshmanaperumalsamy. 2002. Phytochemical and antimicrobial studies of *Begonia malabarica*. J. Ethnopharmacol. 79: 129–132.

Ramesh, N., M.B. Viswanathan, V. Tamil Selvi, and P. Lakshmanaperumalsamy. 2004. Antimicrobial and phytochemical studies on the leaves of *Phyllanthus singampattiana* (sebastine & a.n. Henry) kumari & chandrabose from India, Medicinal Chemistry Research 13(6-7): 348–360.

Ramos, A.D.R., L.L. Falcao, G.S. Barbosa, L.H. Marcellino, and E.S. Gander. 2007. Neem (*Azadirachta indica* a. Juss) components: Candidates for the control of *Crinipellis perniciosa* and *Phytophthora* ssp. Microbiological Research 162: 238–243.

Rani, P. and N. Khullar. 2004. Antimicrobial evaluation of some medicinal plants for their anti-enteric potential against multi-drug resistant *Salmonella typhi*. Phytother Res. 18(8): 670–673.

Rao, P.S., K.V.P. Rao, and K.R. Raju. 1987. Synthesis and Antibacterial Activity of Some New Embelin Derivatives. Fitoterapia 58: 417–418.

Ray, P.G. and S.K. Majumdar. 1975. New antifungal substance from *Alpinia officinarum*. Indian J. Exp. Biol. 13(4): 489.

Ray, P.G. and S.K. Majumdar. 1976a. Antifungal flavonoid from *Alpinia officinarum*. Indian J. Exp. Biol. 14(6): 712–714.

Ray, P.G. and S.K. Majumdar. 1976b. Antimicrobial Activity of Some Indian Plants. Economic Botany 30(4): 317–320.

Rees, L.P., S.F. Minney, N.T. Plummer, J.H. Slater, and D.A. Skyrme. 1993. A quantitative assessment of the antimicrobial activity of garlic (*Allium sativum*). World Journal of Microbiology and Biotechnology 9(3): 303–307.

Saadabi, A.M.A. 2006. Antifungal activity of some Saudi plants used in traditional medicine. Asian Journal of Plant Sciences 5: 907–909.

Samuel Karunyal, K., B. Andrews, and H. Shyla Jebashree. 2000. *In vitro* evaluation of the antifungal activity of *Allium sativum* bulb extract against *Trichophyton rubrum* a human skin pathogen. World Journal of Microbiology and Biotechnology 16(7): 617–620.

Samy, R.P., S. Ignacimuthu, and A. Sen. 1998. Screening of 34 Indian medicinal plants for antibacterial properties. Journal of Ethnopharmacology 62: 173–182.

Sangameswaran, B., B.R. Balakrishnan, B. Arul, and B. Jayakar. 2004. Antimicrobial activity of aerial parts of *Argemone mexicana* Linn. Journal of Ecotoxicology and Environmental Monitoring 14(2): 143–146.

Sasidharan, V.K. 1997. Search for antibacterial and antifungal activity of some plants of Kerala. Acta Pharmaceutica (Zagreb) 47: 47–51.

Sato, M., S. Fujiwara, H. Tsuchiya, T. Fujii, M. Iinuma, H. Tosa, and Y. Ohkawa. 1996. Flavones with antibacterial activity against cariogenic bacteria. Journal of Ethnopharmacology 54: 171–176.

Satya, V.K., R. Radhajeyalakshmi, K. Kavitha, V. Paranidharan, R. Bhaskaran, and R. Velazhahan. 2005. *In vitro* antimicrobial activity of zimmu (*Allium sativum* L.) leaf extract. Archives of Phytopathology and Plant Protection 38(3): 185–192.

Sedigheh, M., M. Ahmad, and M. Iman. 2000. Antimicrobial effects of three plants (*Rubia tinctorum, Carthamus tinctorius* and *Juglans regia*) on some airborne microorganisms. Aerobiologia 16(3-4): 455–458.

Shamim, S., S.W. Ahmed, and I. Azhar. 2004. Antifungal activity of Allium, Aloe, and Solanum species. Pharmaceutical Biology 42: 491–498.

Sharma, B.K. 1998. Antifungal properties of biocontrol agents and plants extracts against causal fungi of yellows and rhizomes rot of ginger. Journal of Biological Control 12: 77–80.

Sharma, K.S., K.M. Porwal, and B.K. Metha. 1989. *In-Vitro* Antimicrobial Activity of *Musa paradisiaca* Root Extracts. Fitoterapia 60(2). 157–158.

Sheldon, J.W, M.J. Balick, and S.A. Laird. 1997. Medicinal plants: can utilization and conservation coexist? In: Advances in Economic Botany, 12. The New York Botanical Garden, New York.

Shirwaikar, A., A.P. Somashekar, A.L. Udupa, S.L. Udupa, and S. Somashekar. 2003. Wound healing studies of *Aristolochia bracteolate Lam.* With supportive action of antioxidant enzymes. Phytomed. 10: 558–562.

Shoba, F.G. and M. Thomas. 2001. Study of antidiarrhoeal activity of four medicinal plants In castor-oil induced diarrhea. J. Ethnopharmacol. 76: 73–76.

Shukla, Y.N., A. Srivastava, S. Kumar, and S. Kumar. 1999. Phytotoxic and antimicrobial constituents of *Argyreia speciosa* and *Oenothera biennis.* Journal of Ethnopharmacology 67(2): 241–245.

Siddiqui, M.A., A. Haseeb, and M.M. Alam. 1992. Control of plant-parasitic nematodes by soil amendments with latex bearing plants. Indian Journal of Nematology 22: 25–28.

Silva, O., S. Barbosa, A. Diniz, M.L. Valdeira, and E. Gomes. 1997. Plant extracts antiviral activity against Herpes simplex virus type 1 and African swine fever virus. International Journal of Pharmacognosy 35: 12–16.

Simonsen, H.T., J.B. Nordskjold, U.W. Smitt, U. Nyman, P. Palpu, P. Joshi, and G. Varughese. 2001. *In vitro* screening of Indian medicinal plants for antiplasmodial activity. Journal of Ethnopharmacology 74: 195–204.

Singh, A.K., A.S. Raghubanshi, and J.S. Singh. 2002. Medical ethnobotany of the tribals of Sonaghati of Sonbhadra district, Uttar Pradesh, India. J. Ethnopharma. 81: 31–41.

Singh, R., A. Jain, S. Panwar, D. Gupta, and S.K. Khare. 2005a. Antimicrobial activity of some natural dyes. Dyes and Pigments 66: 99–102.

Singh, S., M. Malhotra, and D.K. Majumdar. 2005b. Antibacterial activity of *Ocimum sanctum* L. fixed oil. Indian Journal of Experimental Biology 43: 835–837.

Singh, U.P., B.P. Srivastava, K.P. Singh, and V.B. Pandey. 1992. Antifungal activity of steroid saponins and sapogenins from *Avena sativa* and *Costus speciosus.* Naturalia (Rio Claro) 17: 71–77.

Sinha, A.K., K.P. Verma, K.C. Agarwal, N.K. Toorray, and M.P. Thakur. 2004. Antifungal activities of different plant extracts against *Colletotrichum capsici.* Advances in Plant Sciences 17: 337–338.

Sivakumar, T., P. Perumal, R.S. Kumar, M.L. Vamsi, P. Gomathi, U.K. Mazumder, and M. Gupta. 2004. Evaluation of analgesic, antipyretic activity and toxicity study of *Bryonia laciniosa* in mice and rats. Am. J. Chin. Med. 32(4): 531–539

Southwell, I.A. and A.B. Christopher. 2001. Seasonal variation in hypericin content of *Hypericum perforatum* L. (St. John's Wort). Phytochemistry 56(5): 437–441.

Srikrishna, L.P., H.M. Vagdevi, B.M. Basavaraja, and V.P. Vaidya. 2008. Evaluation of antimicrobial and analgesic activities of *Aporosa lindleyana* (Euphorbiaceae) bark extract. Int. J. Green Pharm 2: 155–157.

Srinivasan, D., S. Nathan, T. Suresh, and P.L. Perumalsamy. 2001. Antimicrobial activity of certain Indian medicinal plants used in folkloric medicine. J. Ethnopharmacol. 74: 217–220.

Srivastava, M., S. Srivastava, K. Sayyada, A.K.S. Rawat, S. Mehrotra, and P. Pushpangadan. 2006. Pharmacognostical evaluation of *Cassia angustifolia* seeds. Pharmaceutical Biology 44(3): 202–207.

Sule, A.M. and O. Olusanys. 2001. Studies of the antibacterial activities of some Nigerian medicinal plants. International Journal of Antimicrobial Agents 17.

Sunitha, R.K. and A. Chandrashekar. 1994. Isolation and purification of three antifungal proteins from sorghum endosperm. Journal of the Science of Food and Agriculture 64(3): 357–364.

Svarstad, H. and S.S. Dhillion. 2000. Responding to bioprospecting: From biodiversity in the South to medicines in the North. Spartacus Forlag As, Oslo. ISBN 82-430-0163-8.

Swain, S.R., B.N. Sinha, and P.N. Murthy. 2008. Antiinflammatory, diuretic and antimicrobial activities of *Rungia pectinata* linn. and *Rungia repens* nees, Indian J. Pharm. Sci. 70(5): 679–683.

Swami, C.S. and D.S. Mukadam. 2006. Antifungal property of some plant extracts against tomato fungi. Geobios (Jodhpur) 33: 261–264.

Sydiskis, R.J., D.G. Owen, J.L. Lohr, K.H. Rosler, and R.N. Blomster. 1996. Inactivation of enveloped viruses by anthraquinones extracted from plants. Antimicrobial Agents and Chemotherapy 35(12): 2463–2466.

Tabuti, J.R.S., S.S. Dhillion, and K.A. Lye. 2003. Traditional medicine in Bulamogi county, Uganda: its practitioners, users and viability. Journal of Ethnopharmacology 85: 119–129.

Tahir, F. and F.M. Chi. 2002. Phytochemical and antimicrobial screening of the leaf and root bark extracts of *Calotropis procera* (Ait). Pakistan Journal of Scientific and Industrial Research 45: 337–340.

Teh, R.N. 1998. The role of traditional medical practitioners: in the context of the African traditional concept of health and healing. International Mental Health Workshop. http://www.globalconnections.co.uk/pdfs/HealersMentalHealth.pdf accessed 01 January 2002.

Thevissen, K., I.E.J.A. Francois, L. Sijtsma, A.A. van, W.M.M. Schaaper, R. Meloen, T. Posthuma-Trumpie, W.F. Broekaert, and B.P.A. Cammue. 2005. Antifungal activity of synthetic peptides derived from *Impatiens balsamina* antimicrobial peptides Ib-AMP1 and Ib-AMP4. Peptides (New York) 26: 1113–1119.

Tripathi, R.D. and S.N. Dixit. 1981. Fungitoxicity of lawsone, isolated from leaves of *Lawsonia inermis* and some related napthaquinones of angiosperm origin, (abstract), paper presented at 3rd Int Symp on Plant Pathol, IARI, New-Delhi, Dec 14–18.

Tsey, K. 1997. Traditional medicine in contemporary Ghana: A public policy analysis. Social Science & Medicine 45(7): 1065–1074.

Ugent, D. 2000. Medicine, myths and magic: the folk healers of a Mexican market. Economic Botany 54(4): 427–438.

Uma Reddy, B., T. Sandeep, D. Kumar, G. Prakash, M.O. Farooque, and H.B. Nayaka. 2008. Antimicrobial activity of *Tagetes erecta* Linn. and *Argemone mexicana* Linn. Aryavaidyan 21(3): 171–174.

Upadhyay, O.P., Kumar Kaushal, and R.K. Tiwari. 1998. Ethnobotanical Study of Skin Treatment Uses of Medicinal Plants of Bihar. Pharmaceutical Biology 36(3): 167–172.

Uzun, E., G. Sariyar, A. Asersen, B. Karakoc, G. Otuk, E. Oktayoglu, and S. Pirildar. 2004. Traditional medicine in Sakarya province (Turkey) and antimicrobial activities of selected species. J. Ethnopharmacol. 95: 287–296.

Valsaraj, R., P. Pushpangadan, U.W. Smitt, A. Adsersen, and U. Nyman. 1997. Antimicrobial screening of selected medicinal plants from India. J. Ethnopharmacol. 58: 75–83.

Venkitraman, S. 1978. Anti Fungal Activity of Certain Rhizomes *Curcuma longa, Curcuma amada* etc. Indian Journal of Physiology and Pharmacology 22.

Verastegui, M.A., C.A. Sanchez, N.L. Heredia, and J.S.G. Alvarado. 1996. Antimicrobial activity of extracts of three major plants from the Chihuahuan desert. J. Ethnopharmacol. 52: 175–177.

Viyoch, J., N. Pisutthanan, A. Faikreua, K. Nupangta, K. Wangtorpol, and J. Ngokkuen. 2006. Evaluation of *In vitro* antimicrobial activity of Thai basil oils and their micro-emulsion formulas against Propionibacterium acnes. International Journal of Cosmetic Science 28: 125–133.

Vohora, S.B., I. Kumar, S.A.H. Naqvi, and S.H. Afaq. 1972. Pharmacological investigations on *Picrorhiza kurroa* roots (kutki) with special reference to its chloretic and antimicrobial properties. Indian Journal of Pharmacy 34(1): 17–19.

Voravuthikunchai, S.P. and S. Limsuwan. 2006. Medicinal plant extracts as anti-*Escherichia coli* O157: H7 agents and their effects on bacterial cell aggregation. Journal of Food Protection 69: 2336–2341.

Vrushabendra Swamy, B.M., K.N. Jayaveera, G.S. Kumar, C.K. Ashok Kumar, and C. Sreedhar. 2006. Anti diarrheal activity of root extracts of *Elephantopus scaber* L. Planta Medica 72.

Wang, Y.C. and T.L. Huang. 2005. Anti-*Helicobacter pylori* activity of *Plumbago zeylanica* L. FEMS Immunology and Medical Microbiology 43: 407–412.

World Bank. 1992. World Development Report 1992: Development and the environment. The World Bank, Washington, D.C.

World Health Organization (WHO) 1995. Traditional practitioners as primary health care workers: Guidelines for training traditional health practitioners in primary health care. WHO, Geneva.

Young, R.N. 1999. Importance of biodiversity to the modern pharmaceutical industry. Pure Appl. Chem. 71(9): 1655–1661.

Zafar, R., P. Singh, and A.A. Siddiqui. 1989. Antimicrobial and Preliminary Phytochemical Studies on Leaves of *Butea-Monosperma* Linn. Indian Journal of Forestry 12: 328–329.

Zhu, N., M.R. Mohamed, S.D. Robert, J. Xin, C.V. Chee-Kok, G. Badmaev, R.T.R. Geetha, and Chi-Tang Ho. 2001. Bioactive constituents from gum guggul (*Commiphora wightii*), Phytochemistry 56(7): 723–727.

Chapter 9

Medicinal Usefulness of *Woodfordia fruticosa* (Linn.) Kurz

Shandesh Bhattarai and *Dinesh Raj Bhuju*

Introduction

Since long, *Woodfordia fruticosa* has been considered as an important medicinal plant. Its flowers and leaves have been commonly used in traditional medicinal practices in South East Asian countries, including Nepal. The other parts of the plant such as twigs and the bark have also been reported for other uses.

Many researchers have been attracted by this plant. A total of 87 research papers were reviewed related to research on *Woodfordia fruticosa*, of which, 79 research papers describe the ethnobotanical research including medicinal uses (Anonymous 1970, Rajbhandari 2001, Joshi and Joshi 2001, Manandhar 2002, Prajapati et al. 2003, Prakesh and Chanda 2007, etc.), chemical constituents (Yoshida et al. 1989a, 1990, Kadota et al. 1990, etc.), biological activity (Kuramochi-Motegi et al. 1992, Kroes et al. 1993, Tewari et al. 2001, Chandan et al. 2008, Kaou et al. 2008, etc.), commercial application (Tsuji et al. 1996, etc.), and other uses (Agarwal and Hasija 1961, Chadha 1976, Bhagat et al. 1992, Krishnan and Seeni 1994, Samantaray et al. 1999, Kala et al. 2004, etc.), followed by three on taxonomic study (Rajbhandari et al. 1995, etc.), and five on ecological study and distribution (Bhuju and Yonzon 2000, Bhuju 2003, Bhuju et al. 2004, etc.). A survey of literature revealed that *Woodfordia fruticosa* has been identified and used in the process of drug development encouraging

Nepal Academy of Science and Technology, GPO Box 3323, Kathmandu.
E-mail: *bhattaraishandesh@yahoo.com*

many researchers to find its effectiveness to treat several diseases. Thus, the plant can be a source of potentially important new pharmaceutical substances.

Nepal is a multiethnic and multilingual country with more than 60 different ethnic groups speaking about 75 languages (CBS 2002, Bista 2004). Several vernacular names of *Woodfordia fruticosa* exist but the most common is 'Dhainyaro, Anarephul' in Nepali. Some vernacular names are 'Daring' Chepang; 'Dhayar, Dhanyar, Dhurseli' Gurung; 'Birukanda, Jamjasa, Dhayaro, Syakte' Tamang; 'Ghayaro' Tharu; 'Bajhiya, Chyuhuwa, Danuwar, Dhainra' Magar; 'Dhauli' Majhi; 'Chenchev' Rai; 'Dhaya' Raute; and 'Dha-ta-ki' Tibetan. The Sanskrit names are 'Dhataki, Dhatupuspi, Agnijwala, Tamra-pushpi'. In Hindi, it is called 'Dawi, Dhataki, Jargi, Darwari, Phulsatti, Dhavdi, Thawi' etc. The common English name is Fire Flame Bush (Anonymous 1970, Chadha 1976, Shome et al. 1981, Rajbhandari et al. 1995, Rajbhandari 2001, Joshi and Joshi 2001, Manandhar 2002, Prajapati et al. 2003, Khare 2004, Das et al. 2007).

Considering the above facts, the ethnobotanical research of Nepalese *Woodfordia fruticosa* is continuing to follow the methods adopted by Taylor et al. (1995, 1996, 2002) and Martin (1995). Information of *Woodfordia fruticosa* on the traditional practices of the indigenous Nepalese people is fragmented and scattered in old literature, but there are a few accurate details of plants and their uses and some of them contain interesting observations on medicinal practices (Manandhar 2002). Taking both the ethnobotanical and pharmacological applications of the plant, the aim was to compile the scattered information (focusing mostly on Nepalese research) and to present our preliminary research findings.

General characters (Taxonomy, Distribution and Habit)

Woodfordia fruticosa (Linn.) Kurz belongs to the family Lythraceae and its taxonomy is well described. It is a beautiful shrub with spreading branches, ca. 1–3 m tall (see Fig. 9.1). Leaves are sessile, opposite, stalked, lanceolate, oblong-lanceolate or ovate-lanceolate, 5–9 cm long and 1.3–2.5 cm abroad, upper surface green, lower surface velvety, with black dots. Flowers are numerous in a single plant, short-stalked which are arranged in 2–15 flowered cymes. The color of the flowers is brilliant red in dense axillary clusters. Fruits are defined as capsules, ovoid in shape and 1 cm in length. The seeds are brown, minute, smooth and ovate. The flowering and fruiting seasons range from March–June and propagation is done by seeds or root upshoots.

Rajbhandari et al. (1995) studied the microscopical characters of Nepalese *Woodfordia fruticosa*. Some of the characters are recognized to be important. The color of the powdered drug is brown. The parenchyma

Fig. 9.1. Flowering twigs of *Woodfordia fruticosa* (Linn.) Kurz.
Color image of this figure appears in the color plate section at the end of the book.

cells contain minute clusters of calcium oxalate, with annular vessel and unicellular trichome with a pointed tip. Pollen grains are spherical in shape and the diameter is 15 mm. Palillose epidermis of the stigma and of the epidermal cell in surface view is polygonal but in sections almost rectangular.

Woodfordia fruticosa has wide range of ecological distribution in Nepal, mainly in altitudes of 200–1800 m. It is also widely distributed throughout India up to an elevation of about 1,500 m, and also in most countries of subtropical Himalayas, South and South East Asia such as India, Myanmar, Pakistan, Nepal, Bhutan, Sri Lanka, Malaysia, Indonesia and also in China. The plant has also been reported from tropical Africa as well as Australia (Kirtikar and Basu 1935, Press et al. 2000).

Research on *Woodfordia fruticosa*

Ethnobotanical information

The traditional uses of *Woodfordia fruticose* have long been recognized in traditional medicinal practices in India. It is used a lot in Ayurvedic and Unani systems of medicines (Chopra et al. 1956, Watt 1972, Dymock et al. 1995). Literature reveals that flowers, leaves, stems and barks have an exceptionally wide diversity of medicinal uses to treat more then 60 ailments (Table 9.1). However, in the case of medicinal plants, it is important to know as much as possible about their local use in order to help assure their rational exploitation (Filho and Sor 1997).

Table 9.1. Medicinal uses of *Woodfordia fruticosa*

Medicinal Uses	Reference
Acrid, alexeteric, antibacterial, bile diseases, depurative, erysipelas, haemoptysis, hepatopathy, refrigerant, styptic, verminosis, vulnerary	Prajapati et al. 2003
Antipyretic	Mhasker et al. 2000, IUCN-Nepal 2004
Anti-inflammatory	Mhasker et al. 2000, Oudhia 2003, Prajapati et al. 2003, IUCN-Nepal 2004
Antitumor	Kuramochi-Motegi et al. 1992
Astringent	Burkil 1966, Dey 1984, Aminuddin Girach 1996, Mhasker et al. 2000, Prajapati et al. 2003, IUCN-Nepal 2004, GoN 2007
Boils, gastric trouble, angular stomatitis, sprains, swellings	Manandhar 2002
Burns, burning sensation	Pandey 1992, Mhasker et al. 2000, Prajapati et al. 2003, IUCN-Nepal 2004
Cough	Aminuddin Girach 1996, Oudhia 2003, Prajapati et al. 2003
Cuts, bleeding injuries	Chadha 1976, Pandey 1992, Rajbhandari et al. 1995, Oudhia 2003
Derangements of the liver, disorders of the mucous membrane	Mhaskar et al. 2000
Diabetes, leprosy	Prajapati et al. 2003, Prakesh and Chanda 2007
Diarrhea, dysentery	Burkill 1966, Dey 1984, Borthakur 1992, Rajbhandari et al. 1995, Manandhar 2002, Prajapati et al. 2003, IUCN-Nepal 2004, GoN 2007
Fever	Pandey 1992, Manandhar 2002, Prajapati et al. 2003, IUCN-Nepal 2004
Headache	Mhasker et al. 2000
Hemorrhoids	Chadha 1976, Rajbhandari et al. 1995, Mhaskar et al. 2000, Prajapati et al. 2003, IUCN-Nepal 2004
Immunomodulatory	Labadie et al. 1989, Kroes et al. 1993
Indigestion, constipating	Manandhar 2002, Prajapati et al. 2003, IUCN-Nepal 2004
Leucorrhoea	Prajapati et al. 2003, IUCN-Nepal 2004
Menorrhagia	Rajbhandari et al. 1995, Prajapati et al. 2003, IUCN-Nepal 2004
Menstruation disorders, pregnancy	Rajbhandari et al. 1995, Manandhar 2002, Prajapati et al. 2003, IUCN-Nepal 2004, GoN 2007
Skin diseases	Prajapati et al. 2003, Prakesh and Chanda 2007
Small pox	Chadha 1976, Oudhia 2003
Seminal weakness, antibiotic, otorrhoea	Oudhia 2003

Medicinal Uses	Reference
Spruce, rheumatism, hematuria, infertility	Burkil 1966, Dey 1984
Ulcers	Khorya and Katrak 1984, Oudhia 2003, Prajapati et al. 2003
Urinary disorders, urinary troubles, uterine sedative, correction of urinary pigments, dysuria	Burkil 1966, Dey 1984, Pandey 1992, Manandhar 2002, Prajapati et al. 2003, IUCN–Nepal 2004
Wounds	Khorya and Katrak 1984, Pandey 1992, Oudhia 2003

Besides its use in traditional medicine, the plant has long been used for miscellaneous purposes in India. Chadha (1976), Kroes et al. (1993) and Prajapati et al. (2003) reported that the use of flowers in the process of alcoholic fermentation significantly raised alcohol content. Chadha (1976) reported the miscellaneous use of the plant, i.e., in the preparation of axe-handles, cooling drink, dye, tanning material in the industry, to check landslides and as a good source of fuelwood (Table 9.2).

Table 9.2. Other usage of *Woodfordia fruticosa*

Other Uses	Reference
Alcoholic fermentation	Chadha 1976, Kroes et al. 1993, Prajapati et al. 2003
Axe-handles, tanning material, cold drink, landslides	Chadha 1976
Dye	Chadha 1976, Manandhar 2002, IUCN-Nepal 2004
Fuel	Chadha 1976, Manandhar 2002
Narcotic (smoking)	Manandhar 2002

Chemical Constituents

The use of *Woodfordia fruticosa* in the management of various illnesses is due to its phytochemical constituents. Several studies from South East Asian countries have been submitted to numerous chemical investigations (Desai et al. 1971, Nair et al. 1976, Srivastava et al. 1977, Chauhan et al. 1979a, 1979b, Yoshida et al. 1989a, 1990, 1991, 1992, etc., see Table 9.3 for details). Such phytochemical studies revealed the presence of numerous bioactive compounds: including tannins (especially those of macrocylic hydrolysable class), flavonoids, anthraquinone glycosoides, and polyphenols (Table 9.3) from flowers, leaves, and stem.

Biological activity

Biologically active flowers and leaves of *Woodfordia fruticosa* are storehouses of several chemical compounds (Sharma 1956, Yoshida et al. 1989b, Choe et al. 1990, Kadota et al. 1992, Das et al. 2006, Dabur et al. 2007) and possess immunomodulatory, anti-inflammatory, antileucorrhoeic, antitumor activity (see Table 9.4 for details). Dudov et al. (1994) and Qin and Sun (2005) stated that flavonoides of plants are known by their immunomodulatory and anti-inflammatory activities. Roldao et al. (2008) explained that plant extracts with anti-inflammatory activity could represent a promising and practical approach for the management of inflammatory disease, especially if these compounds do not harm the gastrointestinal tract.

Commercial applications

This medicinal plant has a long history of traditional and commercial applications in South Asian countries where the drugs are produced (Table 9.5). Many marketed drugs comprise of flowers, leaves, fruits, and

Table 9.3. Chemical constituents of *Woodfordia fruticosa*

Compounds	Plant Parts	Reference
β-sitosterol, triterpenoids, lupeol, betulin, betulinic acid, oleanolic acid, ursolic acid, gallic acid, ellagic acid, naphtoquinone lawsone, 3-β-L-arabinoside (polystachoside), D 3-O-α-L-arabinopyranoside, 3-O-β-D-xylopyranoside, 3-O-6"-galloyl)-β-D-glucopyranoside, 3-O-(6"-galloyl)-β-D-lactopyranoside, myricetin glycosides [3-O-β-D-galactoside], 3-O- α-L-arabinopyranoside, 3-(6"-galloyl)-β-D-galactopyranosides, woodfordisin	Leaves	Saoji et al. 1972, Nair et al. 1976, Kalidhar et al. 1981, Dan and Dan 1984, Kadota et al. 1990
Octacosanol, β-sitosterol, nor-bergenin, bergenin, gallic acid	Stem	Desai et al. 1971, Kalidhar et al. 1981, Kadota et al. 1990
Sapogenin, hecogenin, β-sitosterol, meso-inositol, octacosanol, ellagic acid, myricetin glycosides [3-O-β-D-galactoside], anthocynanidin pigment cyaniding-3,5-diglucoside, pelargonidin-3,5-diglucoside, Chrysophanol-8-O-β-D-glucopyranoside, 6 quercetin glycosides [3-rhamnoside, 3-β-L-arabinoside (polystachoside) naringenin 7-glucoside, keampferol 3-O-galacoside, 1,2,3,6-tetra-O-galloyl-β-D-glucose, 1,2,4,6-tetra-O-galloyl-β-D-glucose, 1,2,3,4,6-penta-O-galloyl-β-D-glucose, tellimagrandin, gemin D, heterophylliin A, oenothein β, woodfordin A-C, woodfordin D, oenothenin A, isoschimawalin A, woodfordins E-I	Flowers	Nair et al. 1976, Srivastave et al. 1977, Chauhan et al. 1979a, 1979b, Yoshida et al. 1989a, 1990, 1991, 1992

Table 9.4. Biological Activity of *Woodfordia fruticosa*

Activity	Parts used	Drug produced	Reference
Anti-inflammatory	Flowers	*Balarishta*	Anonymous 1978, Alam et al. 1998
Antileucorrohoeic	Flowers		Tewari et al. 2001
Antitumor	Flowers, Leaves	Woodfordin	Kuramochi-Motegi et al. 1992
Immunomodulatory	Flowers	*Nimba Aristha*	Kroes et al. 1993

buds mixed with pedicels and thinner twigs of the plant (Nadkarni 1954, Chopra et al. 1956, Ahuja 1965, Dutt 1992). Atal et al. (1982) explained that the flowers are used in the preparation of Ayurvedic fermented drugs called '*Aristhas* and *Asavas*' and are very popular in the Indian subcontinent as also in other South Asian countries (Jayaweera 1981, Kroes et al. 1993). Burkill (1966) stated that a popular drug of Indonesia and Malaysia '*Sidowayah* or *Sidawayah*' contains mainly dried flowers.

Chadha (1976) and Ambasta (1986) stated its commercial application in the preparation of cold drinks. Dastur (1951) and Watt (1972) explained the use of the plant in dyeing of fabrics, and as an adjunct/mordant. Similarly, Ambasta (1986) reported its use as gum for coating paste of the fabric. It is also used as an ointment for skin whitening and preservatives of plant fruits and vegetables (Table 9.5). Chadha (1976) recorded the use of the plant in the tanning industry and Oshima and Mitsunga (1999) explained its use in prevention and treatment of dental plaque formation.

Das et al. (2007) reported the regular demand of flowers amongst practitioners of traditional medicines in different South East Asian countries. Oudhia (2003) recorded the high demand for flowers in national and international markets for the preparation of traditional herbal medicines. Therefore, unique commercial application of the plant increases its demand in international markets.

Study of Nepalese *Woodfordia fruticosa*

The traditional uses of *Woodfordia fruticose* have long been recognized in Nepal (Manandhar 2002, GoN 2007, etc.). Most often, flowers, leaves, stems and barks have exceptionally wide diversity of ethnobotanical uses among the various ethnic groups in different parts of the country. Many ethnic communities in Nepal have elaborated plant knowledge and most of the knowledge on plants is transferred orally and therefore, there is the risk of losing this precious cultural heritage. To overcome the rapid loss of cultural diversity and traditional knowledge of plants, an increasing numbers of ethnobotanical inventories need to be established (Van Wyk et

Table 9.5. Commercial Applications of *Woodfordia fruticosa*

Application	Plant Parts	Reference
Cold drinks	Flowers	Chadha 1976, Ambasta 1986
Dyeing of fabrics, an adjunct/mordant	Flowers	Dastur 1951, Watt 1972
Gum used for coating the paste of the fabric	Whole plant	Ambasta 1986
Ointment for skin whitening	Flowers, Leaves	Nanba et al. 1995, Ueda and Shimomura 1995, Suzuki et al. 2004, Adachi et al. 2005, 2006
Preservatives of plant fruits and vegetables	Flowers	Tsuji et al. 1996
Prevention and treatment of dental plaque formation	Whole plant	Oshima and Mitsunga 1999
Tanning Industry	Bark, Leaves, Flowers, Fruit	Chadha, 1976

al. 2002), in different parts of Nepal where the plant has been significantly used in the day to day life by indigenous people.

In connection with the ethnobotanical research of *Woodfordia fruticosa*, we visited the Makawanpur and Nawalparasi districts of Nepal in 2009 (Fig. 9.2), where the wide uses of *Woodfordia fruticosa* were also documented. Local people have been using different parts of the plant for various purposes including medicinal, fuel and fodder. In Makawanpur district, about two tea spoonfuls of fresh flower juice have traditionally been used three times a day to treat diarrhoea and dysentery whereas the Tharu people of Nawalparasi district have been using about five tea spoonfuls of flowers and bark juice twice a day for stomachache, diarrhoea, dysentery and typhoid fever until recovery.

Besides medicinal purposes, the people of Makawanpur, and Nawalparasi districts also use the plant as a good source of fuelwood and fodder. Fodder from this plant is thought to be highly nutritious. In addition, during the peak months of flowering (May-July), people suck the nectar deposited in the red flowers (Fig. 9.3). The nectar which remains in the flowers has a sweet taste. However, research on analysis of nutritive values of the flowers and leaves of the plant has not yet been conducted in Nepal and therefore the nutritive constituents from Nepalese samples need to be identified.

The research by Manandhar (2002) and IUCN-Nepal (2004) also showed the multiple use of *Woodfordia fruticosa* in dye, fuel, and narcotic (smoking) (Table 9.2).

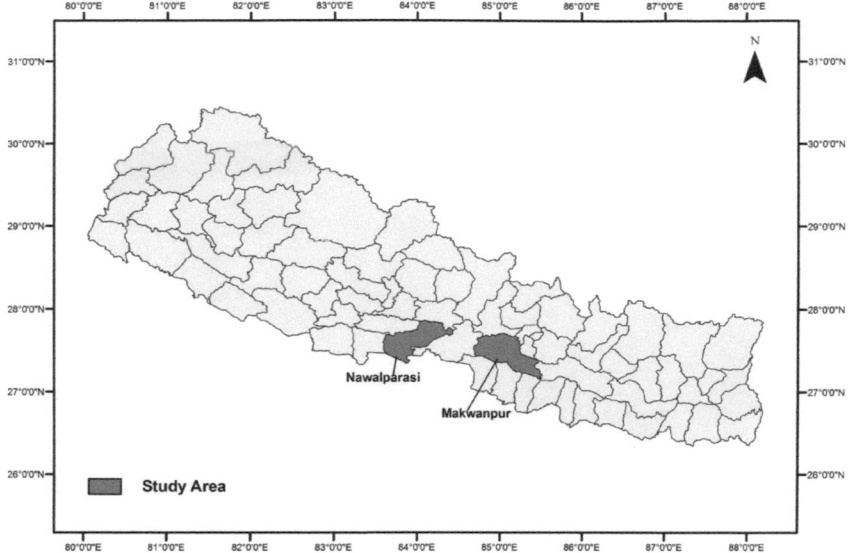

Fig. 9.2. Map of Nawalparasi and Makawanpur Districts of Nepal.

Color image of this figure appears in the color plate section at the end of the book.

We observed the flowers of *Woodfordia fruticosa* in the market in major cities of Nepal (Chitwan, Parsa, Kathmandu and Makawanpur districts), in the peak seasons (June -August), along with other trading medicinal

Fig. 9.3. Sucking of nectars of flowers of W*oodfordia fruticosa* (Linn.) Kurz.

Color image of this figure appears in the color plate section at the end of the book.

plants. One kg of flowers was sold for NRs 80-130 (about US $ 1–1.5) in different districts (see Fig. 9.4).

Fig. 9.4. *Woodfordia fruticosa* flowers in the market (in circle).
Color image of this figure appears in the color plate section at the end of the book.

Few studies have been designed to test the hypothesis related to the knowledge and use of medicinal plants in Nepal (Bhattarai et al. 2008a,b). Bhattarai et al. (2009) preliminary antibacterial research on methanol extracts of Nepalese flower samples of *Woodfordia fruticosa*, observed positive activity against two Gram-positive bacteria (*Stphylococcus aureus* and *Bacillus subtilis*) and two Gram-negative bacteria (*Pseudomonas aeruginosa* and *Escherichia coli*) in *in vitro* testing by disc diffusion method.

Bajrachraya et al. (2008) obtained the antimicrobial activity of the ethanol extracts of leaves against *Escherichia coli*, *Salmonella typhi*, *Salmonella paratyphi*, *Proteus vulgaris*, *Proteus mirabilis*, *Klebsiella* spp., *Citrobacter* spp., *Enterobacter* spp., *Shigella* spp., and *Pseudomonas* spp. Similarly, Timsina (2003) obtained the antimicrobial effect of methanol extracts of leaves against *Bacillus subtilis*, *Candida albicans*, *Proteus vulgaris*, *Pseudomonas aeruginosa*, *Salmonella typhi*, *Salmonella paratyphi*, *Shigella dysenteriae*, *Staphylococcus aureus*, and *Vibrio cholera* but showed inactivity against *Escherichia coli*. Based on the traditional ethnomedicinal use of *Woodfordia fruticosa* and the above findings, we are continuing the antimicrobial research and identifying the Minimum Inhibitory Concentration (MIC) of Nepalese flower and leaves samples in methanol, chloroform and hexane against 15 microorganisms.

Information regarding the important chemical constituents of the plant justified its significance in traditional remedies. However, there is no study carried out to evaluate the chemical composition from Nepalese samples. In our study we considered investigating the active chemical constituents of Nepalese samples for isolation and structural elucidation of the chemical constituents. After the correct identification of the active chemical constituents of the plant along with the detailed ethnobotanical, and pharmacological findings we hope to further our research towards the commercialization of the Nepalese *Woodfordia fruticosa*.

Outlook

The utilization of *Woodfordia fruticosa* in combating diseases is a common practice among the people but in the majority of cases, there is no evidence of efficacy of treatment in popular use, or there has not been an adequate evaluation for possible adverse effects. Ethnobotanical data provide a basis for further validation of practices and plant uses in the context of a professional approach to ethnomedicine. The present chapter is intended to provide a concise source of information of *Woodfordia fruticosa*. Review of literature justified the traditional usage of the plant in remote populations and scientifically correlates its promising usage in the laboratory.

The plant has long been proved to be a source of several chemical compounds including tannins, flavonoids, anthraquinone glycosides and polyphenols which are considered useful against immunomodulatory, tumor, inflammatory and leucorrhoea. To date, research on the Nepalese *Woodfordia fruticosa* has not yet been conducted, although the plant is distributed vigorously in tropical and subtropical zones. We recommend further investigation of new natural products from the Nepalese *Woodfordia fruticosa* although research has started on antimicrobial activity, nutritive values, endophytes and ecological distribution. We hope that the results of this study play a significant role in the conservation of traditional medicine knowledge of the Nepalese *Woodfordia fruticosa* and encourage the scientific community for further investigations by extracting and identifying the active chemical compounds responsible for the antibacterial effect that was observed.

References

Adachi, K., T. Tada, and K. Sakaida. 2005. Skin composition containing *Punica granatum* flower extracts and other active components. Japanese Patent 2(5): 306–832.

Adachi, K., T. Tada, and K. Aramaki. 2006. Elastase inhibitors or Maillard reaction inhibitors for antiwrinkle cosmetics. Japanese Patent 2(006): 062–989.

Agarwal, G.P. and S.K. Hasija. 1961. Fungi causing plant diseases at Jabalpur (Madhya Pradesh)-VII. Some Cercosporae. Journal of the Indian Botany Society 44: 542–547.

Ahuja, B.S. 1965. Medicinal Plants of Saharanpur. Gurukul Kangri Printing Press, Hardwar, Delhi (India).

Alam, M., K.K.S. Dasan, S. Thomas, and J. Suganthan. 1998. Anti-inflammatory potential of Balarishta and Dhanvantara Gutika in albino rats. Ancient Science of Life 17: 305–312.

Ambasta, S.P. 1986. The Useful Plants of India. Publication and Information Directorate. CSIR, New Delhi (India).

Aminuddin Girach, R.D. 1996. Native phytotherapy among the Paudi Bhuinya of Bonai hills. Ethnobotany 8: 66–70.

Anonymous. 1970. Medicinal Plants of Nepal, Bulletin No. 3. Department of Medicinal Plants, Thapathali, Kathmandu (Nepal).

Anonymous. 1978. The Ayurvedic Formulary of India. Part I. Ministry of Health and Family Planning, New Delhi (India).

Atal, C.K., A.K. Bhatia, and R.P. Singh. 1982. Role of *Woodfordia fruticosa* Kurz. (*Dhataki*) in the preparation of *Asavas* and *Aristhas*. Journal of Research in Ayurveda and Siddha III, 193–199.

Awasthi, A.K. and A. Sharma. 2002. Phenological Studies on *Woodfordia fruticosa* Kurz. Environment and Ecology 20(4): 880–884.

Bajrachraya, A.M., K.D. Yami, T. Prasai, S.R. Basnyat, and B. Lekhak. 2008. Screening of some medicinal plants used in Nepalese traditional medicine against Enteric Bacteria. Scientific World 6(6): 107–110.

Bhagat, S., V. Singh, and O. Singh. 1992. Studies on germination behavior and longevity of *Woodfordia fruiticosa* Kurz. Seeds. Indian Forester 118: 797–799.

Bhattarai, S., R.P. Chaudhary, and R.S.L. Taylor. 2008a. Screening of selected ethnomedicinal plants of Manang district, Central Nepal for antibacterial activity. Ethnobotany 20: 9–15.

Bhattarai, S., R.P. Chaudhary, and R.S.L. Taylor. 2008b. Antibacterial Activity of Selected Ethnomedicinal Plants of Manang District, Central Nepal. Journal of Theoretical and Experimental Biology 5(1&2): 01–09.

Bhattarai S., R.P. Chaudhary, R.S.L. Taylor and S.K. Ghimire. 2009. Biological activities of some Nepalese Medicinal Plants used in treating bacterial infections in Human beings. Nepal Journal of Science and Technology, 10: 83–90.

Bhuju, D.R. 2003. Inventory of vegetation for biodiversity monitoring in the key area of Terai Arc Landscape. A preliminary report on Mahadevpuri, Lamahi, and Dovan prepared for the World Wildlife Fund Nepal Program.

Bhuju, D.R. and P.B. Yonzon. 2000. Floristic composition, forest structure and regeneration in the Churiya forests, Eastern Nepal. In: P.B. Yonzon, and D.R. Bhuju (eds.). Ecology of Nepal Churiya Part I: Mechi-Saptakoshi. A report submitted to Nature Conservation Society, Japan Resource Himalaya, Kathmandu (Nepal), pp. 24–52.

Bhuju, D.R., A. Rijal, and P.B. Yonzon. 2004. Strategic planning to maintain the ecology of the Western Terai—Churiya Forests, Nepal. A report submitted to the World Wildlife Fund Nepal Program.

Bista, D.B. 2004. People of Nepal. Ratna Pustak Bhandar, Kathmandu (Nepal).

Borthakur, S.K. 1992. Native phytotherapy for child and women diseases from Assam in Northeastern India, Fitoterapia, 63: 483–388.

Burkill, I.H. 1966. A Dictionary of Economic Products of the Malay Peninsula. Ministry of Agriculture and Co-operatives, Kualalumpur.

CBS. 2002. Nepal population census. 2001. Main report. Central Bureau of Statistics, Kathmandu (Nepal).

Chadha, Y.R. 1976. The Wealth of India: A Dictionary of Indian Raw Materials and Industrial Products-Raw Materials, vol X. Publications and Information Directorate, CSIR, New Delhi (India).

Chandan, B.K., A.K. Saxena, S. Shukla, N. Sharma, D.K. Gupta, K. Singh, J. Suri, M. Bhadaurie, and G.N. Qazi. 2008. Hepatoprotective activity of *Woodfordia fruticosa* Kurz. flowers against carbon tetrachloride induced hepatotoxicity. Journal of Ethnopharmacology 119: 218–224.

Chauhan, J.S., S.K. Srivastava, and S.D. Srivastava. 1979a. Phytochemical investigation of the flowers of *Woodfordia fruticosa*. Planta Medica. 36: 183–184.

Chauhan, J.S., S.K. Srivastava, and S.D. Srivastava. 1979b. Chemical constituents of *Woodfordia fruticosa* Linn. Journal of the Indian Chemical Society 56: 1041.

Cho, T., R. Koshiur, K. Miyamot, A. Nitt, T. Okud, and T. Yoshid. 1990. Woodfordin C, a macro-ring hydrolysable tannin dimmer with antitumor activity, and accompanying dimmers from *Woodfordia fruticosa* flowers. Chemical and Pharmaceutical Bulletin 38: 1211–1217.

Chopra, R.N., S.L. Nayar, and I.C. Chopra. 1956. Glossary of Indian Medicinal Plants. CSIR, Delhi (India).

Dabur, R., A. Gupta, T.K. Mandal, D. Singh, A.M. Bajpai, and G. Lavekar. 2007. Antimicrobial activity of some Indian medicinal plants. African Journal of Traditional and Complementary Medicine 4(3): 313–318.

Dan, S. and S.S. Dan. 1984. Chemical examination of the leaves of *Woodfordia fruticosa*. Journal of the Indian Chemical Society 61: 726–727.

Das, P.K., N.P. Sahu, S. Banerjee, S. Sett, S. Goswami, and S. Bhattacharya. 2006. Anti-peptic ulcer activity of an extract of a plant flower *Woodfordia fruticosa*. US Patent 20,060,040,005.

Das, P.K., S. Goswami, A. Chinniah, N. Panda, S. Banerjee, N.P. Sahu, and B. Achari. 2007. *Woodfordia fruticosa*: Traditional uses and recent findings. Journal of Ethnopharmacology 10: 189–199.

Dastur, J.F. 1951. Medicinal Plants of India and Pakistan. Taraporevala Sons and Co. Ltd., Mumbai (India).

Desai, H.K., D.H. Gawad, T.R. Govindachari, B.S. Joshi, V.N. Kamat, J.D. Modi, P.C. Parthasarathy, S.J. Patamkar, A.R. Sidhaye, and N. Viswanathan. 1971. Chemical investigation of some Indian plants. VI. Indian Journal of Chemistry, 9: 611–613.

Dey, K.L. 1984. The indigenous drugs of India. International Book Distributors, Dehradun (India).

Dudov, I.A., A.A. Morenets, V.P., Artiukh, and N.F. Starodub. 1994. Immunomodulatory effect of honeybee flower pollen load. Ukrainskii Biokhimicheskii Zhurnal. 66: 91–93.

Dutt, U.C. 1992. The materia Medica of the Hindus. Adi Ayurveda Machine Press, Kolkata (India).

Dymock, W., C.J.H. Warden, and D. Hooper. 1995. A history of the principal drugs of vegetable origin met within British India Republic. Pharmacographia Indica, vol I. Vivek Vihar, Delhi (India).

Filho, C.J.M. and J.L. Sor. 1997. Recursos naturais e meio ambiente: uma visao do Brasil IBGE, Departamento de Recursos Naturais e Estudos Ambientais, Rio de Janeiro.

GoN. 2007. Medicinal Plants of Nepal (Revised). Bulletin of the Department of Plant Resources No 28. Government of Nepal, Ministry of Forests and Soil Conservation, Department of Plant Resources, Thapathali, Kathmandu (Nepal).

IUCN-Nepal. 2004. National register of medicinal and aromatic plants. IUCN-Nepal Country Office for His Majesty's Government of Nepal, Ministry of Forest and Soil Conservation.

Jayaweera, D.M.A. 1981. Medicinal Plants used in Ceylon. Part III. The National Science Council SriLanka, Colombo (SriLanka).

Joshi, K.K. and S.D. Joshi. 2001. Genetic Heritage of Medicinal and Aromatic plants of Nepal Himalayas. Buddha Academic Publishers and Distributors Pvt., Ltd., Kathmandu (Nepal).

Kadota, S., Y. Takamori, K.N. Nyein, T. Kikuchi, K. Tanaka, and H. Ekimoto. 1990. Constituents of the leaves of *Woodfordia fruticosa* Kurz. L. Isolation, structure, and proton and carbon-13 nuclear magnetic resonance signal assignments of Woodfruticosin (Woodfordin C), an inhibitor of deoxyribonucleic acid Topoisomerase II. Chemical and Pharmaceutical Bulletin 38(10): 2687–97.

Kadota, S., Y. Takamori, and I. Kikuchi. 1992. Woodfruticosin (Woodfordin C) a new inhibitor of DNA topoisomerase II: Experimental antitumor activity. Biochemical Pharmacology 44: 1961–1965.

Kala, C.P., N.A. Farooquee, and U. Dhar. 2004. Prioritization of medicinal plants on the basis of available knowledge, existing practices and use value status in Uttaranchal (India). Biodiversity and Conservation 13: 453–469.

Kalidhar, S.B., M.R. Parthasarathy, and P. Sharma. 1981. Norbergin, a new C-Glycosoide from *Woodfordia fruticosa* Kurz. Indian Journal of Chemistry 20: 720–721.

Kaou, A.M., V. Mahiou-Leddet, S. Hutter, S. Ainouddine, S. Hassani, I. Yahaya, N. Azas, and E. Ollivier. 2008. Antimalarial activity of crude extracts from nine African Medicinal plants. Journal of Ethnopharmacology 116: 74–83.

Khare, C.P. 2004. Encyclopedia of Indian Medicinal Plants. Rational Western Therapy, Ayurvedia and other Traditional Usage. Botany, Springer, Berlin, pp. 483–484.

Khorya, R.N. and N.N. Katrak. 1984. Materia Medica of India and their therapeutics. Neeraj Publishing House, Ashok Vihar, Delhi (India), pp. 278 – 279.

Kirtikar, K.R. and B.D. Basu. 1935. Indian Medicinal plants. Part I-3. L.M., Basu, Allahabad (India).

Kroes, B.H., A.J.J. Van-den Berg, A.M., Abeysekera, K.T.D. de Silva, and R.P. Labide. 1993. Fermentation in traditional medicine: the impact of *Woodfordia fruticosa* flowers on the immunomodulatory activity, and the alcohol and the sugar content of *Nimba Aristha*, Journal of Ethnopharmacology 40: 117–125.

Krishman, P.N. and S. Seeni. 1994. Rapid micropropagation of *Woodfordia fruticosa* (L.) Kruz (Lythraceae), a rare medicinal plant. Plant Cell Report 14: 55–58.

Kuramochi-Motegi, A., H. Kuramochi, F. Kobayashi, H. Ekimoto, and K. Takahashi. 1992. Woodfruticosin (Woodfortin C), a new inhibitor of DNA topoisomerase. Part II. Experimental antitumor activity. Biochemical Pharmacology 17: 1961–1965.

Labadie, R.P., J.M. Van der Nat, J.M. Simons, B.H. Kroes, S. Kosasi, A.J.J. Van den Berg, L.A. 't Hart, W.G. Van der Sluis, A. Abeysekera, A. Bamunuarachchi, and K.T.D. De Silva. 1989. An ethnopharmacognostic approach to the search for immunomodulators of plant origin. Planta Medica. 55: 339–348.

Manandhar, N.P. 2002. Plants and People of Nepal. Timber Press, Inc. Portland, Oregon, U.S.A.

Martin, G.J. 1995. Ethnobotany: A Methods Manual. Chapman and Hall, London.

Mhaskar, K.S., E. Blatter, and J.F. Caius. 2000. Kiritikar and Basus illustrated Indian medicinal plants, their usages in Ayurveda and Unani medicines. vol 5, Sri Satguru Publications, New Delhi (India), pp. 1494–1499.

Nadkarni, K.M. 1954. Indian Materia Medica. vol 2, 3rd ed.. Popular Book Depot, Mumbai (India), p. 489.

Nair, A.G.R., J.P. Kotiyal, P. Ramesh, and S.S. Subramanian. 1976. Polyphenols of the flowers and leaves of *Woodfordia fruticosa*. Indian Journal of Pharmacy 38: 110–111.

Nanba, T., Y. Hattori, K. Shimomura, and S. Takamatsu. 1995. Cosmetic. Japanese Patent, 7: 126–144.

Oshima, K. and T. Mitsunga. 1999. Glucosyltransferase inhibitor. Japanese Patent 11: 343–347.

Oudhia, P. 2003. Interaction with the herb collectors of Gandai Region, Chhatisgarh, MP, India. www.botanical.com.

Pandey, G. 1992. Medical flowers, Puspayurveda medical flowers of India and adjacent regions, Indian Medical Sciences Series No. 14, Sri Satguru Publications, New Delhi (India), pp. 158–159.

Prakesh, J. and S. Chanda. 2007. *Invitro* screening of antibacterial activity of aqueous and alcoholic extracts of various Indian plant species against selected pathogens from Enterobacteriaceae. African Journal of Microbiology Research 1(6): 092–099.

Prajapati, N.D., S.S. Purohit, A.K. Sharma, and T. Kumar. 2003. A Handbook of Medicinal Plants: A complete source book. Agrobios (India), Agro House, Jodhpur (India).

Press, J.R., K.K. Shrestha, and D.A. Sutton. 2000. Annotated Checklist of the flowering Plants of Nepal. Natural History Museum, London and Central Department of Botany, Tribhuvan University, Kathmandu (Nepal).

Qin, F. and H.X. Sun. 2005. Immunosuppressive activity of pollen Typhae ethanol extract on the immune responses in mice. Journal of Ethnopharmacology 102: 424–429.

Rajbhandari, K.R. 2001. Ethnobotany of Nepal. Ethnobotanical Society of Nepal, Kathmandu (Nepal).

Rajbhandari, T.K., N.R. Joshi, T. Shrestha, S.K.G. Joshi, and B. Achraya. 1995. Medicinal Plants of Nepal for Ayurvedic Drugs. HMGN, Natural Products Development Division, Thapatali, Kathmandu (Nepal).

Roldao, E.F., A. Witaicenis, L.N. Seito, C.A. Hiruma-Lima, L.C.D. Stasi. 2008. Evaluation of the antiulcerogenic and analgesic activities of Cordia Verbenacea DC. (Boraginaceae). Journal of Ethnopharmacology 119 (1&2): 94–98.

Samantaray, S., G.R. Rout, and P. Das. 1999. Studies on the uptake of heavy metals by various plant species on chromite minespoils in sub-tropical regions of India. Environmental Monitoring and Assessment, 55: 389–399.

Saoji, A.G., A.N. Saoji, and V.K. Deshmukh. 1972. Presence of lawsone in *Ammania baccifera* Linn. and *Woodfordia fruticosa* Salisb. Current Science 41: 192.

Sharma, P.V. 1956. Dravyagun Vigyan. The Chowkhamba, Varanasi (India).

Shome, U., S. Mehrotra, and H.P. Sharma. 1981. Pharmacognostic studies on the flower of *Woodfordia fruticosa* Kurz. Proceedings of the Indian Academy of Sciences (Plant Science) 90(4): 335–352.

Srivastava, S.K., M. Sultan, and J.S. Chauhan. 1977. Anthocyanin pigment from the flowers of *Woodfordia fruticosa*. Proceedings of the National Academy of Sciences, India 47(A): 35–36.

Suzuki, R., K. Umishio, K. Hasegawa, and K. Moro. 2004. External preparation for skin. Japanese Patent 2:004: 352–658.

Taylor, R.S.L., NP. Manandhar, and G.H.N. Towers. 1995. Screening of selected medicinal plants of Nepal for antimicrobial activities. Journal of Ethnopharmacology 46: 153–159.

Taylor, R.S.L., F. Edel, N.P. Manandhar, and G.H.N. Towers. 1996. Antimicrobial activities of southern Nepalese medicinal plants. Journal of Ethnopharmacology 50: 97–102.

Taylor, R.S.L., S. Shahi, and R.P. Chaudhary. 2002. Ethnobotanical research in the proposed Tinjure-Milke-Jaljale Rhododendron conservation area, eastern Nepal. In: R.P., Chaudhary, BP. Subedi, OR. Vetaas, and TH. Asae (eds.). Vegetation and Society their Interaction in the Himalayas. Tribhuvan University, Kathmandu (Nepal) and University of Bergen, Norway.

Tewari, P.V. and K.S. Neelam Kulkarni. 2001. A study of Lukol in leucorrhoea, pelvic inflammatory diseases and dysfunctional uterine bleeding. Ancient Science of Life XXI, 139–149.

Timsina, G. 2003. Evaluation of antimicrobial activities of some medicinal plants used in traditional medicine in Nepal. Central Department of Botany (M.Sc. Dissertation), Tribhuvan University, Kirtipur, Kathmandu (Nepal).

Tsuji, T., U. Seitai, T. Shu, and T. Yoshida. 1996. Polygalacturonase inhibitors isolation from plant. Japanese Patent 08: 481–489.

Ueda, K. and K. Shimomura. 1995. Cosmetic. Japanese Patent 7: 157–420.

Van Wyk, B.E., B. Van Oudtshoorn, and N, Gericke. 2002. Medicinal Plants of South Africa. Briza Publications, Pretoria, South Africa.

Watt, G. 1972. A Dictionary of Economic products of India. III. Periodical Expert. Shahdara, Delhi (India).

Yoshida, T., T. Chou, K. Haba, Y. Okano, I. Shingu, K. Miyamot, R. Koshiur, and T. Okuda. 1989a. Camellin B and nobotanin I, macrocyclic ellagitannin dimmers and related dimmers, and their antitumor activity. Chemical and Pharmaceutical Bulletin 37: 3174–3176.

Yoshida, T., T. Chou, A. Nitta, and T. Okuda. 1989b. Woodfordins A, B and C, dimeric hydrolyzable tannin from *Woodfordia fruticosa* flowers. Heterocycles 29: 2267–2271.

Yoshida, T., T. Chou, A. Nitt, K. Miyamot, and R.T. Koshiur. 1990. Woodfordin C, a macro-ring hydrolysable tannin dimmer with antitumor activity, and accompanying dimmers from *Woodfordia fruticosa* flowers. Chemical and Pharmaceutical Bulletin 38: 1211–1217.

Yoshida, T., T. Chou, M. Matsuda, T. Yasuhara, K. Yazaki, T. Hatano, A. Nitta, and T. Okuda. 1991. Woodfordin D and Oenothein A, trimeric hydrolysable tannins of macro-ring structure with anti-tumor activity. Chemical and Pharmaceutical Bulletin 39: 1157–1162.

Yoshida, T., T. Chou, A. Nitta, and T. Okuda. 1992. Tannins and related polyphenols of Lythraceous plants. III. Hydrolyzable tannin oligomers with macrocyclic structures, and accompanying tannins from *Woodfordia fruticosa* Kurz. Chemical and Pharmaceutical Bulletin 40: 2023–2030.

Chapter 10

Chemistry and Pharmacology of *Azadirachta indica*

Rosa Martha Perez Gutierrez

Introduction

The neem tree, *Azadirachta indica* A. Juss. syn. *Melia azadirachta* Linn. (Meliaceae), is a tropical and subtropical species, native of Myanmar and habituated in the Indian sub-continent and now cultivated widely in tropical areas of the world. Neem is moderate to large sized, 15–20 meters in height, usually an evergreen, tree, with a fairly dense rounded crown. Leaves are glabrous imparipinnate, leaflets, sub-opposite, alternate, estipalate, 22.5 to 37.5 cm long on long slender petiole. The honey scented flowers are small cream or yellowish white in color in auxiliary pennicdes, staminal tubes constituents cylindrical, widening above 9–10 lobed at the apex. The neem tree yields 50 kg of fruit per year (Gupta 1993).

Neem derives from from the Sanskrit Nimba and known as the Sarva Roga Nivarini or curer of all illnesses. The Sanskrit treatises Charaka Samhita, Susrutha Samhita and Brihat Samhita, compiled between 6th century BC and 6th century AD, mention the properties of neem many times. In India, the neem tree is sacred, with thousands of uses since ancient times. The first indication that neem was used in medical treatments dating back some 4500 years. It was the main feature of the Harappan culture, one of the greatest civilizations of the old world. The neem tree

Laboratorio de Investigación de Productos Naturales. Escuela Superior de Ingeniería Química e Industrias extractivas IPN. Punto Fijo 16, Col. Torres Lindavista, cp 07708, México D.F. México.

E-mail: *rmpg@prodigy.net.mx*

Corresponding address: Dra Rosa Martha Perez Gutierrez, Punto Fijo 16, Col. Torres Lindavista, Cp 07708, D.F. Mexico.

was intimately connected with the everyday life of Indians. Neem oil extraction in India was a specialized profession and people undertaking these jobs were called 'teli', or oilmen. A traditional oil extractor was called a kohlu. This was originally made of wood and looked like a pestle and mortar connected to a wooden plank.

With the advent of the British, the colonizers systematically discouraged traditional practices like using neem leaves to protect crops and stored grains and over time, these came to be regarded as backward. There was a tendency on the part of the colonial rulers to encourage people to abandon their ecologically sound practices in favor of modern chemical products imported from the West. With the end of the colonial era, interest in the neem tree was revived. Work on the possible commercial use of Neem oil and cake was done by the Indian Institute of Science in Bangalore as early as the 1930s. Recalling the insecticidal properties of neem, researchers began programs in the early 60's to identify the active principles and screen them against major crop pests.

In the last two decades research on neem has been intensified and many agricultural and medical properties of the tree were rediscovered. Neem is widely used in reforestation agroforestry programs in South and South East Asia, West Africa and America. The tree is also planted for shade and shelter, erosion control and soil improvement, and as windbreak. Today, neem plays a major role in the rural industry of India and projects for the commercial use of neem have been successfully introduced in such places as Kenya, Saudi Arabia, Australia, and the USA (Gupta 1993).

There has been a tremendous interest in this plant as can be evidenced by the voluminous work on it. Therefore, we aimed to compile an up to date and comprehensive review of *Azadirachta Indica* that covers its traditional and folk medicine uses, phyochemistry and pharmacology.

Uses of *Azadirachta indica*

Use in traditional medicine

The medicinal uses of neem were first described in classical Hindu texts, especially in the Ayurveda which had been compiled in 500 AD, but originated long before. Neem is one of the most important trees of India and a holy tree of the Hindus. Called 'the village pharmacy' in India, neem is one of the most ancient and widely used herbs in the world (Deepak and Anshu 2008).

In the Ayurvedic medical tradition, Neem is considered a useful therapy for ulcers and gastric discomfort. Throughout India, people take Neem leaves for all sorts of stomach ailments. Peptic ulcers and duodenal ulcers are treated with neem leaf extracts. Neem is also useful in treating other problems in the stomach and bowels such as gastritis. Its leaves are often used to treat heartburn and indigestion (Shodini 1997).

Azadirachta excelsa known as 'tiem' in Thailand and 'sentang' in Malaysia, is fast growing and economically valuable not only for timber but also for its medicinal and pesticidal properties. The tender leaves and flowers are said to be used as a vegetable in Thailand, and the seeds are used as a natural insecticide (Kurose and Yatagai 2005).

Neem twigs contain antiseptic ingredients which provide dental hygiene and has been used for this purpose by people from rural areas in India and parts of Africa. The practice has inspired the use of neem bark in extracts in commercial toothpastes and mouthwashes (Shodini 1996).

Insecticidal uses

The dried leaves are kept in woolen and cotton fabrics as well as books to ward off moths and insects. This practice has been immemorial in India. Dried leaves when mixed with stored grain (wheat, sorghum, rice and legumes and pulses) protects against stored grain pests for a period of 4–5 mon. It protects against *Ephestia cautella* (moth) *Calandra oryzae* (rice weevil), *Sitotroga cerealella* (grain moth) and *Tribolium catannelem* (red flower beetle) (NRC 1992).

Limonoids are responsible for pesticide properties. Neem seed kernel extract (5%) in 1 l parts of water with 2% soap or teepol as a surfactant protects against major pests on all crops including horticultural and greenhouse plants. It also safeguards from pathogens that attack crops. Neem oil mixed with urea (5% oil) is also effective as a nitrification inhibitor. Today neem oil (EC) 300 ppm aradirachtin content are commercially available.

Neem cake used as nitrification inhibitor when broadcast in the field also serves as a pesticidal, antifeedant repellent, insecticide, nematicide, and antimicrobial. For soil application it acts as systemic insecticide. It acts as a soil amendment, and it is believed to enhance the efficiency of nitrogen fertilizers by reducing the rate of nitrification and to inhibit soil pests. Neem leaves and small twigs are also used as mulch and green manure. The cake extract (7% aqueous extract) is also used as a spray to control pests. The cake when mixed with grains (3%) is also useful as stored grain protectants (Randhawa and Parmar 1993).

Another traditional agricultural practice involves the production of 'neem tea'. The seeds are dried, crushed and soaked in water overnight to produce a liquid pesticide that can be applied directly to crops. Crushed seed kernels are also sometimes used as a dry pesticide application, especially to control stem borers on young plants. These home-made remedies are often very effective at repelling pests or as a feeding deterrent on insects. The active compounds break down quickly, so an application of neem tea can generally provide protection for only about 1 wk. Commercial preparations contain sunscreens which maintain their effectiveness for 2–3

Table 10.1. Ethnomedicinal uses of *Azadirachta indica*

Place, country	Part(s) used	Ethnomedical uses	Preparation(s)	Reference(s)
Mexico	Leaves	Diabetes, wounds	Decoction	Aguilar et al. (1994)
Kenya	Leaves, bark	Ulcers, diabetes, malaria, bronchitis	Infusion	Sharma (1996)
India	Leaves	Body heat, painful periods, infections, worms, fever, hemorrhage, jaundice, heart diseases, vaginal problems, gray hairs, malaria, stomach problems	Infusion or decoction	Jamir et al. (1999)
India	Bark	Fever, vaginal problems, teeth diseases, malaria	Infusion or decoction	Shodini (1997)
India	Seeds	Poisoning, piles, wound	oil	Sharma (1996)
Africa, India, Thailand, Malaysia, Myanmar, Indonesia	All parts	Fumigation as natural pesticides	oil or infusion or decoction	Shodini (1997)
Thailand Myanmar	Leaves, Flowers		Cooking	Siddiqui et al (2006)
Africa	All parts	Diseases of bacterial and fungal origin, ulcers, eczema, jaundice, liver complaints	Infusion or decoction	Siddiqui et al. (2006)
Africa	Fruits	Purgative, emollient, intestinal worms, urinary diseases and piles	Decoction	Siddiqui et al. (2006)
Africa	Bark	Bitter tonic, astringent, anti-periodic	Decoction or infusion	Siddiqui et al. (2006)
Africa	Flowers	Tonic	Infusion or decoction	Siddiqui et al. (2006)
India	Seed	Spermicidal, antifertility	Oil	Garg (1993)
Pakistan	All plant	Anti-inflammatory, malaria, skin ailments anti-pyretic, antitumor, urinary disorders, diabetes, fungi infections and viral diseases, for giving bath to new born infants, protect people and plants from insects	Infusion or decoction	Siddiqui et al. (2000), Shodini (1997), Ganguli (2002)

Place, country	Part(s) used	Ethnomedical uses	Preparation(s)	Reference(s)
India, Bangladesh, Pakistan ,Saudi Arabia	Seeds	Preparing cosmetics (soap, shampoo, balms and creams) for skin care, acne treatment, keeping skin elasticity	Oil	Ganguli (2002)
India, Bangladesh, Pakistan, Saudi Arabia	Twigs	Dental care	Brushing teeth	Shodini (1997)
USA	Leaves	Diabetes	Infusion or decoction,	Kay (1996)
Southeast Asia	Leaves	Prostate cancer	Infusion or decoction	Kumar et al. (2006)

wk. Neem extracts may have toxic effects on fish, other aquatic wildlife, and some beneficial insects (Read and French 1993). Neem compounds do not persist or accumulate in the environment after being applied as pesticides.

Neem oil has been found to be an effective mosquito repellent. Studies have shown that neem compounds are more effective insect repellents than DEET (N,N-diethyl-toluamide), a chemical widely used in commercial mosquito repellents. Oil repels or kills mosquitoes, their larvae and a plethora of other insects including those in agriculture (Mishra et al. 1995, Trongtokit et al. 2005).

Commercial applications

The neem tree comes into production to yield fruit in about 10 yr time, A tree produces 35–50 kg of seeds per year. The seed, which contains a shell with a kernel inside, has 44.7% kernel and 55.3% shell or also known as seed coat. The kernel is obtained from seeds, by lightly pounding and winnowing or by using a decorticator. There are about 14 million neem trees in India yielding about 100,000 tones of Neem oil and 400,000 tones of cake (Gupta 1993).

Neem oil has long been produced in Asia on an industrial scale in soaps, cosmetics, and pharmaceuticals. Eighteen thousand tones of neem oil is used in the soap industry. During the 1980s companies began commercial production and distribution of pest control formulations that use azadirachtin as the principal active ingredient. Interest in pesticides agents has developed along with the environmental and consumer rights movements, and the recognition that strategies are needed to sustain agricultural production. New markets for organically grown produce and 'natural' products also spurred the development of the industry, and azadirachtin was among the first to be commercialized.

In the United States neem-based BPCs were first approved for use on non-food crops in 1985. After subsequent testing, the Environmental Protection Agency (EPA) regulated the use of dihydroazadirachtin (DAZA), a reduced derivative of azadirachtin, for use on food crops. In 1996 the EPA exempted raw agricultural commodities from meeting DAZA residue requirements, as long as the chemical is applied as an insect growth regulator or antifeedant at no more than 20 g/acre with a maximum of seven applications per growing season (EPA 1997). The EPA only allows this exemption if approved commercial products are used; food products treated with home-made extracts would not meet these requirements.

A neem-based contraceptive cream was developed by a pharmaceutical company in India. Tests of its effectiveness showed that it compared favorably with the chemical-based foams and gels. It was safer and easier to use, caused no irritation or discomfort, was nearly 100% effective, and was therefore used more frequently than the foam or gel spermicides. The effect does not appear to be hormonal and is considered a safe and effective alternative to other methods that use hormones (Garg 1993).

Neem and its formulations can be used as general antiseptics and contain antibacterial properties and are highly effective in treating epidermal conditions such as acne, psoriasis and eczema. The neem oil moisturizes and protects the skin while healing the lesions, scaling and irritations. For women, the neem is the mainstay of herbal beauty tradition. It also protects stored grains and pulses throughout the year. For men, the tree provides seeds, leaf and bark, which could be converted into fertilizer and pest control material. The oil from neem seeds can be used for various formulations like cosmetics, toothpaste, disinfectants, and various other industrial products without showing harmful side effects (Hitesh et al. 2007).

In Africa the tree is used for its shade and as a source of fuelwood. In the Sahel countries, neem has been used for halting the spread of the Sahara desert. It is also a preferred tree along avenues, in markets and near homesteads, because of the shade it provides. The relatively hard and heavy wood of neem is not only durable, but also termite resistant. In many developing countries the wood is used for making fence posts, poles for house construction, and furniture. The hard wood of the tree is used as timber. The leaves are used as fodder (Jamir et al. 1999).

Chemical constituents and their biological activities

Various bioactive phytochemicals with medicinal properties have been extracted from various parts of the neem tree and continue to receive the attention of phytochemists, biochemists and medicinal chemists on account of the extraordinary biological properties of extracts and of several individual compounds (Patel and Patel 2004). The neem tree contains more than 100 bio-active ingredients and is rich in proteins. Its bitter taste is due to an array of complex compounds called limonoids. The most important bio-active principal is azadirachtin, neem leaves contain organic compounds that have insecticidal and medicinal properties (Rao et al. 1992). Each part of the Neem tree has some biological activity and is thus commercially exploitable (Tewari 1992). Compounds such as gedunin (antimalarial), nimbin (antiinflammatory, antipyretic), nimbidin (antibacterial), nimbidol (antimalarial, antipyretic), quercetin (antimalarial), salannun (repellent), and sodium nimbinate (spermicide) have been isolated (Keating 1994).

Neem seeds

In recent years, the use of environment-friendly and easily biodegradable natural insecticides of plant origin has received much attention for control of medically important arthropods. Vector borne diseases, such as malaria, still cause thousands of deaths per year. Malaria is by far the most important insect transmitted disease (Gilles and Warrell 1993), remaining a major health problem in many parts of the world and is responsible for high childhood mortality and morbidity in Africa and Asia (Pates and Curtis 2005). *Anopheles stephensi* Liston (Diptera: Culicidae), a major malaria vector, breeds in wells, overhead or ground level water tanks, cisterns, coolers, roof gutters and artificial containers (Herrel et al. 2001). Management of this disease vector using synthetic chemicals has failed because of insecticide resistance, vector resurgence and environmental pollution. Consequently, an intensive effort by members of the Meliaceae plant family possess insect- growth regulating properties against many insect pests (Schmutterer 1990).

It is because of these properties that the family Meliaceae has emerged as a potent source of insecticides. The Indian neem tree, *A. indica* has been found to be a promising source of natural pesticides, several constituents of its leaves and seed show marked insect control potential. Neem seed kernel extract suppresses the feeding, growth and reproduction of insects (Schmutterer 1990).

Butterworth and Morgan (1971) first isolated the tetranortriterpenoid azadirachtin from *A. indica* seeds, which primarily showed antifeedant activity and later, regulatory effects on larval development and metamorphosis. Limonoids were described as modified triterpenes, having a 4,4,8 trimethyl-17 furanyl steroid skeleton. These characteristics include insecticidal, insect growth regulation and insect antifeedant properties (Butterworth and Morgan 1971).

Anti-inflammatory

Nimbidin, a major crude principle extracted from neem oil, has been identified as the active constituent of the oil (Pillai and Santhakumari 1981, Pal et al. 2002). Nimbidin is a mixture of tetranortriterpenes including nimbin, nimbinin, nimbidinin, nimbolide and nimbidic acid and possesseses potent anti-inflammatory, antiarthritic and antiulcer activities (Pillai and Santhakumari 1981, Bandyopadhyay et al. 2002). It significantly inhibits carrageenin and kaolin-induced paw oedema wherein it is much more potent than the standard non steroidal anti-inflammatory drug phenylbutazone (Pillai and Santhakumari 1981). Nimbidin is, as such, a general anti-inflammatory drug and is effective in both acute and chronic states of inflammation. However, the mechanism of its action is ambiguous.

According to Guipreet et al. (2004) nimbidin significantly inhibited some of the functions of macrophages and neutrophils relevant to the inflammatory response *in vivo* exposure. Oral administration of 525 mg/kg of nimbidin to rats for three consecutive days significantly inhibited the migration of macrophages to their peritoneal cavities in response to inflammatory stimuli and also inhibited phagocytosis and phorbol-12-myristate-13-acetate (PMA) stimulated respiratory burst in these cells. The results suggest that nimbidin suppresses the functions of macrophages and neutrophils relevant to inflammation. It is also reported by Kaur et al. (2004) that in the mechanism of anti inflammatory action, nimbidin significantly suppresses some of the functions of macrophages and neutrophils relevant to inflammation.

Anticancer/antitumor

The activity was assessed on the meristematic cells of onion root tips (Santhakumari and Stephen 2005). The effect was almost similar to those of colchicine and vinca alkaloids. Recovery trials showed that the drug induces lethal damage in a considerable proportion of treated cells and may hence have applications in cancer chemotherapy .

Antimalarial

Azadirachta indica has been shown to possess anti-malarial activity. Extracts of neem seeds and purified fractions *in vitro* growth and development of asexual and sexual stages of the human malaria parasite *Plasmodium falciparum* (Dhar et al. 1998). Neem seeds are thus active not only against the parasite stages that cause the clinical infection but also against the stages responsible for continued malaria transmission. In addition, all the maturation stages of gametocytes were also killed by various neem fractions tested. The anti-plasmodial effect of neem components was also observed on parasites previously shown to be resistant to other anti-malarial drugs, i.e., chloroquine and pyrimethamine suggesting a different mode of action.

Hypoglycemic

The crude leaf ethanol extract of neem lowered hyperglycemia in streptozotocin diabetes (Gupta et al. 1993). It hipoglycemic activity is also reported by Bopanna et al. (1997) the administration of neem kernel powder (NP) (500 mg/kg) as well as the combination of NP (250 mg/kg) with glibenclamide (0.25 mg/kg) significantly decreased the concentration of serum lipids, blood glucose and activities of serum enzymes like alkaline phosphatase (alk P), acid phosphatase (acid P), lactate dehydrogenase (LDH), liver glucose 6-phosphatase (G6P) and HMG-CoA reductase activity in liver and intestine of alloxan diabetic rabbits. However, all the treatments produced an increased liver hexokinase activity. The changes observed were significantly greater when the treatment was given in combination of NP and glibenclamide than with NP alone. These data suggest a significant antidiabetic and antihyperlipemic effect of NP in alloxan diabetic rabbits.

Toxicity

The oil of *A. indica* given orally to mice showed low toxicity. It was non-irritant to the skin of rabbits in primary dermal irritation test. In sub-acute dermal toxicity, rabbits exposed daily to *A. indica* seed oil for 21 d showed no significant changes in body weight and organ/body weight ratio. Serum oxalo-acetic transaminase, serum pyruvic transaminase levels, blood glucose and blood urea nitrogen values were found to be unaltered. No treatment-related histopathological changes were observed (Tandan et al. 1995). After its administration to mice neem does not seem to have any side effects.

Essential oil

The components of the essential oils from seeds of *Azadirachta indica*, *Azadirachta siamensis*, and *Azadirachta excelsa* were studied by gas chromatography-mass spectrometry. The main components of *A. indica* oil were hexadecanoic acid (34.0%), oleic acid (15.7%), 5,0-dihydro-2,4,6-triethyl-(4H)-1,3,5-dithiazine (11.7%), methyl oleate (3.8%), and cudesin-7(11)-en-4-ol (2.7%). The major components of A. siamensis oil were hexadecanoic acid (52.2%), tricosane (10.5%), tetradecanoic acid (6.8%), oleic acid (4.9%), and pentacosane (4.9%). *Azadirachta excelsa* oil contained oleic acid (31.3%), hexadecanoic acid (14.2%), octadecanoic acid (13.0%), 4-octylphenol (9.7%), and O-methyloximedecanal (6.8%) as the main constituents. The essential oils from *A. indica, A. siamensis,* and *A. excelsa* were found to contain fatty acids (52.6%–72.3%) as major components. The minor components of the oils were n-alkanes, aromatics, esters, sulfur and nitrogen compounds, and terpenoids. Differences in oil composition were observed between the three species (Kurose and Yatagai 2005).

Insecticidal activity

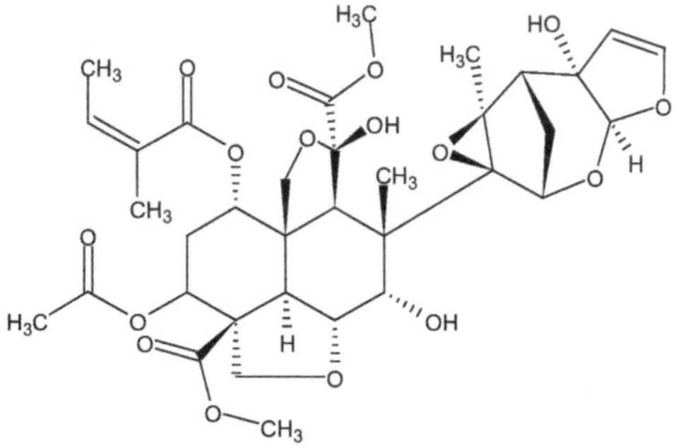

Azadirachtin

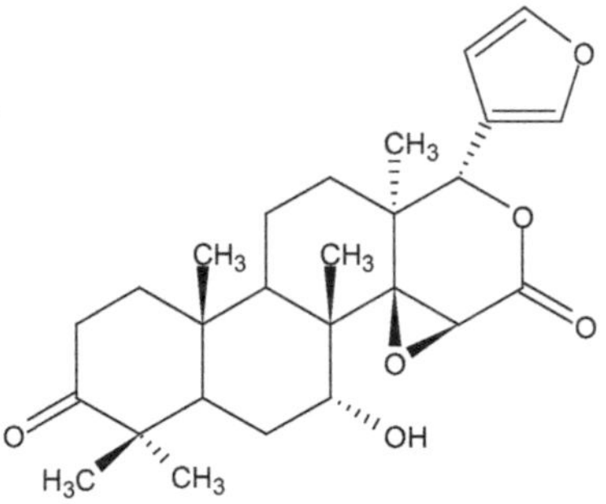

Deacetylgedunin

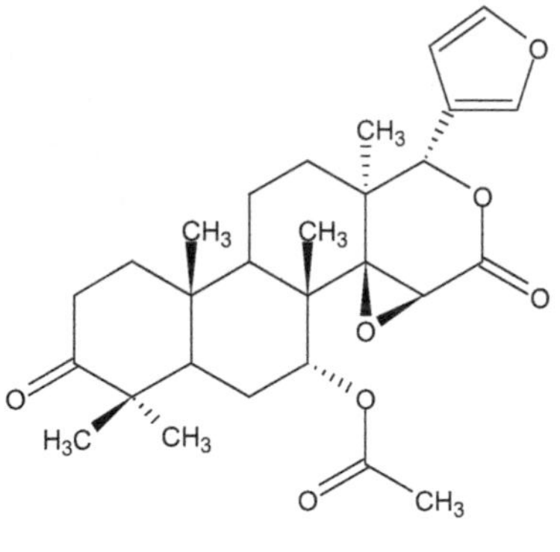

Gedunin

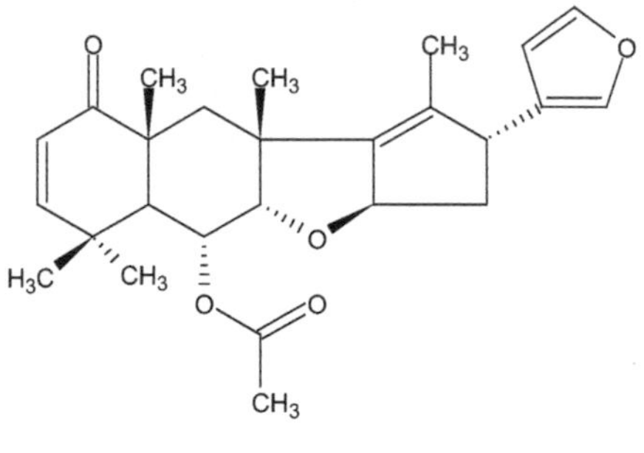

Deacetylnimbin

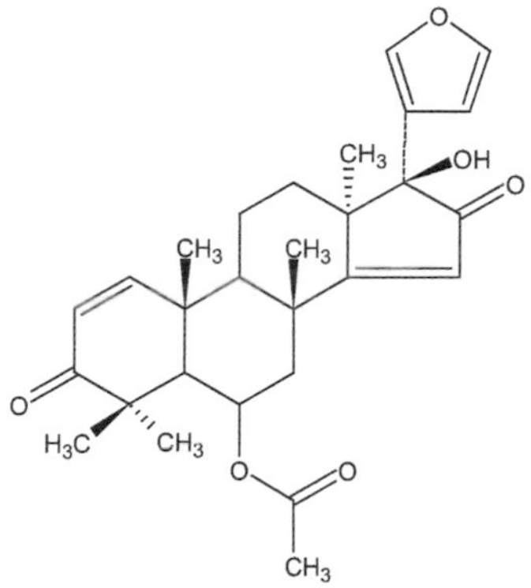

17-Hydroxyazadiradione

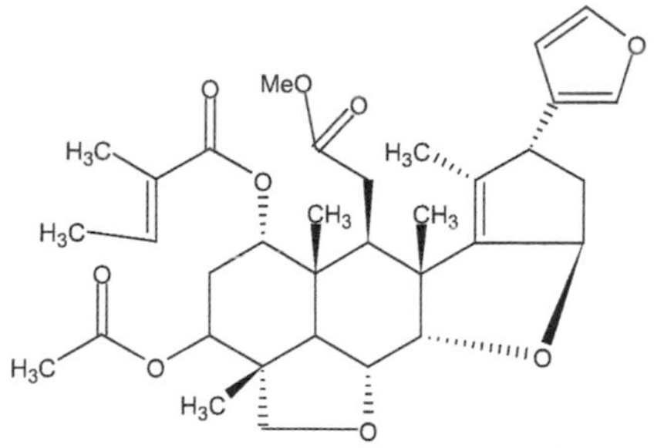

Salannin

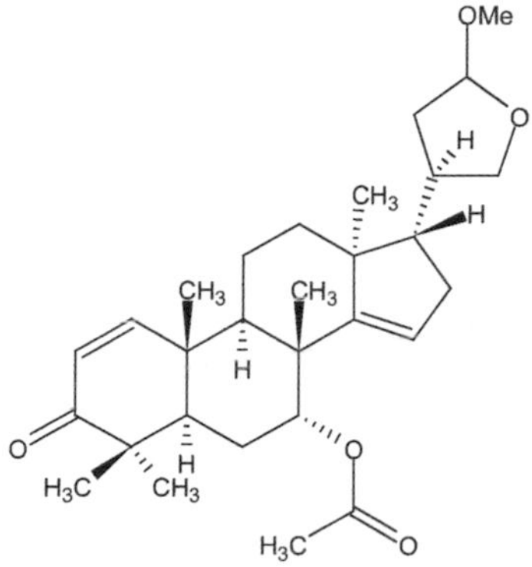

Limocin B

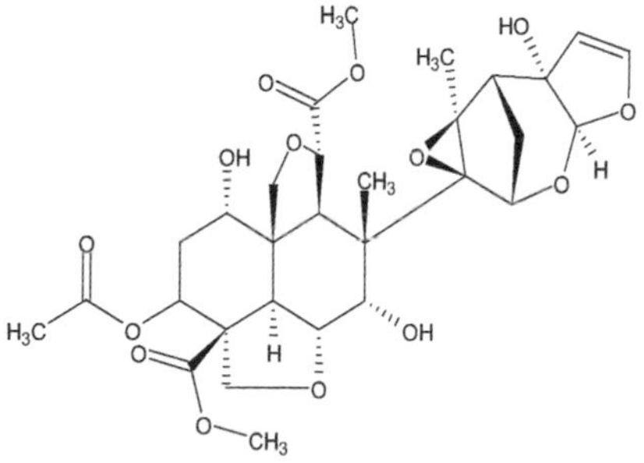

11-Hydroxyazadirachtin B

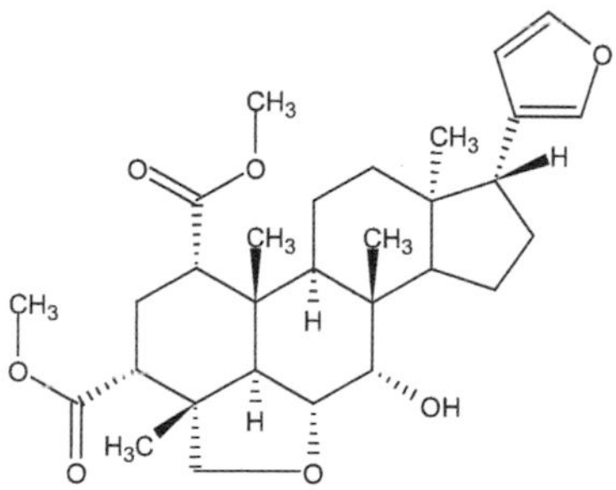

13-Diacetylvilasinin

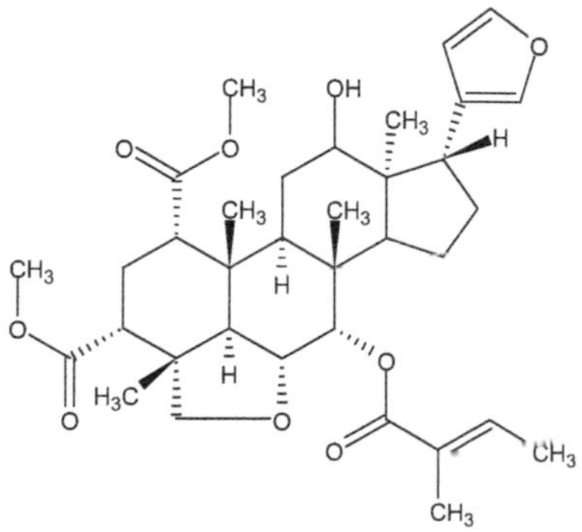

1,3-Diacetyl-7-tigloyl-12-hydroxyvilasinin

23-Desmethyllimocin B

7-Deacetyl-17β-hydroxyazadiradione

1-Tigloyl-3-acetylazadirachtinin

1-Tigloyl-3-acetyl-11-hydroxy-4β-methylmeliacarpin

1- Cinnamoyl-3-feruloyl-11-hydroxymeliacarpin

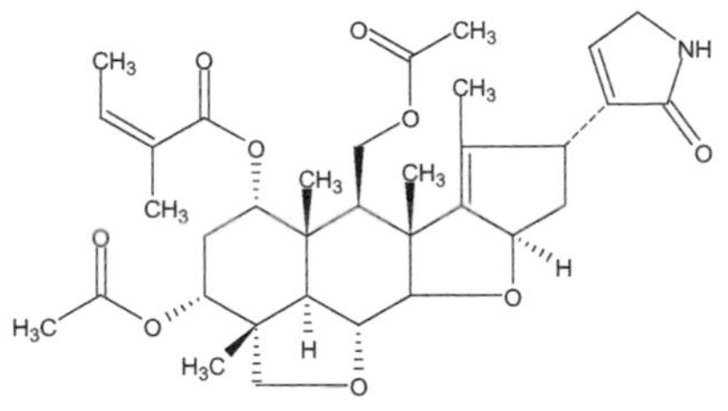

Salannolactam-(21)

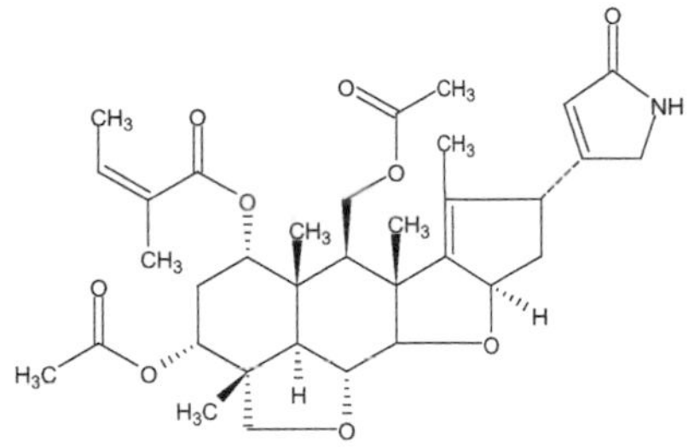

Salannolactam-(23)

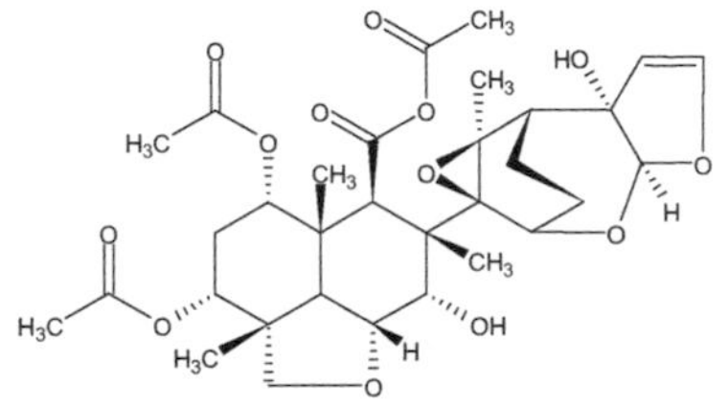

1,3-Diacetyl-11,19-deoxa-11-oxomeliacarpin

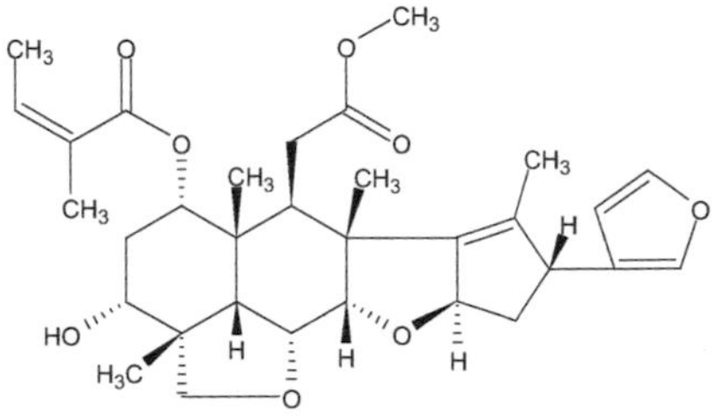

Desacetylsalannin

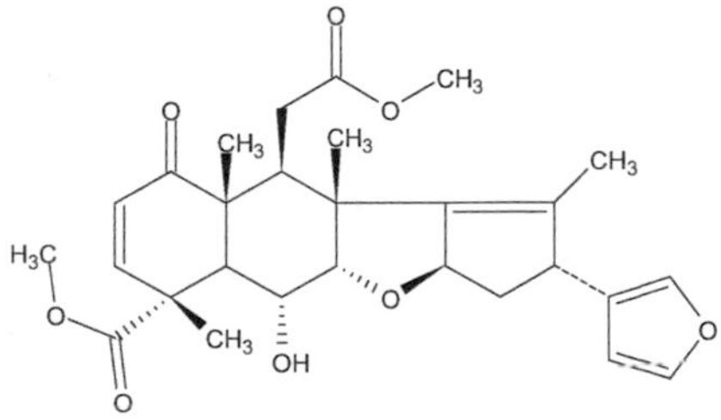

Desacetylnimbin

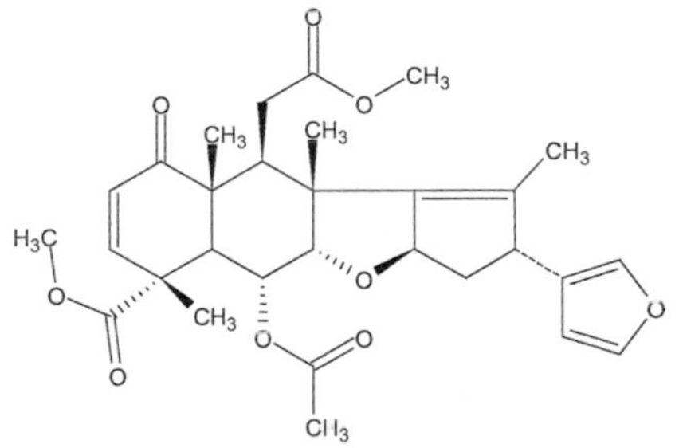

Nimbin

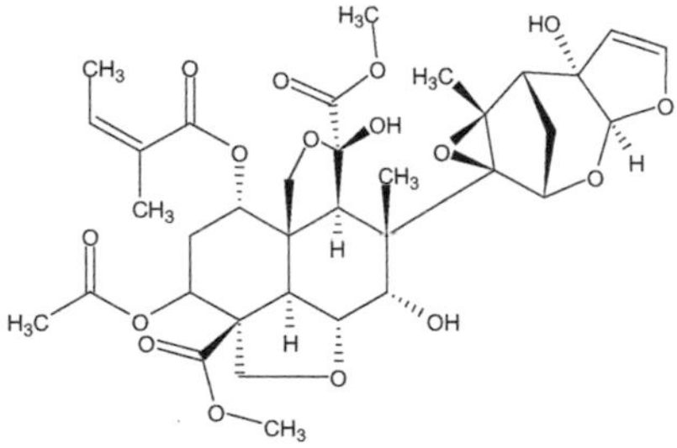

Azadirachtin A

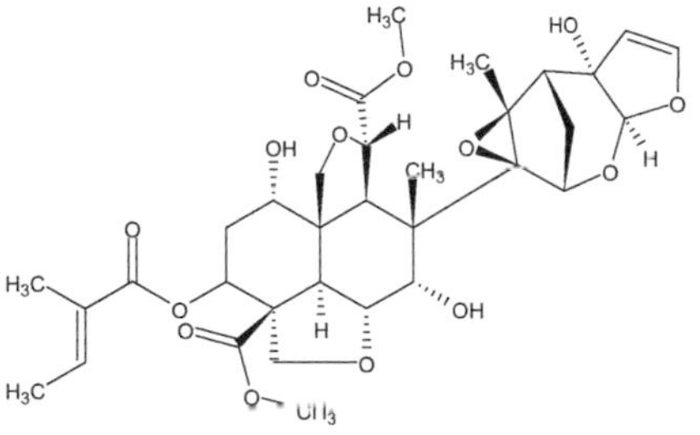

Azadirachtin B

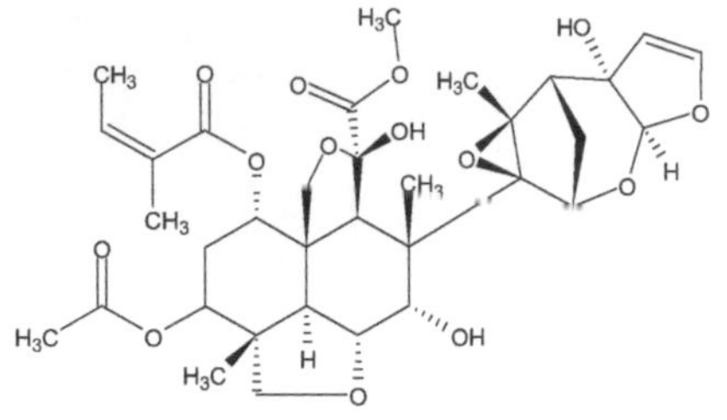

Azadirachtin D

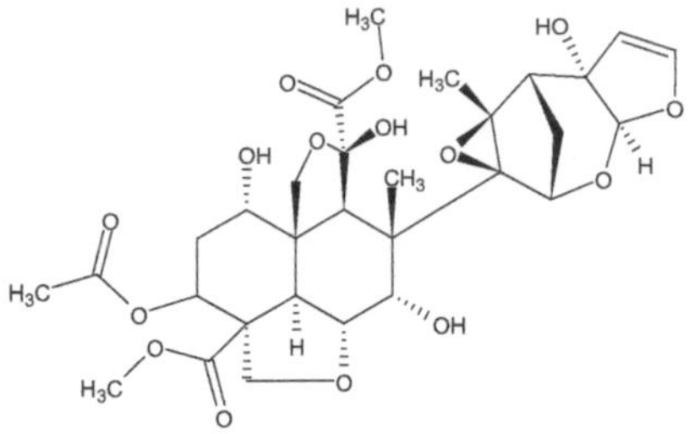

Azadirachtin E

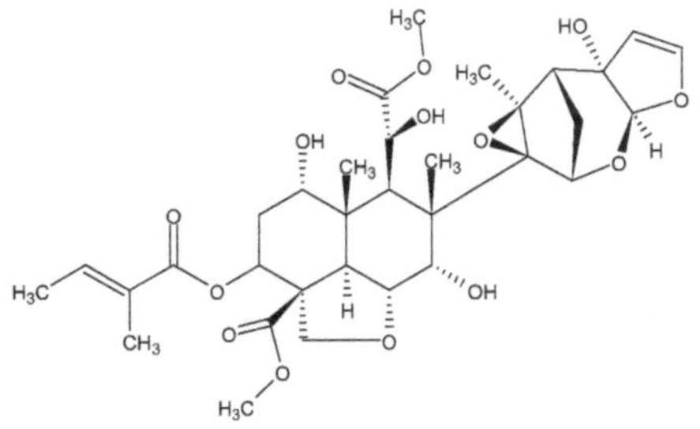

Azadirachtin F

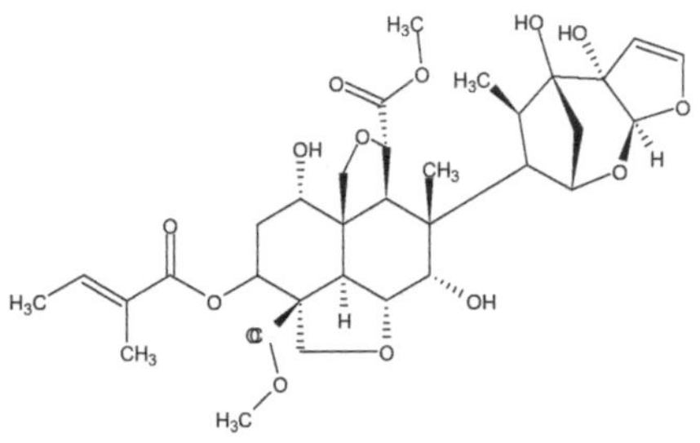

Azadirachtin G

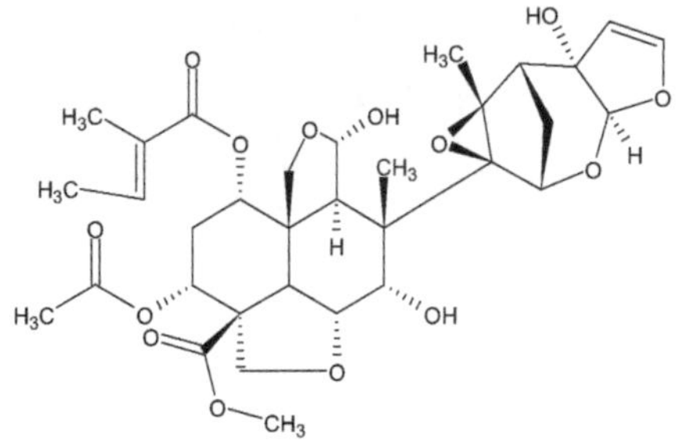

Azadirachtin H

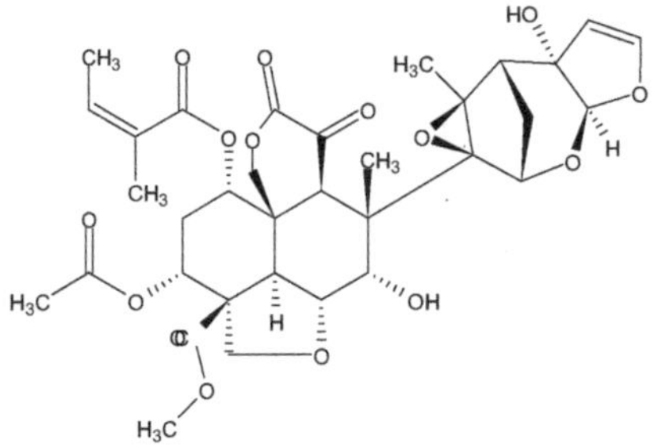

Azadirachtin K

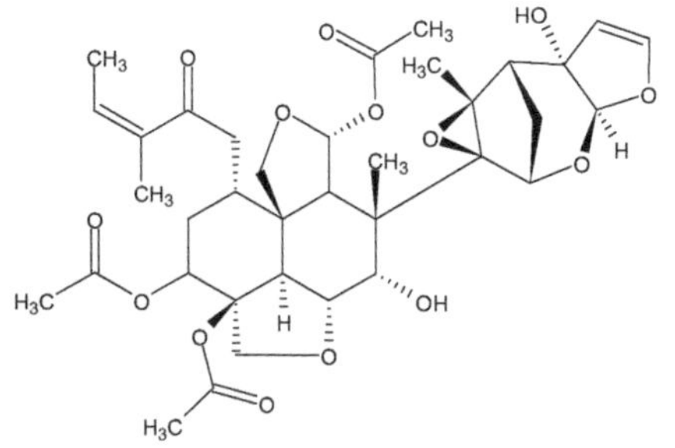

Azadirachtin L

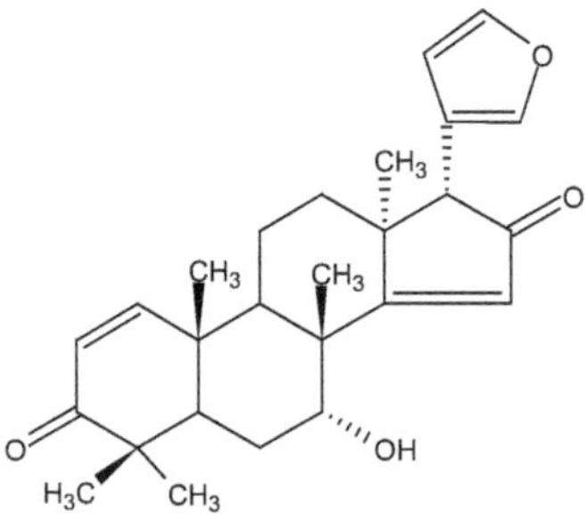

Nimbocinol

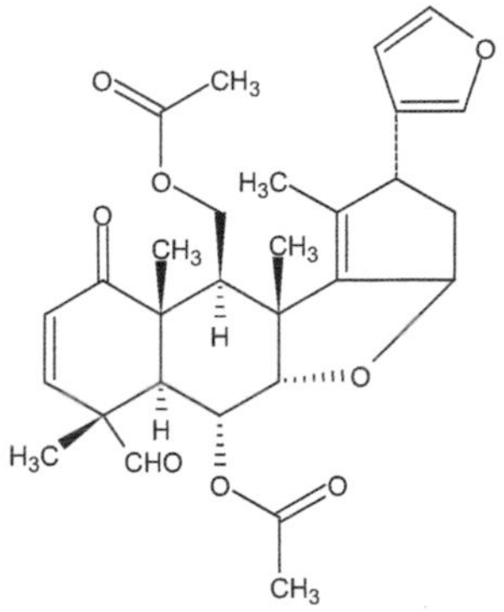

Nimbanal

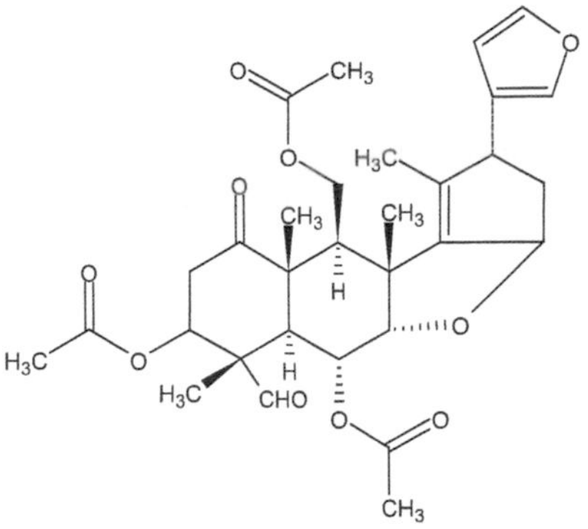

3-Acetyl-nimbanal

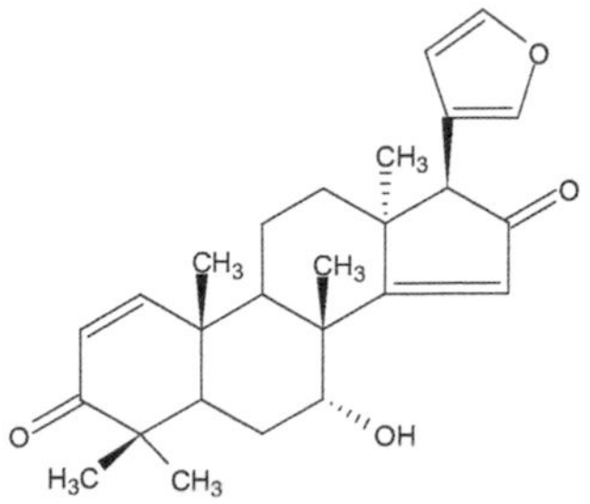

17-Epimimbocinol

Neem insecticidal properties have been widely investigated by several authors. Neem has attracted worldwide attention due to its activity against 400 insect pests. More than 300 compounds have been characterized from neem seeds, one-third of which are tetranortriterpenoids (limonoids). In addition to azadirachtin, several of its analogues with similar biological activity but present in minor quantities, have also been isolated and named as azadirachtins B to L. Purification of neem seed limonoids (azadirachtin A, azadirachtin B, azadirachtin H, desacetylnimbin, desacetylsalannin, nimbin and salannin) has been reported with a counter-current chromatography (CCC) sequentially followed by isocratic preparative reversed-phase high-performance liquid chromatography by Silva et al. (2007).

Azadirachtin is one of the most interesting constituents of neem, because of its influence on insect feeding behavior and insect development. Azadirachtin derivative such as 11-hydroxyazadirachtin B, 1-tigloyl-3-acetylazadirachtinin, 1,3-diacetyl-7-tigloyl-12-hydroxyvilasinin and 23-desmethyllimocin B have also been isolated (Kumar et al. 1996). In other study, Rojatkar and Nagasampagh (1993) isolated others derivatives as 1-tigloyl-3-acetyl-11-hydroxy-4β-methylmeliacarpin and 1-cinnamoyl-3-feruloyl-11-hydroxymeliacarpin with insecticidal properties.

A larger number of limonoids as azadirachtin, salannin, deacetylgedunin, gedunin, 17-hydroxyazadiradione and deacetylnimbin showed toxicity on *Anopheles stephensi* Liston (larvicidal, pupicidal, adulticidal and ovipositional). Azadirachtin, salannin and deacetylgedunin showed high bioactivity, while the rest of the neem limonoids were less active, and were only biologically active at high doses. Azadirachtin was the most potent in all experiments and produced almost 100% larval mortality at 1 ppm concentration. In general, first to third larva instars were more susceptible to the neem limonoids. Neem products may have benefits in mosquito control programs (Nathan et al. 2005).

Kraus et al. (1987) first demonstrated the presence of tetranortriterpenoid lactams in Meliaceae with the isolation of salannolactam-(21) and salannolactam-(23), from methanol extract of *A. indica* seed kernels. Both compounds showed antifeedant activity towards the Mexican bean beetle *Epilachna varivestis*. In addition, a tetranortriterpenoid of the limonoid type, 7-deacetyl-17β-hydroxyazadiradione was isolated from the seeds of *A. indica*. The activity of the compound as an insect growth inhibitor against *Heliothis virescens* was found to be greater than that of azadiradione and 7-deacetylazadiradione (Lee et al. 1988).

From the methanol extract of neem seeds, nimbanal and 3-acetyl-nimbanal derivatives of salannol, were isolated (Supada et al. 1988). In addition, 1,3-Diacetyl-11,19-deoxa-11-oxomeliacarpin, a possible intermediate in the biosynthesis of zadirachtin, was isolated from methanolic extract of *A. indica* seeds (Kraus et al. 1989). In another study, the tetranortriterpene 17-epimimbocinol and nimbocinol were isolated from the methanol extract of neem oil (Balasaheb et al. 1990).

Other constituents

7α-Diacetoxyapotirucall-14-ene-3α,-21,22,24,25-pentaol

Odoratone   2β,3 β,4 β-Trihydroxypregnan-16-one

The other constituents reported included three triterpenoids isolated from methanolic extract of the seed kernels of neem (Luo et al. 2000).

Neem activities

Antioxidant effect

Extracts of leaves of neem from Thailand showed proportionality between total phenolics and radical-scavenging. The extracts were found to be efficient scavengers for reaction with ferrylmyoglobin and showed high efficiencies as chain-breaking antioxidants. This was indicated by lowering of oxygen consumption rates in a peroxidizing lipid emulsion, suggesting a role as dietary antioxidants. Siamese neem tree leaf extracts were found to interact with α-tocopherol in peroxidizing liposomes, resulting in synergistic effects (Pongtip et al. 2007).

Administration of *A. indica* leaf ethanol extract markedly enhanced the hepatic level of glutathione-dependent enzymes and superoxidedismutase and catalase activity, indicating that the hepatoprotective effect of the extract on paracetamol-induced hepatotoxicity may be due to its antioxidant activity. Chemical analyses of the *A. indica* leaf extract revealed that it contains the compounds quercetin and rutin, which may be responsible for the plant's hepatoprotective effect (Chattopadhyay and Bandyopadhyay 2005).

Hypotensive

In an isolated frog heart, there was no noticeable change in amplitude of contraction or rate of the heart at lower doses of leave extract. However, at higher doses, there was temporary cardiac arrest in diastole (Chattopadhyay 1997). In another study a intravenous administration of neem leave alcoholic extract at doses of 100, 300 and 1000 mg/kg resulted in initial bradycardia followed by cardiac arrhythmia in rats. Extract produced a significant and dose-related fall in blood pressure which was immediate, sharp and persistent. Pre-treatment with either atropine or mepyramine failed to prevent the hypotensive effect of extract (Koley et al. 1994).

Anticancer/antitumor

Nimbolide, isolated from leaves of neem was found to be growth-inhibitory in human colon carcinoma HT-29 cells. Nimbolide treatment of cell 2.5 to 10 μM resulted in moderate to large growth inhibition (Kumar et al. 2006).

Enamul and Rathindranath (2006) reported that pretreatment of mice with neem leaf ethanol extract (NLP) causes prophylactic growth inhibition of murine Ehrlich's carcinoma (EC) and B16 melanoma. Using adoptive cell transfer technology, established that NLP-mediated activation of immune cells may be involved in tumor growth restriction. Mononuclear cells from blood and spleen of NLP-activated Swiss and C57BL/6 mice causes enhanced cytotoxicity to murine EC cells *in vitro*. Thus, NLP-activated NK and NK-T cells in mice may regulate tumor cell cytotoxicity by enhancing the secretion of different cytotoxic cytokines.

Antimicrobial

In search of antibacterial activity from neem, ethanol extract was screened against specific multidrug-resistant bacteria. This extract showed MIC values ranged from 0.32–7.5 mg/ml against *Staphyloccocus aureus*, 0.31–6.25 mg/ml against β-lactamases-producing enteric bacteria (Aqil and Ahmad 2007).

The steam distillate of fresh matured leaves of *A. indica* yielded an odorous viscous oil exhibiting antifungal activity against *Tricophyton mentagrophytes, Fusarium oxysporum, F. avenaceum* and *Phytophthora capsici in vitro*. Gas chromatog-mass spectroscopy indicated it to be a mixture of cyclic tri- and tetrasulfides of C_3, C_5, C_6, and C_9 units (Moslem and El-Kholie, 2009).

In another study using the groundnut rust disease (causal agent *Puccinia arachidis* Speg.) as the bioassay system, two limonoids (nimonol and isomeldenin) from the leaves of neem trees showed antifungal activity (Sureash et al. 1997).

Antimalarial

The antimalarial activities of two fractions of an extract from the leaves of the neem tree were compared with those of chloroquine, *in vitro* assays against *Plasmodium falciparum* (Udeinya et al. 2006). Each of the two neem-leaf fractions lysed 50% and 100% of developing gametocytes, at the concentration of 10^{-3} and 1.0 μg/ml, respectively; and 50% and 100% of mature gametocytes at 10^{-3} and 10^{-2} μg/ml, respectively.

In another study a crude acetone/water (50/50) extract of neem leaves showed activity against the asexual (trophozoites/schizonts) and the sexual (gametocytes) forms of the malarial parasite, *Plasmodium falciparum, in vitro.* This extract, if found safe, may provide material for development of new antimalarial drugs that may be useful both in treatment of malaria as well as the control of its transmission through gametocytes (Udeinya et al. 2008).

Recent experiments have shown that several components of the neem are effective against malaria parasites. Irodin A, a substance found in neem leaves, is toxic for resistant strains of malaria. Studies showed a 100% mortality in 72 hr at a ratio of 1:20,000 *in vitro.* Gedunin and quercentin, other compounds found in neem leaves, are at least as effective against malaria as quinine and chloroquine. Because the anti-malarial effects of neem appear to be greater in the body than on the *in vitro* test, there has been some speculation that stimulation of the immune system is a major factor in neem's effectiveness against malaria. In addition to its anti-malarial activity, neem also lowers the fever and increases the appetite. Thus, strengthening the body and speeding up recovery (Buckling et al. 1999).

Insecticidal activity

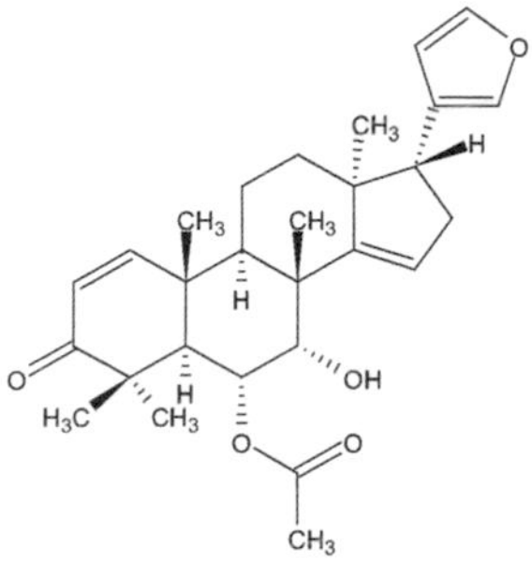

6α-O-Acetyl-7-deacetylnimocinol

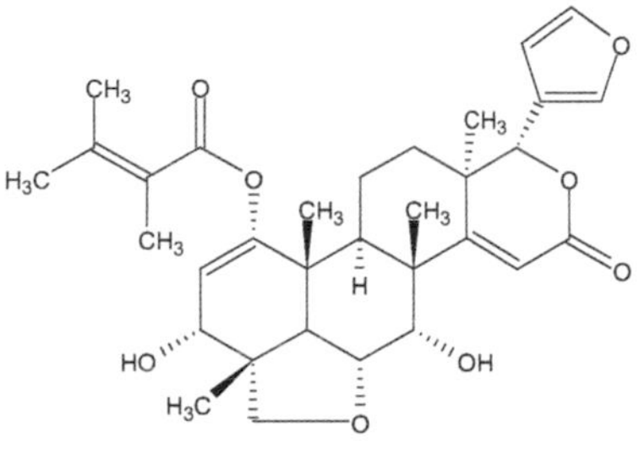

Meliacinol

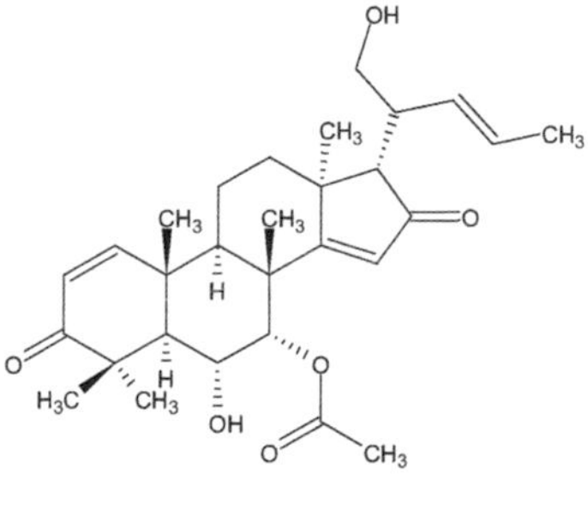

Zeeshanol

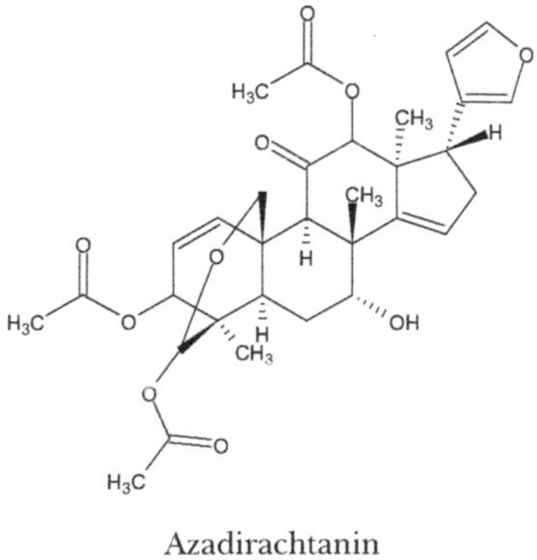

Azadirachtanin

Triterpenoids, 6α-O-acetyl-7-deacetylnimocinol, meliacinol and zeeshanol were isolated from the methanolic extract of the fresh leaves of *A. indica*. Compound showed toxicity on fourth instars larvae of mosquitoes (*Aedes aegypti*) with LC_{50} values of 21, 83, 67 ppm, respectively. Meliacinol had no effect upto 100 ppm (Siddiqui et al. 2000a, Siddiqui et al. 2006). All compounds possessed pesticidal activity. In another study a limonoid of the Δ^{14}-meliacan skeleton, was isolated from the leaves methanol extract by Gurudas and Shashi (1985).

Other constituents

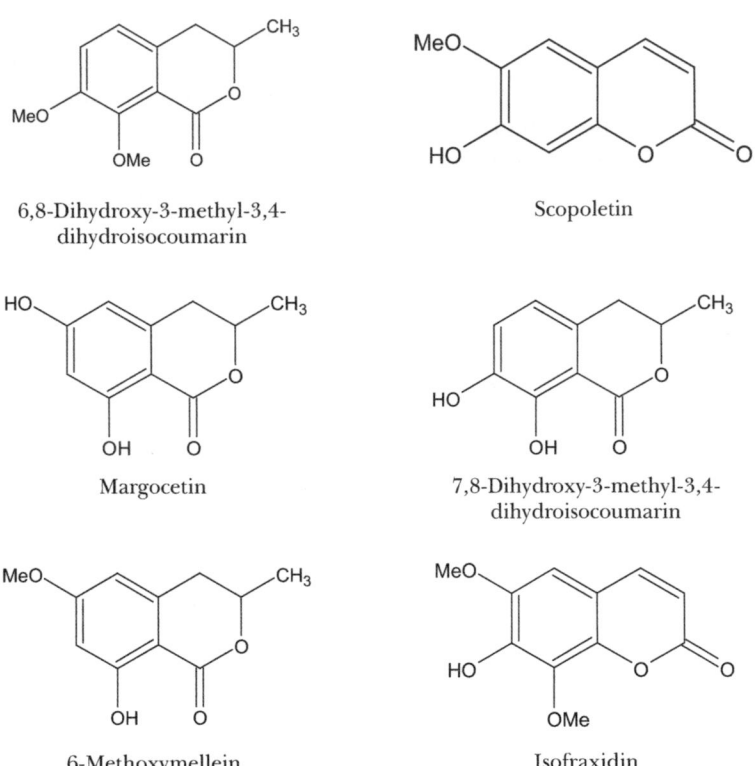

6,8-Dihydroxy-3-methyl-3,4-dihydroisocoumarin

Scopoletin

Margocetin

7,8-Dihydroxy-3-methyl-3,4-dihydroisocoumarin

6-Methoxymellein

Isofraxidin

In addition limonoids and triterpenoids isocoumarins, were isolated from fresh, uncrushed, spring twigs of neem (Siddiqui et al. 1988).

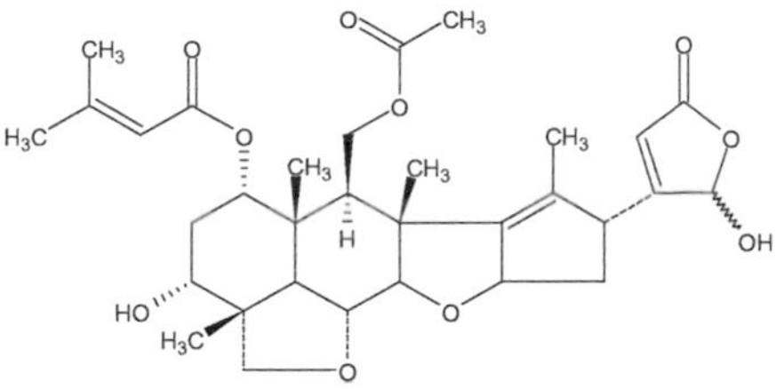

Isoazadirolide

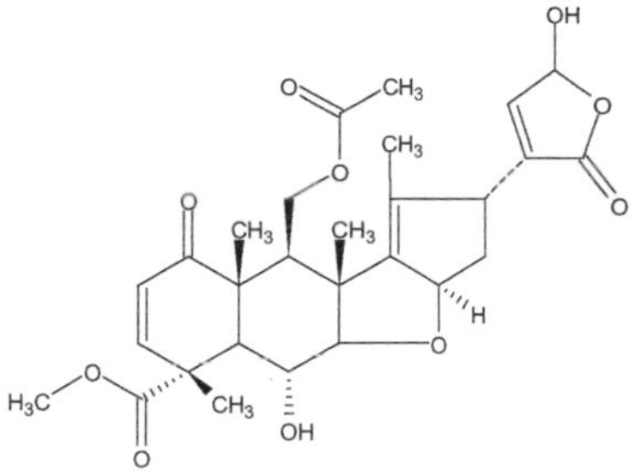

Desacetylnimbinolide

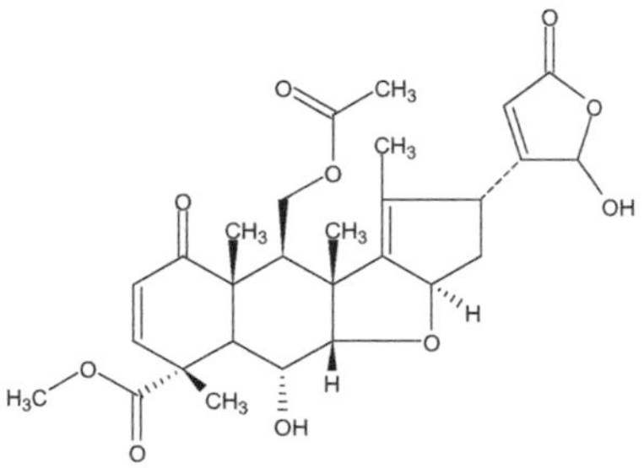

Desacetylisonimbinolide

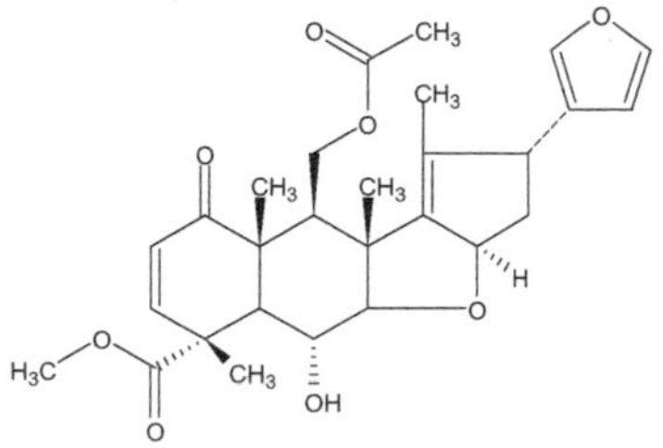

Desacetylnimim

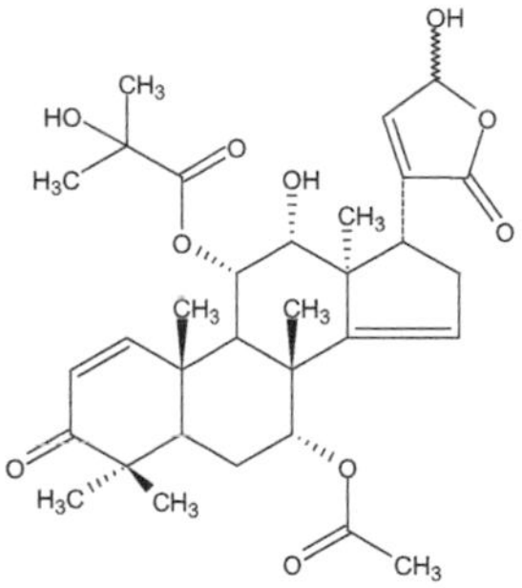

Nimbocinolide

A large number of new and known limonoids have been insolated from the neem leaf and the structures determined mainly by Siddiqui et al. (1986a). A ring-C seco tetranortriterpenoid γ-hydroxybutenolide, (isoazadirolide) was isolated from the acidic fraction of the fresh, winter leaves of *A. indica* (Siddiqui et al. 1986a). In another study from the fresh, green, spring twigs γ-hydroxybutenolides, desacetylnimbinolide and desacetylisonimbinolide, were isolated together with desacetylnimim (Siddiqui et al. 1986b). In addition Siddiqui et al. (1989) isolated from the acidic fraction of fresh, undried leaves of neem nimbocinolide.

Neem fruits

Neem fruits are an important source of food for some wildlife, especially birds and bats, although they only digest the pulp and not the seed. Neem compounds have been judged to be relatively non-toxic to mammals.

Insecticide

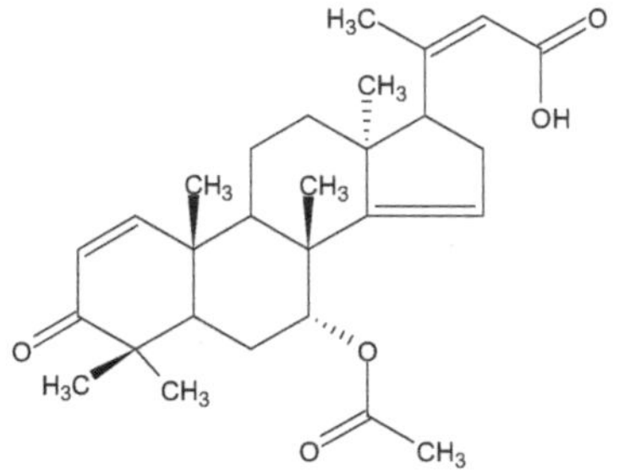

Azadironic acid

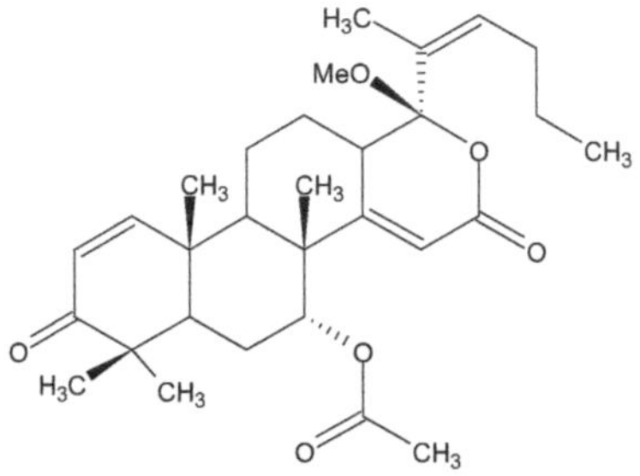

Meliacinin

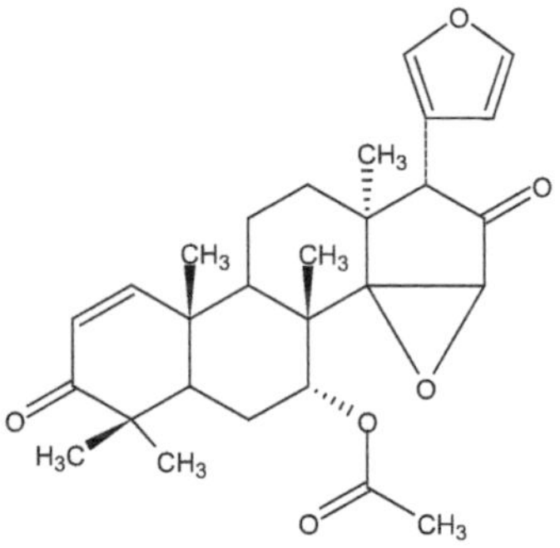

Epoxyazadiradione

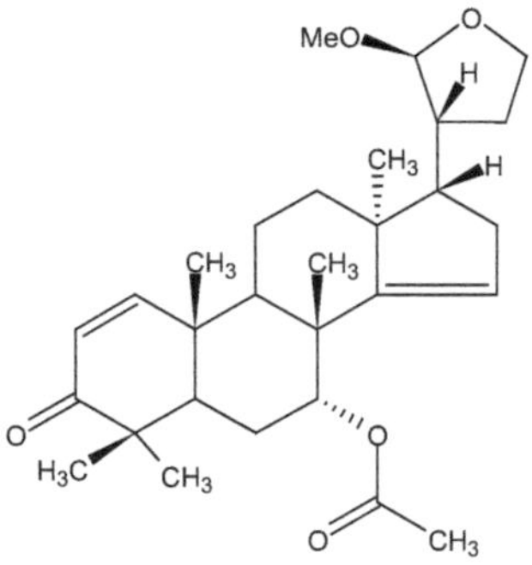

Limocin A

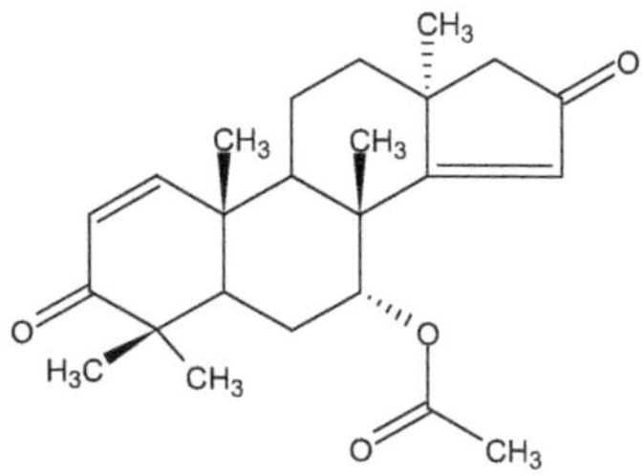

Desfuranoazadiradione

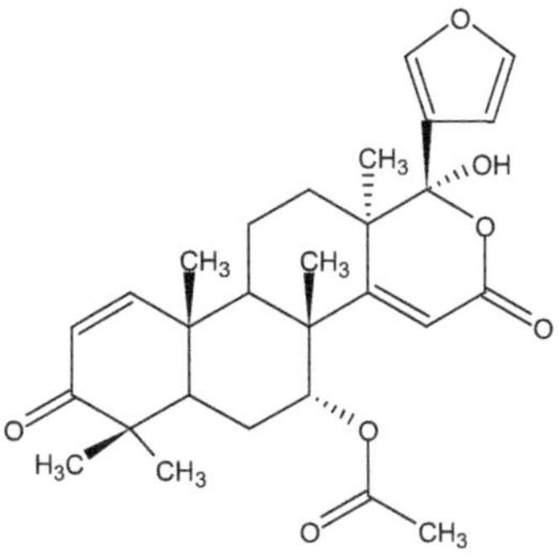

Mimolicinol

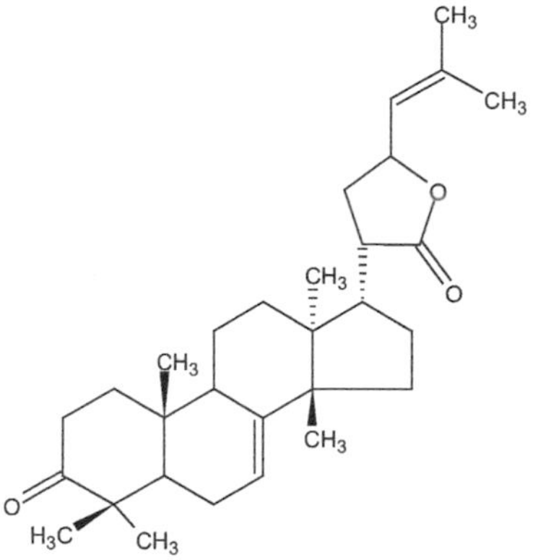

Mimolinone

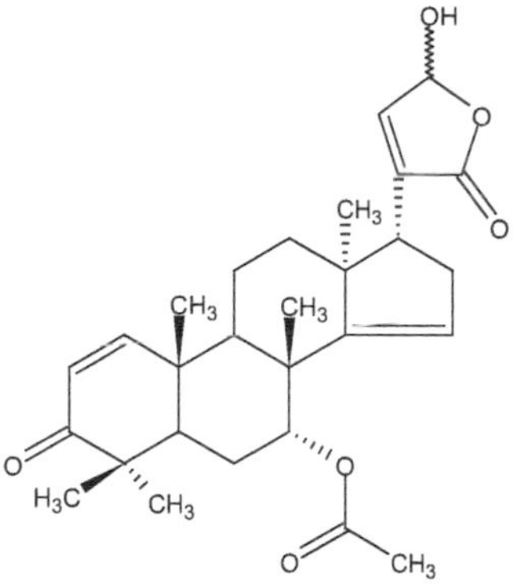

Azadironolide

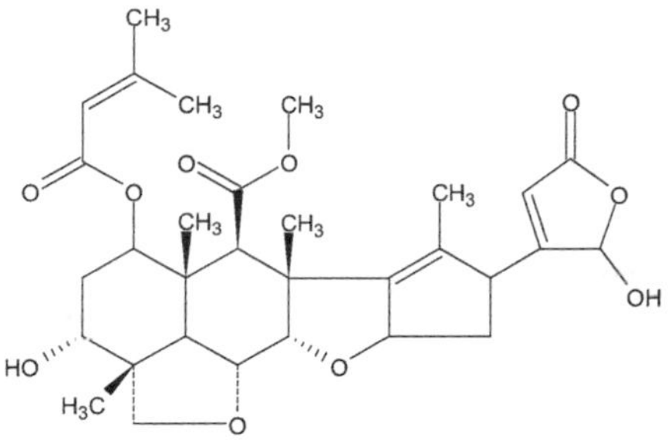

Isoazadironolide

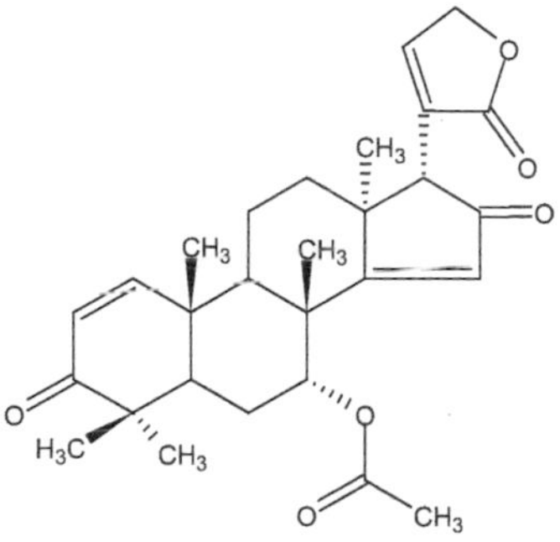

Azadiradionolide

Ripcning neem fruits and expressed neem seed oil give off a strong alliaceous (garlic-like) odor, and some of the reputed medicinal efficacies of neem oil of fruits and seeds have been attributed to the sulfurous compounds that contains *cis* and *trans* 3,5 diethyl-1,2,4-trithiolanes and *cis* and *trans-n*-propyl-1-propenyl-trisulphides (Mubarak and Kulatilleke 1990).

Triterpenoids, meliacinin, azadiradione and azadironic acid have been isolated from the fresh fruit of neem. The insecticidal activity was determined on *Anopheles stephensi* (4th instar larvae). The known constituents azadiradione, epoxyazadiradione, limocin A and B and desfuranoazadiradione that showed insect growth regulating (IGR) effect on these larvae were also isolated Meliacinin and azadironic acid had LC_{50} 13, and 4.5 ppm, respectively, while azadiradione, epoxy-azadiradione, mix of limocin A and B and desfuranoazadiradione showed LC_{50} 15, 18, 19 and 37 ppm, respectively (Siddiqui et al. 2000).

In other studies mimolicinol, mimolinone, azadironolide, isoazadironolide and azadiradionolide also were isolated by Siddiqui et al. (1984, 1986, 1999) from the fresh fruits.

Neem flowers

Anticancer/antitumor

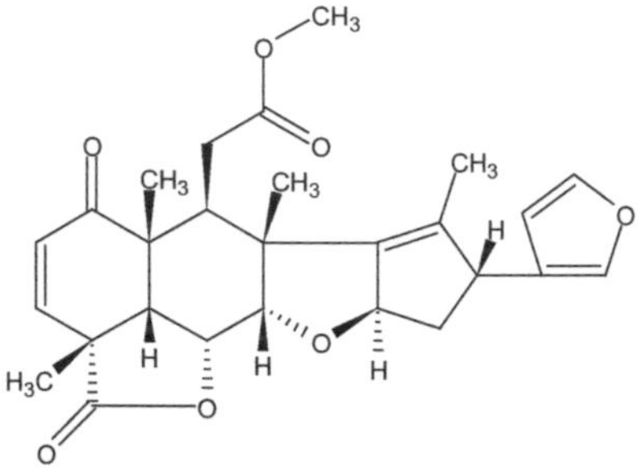

Nimbolide

Nimbolide a triterpenoid extracted from the flowers of the neem tree, was found to have antiproliferative activity in some mammalian cancer cell lines including N1E-155 (murine neuroblastoma cells), 143BK.TK (human osteosarcoma cells), and murine RAW 264.7 macrophages (Cohen et al 1996). The toxicity of the compound is a major issue in its therapeutic use. Nimbolide when given through the intragastric route was not toxic to some experimental animals.

Several studies have been conducted with other limonoids on cancer cell lines as epoxyazadiradione, salannin, nimbin, deacetylnimbin and azadirachtin showed antiproliferative activity with IC_{50} values of 27 μM, 112 μM, each > 200 μM respectively. In contrast nimbolide treatment of cells with 0.5-5.0 μm concentrations resulted in moderate to very strong growth inhibition in U937, HL-60, THP1 and B16 cell lines (Roy et al. 2006a). Nimbolide (5,7,4'-trihydroxy-3',5'-diprenylflavanone) is a potent cytotoxic compound. This has the potential to alter the deregulated cell cycle progression characteristic of the HT-29 colon carcinoma cell line, by arresting the cell cycle at G2/M or GO/G t phase, which involves the modulation of Cdk inhibitors, Cdks, cyclins, and other Cdk regulators (Roy et al. 2006b).

5,7,4'-Trihydroxy-8-prenylflavanone

5,4'-Dihydroxy-7-methoxy-8-prenylflavanone

5,7,4'-Trihydroxy-3',8-diprenylflavanone

5,7,4'-Trihydroxy-3',5'- diprenylflavanone

Prenylated flavanones were isolated from the methanol extract of the flowers of the neem tree as potent antimutagens against Trp-P-1 (3-amino-1,4-dimethyl-5H-pyrido[4,3-b]indole) in the *Salmonella typhimurium* TA98 assay. Antimutagenic IC_{50} values of compounds were 2.7, 3.7, 11.1 and 18.6 μM in the pre incubation mixture, respectively. These compounds similarly inhibited the mutagenicity of Trp-P-2 (3-amino-1-methyl-5H-pyrido[4,3-b]indole) and PhIP (2-amino-1-methyl-6-phenylimidazo[4,5-b] pyridine). All the compounds strongly inhibited ethoxyresorufin O-dealkylation activity of cytochrome P4501A isoforms, which catalyze N-hydroxylation of heterocyclic amines. However, the compounds did not show significant inhibition against the direct-acting mutagen NaN3- Thus, the antimutagenic effect of the flavanones would be mainly based on the inhibition of the enzymatic activation of heterocyclic amines (Nakahara et al. 2003).

Other Constituents

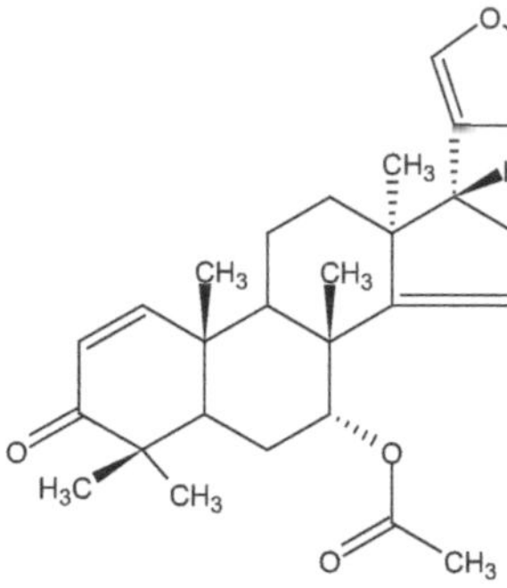

Azadirone

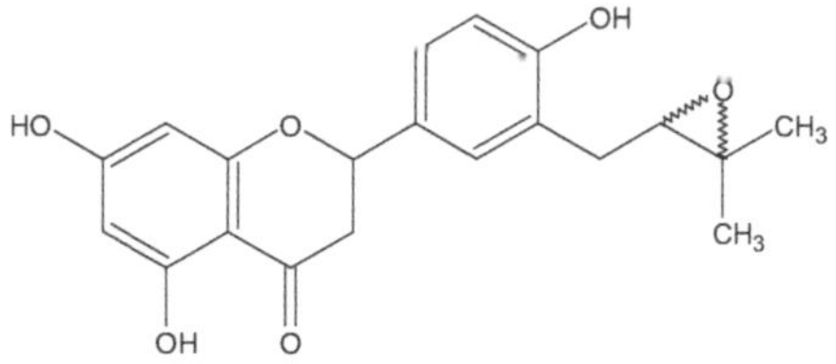

Azharone

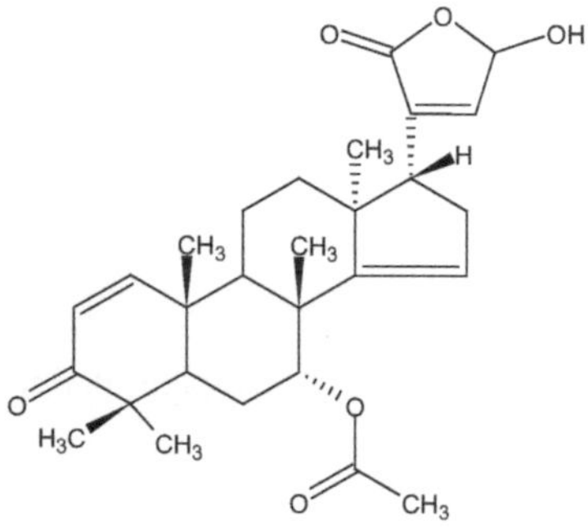

Isoazadironolide

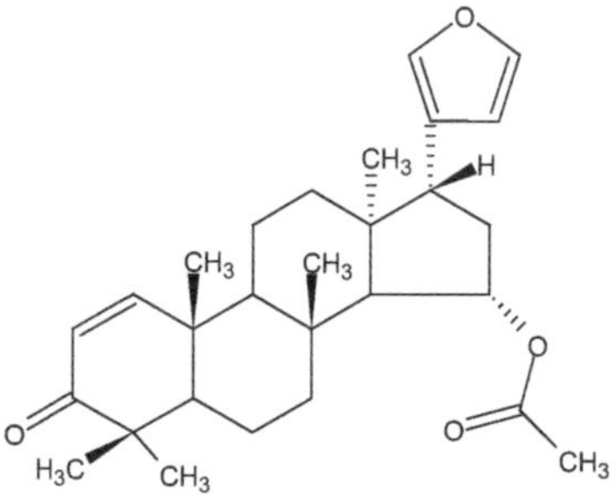

Neoflone

Although different parts of the plant have been studied extensively for its chemical constituents, the study on flowers is lacking. Studies on the chemical constituents of the flowers of *A. indica* have led to the isolation of flavanone as azadirone, azharone and isoazadironolide (Siddiqui et al. 2006). Tetratriterpenoid, neoflone (15-acetoxy-7-deacetoxydihydroazadirona), was isolated from the air-dried flowers of *Azadirachta indica* (Srinivas and Prasad 1995).

Neem stem bark, root bark

Anti-oxidant Neem

The bark of the neem, was found to have high phenolic content of 89.8 mg/kg and high antioxidant effect 96.9% (Dhan et al. 2007).

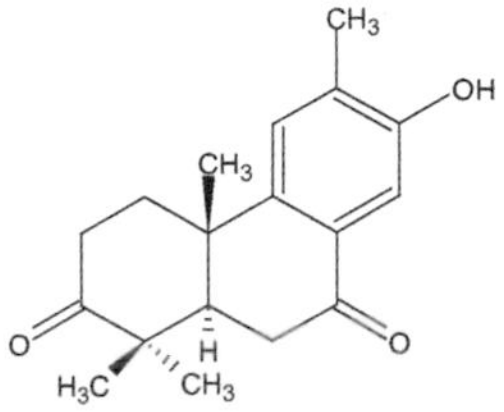

Nimbione

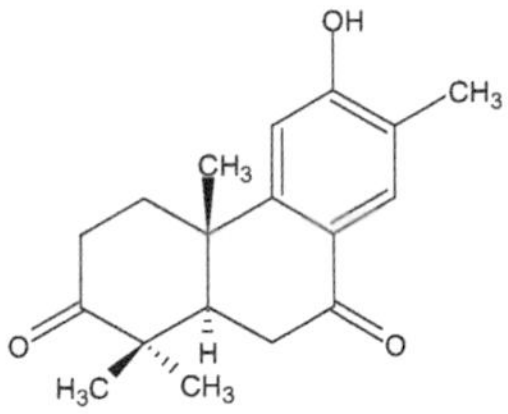

Nimbinone

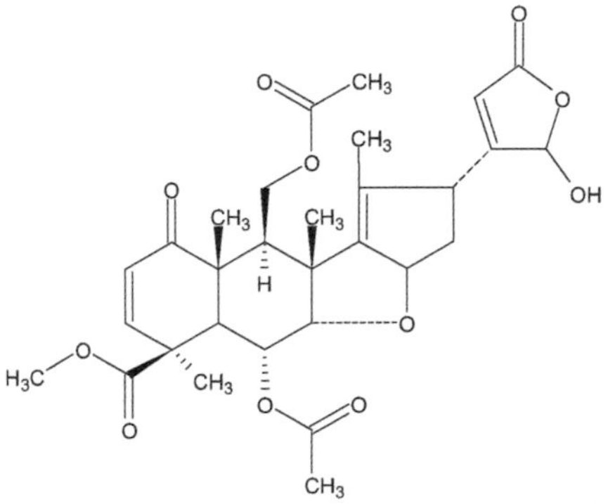

Isonimbinolide

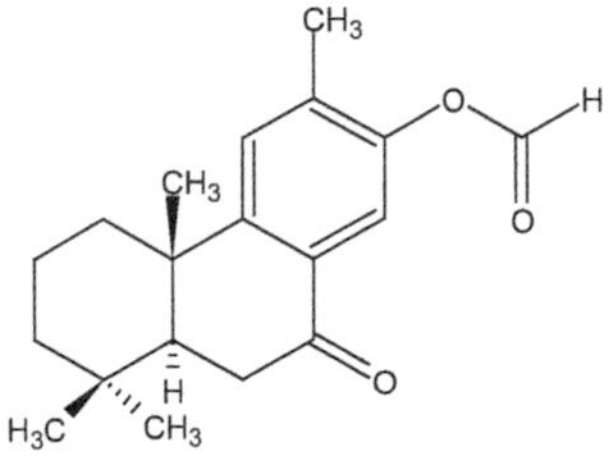

Margolone

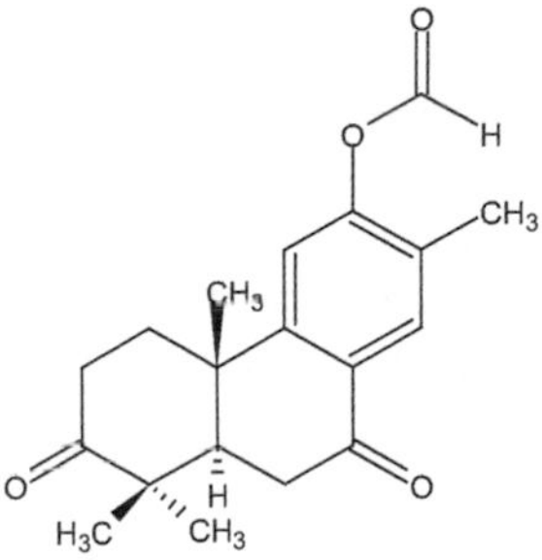

Margolonone

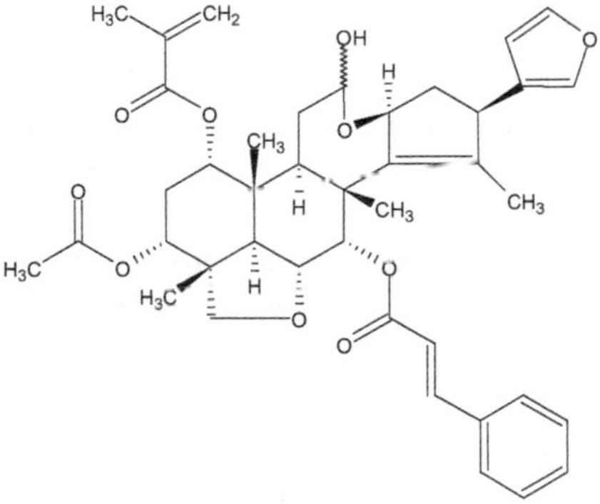

Nimbolicin

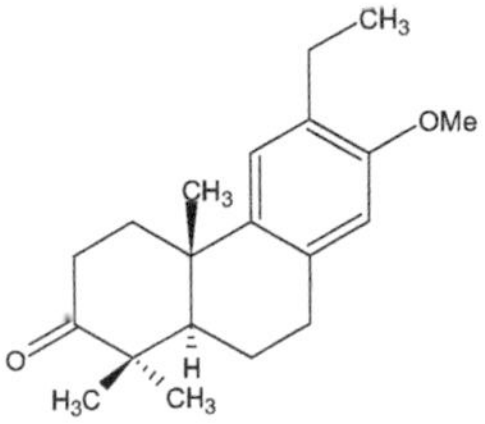

Nimbonone

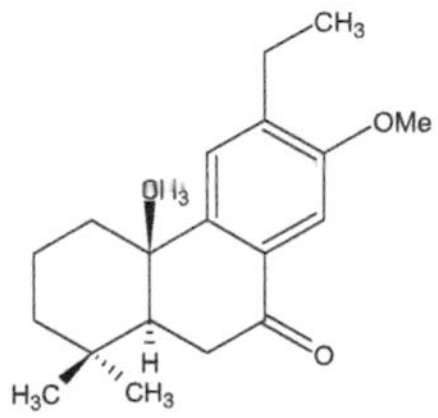

Nimbonolone

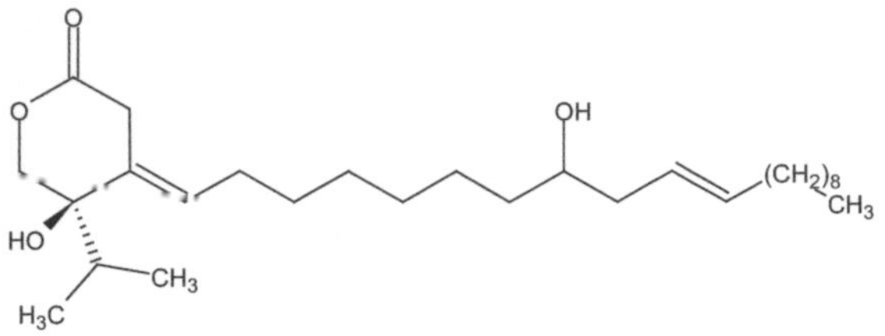

Margosinone

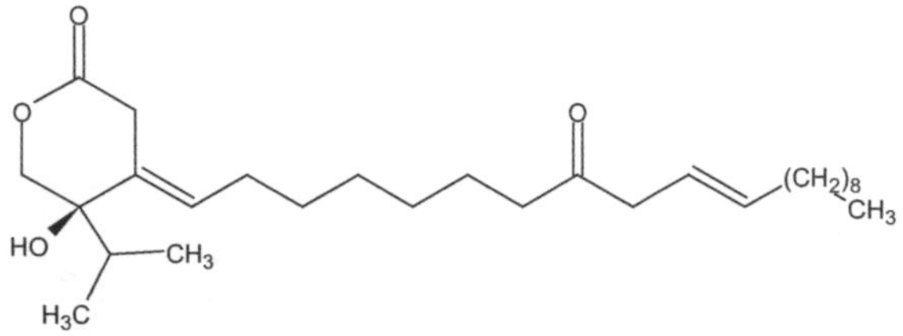

Margosinolone

Other constituents

Isomeric diterpenoids from the stem bark of neem and ring C-seco-tetranortriterpenoid have been isolated from the stem bark of *A. indica* (Iffat et al. 1988). In another study, Iffat et al. (1989) the diterpenoids margolone, margolonone and isomargolonone were isolated from neutral fraction of the stem bark of neem. They also reported that from the neutral fraction of the root bark of *A. indica* nimbolicin was isolated, in addition to nimbolin B which was previously reported from the trunk wood (Iffat et al. 1989a). Continuing with the study of neutral fraction of the stem bark of neem (Iffat et al. 1989b) two isomeric diterpenoids named nimbonone and nimbonolone were isolated. In addition margosinone and margosinolone, two polyacetate derivatives, were isolated from the stem bark of *A. indica* (Siddiqui et al. 1989a).

Conclusions

Azadirachta indica A. Juss, Meliaceae is the well-known traditional medicinal plant in India. Neem is now considered as a valuable source of natural product for development of medicines and also for the development of industrial chemical products. This tree has been identified as one of the most suitable sources of obtaining environmentally friendly pest controlling agents. Many chemical constituents possessing pesticidal activity have been isolated from this plant. The experiments carried out on neem justify, the reputation of the plant as a useful remedy against cancer, diabetes, antioxidant and anti-inflammatory. Recent findings tend to support this ethnomedicinal use of *A. indica*.

References

Aguilar, A., A. Argueta, and L. Cano. 1994. Flora Medicinal Indígena de México. Instituto Nacional Indigenista México.

Aqil, F. and I. Ahmad. 2007. Antibacterial properties of traditionally used Indian medicinal plants. Methods and Finding in Experimental and Clinical Pharmacology 29: 79–92.

Balasaheb, O. R., M.T. Vyas, A. Brahamanand, and B.V. Sujeta. 1990. Nimbocinol and 17–epinimbocinol from the nimbidin fraction of neem oil. Phytochemistry 29: 3963–3965.

Bandyopadhyay, U., Biswas, K., Chatterjee, R., Bandyopadhyay, D., Chattopadhyay, I., Ganguly, C.K., Chakraborty, T., Bhattacharya, K. and R.K. Banerjee. 2002. Gastroprotective effect of Neem (Azadirachta indica) bark extract: possible involvement of H(+)- K(+)-ATPase inhibition and scavenging of hydroxyl radical. Life Sciences 71: 2845–2865.

Bopanna, K.N., J. Kannan, G. Sushma, R. Balaraman, and S.P. Rathod. 1997. Antidiabetic and antihyperlipemic effect of neem seed kernel powder on alloxan diabetic rabbits. Indian Journal of Pharmacology 29:162–167.

Buckling, A., L.C. Ranford-Cartwright, A. Miles, and A.F. Read. 1999. Chloroquine increases *Plasmodium falciparum* gametocytogenesis *in vitro*. Parasitology 118:339–346.

Butterworth, J.H. and E.D. Morgan. 1971. Investigation of the locust feeding inhibition of the seeds of the neem tree, *Azadirachta indica*. Journal of Insect Physiology 17, 969–977.

Chattopadhyay, R.R. 1997. Effect of *Azadirachta indica* hydroalcoholic leaf, extract on the cardiovascular system. General Pharmacology 28: 449–451.

Chattopadhyay, R.R. and M. Bandyopadhyay. 2005. Possible mechanism of hepatoprotective activity of leaf extract against paracetamol-induced hepatic damage in rats. Indian Journal of Pharmacology 37: 184–185.

Cohen, E., G.B. Quistad, and J.F. Casida. 1996. Cytotoxicity of nimbolide, epoxyazadiradione and other limonoids from neem insecticide. Life Science 58: 1075–1081.

Dhan, S., S. Samikaha, P. Garima, and N. Brama. 2007. Total phenol, antioxidant and free radical scavenging activities of some medicinal plants. Journal Food Science Nutrition 58: 18–28.

Dhar, R., K. Zhang, G.P. Talwar, S. Garg, and N. Kumar. 1998. Inhibition of the growth and development of asexual and sexual stages of drug-sensitive and resistant strains of the human malaria parasite *Plasmodium falciparum* by Neem (*Azadirachta indica*) fractions. Journal of Ethnopharmacology 61: 31–39.

Deepak, A. and S. Anshu. 2008. Indigenous Herbal Medicines: Tribal Formulations and Traditional Herbal Practices, Aavishkar Publishers Distributor, Jaipur, India.

Enamul H. and B. Rathindranath. 2006. Neem (*Azadirachta indica*) leaf preparation induces prophylactic growth inhibition of murine Ehrlich carcinoma in Swiss and C57BL/6 mice by activation of NK cells and NK-T cells. Immunobiology 211: 721–731.

EPA 1997. NSW State of the Environment USA.

Ganguli, S. 2002. Neem: A therapeutic for all seasons, Current Science 82: 1304–1312.

Garg, S., V. Taluja, S.N. Upadhyay, and G.P. Talwar. 1993. Studies on the contraceptive efficacy of Praneem polyherbal cream. Contraception 48: 591–596.

Gilles, H.M. and D.A. Warrell. 1993. Bruce-Chwatt's Essential Malariology, 3rd ed. Edward Arnold, London.

Gupta, R.K. 1993. Multipurpose Trees for Agroforestry and Wasteland Utilization. International Science Publisher, New York, NY.

Guipreet, K., A.M. Sarwar, and M. Athar. 2004. Nimbidin suppresses functions of macrophages and neutrophils relevance to its anti inflammatory mechanisms. Phytotherapy Research 18: 419–424.

Gurudas, P. and B. Shashi. 1985. Azadirachtanin, a new limonoid from the leaves of *Azadirachta indica*. Heterocycles 23: 2321–5.

Herrel, N., F.P. Amerasinghe, J. Ensink, M. Mukhtar, W. van der Hoek, and F. Konradsen 2001. Breeding of Anopheles mosquitoes in irrigated areas of South Punjab, Pakistan. Medical Vetetinary Entomology 15: 236–248.

Hitesh, V., P. Pratik, and K. Vimal. 2007. Neem. Indian Pharmacist 6:16–20.

Iffat, A., B.S. Siddiqui, Faizi, F. Salimuzzaman, and S. Siddiqui. 1988. Terpenoids from the stem bark of Azadirachta indica. Phytochemistry. 27: 1801–1804.

Iffat, A., S. Bina, S. Faizi, and S. Siddiqui. 1989. Structurally novel diterpenoid constituents from the stem bark of *Azadirachta indica* (Meliaceae). Journal of the Chemical Society Perkin Transactions 2: 343–345.

Iffat, A., S. Bina, S. Faizi, and S. Siddiqui. 1989a. Isolation of meliacin cinnamates from the root bark of *Azadirachta indica* A. Juss (Meliaceae). Heterocycles 29: 729–735.

Iffat, A., S. Siddiqui, S. Faizi, and S. Siddiqui 1989b. Diterpenoids from the stem bark of *Azadirachta indica*. Phytochemistry 28: 1177–1180.

Jamir, T.T., H.K. Sharma, and A.K. Dolui. 1999. Folklore medicinal plants of Nagaland, India. Fitoterapia 70: 395–401.

Kaur, G., M.S. Alam, and M. Athar. 2004. Nimbidin Suppresses Functions of Macrophages and Neutrophils: Relevance to its antiinflammatory mechanisms. Phytotherapy Research 18: 419–424.

Kay, M.A. 1996. Healing with plants in the American and Mexican West. University of Arizona Press, Tucson.

Keating, B. 1994. Neem: The miraculous healing Herb. The neem association. Winter Park, FL (USA).

Koley, K.M. and J. Lal. 1994. Pharmacological effects of Azadirachta indica (neem) leaf extract on the ECG and blood pressure of rat. Indian Journal of Physiology and Pharmacology. 38: 223–225.

Kraus, W., A. Klenk, M. Bokel, and B. Vogler. 1987. Tetranortriterpenoid lactams with insect antifeeding activity from *Azadirachta indica* A. Juss (Meliaceae). Liebigs Annalen der Chemie 4: 337–340.

Kraus, W., H. Gutzeit, and M. Bokel. 1989. 1,3,Diacetyl-11,19-deoxa-11-oxomeliacarpin, a possible precursor of azadirachtin, from *Azadirachta indica* (Meliaceae). Tetrahedron Letter 30: 1797–1798.

Kumar, M.R., K. Masuko, T. Makiko, N. Kazuhiko, H. Shinmoto, and T. Tojiro. 2006. Inhibition of colon cancer (HT-29) cell proliferation by a triterpenoid isolated from *Azadirachta indica* is accompanied by cell cycle arrest and up-regulation of p21. Planta Medica 72: 917–923.

Kurose, K. and M. Yatagai. 2005. Components of the essential oils of *Azadirachta indica* A. Juss, *Azadirachta siamensis* Velton, and *Azadirachta excelsa* (Jack) Jacobs and their comparison. Journal of Wood Science 51:185–188.

Kumar, C.S., M. Srinivas, and S. Yakkundi. 1996. Limonoids from the seeds of *Azadirachta indica*. Phytochemistry 43: 451–455.

Kumar, S., P.K. Suresh,and M.R.Vijayababu, A. Arimkumar, and J. Arunakaran. 2006. Anticancer effects of ethanolic neem leaf extract on prostate cancer cell line (PC-3). Journal of Ethnopharmacology 105, 246–250.

Lee, S.M., J.I. Olsen, M.P. Schweizer, and J.A. Klocke.1988. 7-Deacetyl-17β-hydroxyazadiradione, a new inhibitor from *Azadirachta indica*. Phytochemistry 27: 2773–2775

Luo, X., S. Wu, Y. Ma, and D. Wu. 2000. A new triterpenoid from *Azadirachta indica*. Fitoterapia 71: 668–672.

Mishra A.K., N. Singh, and V.P. Sharma. 1995. Use of neem oil as a mosquito repellent in tribal villages of Mandla district, Madhya Pradesh. Indian Journal Malariol 32: 99–103.

Mubarak, A.M., and C.P. Kulatilleke. 1990. Sulphur constituents of neem seed volatile: revision. Phytochemistry 29: 3351–3352.

Nakahara, K., M.K. Roy, H. Ono, and I. Maeda, M. Ohnishi-Kameyama, M. Yoshida, and G. Trakoontivakorn. 2003. Prenylated flavanones isolated from flowers of *Azadirachta indica* (the Neem Tree) as antimutagenic constituents against heterocyclic amines. Journal of Agricultural Food Chemistry 51: 6456–6460.

Nathan, S.S., K. Kalaivani, and K. Murugan. 2005. Effects of neem limonoids on the malaria vector *Anopheles stephensi* Liston (Diptera: Culicidae). Acta Tropica 96: 47–55.

National Research Council (NRC). 1992. Neem: A Tree For Solving Global Problems. National Academy Press, Washington, D.C.

Pal, S.K., Biswas, S., Sinharay, K., Banerjee, A. 2002. Rheumatological problems in diabetes mellitus. Indian Medical Association 100: 458–460.

Patel, G.R, and N.K. Patel. 2004. Biochemical and ethnomedicinal uses of neem (*Azadirachta indica*). Plant Sciences 17: 403–406.

Pates, H., and C. Curtis. 2005. Mosquito behavior and vector control. Annual Review Entomology 50: 53–70.

Pillai, N.R. and G. Santhakumari. 1981. Anti-arthritic and anti-inflammatory actions of nimbidin. Planta Medica 43: 59–63.

Pongtip, S., C.U. Charlotte, A. Mogens, G. Wandee, and S.H. Leif. 2007. Antioxidant effects of leaves from *Azadirachta* species of different provenience. Food Chemistry 104: 1539–1549.

Randhawa, N.S. and B.S. Parmar 1993. Neem Research and Development. Society of Pesticide Science, New Delhi, India.

Rao, D.R, Reuben, R., Venugopal, M.S., Nagasampagi, B.A. and H. Schmutterer. 1992. Evaluation of neem, *Azadirachta indica*, with and without water management, for the control of culicine mosquito larvae in rice-field. Medical Veterinary Entomology 6: 318–324.

Read, M.D. and, J.H. French. 1993. Genetic Improvement of Neem: Strategies for the Future. Proceedings of the International Consultation on Neem Improvement held at Kasetsart University, Bangkom, Thailand. Rojarkar, S.R. and B.A. Nagasampagh. 1993. 1-Tigloyl-3-acetyl-II-hydroxy-4p-methylmeliacarpin from *Azadirachta indica*. Phytochemistry 32: 213–214.

Roy, M.K., M. Kobori, M. Takenaka, K. Nakahara, H. Shinmoto, and T. Tsushida. 2006a. Antiproliferative effect on human cancer cell lines after treatment with nimbolide extracted from an edible part of the neem tree (*Azadirachta indica*). Phytotherapy Research 21: 245–250.

Roy, M.K., M. Kobori, M. Takenaka, K. Nakahara, H. Shinmoto, and T. Tsushida. 2006b. Inhibition of Colon Cancer (HT-29) Cell proliferation by a triterpenoid isolated from *Azadirachta indica* is accompanied by cell cycle arrest and up-regulation of p2. Planta Medica 72: 917–923.

Santhakumari, G., and J. Stephen. 2005. Antimitotic effect of nimbidin. Cellular and Mol. Life Science 37: 91–93.

Schmutterer, H. 1990. Properties and potential of natural pesticides from the neem tree, *Azadirachta indica*. Annual Review of Entomology 35: 271–297.

Sharma, P.V. 1996. Classical Uses of Medicinal Plants. Chaukbambha Visvabharati. Varanasi 1, India.

Shodini 1997. Touch Me, Touch-me-not: Women, Plants and Healing. Kali for Women. New Delhi, India.

Siddiqui, S., S. Bina, S. Faizi, and R.M. Tariq. 1984. Studies on the chemical constituents of *Azadirachta indica*. Part I: Isolation and structure of a new tetranortriterpenoid-nimolicinol. Heterocycles. 22: 295–298.

Siddiqui, S., S. Faizi, S. Bina, and R.M. Tariq. 1986. Studies on the chemical constituents of *Azadirachta indica* A. Juss (Meliaceae). Part VI. Journal of the Chemical Society of Pakistan 8: 341–347.

Siddiqui, S., S. Bina, S. Faizi, S. Mahmood, and M. Tariq. 1986a. Isoazadirolide, a new tetranortriterpenoid from *Azadirachta indica* A. Juss (Meliaceae). Heterocycles 24: 3163–3167.

Siddiqui, S., M. Tariq, S. Bina, S. Faizi, and S. Mahmood. 1986b. Two new tetranortriterpenoid from *Azadirachta indica*. Journal Natural Products 49: 1068–1073.

Siddiqui, S., T. Mahmood, B.S. Siddiqui, and S. Faizi. 1988. Non-terpenoidal constituents from *Azadirachta indica*. Planta Medica 54: 457–458.

Siddiqui, S., S. Bina, R.M. Tariq, and S. Faizi. 1989. Tetranortriterpenoids from *Azadirachta indica*. Heterocycles 29: 87–96.

Siddiqui, B.S., S. Faizi, and S. Siddiqui. 1989a. Margosinone and margosinolone, two new polyacetate derivatives from *Azadirachta indica*. Fitoterapia 60: 519–523.

Siddiqui, B. S., S. Ghiasuddin, S. Faizi, L. Rasheed. 1999. Triterpenoids of the fruit Coats of *Azadirachta indica*. Journal Natural Products 62: 1006–1009.

Siddiqui, B.S., F.A. Ghiasuddin, S. Faizi, S.N.H. Naqvi, and R.M. Tariq. 2000. Two insecticidal tetranortriterpenoids from *Azadirachta indica*. Phytochemistry 53: 371–376.

Siddiqui, B.S., M. Rasheed, G.S. Faizi, S.N.H. Naqvi, and R.M. 2000a. Biologically active triterpenoids of biogenetic interest from the fresh fruit coats of *Azadirachta indica*. Tetrahedron 56: 3547–3551.

Siddiqui, B.S., F. Afshan, S. Arfeen, and T. Guizar. 2006. A new tetracyclic triterpenoid from the leaves of *Azadirachta indica*. Natural Product Research 20: 1036–1040.

Silva, J.C., G.N.J. Rosangela, R.D.L. Oliveira, and L. Brown. 2007. Purification of the seven tetranortriterpenoids in neem (*Azadirachta indica*) seed by counter-current chromatography sequentially followed by isocratic preparative reversed-phase high-performance liquid chromatography. Journal of Chromatography A 1151: 203–210.

Srinivas, S.S., and B.K. Prasad. 1995. Neoflone, a new tetratriterpenoid from the flowers of *Azadirachta indica* A. Jus (Meliacea). Organic Chemistry Included Medicinal Chemistry 34B: 1019–1020.

Supada, R.R., B.S. Vidya, K.M. Mandakini, J.S. Vimal, and N.A. Bhimsen. 1988. Tetranotriterpenoids from *Azadirachta indica*. Journal Chemical Society Perkin Transactions 28: 203–205.

Sureash, G., N.S. Narasimhan, S. Masilamani, P.D. Partho, and G. Gopalakrishnan. 1997. Antifungal fractions and compounds from uncrushed green leaves of *Azadirachta indica*. Phytoparasitica 25: 33–39.

Tandan, S.K., S. Gupta, S. Chandra, and J. Lal. 1995. Safety evaluation of *Azadirachta indica* seed oil, a herbal wound dressing agent. Fitoterapia LXVI: 69–72.

Trongtokit, Y.,Y. Rongsriyan, N.L. Komalamisra, and L. Apiwathnasom. 2005. Comparative repellency of 38 essential oils against mosquito bites. Phytotherapy Research 19: 303–309.

Udeinya, I.J., N. Brown, E.N. Shu, F.I. Udeinya, and I. Quakeyie. 2006. Fractions of an antimalarial neem-leaf extract have activities superior to chloroquine, and are gametocytocidal. Journal Annals of Tropical Medicine and Parasitology 100: 17–22.

Udeinya, J.I., E.N. Shu, I. Quakyi, and F.O. Ajayi. 2008. An antimalarial neem leaf extract has both schizonticidal and gametocytocidal activities. American Journal and Therapeutics 15: 108–110.

Chapter 11

Ethnomedicine of Quassia and Related Plants in Tropical America

Rafael Ocampo[1] and Gerardo Mora[2]*

Introduction

Quassi[3] was the name of a Negro slave, a medicine man of Surinam, who acquired a great reputation in the treatment of fevers with a secret bitter-plant remedy. His secret was made public by the Swedes Daniel Rolander and Carl G. Dahlberg in 1756 (Felter and Lloyd 1898) and his name was honored by Linné (Linaeus) in naming the plant *Quassia amara* in 1762.[4] The term 'amara' refers to the bitterness of the plant and its extracts.

The plant, its wood and different preparations thereof, received great attention in Europe and it was commercialized as 'Quassia' or 'Surinam Quassia'. Eventually, however, it was substituted by another plant of the same family, called at that time *Picraena excelsa* (Swartz) Lindley, and today known as *Picrasma excelsa* Planch. This was also commercialized as 'Quassia' or 'Jamaica Quassia'. The reason for substitution was based on similar uses and biological properties. Eventually, it was also proven that the chemical composition was very similar. Surinam Quassia was extensively extracted

[1]Bougainvillea Extractos Naturales, S.A. Apartado Postal 764-3100 Santo Domingo de Heredia, Costa Rica.
E-mail: *quassia@racsa.co.cr*
[2]Centro de Investigaciones en Productos Naturales (CIPRONA); Universidad de Costa Rica. 2060 San José, Costa Rica.
E-mail: *gamora@racsa.co.cr*
[3]The name has been spelled as Kwasi by V. E. Fernand in her Masters Thesis (2003) http://etd.lsu.edu/docs/available/etd-0407103-163959/unrestricted/Fernand_thesis.pdf
[4]According to the database TROPICOS, of the Missouri botanical Garden Tropicos.org. Missouri Botanical Garden. 07 Aug 2009 http://www.tropicos.org/Name/29400114
*Corresponding author

for commercial purposes, endangering the species. Jamaica Quassia was a bigger tree, very abundant in Jamaica so the substitution was only to be expected, even though the distribution of the plants is not the same. To date, *P. excelsa* is an endangered species (Areces-Mallea 2009).

The wood was commercialized, first in Europe and later in the rest of the world. According to Lloyd (1911) the wood became official in the London Pharmacopeia in 1788, though in the German Pharmacopeia of 1872 the origin of the drug was assumed to be the wood of *Quassia amara*, while in the second edition in 1882, *Picraena excelsa* was admitted as the official Quassia.

The bitter principle was readily extracted with water or ethanol and this extract was known as 'quassin' or 'quassiin' when obtained from *Quassia amara* and also as 'picrasmin' when the source was *Picrasma excelsa*. The early chemical studies, presented in the King´s American Dispensatory (Felter and Lloyd 1898), established the similarities and apparent differences of the 'purified' extracts from both sources. These studies, however, were too crude and could not lead to the definition of any structure. The extract known as quassin was shown to be a mixture of mainly two unidentified compounds, one of which was called quassin—causing some confusion in the field— and the other neoquassin (Clark 1937). Later on, other compounds would be shown to be part of the 'bitter principle' of these and other, related plants. The structures and stereochemistry of what we now know as quassin (**1**) and neoquassin (**2**) were established only after the development of modern physical methods and techniques of structural analysis, and the NMR work of a Canadian group in the early 1960s (Valenta et al. 1961, 1962). These, and structurally similar compounds are called quassinoids and, sometimes, simaroulides (Polonsky 1973), simaroubolides, or amaroids (Adams and Whaley 1950, Evans 2002).

1   2

The wood was sold to be extracted and used as a tonic, anthelmintic, and, of course, considering its origin, as a febrifuge. As for the anthelmintic use, the following citation is interesting and probably reminds young readers of some passage of Harry Potter´s: "A very excellent injection for ascarides (thread-worms), is a strong infusion of 3 parts of quassia, and 1 of mandrake root, to every ounce of which a fluid drachm of tincture of asafoetida may be added… For a child 2 years old, 2 fluid ounces may be injected into the rectum twice a day" (Lloyd 1911).

These and other effects will be described for each plant below.

The action on insects is already mentioned in the King´s American Dispensatory (Felter and Lloyd 1898) as a toxic effect on flies and other insects. According to Holman (1940), Quassia, as an extract or in chips, was introduced in the American market in 1850 for the control of aphids and was successfully used as an insecticide in England, in 1884, for the control of hop aphids. Other effects, on many different insects would be reported later on, including antifeedant, repellence, and insecticidal actions (see, for example, Leskinen et al. 1984, and Hilje and Mora 2006).

Eventually, other plants, known to be bitter and having similar properties and uses as either of the two quassias, were given the same or similar names. The most important ones in tropical America are *Simarouba amara* Aubl., *S. glauca* DC., *Picrasma crenata* Engl. in Engl. & Prantl, *Picramnia antidesma* Sw., and *Simaba cedron* Planch., all of them of the same family, except for *P. antidesma* which has been moved from the Simaroubaceae to the Picramniaceae. Each of these plants will be discussed below.

Characteristics of the two plants known as Quassia

1. *Quassia amara* L. ex Blom. Simaroubaceae (Surinam Quassia).

Synonyms: Quassia alatifolia Stokes and *Quassia officinalis* Rich.

Common Names: Kinina (Cabecar Indians, Costa Rica), kini, quiniclu (Bribri Indians, Costa Rica), hombre grande, big man (Costa Rica); cuasia (Mexico); hombre grande, palo grande (Guatemala); cuasia, hombre grande, limoncillo, tru (Honduras); hombre grande, chile de río, chirrión de río (Nicaragua); guabito amargo, crucete, hombre grande (Panama); cuasia, bitter-ash (West Indies); cuasia amarga (Bolivia); pau quassia, quina, falsa quina, murubá, murupa (Brazil); cuasia, creceto morado, contra-cruceto (Colombia); quashiebitters, quassia-bitters (Guyana); amargo, cuasia, simaba (Peru); Surinam Quassia (English) (Ocampo and Balick 2009).

Botanical Description: Quassia amara is a small tree or a big bush, growing up to 7–8 m. The stem can reach 10 cm in diameter; the leaves are pinnately compound with 3–7 leaflets, 5–11 cm long by 4–7 cm wide, obovate to oblong, dark green on the upper surface, slightly pale on the underside with conspicuously winged petiole and rachis. The flowers grow in thin panicles, red with pink base, petals 2.5–4.5 cm long. The fruits, in drupes, up to 1.5 cm, are oblong, green, turning red, and then black, when ripe, each with one seed.

Even though Linné made a mistake in describing the plant, his disciple Carolus M. Blom published its first correct description in 1763 (Lloyd 1911), referring to it as *Lignum quassiae* (Quassia wood).

Distribution and Ecology: The botanical origin of *Quassia amara* is thought to be the northern part of South America and the Caribbean but it is really not known. The plant is now found almost everywhere in the tropics of the world, carried by Nature or Man. It is found from 18° latitude North, in Mexico, down to Ecuador and Brazil (Villalobos et al. 1999). The Missouri Botanical Garden includes descriptions from Mexico down to Venezuela, and the Guianas, and, of course, Surinam (Tropicos. 02 Sep 2009). It has been reported to occur in Argentina by Fernand (2003) and by Cañigueral et al. (1998), but this is probably a result of the confusion with *Picrasma crenata* (*vide infra*). The distribution study of the species in Costa Rica shows denser populations in areas less than 450 m above sea level with unlimited water supply, good drainage and high light levels. In areas with less than 2,500 mm of annual precipitation the shrub is found only in riparian forests. In humid and very humid forests (with up to 5,500 mm of precipitation) the shrub is more prevalent in areas with higher light levels (Villalobos et al. 1999). The plant has been described in Nigeria, Africa (Ajaiyeoba et al. 1999).

Mistaken Identities: Historically, the wood of *Quassia amara* has been confused with another bitter species: *Picrasma excelsa* (Swartz) Planchon, commonly referred to as Jamaica Quassia, quássia-das-Antillas, quássia-nova, and lenho-de-San Martin. It is widely used as a medicinal plant in Jamaica and other Caribbean islands (Trease and Evans 1989).

The vernacular names given to *Q. amara* by indigenous groups of Costa Rica—quiniclu, kini, and kinina—refer to the bitterness of the plant and reminds us of the cinchona plant (*Cinchona* sp.), known as 'quina' in Spanish, well known for its bitterness. The common names of 'quassia amarga' and 'quassia amer' have been used to refer to *Simarouba amara* and *S. glauca* (*vide infra*), both used as amebicides (Taylor 1998). In Argentina, *Quassia amara* has also been confused with *Picrasma crenata* (Vell.) Engl., another species of bitter wood, present in the humid subtropical region of Misiones and which, in Brazil, has been called Quássia-do-Brasil, quássia amarga, pau-tenente, pau-amarelo,and pau-quassia (Oliveira et al. 2005). In Brazil: *Q. amara* is known as 'false quinine', and grows wild in the humid Amazonian region of Belén and Pará (Ocampo and Balick 2009).

The related plant *Picramnia antidesma* Swartz, is a small shrub common to forest undergrowth, medicinally used and known as hombre grande and cáscara amarga in Central America, the Caribbean, and Mexico (Morton 1981).

Ethnomedical Uses: It is important to clarify that, even though the content of quassinoids varies according to the origin of the wood (Villalobos et al. 1999), dosages are normally given in terms of dry wood. Also some confusion arises when the wood is mixed with the bark.

Quassia amara is a traditionally used medicinal plant, known for its bitter properties and its qualities as a tonic by indigenous populations in South America (Standley and Steyermark 1946) and is reputed to have good stomachic, eupeptic, antiamoebic, antihelmintic, antimalarial, and antianemic properties (Barbetti et al. 1987, López Sáenz and Pérez Soto 2008). According to Pittier (1957) the plant is not common in the hot lands of the Costa Rican Pacific region, but is one of the main remedies used by native communities (Indians). "They divide the trunk into 30–60 cm pieces, one of which they take with them on their travels, and sometimes they sell the pieces in the inland markets. An infusion of the rasping of the wood is used for fevers, and as an aperitif".

It is considered to be effective in treating fever, and liver and kidney stones, as well as in treating weakness of the digestive system and even diarrhea (Ocampo and Maffioli 1987, Ocampo and Díaz 2006). In Panama an infusion of the wood is used as a febrifuge, for liver ailments and for snakebites (Duke 1984), and in Brazil it is used to combat dysentery, dyspepsia, intestinal gases, vesicular colic, malaria and fever (Morton 1981, Gupta 1995). In Peru an infusion of the bark is used as a febrifuge and to treat hepatitis, it is also macerated in water or alcohol and used as a tonic (Brack 1999); and in Colombia people use it as a bitter for dyspepsia, anorexia, and malaria (García-Barriga 1975). In Honduras the bark, boiled in water, is used for stomachaches, diabetes, urinary problems, diarrhea, and migraine, and to fortify the blood (House et al. 1995), and in Nicaragua the root is used for snakebites. For this purpose a 20 cm piece of root is crushed, water is added, and then the liquid strained and drunk; while for malaria, two ounces of the bark are cut, boiled in water and drunk three times daily (Morales and Uriarte 1996, Querol et al. 1996, López Sáenz and Pérez Soto 2008). Barnes and coworkers (Barnes et al. 2002), refer to its being used as a gastric stimulant and as having anthelmintic properties. It has been traditionally used for anorexia, dyspepsia, and nematode infestations (taken orally or rectally). A dose of 0.3–0.6 g of dry wood in an infusion, three times a day, is recommended.

Fernand (2003) mentions that, in Surinam, Quassia has prophylactic activity against lice and is effective in chronic diseases of the liver and the powdered stem is a useful remedy to enhance appetite and aid in digestion because of stimulation of the secretion of gastric juices. In French Guiana, the plant is the most frequently used as antimalarial (Vigneron et al. 2005).

Chemical Constituents: Quassin (**1**) and neoquassin (**2**) are the main constituents of the wood and bark of *Quassia amara*, followed by 18-hydroxyquassin and 14,15-dehydroquassin (Robins and Rhodes 1984). Together these comprise about 60% of the crude extract. Other minor constituents are: parain, 11-α-acetylparain, 12-α-hydroxy-13,18-dehydroparain, isoparain, 1-α-*O*-methylquassin (1-dihydro-α-methoxyquassin), 12-hydroxyquassin, isoquassin, 11-dihydro-12-norneoquassin, 16-α-*O*-methylneoquassin, 11-α-*O*-(β-*D*-glucopiranosyl)-16-α-*O*-methylneoquassin, 1-hydroxy-12-α-hydroxyparain, 12-dihydro-α-hydroxyparain, 1-dihydro-α-hydroxyquassin, quassilactol, and nigakilactone A (López Sáenz and Pérez Soto 2008, Barbetti et al. 1993). Quassimarin and simalikalactone D were isolated from the sap (Kupchan and Streelman 1976) and the latter one from the leaves (Bertani et al. 2006), as well as simalikalactone E (Cachet et al. 2009). Excellent reviews of the chemical aspects of quassinoids can be found elsewhere (Polonsky 1973, 1985, Guo et al. 2005, Almeida et al. 2007).

Other, non-quassinoid bitter constituents are the following indole alkaloids: 1-vinyl-4,8-dimethoxy-β-carboline, 1-methoxycarbonyl-β-carboline and 3-methylcanthin-2,6-dione (Barbetti et al. 1987); 4-methoxy-5-hydroxycanthin-6-one (Grandolini et al. 1987); 4-methoxy-5-hydroxycanthin-6-one-3-N-oxyde, 3-methyl-4-methoxy-5-hydroxycanthin-2,6-dione and 3-methylcanthin-5,6-dione (Barbetti et al. 1990); and, 2-methoxycanthin-6-one (Njar et al. 1993).

Campesterol, stigmasterol, β-sitosterol, some aminoacids and mineral salts are also present (Duke 1984, Germonsén-Robineau 1998).

Pharmacology and Biological Activity Studies: Two recent publications cover this area in a specific way; both in Spanish (Díaz et al. 2006, López Sáenz and Pérez Soto 2008). One review in English (Guo et al. 2005) and one in Portuguese (Almeida et al. 2007) cover the topic in a generic way, concerning quassinoids. The following information comes from more specific papers.

Gastrointestinal effects: The dried aqueous extract of *Quassia amara* wood, in doses of 1,000 mg/kg, caused an increase in intestinal motility of male NGP mice as compared with the control group (García et al. 1997). Another study shows that oral administration of a dried aqueous extract to mice produces an increase of gastrointestinal transit at doses of 500 and 1,000 mg/kg and an antiulcerogenic activity was observed in animals treated with the extract at doses of 500 and 1,000 mg/kg. Animals treated with 1,000 mg/kg showed a reduction in acidity and peptic activity and those given 1,500 mg/kg showed an increase in non-protein sulfhydryl group production (Badilla et al. 1998). Another study, conducted by Toma and coworkers, analyzed the antiulcerogenic activities of four extracts of different polarities: 70% ethanol, 100% ethanol, 100% dichloromethane, and 100% hexane. All extracts, administered in oral doses of 100 mg/kg, inhibited the formation of indomethacin/bethanechol-induced gastric ulcer (23, 23, 51, and 47%, respectively) and reduced the gastric injury induced by the hypothermic restraint–stress test in mice (71, 80, 60, and 83%, respectively). In the ligated pylorus of the mouse stomach, following pre-treatment with a single intraduodenal administration of 100 mg/kg of each extract, all but 70% EtOH showed decreased gastric juice content, increased pH values and decreased acid output. All extracts showed significant increases of free mucous in the gastric stomach content. Prostaglandin synthesis was significantly increased by the administration of the hexane extract by the oral route (100 mg/kg) (Toma et al. 2002). This latter effect is interesting because it is probably not related to the quassinoids as these are polar compounds.

Anticancer and Antiviral Activities: Many quassinoids display antitumor activity in different potencies; but bruceantin, bruceantinol, glaucarubinone, and simalikalacton D are amongst the most potent ones. The mechanism of the action is believed to be related to blockade of protein synthesis by inhibition of the ribosomal peptidyl transferase activity leading to the termination of the chain elongation (Guo et al. 2005). Kupchan and Lacadie (1975) proposed a mechanism in which the A-ring enone of quassinoids acts as a Michael acceptor for biological nucleophiles. Recent studies have provided evidence in support of this hypothesis. A free hydroxyl group at C-1 or C-3 was found to enhance biological activity, presumably due to intramolecular hydrogen bonding between the hydroxyl and the oxygen of the enone, thus further activating the enone towards nucleophilic attack (Guo et al. 2005[1]). The work of Fukamiya et al. (2005) provided the following structure-activity relationships for quassinoids regarding translation inhibition in protein synthesis, which is related to antitumor activity: (i) the nature of the C-15 side chain, (ii) the nature of A ring modifications, (iii) the presence or absence of a sugar moiety, and (iv) the presence of an epoxymethano bridge carry a strong effect on the activity .The sap of *Q. amara* showed significant activity *in vivo* against the P-388 leukemia in mice, and *in vitro* against cells derived from human carcinoma of the nasopharinge. The activity was attributed to quassimarin and simalikalactone D (Kupchan and Streelman 1976). These authors mention that the latter compound had been previously described as having antineoplastic activity.

Simalikalactone D is active against the oncogenic Rous sarcoma virus only at concentrations as high as 0.10 mg/mL (Pierre et al. 1980). An aqueous extract of *Q. amara* showed HIV inhibitory activity on lymphoblastoid cells MT-2 (Abdel-Malek et al. 1996). Quassin and simalikalactone D

---

[1]The same paper has been recently published in Frontiers in Medicinal Chemistry. 2009. 4: 285–308.

(isolated from *Quassia africana* Baill.), were studied for antiviral activity. Simalikalactone D was responsible, at least in part, for the high antiviral activity observed against Herpes simplex, Semliki forest, Coxsackie and Vesicular stomatitis viruses for the chloroform crude extract. Quassin showed no activity (Apers et al. 2002).

Antiparasitic activity: There has been a long search for anti-malarial activity for the quassinoid type of compounds. The reviews mentioned above extensively cover this area. Most of the publications examine several types of quassinoids or several plants. One specific case is the one of Teixeira et al. (1999). The growth of *Plasmodium falciparum, in vitro*, was markedly inhibited by several quassinoids. Simalikalactone D was the most active with complete inhibition at 0.002 µg/mL. Glaucarubinone and soularubinone equally effective at 0.006 µg/mL. Chaparrinone and simarolide showed little effect at 0.01 µg/mL. Relative activities parallel antineoplastic activities (Trager and Polonsky 1981). Simalikalactone E, a new addition to the quassinoid armamentarium, extracted from a widely used Amazonian antimalarial remedy made out of *Quassia amara* leaves, inhibited the growth of *Plasmodium falciparum* cultured *in vitro* by 50%, in the concentration range from 24 to 68 nM, independent of the strain sensitivity to chloroquine. It was also shown that this compound was able to decrease gametocytemia with an IC_{50} 7-fold lower than primaquine. Simalikalactone E was found to be less toxic than Simalikalactone D, and its cytotoxicity on mammalian cells was dependent on the cell line, displaying a good selectivity index when tested on non-tumorigenic cells. *In vivo*, Simalikalactone E inhibited murine malaria growth of *P. vinckei petteri* by 50% at 1 and 0.5 mg/kg/day, by the oral or intraperitoneal routes, respectively (Cachet et al. 2009). The administration of a decoction as enema acted as a vermifuge against *Oxiurus* (Martindale 1982, Martínez 1992).

Sedative and Antiedematogenic Activities: The analgesic and antiedematogenic activities of *Q. amara* bark were studied using polar and non-polar extracts. Both activities were demonstrated for the non-polar extract (hexane) when administered intraperitoneally (Toma et al. 2003).

Antifertility Activity: Njar et al. (1995) initiated a series of studies related to antifertility activity of *Q. amara*. They found that the crude methanol extract of the stem wood inhibited both the basal and LH-stimulated testosterone secretion of rat Leydig cells in a dose-dependent fashion. Fractionation of the extract proved quassin to be the responsible for this action. In a continuation of this study, it was found that the same extract also caused a significant reduction in the weight of testis, epididymis and seminal vesicles, but an increase in that of the anterior pituitary gland. A reduction was also found in the epididymal sperm count, serum levels of testosterone, luteinizing hormone and follicle-stimulating hormone.

All changes were reversed eight weeks after withdrawal of the treatment (Raji and Bolarinwa 1997). More recently, a chloroform extract of bark showed antifertility activity on male albino rats by decrease in the sperm count, mobility and viability (Parveen et al. 2003). Aside from the already described changes, a reduction in both lobes of the prostrate was observed. A number of abnormalities like double heads, double tails, detached heads and fragile tails were frequently seen. Epididymal α-glucosidase activity was drastically reduced. However, prostatic acid phosphatase activity and citric acid levels and seminal vesicle fructose concentrations remained unchanged following treatment. Thus, it appears that the prime site of action is at the level of both the testis and the epididymis. Blood cell counts and hemoglobin levels were in the normal range. Bilirubin, SGPT, SGOT, protein and urea were also not altered by the herbal extract and the authors suggested, from the selective action on the male reproductive tract, that the chloroform extract of the bark of *Q. amara* has potential for use as an antifertility agent (Parveen et al. 2003).

Other Biological Activities: Extracts of *Q. amara* showed a good antibacterial and antifungal activity if the material comes from a sunny area, but not if it is from a shady place (Cáceres 1996). Extracts of *Q. undulata* and *Q. amara* showed antibacterial and antifungal activities against six clinical strains of bacteria and five fungi (Ajaiyeoba and Krebs 2003).

Insect-related Uses and Activities: The potential of *Q. amara* as a natural insecticide was the topic for a meeting held at CATIE (Turrialba, Costa Rica) in 1995 and for which there is a written report in Spanish (Ocampo 1995). Crushed aqueous extracts of leaf, wood, bark and flowers of *Q. amara* showed antilarval activity against *Culex quinquefasciatus*. Quassin was the antilarval principle and was effective against mosquito larvae at a concentration of 6 ppm. Quassin was present to the extent of 0.1 to 0.14% (average 0.12%) on a dry weight basis in wood of *Q. amara* (Evans and Raj 1991). This effect was shown to be caused by inhibition of tyrosinase, which is involved in the sclerotization of insect cuticula (Evans and Kaleysa 1992).

The aqueous wood extract showed toxicity towards the following insects: *Acyrthosiphum pisum, Aphis dabae, Bemisia tabaci, Bombyx mori, Chaitophorus populicola, Hoplocampa flava, H. minuta, Macrosiphum ambrosiae, M. liriodendri, M. rosae, Phyllapis fagi, Phymatocera aterrima*, and *Porosagrotis othogonia*. On the other hand, the extract of roots was toxic to *Attagenus piceus* and *Diaphania hyalinata* (Ocampo 1995). Several quassinoids were tested for antifeedant activity against the aphid *Myzus persicae* (Hemiptera, Aphididae). Isobrucein B, brucein B and C, glaucarubinone, and quassin decreased feeding at concentrations down to 0.05% and isobrucein A was effective at 0.01%. Only quassin showed no phytotoxic effects and is therefore the most promising compound for further development (Polonsky et al. 1989). The effect of a standardized extract from the wood of *Q. amara* was analyzed on the cereal aphids *Sitobion avenae* F., *Rhopalosiphum padi* L. and *Metopolophium dirhodum* WLK. A root application with quassin (1 mg/L) on oat plants in the greenhouse showed an efficacy of 90% after 72 h on *S. avenae*, 50% on *R. padi* and 28% on *M. dirhodum*. One week after the treatment, an efficacy of 42% was obtained for *S. avenae* whereas *R. padi* and *M. dirhodum* showed no comparable reaction. Possible effects on larvae of the ladybeetle *Coccinella septempunctata* L. were checked in glass plate tests and in feeding tests with treated *R. padi*. No effects could be seen both on vitality and pupation of the larvae and on hatching and fertility of the adult beetles (Holaschke et al. 2006).

A more complete list of insects studied for their susceptibility to *Quassia amara* is available in the booklet by Ocampo and Díaz (2006) of which an English version can be obtained upon request.

2. *Picrasma excelsa* (Sw.) Planch. (Surinam Quassia).

Synonyms: *Aeschrion excelsa* (Sw.) Kuntze, *Quassia excelsa* Sw., *Simarouba excelsa* (Sw.) DC, and *Picraena excelsa* Lindl.

Common Names: Bitter ash, bitterwood, cáscara amarga, coralito, cuasia, cuasia elevada, fresno amargo, gorie frene, Jamaica Quassia, leña amarga, palo amargo, Quassia, and Quassia de Jamaica (Morton 1981).

Distribution and Ecology: It is said to be native to Jamaica, Hispaniola and Puerto Rico and also found in northern Venezuela (Morton 1981); however, the Missouri Botanical Garden has reports from Central America, the Caribbean and northern South America, including Bolivia (Tropicos. 03 Aug 2009). The USDA provides a more ample distribution for the Caribbean: Antigua and Barbuda, Barbados, Cuba, Dominica, Dominican Republic, Guadeloupe, Haiti, Jamaica, Martinique, Montserrat, Puerto Rico, St. Kitts and Nevis, St. Lucia, St. Vincent and Grenadines, Virgin Islands (British) and Virgin Islands (U.S.) (GRIN 2009). This plant is considereded an endangered species by the IUCN (Areces-Mallea 2009).

Description: A tree, 6 to 25 m high, with furrowed gray bark and downy twigs. Leaves are evergreen, alternate, 15 to 35 cm long, compound, with 9 to 13 mainly opposite leaflets, lanceolate, pointed, 5 to 13 cm long, up to 4.5 cm wide, thin, smooth on the upper surface, pale and minutely downy on the underside. Flowers are fragrant, small, yellowish green, with 4 to 5 petals, minute hairy sepals, the male with prominent stamens; born profusely in long-stalked axillary clusters; usually male and female (or hermaphrodite) on separate trees. Fruit is blue-black, oval, 3 mm long, fleshy, single-seeded, in branched clusters (Morton 1981).

Ethnomedical Uses: In Jamaica, a decoction of the wood is used for threadworms in children, as a tonic, appetite stimulant and as a remedy for malaria. In Cuba, the same infusion is used as a bitter tonic, stomachic, digestive, febrifuge and for dysentery (Morton 1981).

Chemical Constituents: A German group reported on several alkaloids isolated from the wood of *P. excelsa*: Canthin-6-one, 5-methoxy-canthin-6-one, 4-methoxy-5-hydroxy-canthin-6-one, scopoletin, and a new β-carboline alkaloid, N-methoxy-1-vinyl-β-carboline (Wagner et al. 1978, 1979). In a recent study in Japan, the differences in composition of the constituents among four Jamaica Quassia extract products were analyzed by LC/MS. The results showed that the four products have similar compositions of all constituents. Four of the minor constituents that were commonly included in the four products were isolated and identified as 11-dihydro-12-norneoquassin, canthin-6-one, 4-methoxy-1-vinyl-β-carboline and 4,9-dimethoxy-1-vinyl-β-carboline. Hot water extracts from two other species of plants, *Quassia amara* and *P. quassioides* ('Nigaki' in Japanese) were prepared and investigated. The results showed that the compositions of the constituents in the Jamaica Quassia extract products resembled those in the extract derived from *Q. amara* (Tada et al. 2009).

Pharmacological and Biological Activity Studies: A study in India, with decoctions of 40 medicinal plants, showed that, while several of these inhibited pancreatic lipase, *P. excelsa* activates it (Gowadia and Vasudevan 2000). A study was run on the effect of the Jamaican bitterwood tea, commonly consumed to lower blood sugar levels in diabetics, on a panel of cytochrome P_{450} (CYP) enzymes, which are primarily responsible for the metabolism of a majority of drugs on the market. The two major ingredients, quassin and neoquassin, were then isolated and used for further studies. Inhibition of the activities of heterologously expressed CYP microsomes (CYPs 2D6, 3A4, 1A1, 1A2, 2C9, and 2C19) was monitored, and the most potent inhibition was found to be against CYP1A1, with IC_{50} values of 9.2 μM and 11.9 μM for quassin and neoquassin, respectively. The moderate inhibition against the CYP1A1 isoform by quassin and neoquassin displayed partial competitive inhibition kinetics, with inhibition constants (K_i) of 10.8 ± 1.6 μM, for quassin and 11.3 ± 0.9 μM, for neoquassin (Shields et al. 2009).

The β-carboline alkaloids from *P. excelsa* exhibited positive inotropic effects in hearts isolated from rats (Wagner et al. 1979).

While the above biological activities have been performed with *Picrasma excelsa*, it should be kept in mind that many of the activities reported for *Quassia amara* or its components, apply to *P. excelsa*, since the compositions of the two plants are very similar (Tada et al. 2009).

Characteristics of other plants known or used as Quassia

1. *Simarouba amara* Aubl. and/or *Simarouba glauca* DC. Simaroubaceae. These are conflictive species. They have been described as both *S. amara* and *S. glauca*, it has also been claimed that both are the same species. For this reason they are included together here.

Synonyms: *Simarouba medicinalis* Endl., and *S. officinalis* DC (Cáceres 1996). The Missouri Botanical Garden considers three synonyms for *S. amara*: *Quassia simarouba* L. f., *Simarouba glauca* DC., and *Zwingera amara* (Aubl.) Willd. (Tropicos.org. Missouri Botanical Garden. 10 Sep 2009).

Common Names: Dysentery bark (English); xpaxakil, aceituno (Mexico); negrito, acetuna, pa-sak (Maya–Belize) (Arvigo and Balick 1998); aceituno (both species), chookuolit (cabécar ethnia), refers to *S. amara* (Costa Rica) (Ocampo and Maffioli 1987); aceituno, negrito (Honduras) (House et al. 1995), refers *S. glauca*; aceituno, jocote de mico, negrito, olivo, pasac, zapatero (Guatemala) (Cáceres 1996); Juan Primero, daguilla (Dominican Republic) , refers to *S. glauca*; gavilán, palo blanco, roblecillo (Cuba) (Gupta 2008); marupa, maripa, palo blanco (Perú) (Brack 1999) for both species; cedro blanco, capuli (Ecuador) refers to *S. amara*. In Brazil *S.* amara is called Calunga, Marubá, Marupá, Dysentery bark, Bitterwood, Slave wood, Bitter damson (Botsaris 2007). In India it is called paradise tree, oil tree or aceituno, and Lamix Taru (Indian Biofuels Awareness 2009).

Botanical Description: Morton (1981) describes *S. glauca* as: "A tree, to 16 or even 30 m, with a straight trunk and slender, spreading branches. Leaves are evergreen, alternate, 15 to 25 cm long, compound, with 5 to 21 alternate or opposite leaflets, oblong, oval, or obovate, 5 to 10 cm long, rounded and at times minutely spine-tipped at the apex, oblique at the base, leathery, often with recurved margins, dark green, glossy above, pale beneath. Flowers (mostly male) are greenish yellow, with 4 to 6 oblong or obovate petals, 8 to 9 mm wide; born in axillary panicles 30 to 40 cm long, 30 to 60 cm wide. Fruit is ovoid, turning from green to red and finally purple-black, 1.5 to 2.5 cm long, with white, juicy, slightly astringent, insipid flesh and a single orange-brown seed".

Distribution and Ecology: The genus *Simarouba* comprises five species, characterized for having slightly bitter organs (Gentry 1993). These are trees normally grown in the tropical forests. The distribution of the two ethnomedically more important species, *S. glauca* and *S. amara* is very specific for each one: *S. amara* is found in humid forests, from Belize down to South America, while *S. glauca* is found in dry forests, in Central America, the Antilles and parts of South America. In Costa Rica, *S. glauca* is found in the Pacific and the Central Valley and *S. amara* grows more towards the Caribbean coast.

Ethnomedical Uses: In Nicaragua, *S. glauca* is used for malaria (Grijalva 1992). It is also used as food (Querol et al. 1996). In Costa Rica, an infusion of the bark of *S. glauca* is used as a tonic and for amebas, diarrhea and fevers. In the past, a remedy for amebas was prepared by a pharmacist named Doryan, under the name 'Amebicida Doryan'. The same procedure is used with *S. amara* for the treatment of intermittent fevers (malaria), amebas and trichocephalus (Quirós 1945, Ocampo and Maffioli 1987). According to Segleau (2001) *S. amara*, which grows naturally in forests of the Caribbean, is used for internal parasites, including amebas, and, for this purpose, stripes of the bark must be cut on the east and west sides of the tree, cooked, and the decoction must be drunk, two cups three times a day for three or four days. In El Salvador, a wine made from the fruit juice is used as stomachic and a decoction of the bark is taken as febrifuge. A commercial preparation, with a purified extract, is known as 'Glaumeba' (Morton 1981). In Honduras, *S. amara* is found in regions of low precipitation and the bark is used for dysentery, diarrhea, to expel the placenta, for acne, and parasites. The seeds are used to prepare soap for the treatment of dandruff. In Belize, the bark and roots of *S. glauca* are used for dysentery, diarrhea and uterine hemorrhage (Arvigo and Balik 1998). In Guatemala an infusion of the bark or roots is used for malaria, gastrointestinal ailments (diarrhea, dyspepsia, weakness, amebiasis, intestinal worms, trichocephalus, vomit), nervousness and intermittent fevers; the leaves are used as tincture, for amebiasis, or as poultice, for skin ailments, pruritus or even some forms of cancer (Cáceres 1996). In the Dominican Republic, the leaves are used for skin ailments and pruritus. In Peru, the bark of *S. amara* is used for dysentery, as emmenagogue, febrifuge and anthelmintic; *S. glauca* is used to extract the oil from the seeds and *S. versicolor* is used as a febrifuge (Brack 1999). In Brazil, the bark and roots of *S. amara* are used for intestinal infections, verminosis, fever, wounds, infected ulcers, abdominal pain and as antidiarrheal, antispasmodic, and in healing. It is used by Amazonian Indians to treat fever and malaria (Botsaris 2007).

Chemical Constituents: Early studies distinguish between the two plants and *S. amara* is reported to give simaroubidin ($C_{22}H_{32}O_9$) (believed to be a glycoside), simaroubin ($C_{22}H_{30}O_9$) and another, undetermined compound; *S. glauca* provides glaucarubin, also known as simarubidin ($C_{25}H_{35}O_{10}$; notice the absence of the 'o'; it is not the same compound) (Elslager 1970). According to Gupta (2008) the leaves and bark of this plant, considered as one single species, contain flavonoids, polyphenols, sesquiterpene lactones, tannins, alkaloids and quassinoids and the seeds contain triterpenes (glaucarubin, glaucarubinone, and glucopiranosides of glaucarubol and glaucarubolone). It has been reported that the bark of *Simarouba amara* contains 2,6-dimethoxybenzoquinone (Polonsky and

Lederer 1959). An investigation of *S. amara* for antineoplastic quassinoids led to isolation and structural determination of 2'-acetylglaucarubine and 13,18-dehydroglaucarubinone together with 2'-acetylglaucarubinone and glaucarubinone (Polonsky et al. 1978). Glaucarubinone, 2'-acetylglaucarubinone, ailanthinone, and lialocanthone were reported from extractions of *S. amara* fruits (O´Neill et al. 1987). A chemical investigation of the bark of *S. amara*, collected in Barbados, resulted in the isolation of eight compounds, 3-oxatirucalla-7,24-dien-23-ol; niloticin; a tirucallane triterpene, two apotirucallane derivatives containing an epsilon-lactone in ring A, a mixture of two compounds that could not be separated, and an octanorapotirucallane derivative that lacks the C(8) side chain (Grosvenor et al. 2006). Activity-guided fractionation of a chloroform-soluble extract of *S. glauca* twigs from southern Florida, monitored with a human tumor cell line, afforded six canthin-6-one type alkaloid derivatives, canthin-6-one; 2-methoxycanthin-6-one; 9-methoxycanthin-6-one; 2-hydroxycanthin-6-one; 4,5-dimethoxycanthin-6-one; and 4,5-dihydroxycanthin-6-one; plus a limonoid (melianodiol), an acyclic squalene-type triterpenoid (14-deacetyleurylene), two coumarins (scopoletin and fraxidin), and two triglycerides (triolein and trilinolein) (Rivero-Cruz et al. 2005). An interesting addition to the chemical composition comes from a more bromatologycal point of view, from Mysore, India, in consideration of the potential use of the seeds *S. glauca* as food or feed, in the perspective of its high protein content (47.7 g/100 g). Simarouba meal contained high calcium (143 mg/100 g) and sodium (79 mg/100 g). Saponins with triterpenoid aglycone (3.7 g/100 g), alkaloids (1.01 g/100 g), phenolics (0.95 g/100 g) and phytic acid (0.73 g/100 g) were the major toxic constituents identified. The proteins of simarouba meal recorded high *in vitro* digestibility (88%). SDS-PAGE revealed four major protein bands in molecular weight ranges of 20–24, 36–45 and 55–66 kDa. Apart from glutamic acid (23.43 g/100 g protein) and arginine (10.75 g/100 g protein), simarouba protein contained high essential amino acids like leucine (7.76 g/100 g protein), lysine (5.62 g/100 g protein) and valine (6.12 g/100 g protein) (Govindaraju 2009).

Pharmacological and Biological Activity Studies: Both *Simarouba amara* and *S. glauca* have been of interest for the treatment of amebiasis since about 1915 (Elslager 1970). Cáceres (1996) indicates that a tincture of the leaves of *S. glauca* has partial antibacterial activity and an infusion of the bark has antimalarial activity similar to artemisinin. An infusion of the leaves, given orally, has bronchoconstriction activity in rabbits and the aqueous extract when given subcutaneously, has antiulcerogenic activity, but not when given orally. Morton (1981), reports a clinical trial in Costa Rica, in

1951, with an alcoholic tincture of the bark, steeped for 15 d, taken at the rate of 15 to 30 drops, according to age, three times a day after meals as a complete cure of amebiasis. Interestingly, glaucarubinone was tested for its *in vivo* therapeutic action on mice infected with a *Plasmodium berghei* strain. At low doses, glaucarubinone retarded mortality by exerting a partial, temporary inhibition of parasitemia; its toxicity, however, precluded further applications at the time (Monjour et al. 1987). Another report shows that glaucarubinone and 2´-acetylglaucarubinone have very good antiplasmodial activity and similar antiamebic activity to halocanthone and better than metronidazole. A chloroform extract of stems of *S. amara* showed very good antiamebic activity (Wright et al. 1988). Working with the fruits of *Simarouba amara*, O'Neill and co-workers isolated four quassinoids that were tested *in vitro* against a multidrug-resistant strain (K1) of *P. falciparum* and *in vivo* against *P. berghei* in mice. Although the *in vitro* tests indicated activity in the region of 23–52 times greater than that for chloroquine, the toxicity was greater. Their results also substantiated the fact that antimalarial activity of specific quassinoids and their toxicity are not directly related (O´Neill et al. 1988). A study demonstrated that an aqueous extract of *S. amara* increases human keratinocyte differentiation. After 4 wk of treatment on the hemiface of volunteers, the capacitance and transepidermal water loss evaluation revealed the potential interest of this extract for improvement of skin hydration and antipsoriatic action (Bonté et al.1996). This idea ended up in a patent for the use of a polar extract of four different species of *Simarouba* for reducing patchy skin pigmentation (US patent 5,676,948). In a study of Guatemalan plants, a methanol extract of *Simarouba glauca* significantly reduced parasitemias in *Plasmodium berghei*-infected mice. A dichloromethane fraction was screened for cytotoxicity on *Artemia salina* (brine shrimp) larvae, and 50% inhibitory concentrations were determined for *Plasmodium falciparum in vitro* cultures. Both chloroquine-susceptible and -resistant strains of *P. falciparum* were significantly inhibited by the extract. The dichloromethane extract of *S. glauca* cortex was considered to be toxic to nauplii of *A. salina* in the brine shrimp test (Franssen et al.1997).

In a recent study, an extensive *in vitro* antimicrobial profiling was performed for *S. glauca* and two other plants. Ethanol extracts were tested for their antiprotozoal potential against *Trypanosoma b. brucei*, *Trypanosoma cruzi*, *Leishmania infantum* and *Plasmodium falciparum*. Antifungal activities were evaluated against *Microsporum canis* and *Candida albicans* whereas *Escherichia coli* and *Staphylococcus aureus* were used as test organisms for antibacterial activity. Cytotoxicity was assessed against human MRC-5 cells. Although *S. glauca* exhibited strong activity against all protozoa, the study concludes that it must be considered non- specific. (Valdés et al. 2008).

2. *Picrasma crenata* Engl. in Engl. & Prantl

Synonyms: *Aeschrion crenata* Vell., *Picraena palo-amargo* (Speg.) Speg., *Picraena vellozii* (Planch.) Engl., *Picramnia crenata* (Vell.) Hassl., *Picrasma palo-amargo* Speg., and *Picrasma vellozii* Planch. (Flora Brasiliensis revisitada 2007, SIB-APN 2009).

Common Names: In Argentina it is called palo amargo, cuasia and paraih (Roldán et al. 2007). In Brazil, known as Quássia-do-Brasil, quássia amarga, pau-tenente, pau-amarelo,and pau-quassia (Oliveira et al. 2005).

Botanical Description: Tree or bush, 2 to 6 m high in Argentina and 2–15 m high in Brazil, with a brownish, rough bark and many lenticels. The leaves are compound with a stalk 2–9 cm long; 9–19 oval-elliptical, slightly serrate or crenate leaflets, 4–10 cm long, with petioles of 1–5 mm. Axillary corymbous inflorescences, 7–12 cm with the axis longitudinally striated and pubescent. Flowers masculine and feminine are greenish with 4–5 sepals, oblong and free, 1 mm long; 4–5 petals, oblong with acute apex, 3–4 mm long. Masculine flowers have 4–5 stamens. A pharmacognostic description of the bark and leaves has been made available, with the possibility of inclusion in the Brazilian Pharmacopoeia (Nunes et al. 2002). A proper botanical description has not been accessible.

Distribution and Ecology: *Picrasma crenata* is a tree that grows on the east of Brazil and the northwest of Argentina, where it is found on river and creek banks in forests of almost all the Misiones province (Roldán et al. 2007, Rodríguez et al. 2008). It is said to be found between 0 and 500 meters above sea level in tropical America, in Bolivia, Brazil and Paraguay (Pirani 1997, Xifreda and Seo 2006).

Ethnomedical Uses: The wood of *P. crenata* is used in Argentina for intestinal problems and the control of lice (Rodríguez et al. 2008). Infusions of *Picrasma crenata* wood are used in folk medicine against lice and as a non astringent bitter tonic. In traditional Argentinean medicine, the whole stem is used in an infusion for the control of periods, for syphilis and as a tonic. The stem is used as bitters in the preparation of alcoholic beverages and appetizers; as an insecticide and in tinctures, as a substitute of *Quassia amara* and *Picrasma excelsa* for the treatment of lice (Roldán et al. 2007). In Brazil *Picrasma crenata* is utilized to treat *Diabetes mellitus*, gastric perturbation, and hypertension. Quassinoids are the main constituents obtained from the wood of this plant (Cardoso et al. 2009).

Chemical Constituents: The wood of *Picrasma crenata* contains 2,6-dimethoxybenzoquinone and quassin (Polonsky and Lederer 1959). Several alkaloids of the 2-carboline structure were isolated: 1-carbomethoxy-2-carboline; 1-ethyl-4-methoxy-2-carboline (crenatine) and 1-ethyl-4,8-dimethoxy-2-carboline (crenatidine) (Sánchez and Comin 1971). Three quassinoids were also isolated: 12-norquassin, parain and isoparain (Vitagliano and Comin 1972). A German group isolated two new quassinoids 16-β-O-methylneoquassin and 16-β-O-ethylneoquassin, and four known ones, together with coniferyl aldehyde, coniferin, cantin-6-one, 4,5-dimethoxycantin-6-one and (+)-neo-olivil (Krebs et al. 2001).

Pharmacology and Biological Activity Studies: Several groups in South America have been working on the biological activities of *P. crenata* (Márquez et al. 1999, Rodríguez et al. 2004). The hydroalcoholic extract from wood (without the bark) showed low toxicity in mice. Additionally, used to treat gastric ulcer and diabetes, anti-ulcer and blood glucose lowering effect were detected in rats. However, anti-inflammatory, anti-hyperglycemic and anti-hyperlipidemic effects were not observed in these animals (Novello et al. 2008). In a study of antiplasmodial activity of extracts of *P. crenata* (parts not specified) the methanol extract showed weak activity; the other extracts were inactive (Debenedetti et al. 2002). The insecticide effect of the acetone and ethyl acetate extracts of *P. crenata* were tested on the Rice Weevil *Sitophilus oryzae* L. One milliliter, of six different concentrations for each extract, was applied to filter paper disks in glass Petri dishes. Ten adult beetles were placed on these papers. The mortality was examined at 0.5 h, 6 h, 12 h, 18 h, 24 h, 48 h and 72 h. The highest ethyl acetate extract concentrations showed 100% efficiency at 6 h. All the concentrations of the acetone extract ranged between 80 and 100% of efficiency at 6 h, reaching 100% after 18 h (Rodríguez et al. 2008).

3. *Picramnia antidesma* Swartz. Simaroubaceae or Picramniaceae

Synonyms: The Missouri Botanical Garden recognizes nine synonyms: *Picramnia allenii* D.M. Porter, *Picramnia andicola* Tul., *Picramnia antidesma* var. *normalis* Kuntze, *Picramnia bonplandiana* Tul., *Picramnia brachybotryosa* Donn. Sm., *Picramnia cooperi* D.M. Porter, *Picramnia fessonia* DC., *Picramnia quaternaria* Donn. Sm., and *Picramnia triandra* Stokes. A new family called Picramniaceae has been proposed, of which the genus *Picramnia* would be one of the constituents (Fernando and Quinn 1995). Chemotaxonomics seems to favor this separation (Jacobs 2003, Almeida et al. 2007).

Common Names: In Costa Rica *P. antidesma* it is known as caregre, corteza amarga, cafecillo, quina. In Honduras it is called Quina, and Corteza de Honduras. Morton (1981) provides the following common names: Bitter wood, brasilete bastardo, brasilete falso, bresillet batard, bresillet d´Amerique, cáscara amarga, chilillo, coralito, cuasia de Jamaica, fresno amargo, hombre grande, Honduras bark, macary bitter, majoe bitters, old woman´s bitter, quinine, and Tom Bontrin´s bush.

Botanical Description: Morton (1981) describes *P. antidesma* as : "A shrub 2 to 3 m high, or a tree 6 to 9 m tall, the young branches often downy. Leaves are alternate, compound, with 7 to 13 leaflets, lanceolate-oblong to oblong-ovate, pointed, 2.5 to 14 cm long, 1.5 to 6 cm wide, semileathery. Flowers are tiny, whitish, greenish or yellowish, 3-parted; male short stalked, in long slender racemes; female long-stalked, in hanging racemes to 26 cm long. Fruit is oval or elliptical, 1 to 1.5 cm long, on stalks 5 to 20 cm long, red at first, turning black."

Distribution and Ecology: The genus *Picramnia* comprises 40 species, and all of them are bushes or small trees and, apparently, it is native of tropical America (Gentry 1993). In Costa Rica this genus is represented by five species: *P. latifolia* Tul., known as 'Coralillo', *P. antidesma* Swartz; *P. carpinterae* Pol., *P. quaternaria* Donn.Sm., and *P. teapensis* Tul., all of them with the common name 'caregre' (Holdridge and Poveda 1975, León and Poveda 2000). The Missouri Botanical Garden reports it from Mexico, Central America and Panama (Tropicos.org. Missouri Botanical Garden. 08 Sep 2009). Morton (1981) extends the distribution to Jamaica and Hispaniola.

Ethnomedical Uses: Morton (1981) refers to the traditional use of *Picramnia antidesma* in Central America and the West Indies as antimalarial and for stomach ailments. In Jamaica, a decoction is used to treat teething babies and for yaws,[1] venereal disease, colic, intermittent fevers and ulcers of the skin. All species found in Costa Rica are generally used as a decoction of the bark, trunk and roots, as a bitter tonic and stimulant for appetite and digestion (Núñez 1975, León and Poveda 2000). A decoction of the bark is used in Honduras to treat skin wounds and other ailments (House et al. 1995). In Mexico the branches are used to help labor and for venereal diseases. A decoction of the bark is used for erysipelas and malaria (Biblioteca de la Medicina Tradicional Mexicana 2009).

Chemical Constituents: Until recently, the chemistry of *Picramnia* was quite unknown. Several groups have reported the presence of anthraquinone derivatives of the anthrone type, with very interesting biological activities. Some of these derivatives are O- and C-glycosides. A bioactivity guided fractionation, using KB cells and brine shrimp assays, of the methanolic extract from the leaves yielded two known anthraquinones, aloe-emodin and aloe-emodin anthrone, and three new aloe-emodin C-glycosides, named picramnioside A, picramnioside B and picramnioside C (Solís et al. 1995). A chloroform extract of roots provided a new chrysophanol anthrone C-glycoside, named uveoside plus four known compounds (Hernández-Medel et al. 1998). Similarly, nine compounds were obtained from the stem material. Two of the isolated compounds, mayoside and saroside, which are new, are diastereoisomeric oxanthrones (Hernández-Medel et al. 1999).

---

[1]Interestingly, *Picramnia antidesma* has been used in Africa for the same purpose. See http://www.nhm.ac.uk/resources-rx/files/chapter-8-medicines-19114.pdf

Activity-guided fractionation of extracts prepared from the root bark afforded 10-epi-uveoside (Hernández-Medel and Pereda-Miranda 2002).

Pharmacology and Biological Activity Studies: One study with crude extracts found antiplasmodial activity which proved to be a result of the high levels of cytotoxicity displayed by the anthraquinone derivatives, suggesting that infusions from this crude drug lack the selectivity index needed to be considered an effective antimalarial agent (Hernández-Medel and Pereda-Miranda 2002).

4. *Simaba cedron* Planch. Simaroubaceae.

Synonyms: *Aruba cedron* (Planch.) O.Ktze.

Common Names: In Costa Rica, Colombia (Chocó and Magdalena), Peru and Panama *S. cedron* is known as Cedrón. In Brazil is called Pau pa todo. (Brack 1999). Morton (1981) provides the following names: cedrón, hikuribianda, paratudo, pastileño, pau de gafanhoto, pau para-tudo, and vela de muerto.

Botanical Description: *Simaba cedron* is a tree, 5 to 8, or even 15 m high; all parts exceedingly bitter. Leaves are alternate, stemless or very short-stemmed, to 1 m or more in length, compound, with 20 to 30 alternate (rarely opposite) leaflets, narrow-oblong or narrow-elliptic-oblong; 12 to 15 cm long and 4 to 5 cm wide, abruptly pointed at both ends, pale on the underside. Flowers are slightly fragrant, dark yellow, with 5 brown-hairy petals 2 to 3.5 cm long and to 2.5 mm wide, 10 purplish stamens; born in panicles 15 to 25 cm long. Fruits are oval, to 4 cm long, with 1 seed 2.5 cm long, having 2 white cotyledons, which turn yellowish after exposure (Morton 1981).

Distribution and Ecology: According to Cronquist (1981), *S. cedron* is native to the region of the Amazon. Morton (1981) says it is native to the dry plains from the Pacific coast of Costa Rica (cultivated in Palmar Norte) to Colombia and Brazil and grown in parks and gardens in Peru. León and Poveda (2000) describe it as a tree of the Costa Rican Pacific slope, sometimes planted by the natives because of its medicinal properties. Ocampo (2009) has personally collected seeds from Tortuguero, in the Costa Rican Caribbean coast, and Chiriquí, in the Panamanian Pacific coast, both very humid regions.

Ethnobotanical Uses: In Costa Rica, the cotyledons of *S. cedron* are used for fevers and snakebites (Núñez 1975). In Colombia, García-Barriga (1975) reports that cedron was introduced as a subtitute for quina. The bitter principles of cedron were called 'cedrina'. Both cedrina and a liquor made from the seeds of cedron were used for hepatic colic. The fruits and bark are used as antispasmodic, febrifuges, and tonics. In Panama, the Cuna Indians use the ground fruits or an infusion of leaves and roots as a protection against snakes (Duke 1984). In Peru, the plant is used for dysentery and as tonic and febrifuge (Brack 1999). In Brazil, the seeds are used for snakebites, dysentery and anemia, as a substitute for quinine (Morton 1981). The plant is used as antimalarial in South America.

Chemical Constituents: The following compounds have been reported as bitter principles of *S. cedron*: glaucarubolone, chaparrinone and klaineanone, cedrin, 7-dihydrosamaderine B; cedroniline (7-dihydrosamaderine C), cedronin and cedroniline (Gupta 1995). A Japanese group reported four C(19) quassinoids: cedronolactones A-D, together with the known compounds: simalikalactone D, chaparrinone, chaparrin, glaucarubolone, glaucarubol, samaderine Z, guanepolide, ailanquassin A, and polyandrol, isolated from the wood (Ozeki et al. 1998). Cedrenolactone E was isolated subsequently, also from the wood (Hitotsuyanagi et al. 2001). Other compounds isolated from *Simaba cedron* include nilocitin, dihydronilocitin, piscidinol, bourjutinolone A, glaucarubolol, and glaucarubolone (Vieira et al. 1988).

Pharmacology and Biological Activity Studies: The early work on *Simaba cedron* reports a weak anti-inflammatory effect (Hammarlund 1963). An aqueous extract of the seeds, at a dose of 50 mg/kg, and a chloroform extract at a dose of 34 mg/kg (both s.c.) in chickens, showed strong antiplasmodial activity. Wood, roots and bark showed little activity (O´Neill et al. 1985).

Cedronin was examined for *in vitro* and *in vivo* antimalarial activities and for cytotoxicity against KB cells. Experimental results showed that it was active against chloroquine-sensitive and -resistant strains, with an IC_{50} of 0.25 μg/mL (0.65 μmol/mL). Because its IC_{50} values were similar for both strains, this suggested that quassinoids may act upon malarial parasites by means of a fundamentally different mechanism from that of chloroquine. It was also found to be active *in vivo* against *Plasmodium vinkei* with an IC_{50} of 1.8 mg/kg (4.7 nM/kg) in the classic 4-day test. Cedronin shows a rather low cytotoxicity against KB cells (IC_{50} = 4 μg/mL, 10.4 μM) as compared with C(20) biologically active quassinoids; however, its toxic/therapeutic ratio (10/1.8) remains lower than chloroquine (10/0.5) (Moretti et al. 1994).

General aspects of toxicity of the plants known or used as Quassia

Some concern has been expressed by the European Scientific Committee on Food (SCF) about the possible toxicity of pure quassin (SCF 2002). However, the FDA considers Quassia as generally regarded as safe (GRAS). In doses of 250 to 1,000 mg/kg, no signs of acute toxicity were observed in rats given Quassia extract and no adverse reactions were reported upon topical application of the scalp preparation in the 454 patients in the head lice study (Duke 1984). Large amounts, given orally, have been known to irritate the mucus membrane in the stomach and may lead to vomiting; also, high doses of quassin produce vertigo, a decrease of visual accuracy, colic, decrease of cardiac frequency, muscular tremors, fever and paralysis (Cañigueral et al. 1998). Excessive use may also interfere with existing cardiac and anticoagulant regimens. Because of its cytotoxic and emetic properties, its use during pregnancy should be avoided (Duke 1984).

Quassia extract is regarded safe as a food flavoring agent by the Council of Europe and the international regulatory agencies. In foods it can be included at doses not exceeding 5 mg/kg body mass. The permitted concentrations in alcoholic beverages are up to 50 mg/kg (Newall et al., 1996). Acute toxicity of the orally administered aqueous extracts of Quassia has not been observed in standard tests with mice and rats (García et al. 1997, Badilla et al. 1998). However, the intraperitoneal administration of 1 g/kg appeared to be lethal (García et al. 1997).

The bark of *Simarouba amara* and the wood of *Picrasma crenata* contain 2,6-dimethoxybenzoquinone (Polonsky and Lederer 1959), which has been associated with a case of dermatitis and possibly with a report regarding Negroes stripping the root bark of some *Simarouba* (perhaps not *S. amara*) who had to wear protective clothing because the sap would cause a reaction severe enough to prevent them from working for some days. (Woods and Calnan 1976). Similarly, an extract of *Quassia amara* wood, used as an insecticide, has been reported to cause dermatitis (BoDD 2009).

Cedronolactone A, present in *Simaba cedron*, was shown to exhibit a significant *in vitro* cytotoxicity (IC_{50} 0.0074 μg/mL) against P-388 murine leukemia cells (Ozeki et al. 1998). Cedronolactone E , from the wood of *S. cedron*, was found to show a weak cytotoxic activity (IC_{50} 51 μg/mL) against P-388 cells (Hitotsuyanagi et al. 2001).

Infusions of *Picrasma crenata* (2.5 mg%; 5.0 mg%, 10.0 mg%, 20.0 mg% and 40.0 mg%), showed a statistically significant correlation between the concentration and the length of roots and macroscopic aberrations in the *Allium cepa* test, suggesting the possibility of genotoxic activity (Roldán et al. 2007). A hydroalcoholic extract of the nude stem showed low toxicity in mice (Novello et al. 2008).

Dayan et al. (1999) studied several quassinoids, to establish some correlation between the structure and phytotoxic activity. The presence of the oxymethylene ring group had a great effect on the three-dimensional conformation and biological activity of these natural products. In the absence of the oxymethylene ring, the quassinoids were more planar and had little phytotoxicity. In addition, this bridging function introduced a new reactive center that caused the terpene backbone to bend. Molecules with such conformation were highly phytotoxic, reducing root growth of lettuce (*Lactuca sativa*) and affecting all stages of mitosis in onion (*Allium cepa*) root tips.

Aspects regarding conservation and domestication

Serious studies and practices on the domestication or cultivation of plants for this group are not very common. The following two cases are exceptional.

Case 1. For *Quassia amara* the effort has been sustained for several years and some of the results have been published in Spanish (Ocampo and Diaz 2006). Written in a relatively simple manner, the booklet covers several aspects of the natural distribution and the requirements for the natural growth of the plant. It also covers some aspects of the management plan designed by CATIE (see Villalobos et al. 1999 and references therein). The most important aspects have to do with the cultivation and its requirements.

The propagation of this species can be successfully achieved sexually (seeds) or asexually (air layers, cuttings, naked roots).

The seeds die when the humidity lowers to critical values, which implies that adequate handling has to be used during transportation and storage. High temperatures induce negative effects in the viability of the seeds and, because of this, transportation and storage must have fresh air and good ventilation.

Germination takes a period of about 10 weeks after settling. Under greenhouse conditions (80% shades, river sand, automatically controlled irrigation) between 90 and 95% germination can be obtained. However, under more intensive field conditions, with 50% shade and manual irrigation, germination is about 67%. *Q. amara* germinates better in sand or sandy soils than in clayey ones, which confirms the importance of substrate selection.

Some factors that must be taken into consideration for a better development of the plants are: a) Dehydration of the seed by sun exposure diminishes the viability and, so, solar radiation must be controlled during the greenhouse stage. b) Positive germination results have been reached with 50% controlled shade. c) Humidity plays an important role in the greenhouse stage. d) Stirring of the soil in the upper 30 cm of terraces and beds improves the general development of the plants. e) A distance of 10 cm × 15 cm between plants is recommended for a density of 35 plants per square meter. This could vary, according to the time the plants are to remain in the greenhouse (no more than 8 months). f) The seeds should be placed on the surface and then partially covered with sawdust. g) Antracnosis by the fungus *Colletotricum* sp. has also been found during times of high precipitation. Fungicides should be used, if necessary.

Transplantation must be made 8 mon after planting, when plants have reached a size of 40–50 cm. When transplanting, the soil should be humidified and removed, to make the extraction easy without damaging the roots. The planting should be made on naked roots.

Pruning should be done at about 15 cm from the apical end and the foliage should be removed completely, to prevent dehydration. Bacterial damage to the stem has been observed if the planting is done in times of high precipitation, with a mortality of as much as 40%. For this reason it is recommended to protect the transplant material with a bactericide or to do it in times of less precipitation.

The air layers are made by ring-wounding a branch of 1–2 cm width, at a distance of 30–50 cm from the apex, on a mature section (white color). The ring is covered with moss and, on top, with aluminum foil or plastic. Give 7–8 weeks for the development of roots and separate (sever) from the mother branch and let acclimate before the definite planting. Stems larger than 1 cm and up to 1.5 cm in diameter have shown success of up to 98%. However, survival when settling in the field has shown negative results (up to 50% mortality). When planting in terraces or beds the mortality has been only 10%.

The cultivation system of *Q. amara* requires an adequate management of the shade which, to a great extent, determines the optimal growth and a good quality of the raw material, considered in terms of its contents of quassinoids. Agro-ecological systems offer partial shade which, under determinate conditions, is ideal for the development of *Q. amara* as a crop. This system attempts to use the established cultivation systems (such as fruit, cacao, banana, coconut palm, African palm, or timber tree plantations), characterized by allowing manageable shade for the adequate growth of *Q. amara*. For this reason, the preparation of the land consists of trimming the weeds in a mechanical way, which leads to low cost investments.

In general, cultivations have been established with populations ranging from 2,500 to 6,000 plants per hectare. This implies distances between plants varying from 1 m × 1 m to 2 m × 2 m.

These criteria for planting are the first result of the study for conditions of the habitat, in search for cultivation alternatives for *Q. amara* in systems which have been considered to be of little productivity in the medium term. *Q. amara* should be analyzed as one more component (crop) in a productive system, to increase the total yields. As an example, cacao fields with trees—*Theobroma cacao* plus *Cordia alliodora* mainly—are used as a system for the production of *Q. amara*. This implies that cacao trees should be pruned and the higher stratum of the trees for forestry production must be kept. Clearing, to eliminate competence for nutrients or light, is highly recommended, to improve the growth.

A copy of an English version can be made available upon request to Bougainvillea, S.A. (quassia@racsa.co.cr).

Callus and suspension cultures of *Quassia amara* have been established. Analysis of the tissue culture showed that quassin was present in both callus and suspension cultures. The effect of variation in auxin and cytokinins on both callus growth and the presence of quassin were examined. Good yields of quassin were achieved with a suspension culture growing in a 7 liter bioreactor (Scragg et al. 1990).

Case 2. Another species which has been studied is *Simarouba glauca,* in South India, where it was first introduced by the National Bureau of Plant Genetic Resources in the Research Station at Amravati, Maharashtra, in the 1960s. It was brought to the University of Agricultural Sciences, in Bangalore, in 1986 where systematic research and developmental activities began in 1993. The following information has been taken from the website of the Indian Biofuels Awareness Center (2009), with little editing. It is considered an edible oil seed bearing tree, well suited for warm, humid, tropical regions. Its cultivation depends on rainfall distribution, water holding capacity of the soil and sub-soil moisture. It is suited for temperature range of 10 to 40°C. The tree is now found in different regions of India. It can be grown on waste tracts of marginal, fallow lands of southern India. Some of the characteristics are given below: The saplings are sturdy in nature and can survive under all types of terrain, and soils with some depth for the roots to penetrate. The tree survives under rain fed conditions with rainfall around 400 mm. It can grow in all types of degraded soils and waste lands. It is not grazed by cattle, goat or sheep. It sheds large quantities of leaves, which makes soil more fertile. It protects the soil from parching due to hot sun . Its seed contains 65% edible oil. The oilseed cake is of best NPK value. Its fruit is also edible with a sweet pulp. Most of its parts have medicinal value. It has a life of about 70 yr. Its wood is termite resistant. It is not attacked by insects and pests. Its long roots prevent soil erosion. It can grow at elevations from sea level to 1,000 meters. It grows 40 to 50 feet tall and has a span of 25 to 30 feet. It bears yellow flowers, and oval elongated purple colored fleshy fruits.

It can be propagated from seeds, grafting and tissue culture technology. Fruits are collected in the month of April/May, when they are ripe and then dried in the sun for about a week. The skin is separated and the seeds are grown in plastic bags to produce saplings. Sapling 2 to 3 mon old can be transplanted in plantations.

The depth of the soil should be at least 1 meter and the pH of the soil should be from 5.5 to 8. It can grow in any type of soil which is unsuitable for cultivation of other crops. The average yields per hectare of Simarouba are: Seed 4 tons, oil 2.6 tons, and cake 1.4 tons.

The plants can be grown in orchards, as boundary planting or avenue trees. At the onset of the monsoon, the grafts or seedlings of known sex are planted with 5 m (E–W) × 4 m (N–S) spacing (500 plants/ha; 200 plants/acre), in pits 45 × 45 × 45 cm size half filled with the top soil.

Patents related to the plants known or used as quassia

Several patents have been filed. No description is offered here, just the numbers and titles. Most of these patents can be obtained through http://www.patentsonline.com/ or http://www.freepatentsonline.com/.

US 4,731,459: Novel Anti-ulcer Agents and Quassinoids.

US 6,573,296: Therapeutic quassinoid preparations with antineoplastic, antiviral, and herbistatic activity.

US 7,005,298: Micropropagation and Production of Phytopharmaceutical Plants.

Application #: 20,040,005,344 Head Lice Formulation.

Applicaton #: 20,060,281,807 Quassinoid Compositions for the Treatment of Cancer and Other Proliferative Diseases.

Applicaton #: 20,060,281,808 Process to Extract Quassinoids.

Application #: 20,070,122,492 Plant Extracts and Dermatological Uses thereof.

Application #: 20,080,089,958 Topical and Intravaginal Microbicidal and Antiparasitic Compositions Comprising Quassinoids or Quassinoid-containing Plant Extracts.

Acknowledgements

We are extremely grateful to Hector Keller and Beatriz Eibl, from the Facultad de Ciencias Forestales, Universidad Nacional de Misiones, Argentina, and Fatima Chechetto, from the Universidade Estadual Paulista, Botucatu, Saõ Paulo, Brazil and Vanilde C. Zanette, from the Universidade Estadual de Santa Catarina, Santa Catarina, Brazil for the valuable information on the botanical description of *Picrasma crenata* and to James D. McChesney, Founder and Principal, Ironstone Separations, Inc, Etta, MS 38627-9519 USA, for kindly proof-reading the manuscript.

References

Abdel-Malek, S., J.W. Bastein, W.F. Mahler, Q. Jia, M.G. Reinecke, W.E. Robinson, Y.-H. Shu, and J. Zalles-Asin. 1996. Drug Leads from the Kallawaya Herbalist of Bolivia. I. Background, Rationale, Protocol and Anti-HIV Activity. J. Ethnopharmacol. 50: 57–166.

Adams, R. and W.W. Whaley. 1950. The Amaroids of Quassia. I. Quassin, Isoquassin, and Neoquassin. J. Am. Chem. Soc. 72: 375–379.

Ajaiyeoba, E. and H. Krebs. 2003. Antibacterial and Antifungal Activities of *Quassia amara* and *Quassia undulata* Extracts *in Vitro*. Afr. J. Med. Med. Sci. 32(4): 353–356.

Ajaiyeoba, E.O., U.I. Abalogu, H.C. Krebs, and A.M.J. Oduola. 1999. *In Vivo* Antimalarial Activities of *Quassia amara* and *Quassia undulata* Plant Extracts in Mice. J. Ethnopharmacol. 67: 321–325.

Almeida, M.M.B., A.M.C. Arriaga, A.K. Lima dos Santos, T.L.G. Lemos, R. Braz-Filho and I.J.C. Vieira. 2007. Ocurréncia e Atividade Biológica de Quassinóides da Última Década. Quim. Nova. 30(4): 935–951.

Apers, S., K. Cimanga, D. Vanden Berghe, E. Van Meenen, A.O. Longanga, A. Foriers, A. Vlietinck, and L. Pieters. 2002. Antiviral Activity of Simalikalactone D, a Quassinoid from *Quassia africana*. Planta Med. 68(1): 20–24.

Areces-Mallea, A.E. 2009. *Picrasma excelsa*. In: IUCN 2009. IUCN Red List of Threatened Species. Version 2009.1. <http://www.iucnredlist.org/>. Downloaded on 09 September 2009.

Arvigo, R. and M. Balick. 1998. Rainforest Remedies. One Hundred Healing Herbs of Belize. 2nd ed. Lotus Press, Tween Lakes, WI. USA.

Badilla, B., T. Miranda, G. Mora, and K. Vargas. 1998. Actividad Gastrointestinal del Extracto Acuoso Bruto de *Quassia amara* (Simarubaceae). Rev. Biol. Trop. 46(2): 203–210.

Barbetti, P., G. Grandolini, G. Fardella, and I. Chiappini. 1987. Indole Alkaloids from *Quassia amara*. Planta Med. 53: 289–290.

Barbetti, P., G. Grandolini, G. Fardella, I. Chiappini, and A. Mastalia. 1990. New Canthin -6-one Alkaloids from *Quassia amara*. Planta Med. 56(2): 216–217.

Barbetti, P., G. Grandolini, G. Fardella, and I. Chiappini. 1993. Quassinoids from *Quassia amara*. Phytochemistry 32(4): 1007–1013.

Barnes, J., L. Anderson, and D. Phillipson. 2002. Herbal Medicines. 2nd ed. Pharmaceutical Press. London.

Bertani, S., E. Houël, D. Stien, L. Chevolot, V. Jullian, G. Garavito, G. Bourdy, and E. Deharo. 2006. Simalikalactone D is Responsible for the Antimalarial Properties of an Amazonian Traditional Remedy Use Made with *Quassia amara* L. (Simaroubaceae). J. Ethnorpharmacol. 108: 155–157.

Biblioteca de la Medicina Tradicional Mexicana. 2009. Atlas de las Plantas de la Medicina Tradicional Mexicana. *Picramnia antidesma*. http://www.medicinatradicionalmexicana. unam.mx/monografia.php?l=3&t=Soplador&id=7759.

BoDD 2009. Website of the Botanical Dermatology Database. http://bodd.cf.ac.uk/ BotDermFolder/SIMA.html

Bonté, F., P. Barré, P. Pinguet, I. Dusser, M. Dumas, and A. Meybeck. 1996. *Simarouba amara* Extract Increases Human Skin Keratinocyte Differentiation. J. Ethnopharmacol. 53(2): 65–74.

Botsaris, A.S. 2007. Plants Used Traditionally to Treat Malaria in Brazil: the Archives of Flora Medicinal. J. Ethnobiol. Ethnomed. 3: 18–25.

Brack, A. 1999. Diccionario Enciclopédico de Plantas Útiles de Perú. Cusco, Peru. Centro de Estudios Andinos Bartolomé de las Casas.

Cáceres, A. 1996. Plantas de Uso Medicinal en Guatemala. Guatemala, Universidad de San Carlos de Guatemala pp. 51–52.

Cachet, N., F. Hoakwie, S. Bertani, G. Bourdy, E. Deharo, D. Stien, E. Houel, H. Gornitzka, J. Fillaux, S. Chevalley, A. Valentin, and V. Jullian. 2009. Antimalarial Activity of Simalikalactone E, a New Quassinoid from *Quassia amara* L. (Simaroubaceae). [Epub ahead of print]. Antimicrob. Agents Chemother. 53(10): 4393–498.

Cañigueral, S., R. Vila, and M. Wichtl. 1998. Plantas Medicinales y Drogas Vegetales para Infusión y Tisana. OEMF International. Milan, Italy.

Cardoso, M.L.C., M.S. Kamei, R.F. Nunes, N.S. Lazeri, J.R.S. Neto, C.R. Novello, and M.L. Bruschi. 2009. Development and Validation of an HPLC Method for Analysis of *Picrasma crenata*. J. Liq. Chromatogr. Related Technol. 32(1): 72–79.

Clark, E.P. 1937. Quassin. I. The Preparation and Purification of Quassin and Neoquassin, with Information Concerning their Molecular Formulas. J. Am. Chem. Soc. 59: 927–931.

Cronquist, A. 1981. An Integrated System of Classification of Flowering Plants. Columbia University Press. New York.

Dayan, F.E., S.B. Watson, J.C.G. Galindo, A. Hernández, J. Dou, J.D. McChesney, and S.O. Duke. 1999. Phytotoxicity of Quassinoids: Physiological Responses and Structural Requirements. Pestic. Biochem. Physiol. 65: 15–24.

Debenedetti, S., L. Muschietti, C. van Baren, M. Clavin, A. Broussalis, V. Martino, P.J. Houghton, D. Warhurst, and J. Steele. 2002. *In Vitro* antiplasmodial activity of extracts of Argentinean Plants. J. Ethnopharmacol. 80: 163–166.

Díaz, R., L. Hernández, R. Ocampo, and J.F. Cicció. 2006. Domesticación y Fitoquímica de *Quassia amara* (Simaroubaceae) en el Trópico Húmedo de Costa Rica. Lankesteriana. 6(2): 49–64.

Duke, J.A. 1984. CRC Handbook of Medicinal Herbs. CRC Press. Boca Raton, Florida.

Elslager, E.F. 1970. Antiamebic Agents. In: A. Burger (ed.) Medicinal Chemistry. 3rd. ed. Wiley-Interscience. New York. pp. 522–561.

Evans, D.A. and R.K. Raj. 1991. Larvicidal efficacy of Quassin against *Culex quinquefasciatus*. Indian J. Med. Res. 93: 324–327.

Evans, D.A. and R.R. Kaleysa. 1992. Effect of Quassin on the Metabolism of Catecholamines in Different Life Cycle Stages of *Culex quinquefasciatus*. Indian J. Biochem. Biophys. 29(4): 360–363.

Evans, W.C. 2002. Trease and Evans Pharmacognosy. W. B. Saunders. London.

Felter, H.W. and J.U. Lloyd. 1898. King´s American Dispensatory. 18th ed. 3rd Revision. As presented in http://www.henriettesherbal.com/eclectic/kings/picraena.html.

Fernand, V.E. 2003. Initial Characterization of Crude Extracts from *Phyllanthus amarus* Schum. and Thonn. and *Quassia amara* L. Using Normal Phase Thin Layer Chromatography. Master Thesis, Louisiana St. Univ. This thesis is available through http://etd.lsu.edu/docs/available/etd-0407103-163959/unrestricted/Fernand_thesis.pdf.

Fernando, E.S. and Quinn, C.J. 1995. Picramniaceae, a new family, and a recircumscription of Simaroubaceae. Taxon. 44: 177–181.

Flora Brasiliensis revisitada. 2007. http://flora.cria.org.br/taxonCard?id=FBR3127. Set 23, 2009.

Franssen, F.F., L.J. Smeijsters, I. Berger, and B.E. Medinilla Aldana. 1997. *In Vivo* and *in Vitro* Antiplasmodial Activities of Some Plants Traditionally Used in Guatemala Against Malaria. Antimicrob. Agents Chemother 41(7): 1500–1503.

Fukamiya, N., K.H. Lee, I. Muhammad, C. Murakami, M. Okano, I. Harvey, and J. Pelletier. 2005. Structure-activity Relationships of Quassinoids for Eukaryotic Protein Synthesis. Cancer Lett. 220: 37–48.

García, M., S. González, and L. Pazos. 1997. Actividad Farmacológica del Extracto Acuoso de la Madera de *Quassia amara* (Simaroubaceae) en Ratas y Ratones Albinos. Rev. Biol. Trop. 44/45: 47–50.

García-Barriga, H. 1975. Flora Medicinal de Colombia. Botánica Médica: Tomes 1 and 2. 2nd ed. Instituto de Ciencias Naturales, Universidad Nacional, Santa Fé de Bogotá, Colombia. pp. 44–47.

Gentry, A. 1993. Woody Plants of Northwest South America. Conservation International, Washington, DC USA. pp. 783–786.

Germonsén-Robineau, L. 1998. Investigación Científica y Uso Popular de Plantas Medicinales en el Caribe. Farmacopea Caribeña. . Editorial Enda-Caribe. Santo Domingo, República Dominicana.

Govindaraju, K., J. Darukeshwara, and A.K. Srivastava. 2009. Studies on Protein Characteristics and Toxic Constituents of *Simarouba glauca* Oilseed Meal. Food Chem. Toxicol. 47(6): 1327–1332.

Gowadia, N. and T.N.Vasudevan. 2000. Studies on Effect of Some Medicinal Plants on Pancreatic Lipase Activity Using Spectrophotometric Method. Asian J. Chem. 12(3): 847–852.

Grandolini, G., C.G. Casinovi, P. Barbetti, and G. Fardella. 1987. A New Quassinoid Derivative from *Quassia amara*. Phytochemistry. 26(11): 3085–3087.

Grijalva, A. 1992. Plantas Útiles de la Cordillera los Maribios. Implenta UCA Universidad Centroaméricana, FAO, Managua, Nicaragua, pp. 127–128.

GRIN. 2009. USDA, ARS, National Genetic Resources Program. Germplasm Resources Information Network—(GRIN) [Online Database]. National Germplasm Resources Laboratory, Beltsville, Maryland. (09 September 2009)URL: http://www.ars-grin.gov/cgi-bin/npgs/html/taxon.pl?102590

Grosvenor, S.N., K. Mascoll, S. McLean, W.F. Reynolds, and W.F. Tinto. 2006. Tirucallane, Apotirucallane, and Octanorapotirucallane Triterpenes of *Simarouba amara*. J. Nat. Prod. 69(9): 1315–1318.

Guo, Z., S. Vanfapandu, R.W. Sindelar, L.A. Walker, and R.D. Sindelar. 2005. Biologically Active Quassinoids and Their Chemistry: Potential Leads for Drug Design. Curr. Med. Chem. 12: 173–190.

Gupta, M.P., (ed.) 1995. 270 Plantas Medicinales Iberoamericanas. CYTED. Santa Fé de Bogotá, Colombia.

Gupta, M.P. (ed.) 2008. Plantas Medicinales Iberoamericanas. CAB/CYTED. Bogotá.

Hammarlund, E.R. 1963. Occurrence of a Weak Anti-inflammatory Substance in *Simaba cedron* Seed. J. Pharm. Sci. 52: 204–205.

Hernández-Medel M.D.R, and R. Pereda-Miranda. 2002. Cytotoxic Anthraquinone Derivatives from *Picramnia antidesma*. Planta Med. 68(6): 556–558.

Hernández-Medel, M.D.R., I. García-Salmones, R. Santillan, and A. Trigos. 1998. An Anthrone from *Picramnia antidesma*. Phytochemistry. 49(8): 2599–2601.

Hernández-Medel, M.D.R., C.O. Ramírez-Corzas, M.N. Rivera-Domínguez, J. Ramírez-Méndez, R. Santillán, and S. Rojas-Lima. 1999. Diastereomeric C-glycosyloxanthrones from *Picramnia antidesma* Phytochemistry. 50(8): 1379–1383.

Hilje, L. and G. Mora. 2006. Promissory Botanical Repellents/Deterrents for Managing Two Key Tropical Insect Pests, the Whitefly *Bemisia tabaci* and the Mahogany Shootborer *Hypsipyla grandella*. In: M. Rai and M.C. Carpinella (eds.). Naturally Occurring Bioactive Compounds. Elsevier. London.

Hitotsuyanagi, Y., A. Ozeki, H. Itokawa, S. de Mello Alves, and K. Takeya. 2001. Cedronolactone E, a Novel C(19) Quassinoid from *Simaba cedron*. J. Nat. Prod. 64(12): 1583–1584.

Holaschke, M., L. Hua, T. Basedow, and C. Kliche-Spory. 2006. Untersuchungen zur Wirkung eines standardisierten Extraktes aus dem Holz von *Quassia amara* L. ex Blom auf Getreideblattläuse und deren Antagonisten. Mitt. Dtsch. Ges. Allg. Angew. Ent. 15: 269–272.

Holdridge, L.R. and L.J. Poveda. 1975. Arboles de Costa Rica. Centro Científico Tropical. San José, Costa Rica.

Holman, H.J. 1940. A Survey of Insecticide Materials of Vegetable Origin. Imperial Institute. London.

House, P.R., S. Largos-Witte, L. Ochoa, C. Torres, T. Mejía, and M. Rivas. 1995. Plantas Medicinales Comunes de Honduras. Litografía López – UNAH, CIMN-H, CID/CIIR, GTZ. Tegucigalpa, Honduras.

Indian Biofuels Awareness Center. 2009. Website: http://www.svlele.com/simarouba.htm

Jacobs, H. 2003. Comparative Phytochemistry of *Picramnia* and *Alvaradoa*, Genera of the Newly Established Family Picramniaceae. Biochem. Syst. Ecol. 31(7):773–783.

Krebs, H.C., P.J. Schilling, R. Wartchow, and M. Bolte. 2001. Quassinoids and Other Constituents from *Picrasma crenata*. Z. Naturforsch. 56b: 315–318.

Kupchan, S.M. and J.A. Lacadie. 1975. Dehydroailanthinone, a New Antileukemic Quassinoid from *Pierreodendron kerstingii*. J. Org. Chem. 40(5): 654–656.

Kupchan, S.M. and D.R. Streelman. 1976. Quassimarin, a New Antileukemic Quassinoid from *Quassia amara*. J. Org. Chem. 41: 3481–3482.

León, J. and L.J. Poveda. 2000. Nombres Comunes de las Plantas en Costa Rica. P. Sánchez (ed.). Editorial Guayacán. San Jose, Costa Rica.

Leskinen, V., J. Polonsky, and S. Bhatnagar. 1984. Antifeedant Activity of Quassinoids. J. Chem. Ecol. 10:1497–1507.

López Sáenz, J.A. and J. Pérez Soto. 2008. Etnofarmacología y actividad biológica de *Quassia amara* (Simaroubaceae): Estado de la cuestión. Bol. Latinoam. Caribe Plant Med. Aromat 7(5): 234–246.

Lloyd, J.U. 1911. History of the Vegetable Drugs of the Pharmacopeia of the United States. Bulletin of the Lloyd Library of Botany, Pharmacy and Materia Medica. 18: 90–91. This and other magnificent references can be consulted at the Southwest School of Botanical Medicine website: http://www.swsbm.com/homepage/.

Márquez, A., K. Borri, J. Dobrecky, A.A. Gurni, and M.L. Wagner. 1999. New Aspects in Quality Control of 'Palo Amargo' (*Aeschrium crenata* Vell.-Simaroubaceae). Acta Horticulturae. 503: 111–115.

Martindale: The Extra Pharmacopeia. 1982. 28th ed. J.Reynolds. (ed.). The Pharmaceutical Press. London.

Martínez, M. 1992. Las Plantas Medicinales de México. Editorial Botas. Mexico.

Monjour, L., F. Rouquier, C. Alfred, and J. Polonsky. 1987. Therapeutic Trials of Experimental Murine Malaria with the Quassinoid Glaucarubinone. C. R. Acad. Sci. Ser. III. 304(6): 129–132.

Morales, M.E., and A. M. Uriarte. 1996. Plantas Medicinales Usadas en la IV Región de Nicaragua. Tomo 1. CEPA. Managua, Nicaragua.

Moretti, C., E. Deharo, M. Sauvain, C. Jardel, P.T. David, and M. Gasquet. 1994. Antimalarial Activity of Cedronin. J. Ethnopharmacol. 43(1): 57–61.

Morton, J. 1981. Atlas of Medicinal Plants of Middle America: Bahamas to Yucatan. Charles C. Thomas. Springfield, Illinois.

Newall, C.A., L.A. Anderson, and J.D. Phillipson. 1996. Herbal Medicines. A Guide for Healthcare Professionals. The Pharmaceutical Press. London.

Njar, V.C.O., T.O. Alao, J.I. Okogun, and H.L. Holland. 1993. 2-Methoxycanthin-6-one: A New Alkaloid from the Stem Wood of *Quassia amara*. Planta Med. 59(3): 259–261.

Njar, V.C.O., T.O. Alao, J.L. Okugun, Y. Raji, A.F. Bolarinwa, and E.U. Nduka. 1995. Antifertility Activity of *Quassia amara*: Quassin Inhibits the Steroidogensis in Rat Leydig Cells *in Vitro*. Planta Med. 91: 180–162.

Novello, C.R., R.B. Bazotte, C.A. Bersani-Amado, L.C. Marques, and D.A.G. Cortez. 2008. Toxicological and Pharmacological Studies of *Picrasma crenata* (Vell.) Engler (Simaroubaceae) in Mice and Rats. Lat. Am. J. Pharm. 27(3): 345–348.

Nunes, R.F., M.S. Kamei, A. Caliani, J.C.B. Rocha, and M.L.C. Cardoso. 2002. Caracterização Físico-química das Cascas e Folhas da *Picrasma crenata* (Vellozo) Engler—Simaroubaceae —Pau-tenente. XI Encontro Anual de Iniciação Científica—de 1 a 4/10/2002—Maringá —PR. Universidade Estadual de Maringá/Pró-Reitoria de Pesquisa e Pós-Graduação.

Núñez, E. 1975. Plantas Medicinales de Costa Rica y su Folclore. Universidad de Costa Rica. San José, Costa Rica.

Ocampo, R. 1995. Potencial de *Quassia amara* como Insecticida Natural. Minutes of a Meeting held at CATIE. CATIE. Turrialba, Costa Rica.

Ocampo, R. 2009. Personal communication.

Ocampo, R., and A. Maffioli. 1987. El Uso de Algunas Plantas Medicinales en Costa Rica.. Imprenta LIL, S.A. San José, Costa Rica.

Ocampo, R. and R. Díaz. 2006. Cultivo, Conservación e Industrialización del Hombre Grande (*Quassia amara*). Litografía e Imprenta LIL, S.A. San José, Costa Rica.

Ocampo, R. and M.J. Balick. 2009. Plants of Semillas Sagradas: An Ethnobotanical Garden in Costa Rica. R. Goldstein and K. Herrera [eds.]. This book is available for downloading from http://fincalunanuevalodge.com/sacred-seeds-costa-rica.html.

Oliveira, F. G. Akisue, and M. Akisue. 2005. Farmacognosia. Editora Atheneu. Sao Paulo, Brasil.

O'Neill, M.J., D. H. Bray, P. Boardman, C. W. Wright, J.D. Phillipson, and D.C. Warhurst. 1985. Plants as Sources of Antimalarial Drugs Part. 1. *In Vitro* Test Method for the Evaluation of Crude Extracts from Plants. Planta Med. 51(5): 394–398.

O'Neill, M.J., D.H. Bray, P. Boardman, C.W. Wright, J. D. Phillipson, D.C. Warhurst, W. Peters, M. Correya, and P. Solis. 1987. The Activity of *Simarouba amara* Against Chloroquine-Resistant *Plasmodium falciparum in Vitro*. J. Pharm. Pharmacol. 39(Suppl.): 1–80.

O'Neill, M.J., D.H. Bray, P. Boardman, C.W. Wright, J.D. Phillipson, D.C. Warhurst, M.P. Gupta, M. Correya, and P. Solis. 1988. Plants as Sources of Antimalarial Drugs, Part 6: Activities of *Simarouba amara* Fruits. J. Ethnopharmacol. 22: 183–190.

Ozeki, A., Y. Hitotsuyanagi, E. Hashimoto, H. Itokawa, K. Takeya, and S. de Mello Alves. 1998. Cytotoxic Quassinoids from *Simaba cedron*. J. Nat. Prod. 61(6): 776–780.

Parveen, S., S. Das, C.P. Kundra, and B.M. Pereira. 2003. A Comprehensive Evaluation of the Reproductive Toxicity of *Quassia amara* in Male Rats. Reprod Toxicol. 17(1): 45–50.

Pierre, A., M. Robert-Gero, C. Tempete, and J. Polonsky. 1980. Structural Requirements of Quassinoids for the Inhibition of Cell Transformation. Biochem. Biophys. Res. Comm. 93(3): 675–686.

Pirani, J.R. 1997. Simaroubáceas. Flora Ilustrada Catarinense. . Herbario Barbosa Rodríguez. Itajaí, SC, Brasil.

Pittier, H. 1957. Plantas Usuales de Costa Rica. 2nd ed. Editorial Universitaria. San José, Costa Rica.

Polonsky, J. 1973. Quassinoid Bitter Principles. Fortschr. Chem. Org. Naturst. 30: 101–150.

Polonsky, J. 1985. Quassinoid Bitter Principles II. Fortschr. Chem. Org. Naturst. 47: 221–264.

Polonsky, J. and E. Lederer. 1959. Note sur l'Isolement de la Dimethoxy-2,6-benzoquinone des Ecorces et du Bois de Quelques Simarubacees et Meliacees. Bull. Soc. Chim. Fr. 1959: 1157–1158.

Polonsky, J., Z. Varon, H. Jacquemin, and G.R. Pettit. 1978. The Isolation and Structure of 13,18-dehydroglaucarubinone, a New Antineoplastic Quassinoid from *Simarouba amara*. Experientia. 34: 1122–1123.

Polonsky, J., C.S. Bhatnagar, D.C. Griffiths, J.A. Pickett, and C.M. Woodcock. 1989. Activity of Quassinoids as Antifeedants against Aphids. J. Chem. Ecol. 15(3): 993–998.

Querol, D., M. Díaz, J. Campos, S. Chamorro, and A. Grijalva. 1996. Especies Útiles de un Bosque Húmedo Tropical. Río San Juan, Nicaragua. Güises Montaña Experimental. Río San Juan, Nicaragua. p. 28.

Quirós, M. 1945. Botánica Aplicada a la Farmacia. Universidad de Costa Rica. Escuela de Farmacia. San José, Costa Rica. pp. 72–73.

Raji, Y. and Bolarinwa, A.F., 1997. Antifertility Activity of *Quassia amara* in Male Rats—*in Vivo* Study. Life Sci. 61(11): 1067–1074.

Rivero-Cruz, J.F., R. Lezutekong, T. Lobo-Echeverri, A. Ito, Q. Mi, H.B. Chai, D.D. Soejarto, G.A. Cordell, J.M. Pezzuto, S.M. Swanson, I. Morelli, and A.D. Kinghorn. 2005. Cytotoxic Constituents of the Twigs of *Simarouba glauca* Collected from a Plot in Southern Florida. Phytother. Res. 19(2): 136–140.

Robins, R.J. and M. Rhodes. 1984. High-Performance Liquid Chromatographic Methods for the Analysis and Purification of Quassinoids from *Quassia amara* L. J. Chromatogr. 283: 436–440.

Rodríguez, S.M., M.I. Moreira, R.A. Giménez, S. Russo, A.M. Márquez, R.A. Ricco, A.A. Gurni, and M.L. Wagner. 2008. Acción Insecticida de Extractos de *Picrasma crenata* (Vell.) Engl. (Simaroubaceae) en el Gorgojo del Arroz, *Sitophilus oryzae* L. (Coleoptera, Curculionidae). Dominguezia 24(2): 95–101.

Rodriguez, S., R. Giménez, J. Lista, M. Michetti, and M.L. Wagner. 2004. Respuesta de *Tribolium castaneum* (Coleoptera: Tenebrionidae) a la Aplicación de Soluciones Acuosas de *Picrasma crenata* (Vell.) Engl. (Simaroubaceae). IDESIA 22(2): 43–48.

Roldán, R.M., M.F. Noriega, M.L. Wagner, A.A. Gurni, and G.B. Bassols. 2007. Estudio de Genotoxicidad de *Picrasma crenata* (Vell.) Engl.-Simaroubaceae. Acta Toxicol. Argent. 15(2): 39–42.

Sánchez, E. and J. Comin. 1971. Two New 2-carbolide Alkaloids from *Aeschrion crenata*. Phytochemistry. 10(9): 2155–2159.

SCF. 2002. Opinion of the Scientific Committee on Food on Quassin SCF/CS/FLAV/ FLAVOUR/29 Final. European Commission Health & Consumer Protection Directorate-General.

Scragg, A.H., S. Ashton, R.D. Steward, and E.J. Allan. 1990. Growth of and Quassin Accumulation by Cultures of *Quassia amara*. PCTOC: J. Plant Biotechnol. 23(3): 165–169.

Segleau, E.J. 2001. Plantas Medicinales en el Trópico Húmedo. Editorial Guayacán. San José, Costa Rica.

Shields, M., U. Niazi, S. Badal, T. Yee, M.J. Sutcliffe, and R. Delgoda. 2009. Inhibition of CYP1A1 by Quassinoids Found in *Picrasma excelsa*. Planta Med. 75(2): 137–141.

SIB-APN 2009. The website is http://www.sib.gov.ar/ficha/PLANTAE*Picrasma*crenata

Solís, P.N., A.G. Ravelo , A.G. González , M.P. Gupta , and J.D. Phillipson. 1995. Bioactive Anthraquinone Glycosides from *Picramnia antidesma* spp. *fessonia*. Phytochemistry. 38(2): 477–480.

Standley, O. and J. Steyermark. 1946. Flora de Guatemala. Fieldiana Botany. 24(5): 431–432.

Tada, A., N. Sugimoto, K. Sato, T. Akiyama, M. Asanoma, Y.S. Yun, T.Yamazaki, and K. Tanamoto. 2009. Examination of Original Plant of Jamaica Quassia Extract, a Natural Bittering Agent, Based on Composition of the Constituents. Shokuhin eiseigaku zasshi. J. Food Hyg. Soc. Jap. 50(1): 16–21.

Taylor, L. 1998. Herbal Secrets of the Rainforest: The Healing Power of 50 Medicinal Plants You Should Know About. Prima Health. Rocklin, California.

Teixeira, S.C.F., O.G. Rocha Nieto, J.M. Souza, and S.G. Oliveira. 1999. Avaliação da Actividade Anti-malárica da Especie *Quassia amara*: 1—Estudos *in Vivo* com *Plasmodium berguei* em Camundongos (*Mus musculus*)". Seminario de Iniciação Científica da FCAP. Embrapa Amazonia Oriental. Resumos 337–339.

Toma, W., J.deS. Gracioso, F.D.P. de Andrade, C.A. Hiruma-Lima, W. Vilegas, and A.R.M. Souza Brito. 2002. Antiulcerogenic Activity of Four Extracts Obtained from the Bark Wood of *Quassia amara* L. (Simaroubaceae). Biol. Pharm. Bull. 25: 1151–1155.

Toma, W., J.deS. Gracioso, C.A. Hiruma-Lima, F.D.P. de Andrade, W. Vilegas, and A.R.M. Souza Brito. 2003. Evaluation of the Analgesic and Antiedematogenic Activities of *Quassia amara* Bark Extract. J. Ethnopharmacol. 85: 19–23.

Trager, W. and J. Polonsky. 1981. Antimalarial Activity of Quassinoids Against Chloroquine-Resistant *Plasmodium falciparum in Vitro*. Am. J. Trop. Med. Hyg. 30(3): 531–537.

Trease, G.E., and W.C. Evans. 1989. 13th ed. Pharmacognosy. London. Baillière Tindall. London.

Tropicos.org. Missouri Botanical Garden. 03 Aug 2009. http://www.tropicos.org/Name/29400082

Tropicos.org. Missouri Botanical Garden. 02 Sep 2009 http://www.tropicos.org/Name/29400114

Tropicos.org. Missouri Botanical Garden. 08 Sep 2009 http://www.tropicos.org/Name/29400110

Tropicos.org. Missouri Botanical Garden. 10 Sep 2009 http://www.tropicos.org/Name/29400090

Valdés, A.F-C., J.M. Martínez, R.S. Lizama, M. Vermeersch; P. Cos, and L. Maes. 2008. *In Vitro* Anti-microbial Activity of the Cuban Medicinal Plants *Simarouba glauca* DC, *Melaleuca leucadendron* L and *Artemisia absinthium* L . Memórias do Instituto Oswaldo Cruz. 103(6): 615–618.

Valenta, Z., S. Papadopolus, and C. Podesva. 1961 Quassin and Neoquassin. Tetrahedron. 15, 100–110.

Valenta, Z., A.H. Gray, D.E. Orr, S. Papadopolus, and C. Podesva. 1962. The Stereochemistry of Quassin. Tetrahedron. 18: 1433–1441.

Vieira, I.J., E. Rodrigues-Filho, P.C. Vieira, M.F. Silva, and J.B. Fernandes. 1998. Quassinoids and Protolimonoids from *Simaba cedron*. Fitoterapia 69: 88–90.

Vigneron, M., X. Deparis, E. Deharo, and G. Bourdy. 2005. Antimalarial Remedies in French Guiana: A Knowledge Attitudes and Practices Study. J. Ethnopharmacol. 98: 351–360.

Villalobos, R., D. Marmillod, R. Ocampo, G. Mora, and C. Rojas. 1999. Variations in the Quassin and Neoquassin Content in *Quassia amara* (Simaroubaceae) in Costa Rica.

Ecological and Management Implications. Acta Horticult. 502: II World Congress on Medicinal and Aromatic Plants (WOCMAP II). Part 3: Agricultural Production, Post Harvest Techniques, Biotechnology 369–376.

Vitagliano, J.C. and J. Comin. 1972. Quassinoids from *Aeschrion crenata*. Phytochemistry. 11(2): 807–810.

Wagner, H., T. Nestler, and A. Neszmelyi. 1978. N-Methoxy-1-vinyl-β-carbolin, ein Neues Alkaloid aus *Picrasma excelsa* (Swartz) Planchon. Tetrahedron Lett. 19(31): 2777–2778.

Wagner, H., T. Nestler, and A. Neszmelyi. 1979. New Constituents of *Picramnia excelsa*. I. Planta Med. 36: 113–118.

Wright, C.W., M.J. O'Neill, J.D. Phillipson, and D.C. Warhurst. 1988. Use of Microdilution to Assess *in Vitro* Antiamoebic Activities of *Brucea javanica* Fruits, *Simarouba amara* Stem, and a Number of Quassinoids. Antimicrob. Agents Chemother 32(11): 1725–1729.

Woods, B. and C.D. Calnan. 1976. Toxic Woods. Br. J. Dermatol. 95(Suppl. 13): 1–97.

Xifreda, C.C. and M.N. Seo. 2006. Fasc. 138. SIMAROUBACEAE DC. S.l. Flora Fanerogámica Argentina,. Proflora-CONICET, pp. 1–13.

Chapter 12

An Inventory of Ethnomedicinal Plants Used in Tunisia

Borgi Wahida, * *Mahmoud Amor,* and *Chouchane Nabil*

Introduction

Plants have been widely and successfully used as medicines throughout history. Obviously, the use of plants in traditional medicine has contributed to the reduction of excessive mortality, morbidity and disability caused by diseases. Therefore, medicinal plants present an important opportunity to rural communities, as a source of affordable medicine and a source of income. Traditional knowledge related to the health of humans and animals exist in all countries of the world.

Tunisia, the northernmost country on the African continent, has a great geographical and climatic diversity which produces a large plant biodiversity (Le Houerou 1995). In fact, its flora accounts for more than 2150 species (Pottier-Alapetite 1979, 1981) growing on various bioclimatic zones from sub-humid to arid and Saharan. Many wild plants are considered as multipurpose and are susceptible to produce valuable substances, essential oils and original aroma useful for agrofood, pharmaceutical and cosmetic industries. Actually, more than 200 wild species are considered herbal, medicinal and aromatic plants and used in traditional phytotherapy in the treatment of various diseases (Cuenod 1954, Le Floc'h 1983, Boukef 1986, Issaoui et al. 1996, Chaieb and Boukhris 1998). Thus a screen of plants which include phytotherapic use based on the principal of ethnobotanical studies carried out in Tunisia by Le Floc'h (1983) and Boukef (1986) are presented here.

Laboratory of Pharmacology, Faculty of Pharmacy-Monastir 5000, Tunisia.
*E-mail: *wahida_b2002@yahoo.fr*

Description of the study area

Tunisia is a country located in North Africa (Fig. 12.1). It is bordered by the Mediterranen Sea to the north, Algeria to the west and Libya to the southeast. Its size is almost 165,000 km². Tunisia is the northernmost country on the African continent, and the smallest of the nations situated along the Atlas mountain range. Around 40% of the country is composed of the Sahara desert, while the rest consists of particularly fertile soil and a 1300 km coastline.

Fig. 12.1. Map of the study area (Tunisia).

Methodology

According to literature and the principal ethnobotanical studies company in Tunisia (Le Floc'h 1983, Boukef 1986), a systematic search was performed. Information was collected from research gathered by books including ethnomedicinal use of plants in humans and animals in Tunisia. In the following enumeration, plant names have been arranged alphabetically. The botanical name is followed by local name in Tunisia, the family and the parts used with their medicinal uses and constituents.

Tunisian plant biodiversity

Tunisia is a country situated on the Mediterranean coast of North Africa, midway between the Atlantic Ocean and the Nile Valley. Despite its relatively small size (165,000 km²), Tunisia has great geographical and climatic diversity. The Dorsal, an extension of the Atlas Mountains, traverses Tunisia in a northeasterly direction. North of the Dorsal is the Tell, a region characterized by low, rolling hills and plains. The Sahil is a plain along Tunisia's eastern Mediterranean coast and the south of the country is desert. This geographical diversity in Tunisia includes 15679 hectares which is the total area of 14 natural reserves (Dar Fatma, Ain Zana, Majen Djbal Chitane, Ain Chrichira, Djbal Toiati, Khechem of Kaleb, Etella, Djbal Serdj, Djbal Bouramli, El Haouaria, Djbal Khroufa, Chikly, Sebkhat Kelbia, and Kneiss) and 196478 hectares which is the total area of 8 national parks (Zembra-Zembretta, Bouhedma, Chaambi, Ichkeul, Boukornine, El Feija, Jbil and Didi Toui).

Tunisia's climate is humid to sub-humid in the north, with mild rainy winters and hot, dry summers. The terrain in the north is mountainous, which towards the south, gives way to a hot and dry central plain. The south is semiarid, and merges into the Sahara (Nagendra 2000, Stearns and Leonard Langer 2001, Ham et al. 2004). While Tunisia is not cold during winter, when it snows the temperature can go down to below 0°C. In the summer it can go up to 32°C. Most of the areas in Tunisia have four seasons.

Thus, Tunisia is the geographic transition between the Saharan zone in the South and the temperate zone in the North. It is generally accepted that this geographical diversity plays a major role in the diversity of natural resources. Consequently, Tunisia has important resources of biological diversity. The flora is estimated to about 2150 species distributed in 115 families and 742 genera. Tunisia is the home of plant species which are related to livestock, food, fodder crops, and of hundreds of species that are traditional medicinal plants (Lemordant et al. 1977). In reality, more than 200 species of plants are used in the traditional health care system (Table 12.1) to treat ailments. This traditional medical system is characterized by variation and is shaped by the ecological diversities of the country, socio-cultural background of the people as well as historical developments which are related to migration, introduction of foreign cultures and religions throughout history.

In all regions of the world, the history of nations shows that plants have always held an important place in medicine, in the composition of perfumes and in food preparation. Situated in the Mediterranean with large variations of climate from North to South, Tunisia provides a breeding ground for the development of these cultures. In consequence, the ethnobotanical knowledge and the use of plants have been transferred through generations. Essentially, in the XI century, this empirical knowledge was the object of high levels of study in several cities particularly the Islamic Universities of Kairouan and Zeîtouna in Tunisia (Lapidus 2002). This factor determines the oldest and the richest traditions in Tunisia associated with the use of medicinal plants. Thus, medicinal plants are

Table 12.1. List of medicinal plants used in Tunisia, arranged alphabetically with their scientific names

Scientific name	Family	Local name	Main traditional uses in Tunisia	Plant part used	Constituents	References
Acacia raddiana Savi	Fabaceae	Talh	Respiratory diseases, skin ailments, toothache.	Flowers, fruits	Flavonol glycosides, ellagitannin and galloyl	El-Mousallamy et al. 1991
Ajuga iva (L.) Scherb.	Lamiaceae	Chendgoura	Digestive diseases, hypertension, diabetes, rheumatism, nevralgy (stress), otis, cancer, respiratory diseases, sterility in women.	Flowering branches	- Tannins, few essential oils, polyhydroxylated-sterols, phytoecdysteroids.	Wessner et al.1992
Agave americana L.	Amaryllidaceae	Sabbara	Rheumatism, anti-septic, respiratory diseases, skin ailments	Leaves		
Allium cepa L.	Liliaceae	Bsal	Diabetes, sunstroke	Bulbs		
Allium porrum L.	Liliaceae	Kourrath	Digestive diseases	Bulbs		
Allium roseum L.	Liliaceae	Lazoul	Colds, rheumatism	The fresh leaves and inflorescences	- Essential oil	Najjaa et al. 2007
Allium sativum L.	Liliaceae	Thoum	Hypertension, cardiovascular diseases, skin ailments, diabetes, vascular diseases	Bulbs		
Alnus glutinosa Gaertn.	Betulaceae	Oud ahmar	Gastric ulcer, fever, headache	Fruits		
Aloe vera L.	Liliaceae	Marousbar	Fever, kidney diseases	Whole plant		
Ammi visnaga (L.) Lam	Apiaceae	Nounkha	Eye diseases, digestive diseases	Whole plant		
Amygdalus communis	Rosaceae	Louz	Digestive diseases, kidney diseases	Fruit		
Anacyclus valentinus L.	Asteraceae	Gritfa	Digestive diseases	Flowers		
Andrachne telephioides L.	Euphorbiaceae	Hnihna	Respiratory diseases, skin ailments	Stem		
Andropogon hirtus L.	Poaceae	Bourkiba	Diuretic	Roots		
Anethum graveolens L.	Apiaceae	Chibt	kidney diseases	Fruit		
Anthriscus silvestris Hoffm.	Apiaceae	Kochbor	Headache	Aerial part		

Ancillea garcinii ssp. *radiata* Coss. et Dur.	Asteraceae	Nougd	Cold, diabetes, digestive problems, gastro-intestinal troubles, indigestion, pulmonary affections.	Capitules, leaves and seeds	- Tannin, oxalate of calcium, saponine, pectine, lactone sesquiterpenic.	El Hassany et al. 2004
Apium graveolens L.	Apiaceae	Klafez	Fever	Aerial part		
Arbutus unedo L.		Lenj	Hypertension	Roots		
Artemisia absinthium L.	Asteraceae	Chajret Mariem	Digestive diseases			
Artemisia campestris L...	Asteraceae	Dgouft	Anti-poison, skin ailments	Aerial part		
Artemisia herba alba Asso	Asteraceae	Chih	Digestive diseases, rheumatism	Aerial part	—Santonin, lactones of sesquiterpenic acids, flavonoids, coumarins, pentacyclic triterpens, anthracenosids and tannins.	Bezanger-Beauquesne and Pinkas 2000
Arum italicum Mill.	Araceae	Sabat Elghoula	Antiseptic	Leaves		
Arundo donax L.	Poaceae	Ksab	kidney diseases, veterinary medicine	Aerial part		
Asphodelus microcarpus Viv	Liliaceae	Barouag	Ear diseases	Bulbs	- Starch, anthraquinones.	Bonnier 1990
Astragalus marroCcus Del.	Fabaceae	Hachicht agrab	Scorpion bite	Whole plant		
Atractylis gummife L.	Asteraceae	Ded	Antiseptic	Roots		
Atriplex halimus	Chen-opodiaceae	Ktaf	Cicatrizing	Leaves		
Avena sativa L.	Poaceae	Hchichat errih	Antiseptic	Aerial part		
Beta vulgaris L.	Asteraceae	Oqhouwana	Diabetes, constipation	Flowers		
Beta macrocarpa Guss.	Chen-opodiaceae	Salq	Cicatrizing	Leaves		

Table 12.1. contd...

Table 12.1. contd...

Scientific name	Family	Local name	Main traditional uses in Tunisia	Plant part used	Constituents	References
Borago officinalis	Borraginaceae	Boukhrich	Respiratory diseases	Leaves	- Vitamin C, tannins, soluble silicic acid	Bezanger-Beauquesne and Pinkas 2000
Brassica napus L.	Brassicaceae	Left	Fever, laxative	Leaves, flowers, seed		
Calendula arvensis L.	Asteraceae	Karchoun	Rheumatism, fever, headache	Leaves		
Calycotme villosa Link	Fabaceae	Gandoul	Cicatrizing	Leaves		
Capparis spinosa L.	Capparidaceae	Kabbar	Rheumatism, headache	Leaves, flowers, fruits, seeds	Isothiocyanates, thiocyanates, sulphides, flavonoids glycosides, (6S)-Hydroxy-3-oxo-alpha-ionol glucosides, quercetin triglycoside, p-methoxy benzoic acid	Al-Said et al. 1988, Ali-Shtayeh and Abu Ghdeib 1999, Inocencio et al. 2000, Sharaf et al. 2000, Calis et al. 2002, Germano et al. 2002
Capsicum annuum L.	Solanaceae	Felfel	Ear diseases, rheumatism	Fruits		
Carpobrotus edule L.	Aizoaceae	Charbabbou	Cicatrizing	Leaves		
Carthamus tinctorius L.	Asteraceae	Zaafrane	Eye diseases	Leaves, flowers	- Phytosterol	Hamrouni-Sellami et al. 2007
Carum carvi L.	Apiaceae	Karouia	Digestive diseases, respiratory diseases	Fruit	- Essential oil and fatty acid	Laribi et al. 2009
Centaurea calcitrapa L.	Asteraceae	Bounaggar	Fever, veterinary medicine	Roots		
Centaurea contracta	Asteraceae	Arjegnou	Antiseptic	Whole plant		

Species	Family	Local name	Uses	Part used	Chemical constituents	References
Centaurium pulchellum	Gentianaceae	Qosset sbiya	Elimination of stones from the kidney	Whole plant	- Coumarins, alkaloid, oleanolic acid, erythrosterol, xanthones	Khafagy and Mnajed 1970, Ahmed et al. 1971, Khafagy and Metwally 1971, Kostens and Willuhn 1973, Zirvi et al. 1977
Ceratonia siliqua L.	Caesalpiniaceae	Kharroub	Diarrhoea, diabetes, respiratory diseases	Fruit: pericarp (pulp) and seeds	- Sugars, condensed tannins, mucilage, starch, fats	Bruneton 1999
Chenopodium ambrosioides L.	Chenopodiaceae	Lajouma	Digestive diseases	Whole plant-Leaves		
Chenopodium murale L.	Chenopodiaceae	Mazrita	Fever, kidney diseases	Leaves		
Cicer arietinum L.	Fabaceae	Homs	Antiseptic, kidney diseases	Seeds		
Cichorium intybus L.	Asteraceae	Chikouria	Diabetes	Leaves		
Cistus monspeliensis L.	Cistaceae	Oumillia	Skin ailments, dog bite	Whole plant		
Citrullus colocynthis Schrad.	Cucurbitaceae	Handhal	Rheumatism, skin ailments	Fruit, leaves, seeds, roots	- Cucurbitacins, alkaloids, saponins, phytosterols.	Sayed et al. 1973, 1974, Soliman et al. 1985, Yaniv et al. 1991
Citrus aurantium var. amara	Rutaceae	Aranj	Cardiovascular diseases, fever	Flowers		
Citrus sinensis L.	Rutaceae	Bordguene	Skin ailments, gastric ulcer	Leaves		
Citrus medica (L.) Proparte	Rutaceae	Qaress	Respiratory diseases, hypertension	Fruits		
Citrus paradisi Macfad yen.	Rutaceae	Zenbaa	Hypertension, obesity	Fruits		
Clematis flammula L.	Ranunculaceae	Nar berda	Rheumatism, skin ailments	Aerial part		

Table 12.1. contd...

Table 12.1. contd...

Scientific name	Family	Local name	Main traditional uses in Tunisia	Plant part used	Constituents	References
Cleome arabica L.	Capparidaceae	Om jlajel	Constipation, rheumatism	Leaves		
Convolvulus arvensis	Convolvulaceae	Laouia	Hair tonic	Stem		
Coriandrum sativum L.	Apiaceae	Tabel	Digestive diseases, diabetes	Fruits	- Essential oil, fatty acid	Msaada et al. 2007a, 2007b, 2009a, 2009b, 2009c, Neffati and Marzouk. 2008
Crataegus azarolus L.	Rosaceae	Zaarour	Diabetes, spasm, respiratory diseases	Fruits, flowers, leaves	- Procyanidic oligomers, flavonoids, catechin, polyphenol	Bahri-Sahloul et al. 2009
Cornulaca monacantha Del.	Chenopodiaceae	Elhad	Liver problems, veterinary medicine	Leaves	- Gallotannins, lavonol glycoside, flavonoids, triterpenoidal saponins	Amer et al. 1974, Kandil and Husseiny 1998, Kamel et al. 2000, Kandil, and Grace 2001
Cucurbita pepo L.	Cucurbitaceae	Kraa	Fever, fungal infection	Fruits, leaves		
Cuminum cyminum L.	Apiaceae	Kammoun	Digestive and gynaecological disorders	Fruits		
Cupressus sempervirens L.	Cupressaceae	Saroual	Digestive diseases, haemorrhoids	Leaves, cones.	- Biflavonoids, monoterpenic carbides, sesquiterpenics, diterpenics, tannins.	Bruneton 1999
Cydonia vulgaris Persoon	Rosaceae	Sfarjel	Eye diseases, digestive diseases	Fruits, seeds		
Cynanchum acutm L.	Asclepiadaceae	Alligua	Eye diseases, skin ailments	Latex		

Species	Family	Local name	Uses/Diseases	Plant part	Chemical constituents	Reference
Cynara cardunculus L.	Asteraceae	Khorchef	Hepatitis, kidney diseases, diabetes Cardiovascular diseases	Leaves, roots, Stems	- Cynarin, flavonoids-cynarosid,scolymosid, oxydase, inulin.	Bezanger-Beauquesne and Pinkas 2000
Cynara scolymus L.	Asteraceae	Guennaria	Liver problems, kidney diseases	Leaves, roots		
Cynodon dactylon (L.) Pers	Poaceae	Njem	kidney diseases, rheumatism, diabetes	Roots		
Cynomorium coccineum L.	Cynomoriaceae	Tarthouth	Haemorrhoids, diarrhoea	Aerial part	- Anthocyans, cyanidin 3-glucoside.	Harborne et al. 2003
Cyperus longus L.	Cyperaceae	Saada	Cardiovascular diseases	Roots		
Cytinus hypocistis L.	Cytinaceae	Daghmous	Ear diseases	Leaves		
Daphne gnidium L.	Thymeleaceae	Zez	Cicatrizing	Leaves		
Datura metel L.	Solanaceae	Hchichet elfedda	Antiasthmatic	Leaves	- Alkaloids, withanolides	Bruneton 1993
Daucus carota L.	Apiaceae	Sfinnaria	Aphrodisiac	Leaves		
Diplotaxis harra Forsk.) Boiss	Brassicaceae	Harra		Aerial part	- Sinigroside, Arachidonic acid, palamitic acid, cholesterol, stigmasterol, B-sitosterol and nonmethylated, fatty acid	Fahey et al. 2001
Ecballium elaterium Rich.	Cucurbitaceae	Faqous h'mir	Liver problems, veterinary medicine	Fruits, roots		
Echinops spinosis Tarra	Asteraceae	Teskra	Cancer; skin ailments	Flowers, leaves, roots	- Sesquiterpene lactones, acetylenic elements	Dawidar et al. 1990
Erica arborea L.	Ericaceae	Bou haddad	kidney diseases	Flowering tips	- Tannins, flavonoids	Harborne and Williams 2001
Erica multiflora L.	Ericaceae	Khlenj	Prostate cancer; pain	Flowering tips	-Proanthocyanidols, flavonoids, cyanogenetic glucosides	Pottier Alapetite 1981

Table 12.1. contd...

Table 12.1. contd...

Scientific name	Family	Local name	Main traditional uses in Tunisia	Plant part used	Constituents	References
Erigeron canadensisi L.	Asteraceae	Jaada	Gastric ulcer, cardiovascular diseases, Rheumatism	Flowering tips		
Erodium cicutarium L.	Geraniaceae	Mochita	Stomach ailments	Aerial part		
Ervum lens L.	Fabaceae	Ades	Fever, skin diseases	Seeds		
Eucalyptus globulus L.	Myrtaceae	Kalatous	Fever, respiratory diseases, toothache	Leaves		
Euphorbia calyptrata Coss. DR	Euphorbiaceae	Bouhaliba	Skin diseases	Latex		
Euphorbia guyoniana	Euphorbiaceae	Lbina	Scorpion bites, snake bites			
Ferula communis L.	Apiaceae	Kelkh	Skin diseases, female sterility, pain, rheumatism, spasm, anti-hysteric	Roots, Latex	-Coumarins, daucane sesquiterpenes, 4-hydroxycoumarins	Lamnaouer et al. 1987, 1989, 1991a, 1991b, Valle et al. 1987, Appendinno et al. 1988
Ficus carica L.	Moraceae	Karma	Haemorrhoids, digestive diseases	Fruits, latex		
Foeniculum vulgare Mill	Apiaceae	Besbes	Digestive diseases	Fruits		
Fumaria capreolata L.	Fumariaceae	Sibana	Skin diseases, ear diseases	Whole plant		
Genista saharae	Fabaceae	Almarkh	Veterinary medicine	Aerial part		
Globularia alypum L.	Globulariaceae	Zriga	Ulcer, cancer, skin diseases, fever	Leaves	-A resin, iridoids (aucubosid) and sterols	Bezanger-Beauquesne and Pinkas 2000
Glycyrrhiza glabra L.	Fabaceae	Erqsus	Inflammatory diseases, constipation, respiratory diseases, digestive diseases, ulcer, cancer, kidney diseases	Roots, underground stem		

Species	Family	Local name	Uses/diseases	Part used	Chemical compounds	Reference
Hammada scoparia (Pomel) Il'jin	Chenopodiaceae	Remth	Eye diseases, headache	Leaves	-Alkaloids: anabasin, aphyllidin, lupinin.	Ben Salah et al. 2002
Helianthemum confertum Dunal	Cytinaceae	Assamhari	Skin diseases	Aerial part		
Heliantus annuus L.	Asteraceae	Oubbidet Echems	Fever	Fruits		
Herniaria fontanesii J. Gay.	Illecebraceae	Qouttiba	kidney diseases	Aerial part		
Hordeum vulgare L.	Poaceae	Chair	kidney diseases, galactagogue	Seeds		
Hyoscyamus albus L.	Solanaceae	Sikran	Skin diseases, haemorrhoids	Whole plant, seeds		
Jasminum fruticans L.	Oleaceae	Boulila	Skin diseases	Leaves		
Juglans regia L.	Juglandaceae	Zouza	Hypertension, toothache, aphrodisiac	Leaves, fruits, flowers		
Juncus maritimus Lamk.	Joncaceae	Smar	Haemorrhoids, cicatrizing	Flowers, seeds		
Juniperus oxycedrus L.	Cupreassaceae	Arar	Skin diseases, diuretic, stimulant, vermifuge, asthma, rheumatism, diabetes	Leaves, fruits	-Flavonoids, flavones, terpenoids,monoterpenoids, sesquiterpenoids, volatile oil, resin, tannin, terpenoids, monoterpenoidsfatty acid	Barrero et al. 1991, 1993, 1995, Salido et al. 2002
Juniperus phoenicea L.	Cupreassaceae	Arar hor	Constipation, rheumatism, diabetes, digestive diseases, pain, acute, gynaecological disorders, dysmenorrhoea, prostatitis, animal's skin diseases	Leaves, fruits	- oligosaccharides, catechic tannins, biflavonoids, leucanthocyanes, alcohol acids, terpenic alcohol (sabinol)	El Akrem et al. 2007
Koniga libyca Viv.	Brassicaceae	Hchichteddhars	Toothache, respiratory diseases	Flowers		
Lactuca sativa L.	Asteraceae	khass	Respiratory diseases	Leaves		
Laurus nobilis L.	Lamiaceae	Rand	Headache, rheumatism, skin diseases	Leaves, roots	- Cineol, linalol, eugenol, sesquiterpenic lactones, isoquinoleic alkaloids	Valnet 2001

Table 12.1. contd...

Table 12.1. contd...

Scientific name	Family	Local name	Main traditional uses in Tunisia	Plant part used	Constituents	References
Lavandula stoechas L.	Lamiaceae	Helhal	Respiratory diseases, digestive diseases, Pain	Leaves, flowers	- Fenchone, camphor, alpha-pinene, beta-pinene, camphene, eucalyptol, para-cymene, linalol, borneol, borneol acetate, carvacrol, iso-eugenol, iso-eugenol-methylether	Ulubelen et al. 1988, Ulubelen and Olcay 1989 Buchbauer et al. 1991, Topcu et al. 2001, Goren et al. 2002
Lavandula vera DC.	Lamiaceae	Khzema	Hypertension, respiratory diseases	Aerial part		
Lawsonia inermis L.	Lythraceae	Henna	Fever, skin diseases, digestive diseases, diabetes	Leaves		
Lepidium sativum L.	Brassicaceae	Hab erched	kidney diseases, respiratory diseases	Seeds		
Limonium pruinosum (L.) O.Kuntze	Plum-baginaceae	Zayata	Ear diseases, gynaecological disorders Genital pain	Aerial part		
Linaria relexa Desf.	Scro-fulariaceae	Om lajrah	Skin diseases	Aerial part		
Linum usitatissimum L.	Linaceae	Ketten	Respiratory diseases, digestive diseases Skin diseases	Seeds		
Lycium europaeum L.	Solanaceae	Sakkoum	Eczema	Flowering tips		
Lycopersicum esculentum Mill.	Solanaceae	Tmatem	Hypertension, kidney diseases	Leaves		

Scientific name	Family	Local name	Uses	Part used	Chemical constituents	Reference
Malva sylvestris L.	Malvaceae	Khobbiza	Renal lithiasis, insect bites, digestive disorders, mouth pain	Leaves, flowery tips	-Acid polysaccharids, flavonoids, tannins, anthocyanosids, anthocyanidines	Valnet 2001
Mandragora autumnalis Bertl	Solanaceae	Bidh elghoul	Rheumatism, haemorrhoids	Roots, leaves	- Alkaloids, cuscohygrine	Tessier 1994
Marrubium vulgare L.	Lamiaceae	Morroubia	Hypertension, diabetes, ear diseases, haemorrhoids, skin diseases	Aerial part		
Matricaria chamomilla L.	Asteraceae	Babounej	Digestive diseases, pain, kidney diseases	Whole plant		
Mentha pulegium	Lamiaceae	Flayou	Respiratory diseases	Leaves	- Essential oil	Karray-Bouraoui et al. 2009
Mentha rotundifolia L.	Lamiaceae	Marsita	Skin diseases, fever; toothache	Leaves		
Mentha spicata L	Lamiaceae	Naanaa	Skin diseases, toothache, digestive disorders, rheumatism, gynaecological disorders	Leaves		
Mercurialis annua L.	Euphorbiaceae	Hbaq edhol	Uro-genital disorders	Whole plant		
Mesembryanthemum nodiflorum L.	Aizoaceae	Ghissoul	Eye diseases	Aerial part		
Mespilus germanica. L.	Rosaceae	Boussaa	Toothache, respiratory diseases, kidney diseases	Leaves		
Morus alba	Moraceae	Toutt	kidney diseases, haemorrhoids	Leaves, latex		
Morus nigra L.	Moraceae	Toutt arbi	Toothache	Roots		
Myrtus communis L.	Myrtaceae	Rihan	Gastric ulcer, digestive disorders, toothache, respiratory diseases, rheumatism	Fruits, flowers	- Terpenes, myrtenol, eucalyptol, tannins, fatty acid	Wannes et al. 2009
Nerium oleander L.	Apocynaceae	Defla	Gangrene, eczema, headaches, colds, toothache, abortion	Root, leaves	- Cardenolides, triterpenoids, a resin, tannins, glucose, a paraffin, ursolic acid, steroids vitamin C, essential oil	Begum et al. 1998

Table 12.1. contd...

Table 12.1. contd...

Scientific name	Family	Local name	Main traditional uses in Tunisia	Plant part used	Constituents	References
Nicotiana gluaca Garham	Solanaceae	Dokkhane	Skin diseases, veterinary medicine, eye diseases	Leaves		
Nigelle damascena L.	Ranunculaceae	Sinouj	Cardiovascular diseases, eye diseases headaches, toothache	Seeds		
Ocimum basilicum L.	Lamiaceae	Hbaq	Aphrodisiac, headaches, gynaecological disorders	Leaves		
Olea europaea L.	Oleaceae	Zitoun	Constipation, respiratory diseases, hypertension, ophthalmic diseases, diabetes	Leaves		
Ononis natrix L.	Fabaceae	Mellita	Toothache	Leaves		
Opuntia ficus-indica (L.) Mill.	Cactaceae	Hindi	Kidney diseases, diabetes, diarrhoea, haemorrhoids, vermifuge	Flowers, fruits		
Orchis sp	Orchidaceae	Elhaya welmiyta	Aphrodisiac	Tubercle		
Origanum majorana L.	Lamiaceae	Mardaqoash	Respiratory diseases, colds, gynaecological disorders	Leaves, flowers	- Essential oil, phenolic fraction	Hamrouni et al. 2009
Papaver somniferum L.	Papaveraceae	Khochkhache	Headaches, hypnotic	Aerial part		
Papaver rhoeas L.	Papaveraceae	Bougaroun	Hypnotic, respiratory diseases	Flowers		
Peganum harmala L.	Zygophyllaceae	Harmel	Rheumatism, diabetes, hypertension, respiratory diseases, toothache	Roots, leaves, seeds	- Alkaloids	Pottier Alapetite 1979
Pelargonium capitatum Ait.	Geraniaceae	Attirchia	Spasm, headaches, diabetes, sunstroke	Flowering branches	- Essential oil, tannins	Van Hellemont 1986
Pergularia tomentosa L.	Apocynaceae	Ghelga	Eczema	Latex		

Species	Family	Local name	Uses/Diseases	Part used	Chemical constituents	References
Periploca laevigata Ait.	Betulaceae	Aroug elhalleb	Diabetes, hypertension	Leaves, roots	- Triterpenes: b sistosterol, a and b amyric, periplocadiol and phenol acid derivatives, particularly caffeic acid	Pottier Alapetite 1981
Persica vulgaris Mill.	Rosaceae	Khoukh	Ear diseases, hypertension, skin diseases	Leaves		
Petroselinum sativum Hoffm.	Apiaceae	Maadnous	Kidney diseases, anti-galactagogue	Leaves		
Phlomis crinita Cav.	Lamiaceae	Khayatta	Skin diseases	Leaves		
Phoenix dactylifera L.	Palmaceae	Nakhla	Syphilis, hair tonic	Seeds		
Pimpinella anisum L.	Apiaceae	Habbet hleoua	Galactagogue, digestive disorders	Fruits		
Pinus halepensis Mill.	Pinaceae	Zgougou	Ulcer, cicatrisant, diarrhoea, Hypertension	Leaves		
Pinus pinaster Soland.	Pinaceae	Snouber	Ulcer, cicatrizing, diarrhoea, respiratory diseases	Leaves		
Pirus communis L.	Rosaceae	Anjas	Cancer, kidney diseases	Flowers		
Pistacia lentiscus L.	Anacardiaceae	Dharou	Ulcer, constipation, hypertension, respiratory diseases, rheumatism	Roots, leaves, fruits	- Flavonoids, essential oil, tannins, fatty oil, gallic acid, digallic acid, 1,2,3,4,6-pentagalloylglucose	Tabazik-Wlotzka et al. 1967, Abdelwahed et al. 2007, Bhouri et al. 2009
Pistacia terebinthus L.	Anacardiaceae	Battoum	Respiratory diseases, rheumatism	Fruits		
Pituranthos scoparius Benth.	Apiaceae	Gozzeh	Respiratory diseases, rheumatism, snake bite, veterinary medicine	Aerial part	-Mannitol,isocoumarines, essential oil	Hamada et al. 2004, Verite et al. 2004
Plantago albicans L.	Plantaginaceae	Zouel	Cicatrizing	Aerial part		

Table 12.1. contd...

Table 12.1. contd...

Scientific name	Family	Local name	Main traditional uses in Tunisia	Plant part used	Constituents	References
Polygonum aviculare L.	Polygonaceae	Gordhab	Digestive diseases, respiratory diseases	Roots		
Portulaca oleracea L.	Portulacaceae	Blibcha	Digestive diseases	Whole plant		
Prunus armeniaca Lamk.	Rosaceae	Michmeche	Constipation	Fruits		
Prunus domestica L.	Rosaceae	Aouina	Constipation	Fruits		
Punica granatum L.	Punicaceae	Rommen	Gastric ulcer, diarrhoea, hypertension, Diabetes	Fruits		
Quercus ilex L.	Fagaceae	Ballout	Haemorrhoids, diarrhoea	Fruits, barks		
Ranunculus macrophyllus Desf	Ranunculaceae	Kef jrana	Skin diseases	Leaves		
Reseda alba L.	Resedaceae	Bassous khrouf	Skin diseases	Leaves, root barks		
Retama raetam Webb	Fabaceae	Rttam	Eye diseases, rheumatism, cicatrizing Skin diseases, digestive diseases respiratory diseases	Stems, leaves, flowers	- Flavonoids, quinolizidine alkaloid	El-Shazly et al. 1996
Rhamnus alaternus L.	Rhamnaceae	Oud elkhir	Liver diseases	Aerial part		
Rhaphanus sativus L.	Brassicaceae	Fjel	Respiratory diseases	Tubercle		
Rhus oxyacantha L.	Anacardiaceae	Jderi	Gastric ulcer	Roots		
Ricinus communis L.	Euphorbiaceae	kharouâa	Cold, respiratory problems, abortion Contraceptive, fever, rheumatism	Leaves, seeds, oil	- Water, proteins, Lipids, ricin	Van Hellemont 1986
Rosa canina L.	Rosaceae	Nisri	Acne, asthenia, cardiopathy, sunstroke, Constipation	Flowers	- Vitamin C, provitamins A, tannins, sugars, citric and malic acids, pectin, D-sorbitol, an essential oil, flavonoids, fatty oil	Larsen et al. 2003

Species	Family	Local name	Uses	Parts used	Constituents	References
Rosa gallica L.	Rosaceae	Wared	Eye diseases, skin diseases	Petals, flower buds	- An essential oil rich in geraniol and citronellol; flavonoids; anthocyanes, tannins	Maire 1980
Rosmarinus officinalis L.	Lamiaceae	Klil	Rheumatism, respiratory problems, fever, digestive diseases, liver problems	Leaves, flowery tips	- Cineol, camphor, a-pinene; tricyclic phenolic diterpenes carnosolic acid, carnosol, tannins, methylated flavons, triterpenes, steroids, lipids, polysaccharides, traces of salicylate	Zaouali and Boussaid 2008
Rubus ulmifolius Schott.	Rosaceae	Allig	Skin problems, headaches, mouth pain	Leaves		
Rumex vesicarius L.	Polygonaceae	Hommidha	Jaundice, respiratory problems, veterinary medicine	Whole plant	- Flavonoids, C-glycosides: vitexin, isovitexin, orientin and iso-orientin and anthraquinones: emdin and chrysophanol, rumicine, lapathine, oxalic acid, tannins, mucilage, mineral salts and vitamin C	Al-Easa et al. 1995
Ruta chalepensis L	Rutaceae	Fijel	Digestive diseases, hypertension, rheumatism, ear diseases	Leaves, flowering stems, roots	- Active alkaloids, furocoumarins, flavonoids, tannins, volatile oil, sterols	Brooker et al. 1967, Ulubelen et al. 1988, El Sayed et al. 2000
Salicornia arabic L.	Chenopodiaceae	Ghodhram	Eczema	Latex		
Sambucus nigra L.	Caprifoliaceae	Okez sidi moussa	Rheumatism, skin problems	Leaves		
Samolus valerandi L.	Primulaceae	Oudhina	Ear diseases	Leaves		
Scabiosa atropurpurea L.	Didsaceae	Om ejelejel	Sterility in women	Whole plant		

Table 12.1. contd...

Table 12.1. contd...

Scientific name	Family	Local name	Main traditional uses in Tunisia	Plant part used	Constituents	References
Scorzonera undulata Vahl.	Asteraceae	Guiz	Skin problems	Roots		
Sesamum indicum De Candolle	Pedaliaceae	Jiljlen	Post partum	Seeds		
Sinapis arvensis L.	Brassicaceae	Khardhel	Rheumatism, respiratory problems	Leaves		
Silybum marianum (L.) Gartner	Asteraceae	Khorshef barri	Eczema	Stems	- Flavonolignan complex, silymarin,dehydrosilybin, 3-desoxysilichristin, deoxysilydianin, siliandrin, silybinome, silyhermin, fixed oil, flavonoids, taxifolin, sterols	Morazzoni amd Bombardelli 1994
Solanum nigrum L.	Solanaceae	Tmatem kleb	Skin diseases	Fruits, whole plant		
Solanum sodomeum L.	Solanaceae	Lim ennsara	Haemorrhoids, eczema	Fruits	- Gluco-alkaloid, saponosides	Cham and Meares 1987
Solanum tuberosum L.	Solanaceae	Batata	Sunstroke, skin problems	Tubercle		
Sonchus oleraceus L.	Asteraceae	Tifef	Ophthalmic diseases, skin diseases, galactogogue, diabetes	Flowery tips		
Suaeda fruticosa (L.) Forsk.	Chenopodiaceae	Souida	Skin diseases	Flowers, seeds		
Tamarix gallica L.	Tamaricaceae	Tarfa	Eye diseases, digestive diseases	Leaves, bark	- Alkaloid: tamarixin, tannin (ellagic and gallic), quercetol (methyllic esther)	Bellakhdar 1997
Teucrium polium L.	Lamiaceae	Guattaba	Cicatrisant, cardiovascular diseases	Flowering branches, leaves	- Diterpenoids	
Thapsia garganica L.	Apiaceae	Diryes	Rheumatism	Roots		
Thymelea microphylla Coss. Et Dur.	Thymeleaceae	Methnene	Skin diseases, sterility in women	Leaves		

Species	Family	Local name	Uses	Part used	Chemical constituents	References
Thymus broussonetii Boiss	Lamiaceae	Zaatar essouiri	Colds, pains, coryzas, rheumathisms, articular pains, throat troubles, jaundice, liver diseases, digestive troubles, aphta, gingivitis, wounds, cuts, furuncles and abscess. General antiseptic, galactogogue, vermifuge, emmenagogue, diuretic, digestive, appetizer.	Flowering branches, leaves, aerial part.	- Flavonoids: luteolin, eriodictyol, thymonin and glycosides: luteolin-7-O-glucoside, luteolin-3'-O-glucuronide, eriodictyol-7-O-glucoside. Ursolic acid and oleanolic acid	Ismaili et al. 2002
Thymus capitatus (L.) Hoffm.	Lamiaceae	Zaatar	Skin diseases, respiratory problems, digestive diseases, sterility in women	Leaves		
Traganum nudatum Del.	Chenopodiaceae	Dhemran	Rheumatism, skin diseases, digestive diseases	Aerial part		
Trigonella foenum-graecum L.	Fabaceae	Helba	Cancer, fever, digestive diseases, respiratory problems, eye diseases	Seeds		
Urginea maritima (L.) Bak	Liliaceae	Onsol	Ear diseases, veterinary medicine		- Steroid glycosides,glucoscillaren A, proscillaridin A, scillaridin A, scillicyanoside, scilliglucoside, scilliphaeoside, glucoscilliphaeoside	Kopp et al. 1996, Krenn et al. 1991
Urtica pilulifera L.	Urticaceae	Horriqua	Cold, kidney diseases	Whole plant		
Vaccinum myrtillus L.	Ericaceae	Adherra	Cardiovascular diseases, digestive diseases			
Verbascum sinuatum L.	Scrofulariaceae	Saleh landhar	Eye diseases	Flowers		
Verbena officinals L.	Verbenaceae	Tronjia	Respiratory problems, digestive diseases, fever, tranquillizer	Leaves		
Vicia faba L.	Fabaceae	Foul	Digestive diseases, skin diseases	Fruits		

Table 12.1. contd...

Table 12.1. contd....

Scientific name	Family	Local name	Main traditional uses in Tunisia	Plant part used	Constituents	References
Vitis vinifera L.	Ampelidaceae	Anba	Antiseptic, constipation, eye diseases Gynaecological diseases	Fruits, leaves		
Zea mays L.	Poaceae	Ktania	Constipation	Fruits		
Ziziphus lotus (L.) Desf.	Rhamnaceae	Sedra	Respiratory problems, eye diseases, digestive diseases, hair tonic	Fruits, leaves, roots	- Cyclopeptide alkaloids, flavonoids, dammarane saponins	Ghedira et al. 1995, Renault et al. 1997, Maciuk et al. 2004, Borgi et al. 2007a, 2007b, 2008, Borgi and Chouchane 2009
Zygophyllum cornutum Coss.	Zygophyllaceae	Bougriba	Diabetes, hypertension	Leaves, roots		

important for the people of the region, especially in rural areas. Even in many urban areas, the prices of modern medicines have increased, and people are turning back to traditional plant remedies. Unfortunately, this traditional system of plant use is not very popular with the younger generation. In view of this, it is necessary to conserve natural resources. Many scientific studies have been conducted in different institutions.

Medicinal plants and uses

In Tunisian traditional medicine, usually parts of the plant or complete plants are used. These medicines initially took the form of crude drugs such as tinctures, teas, poultices, powders, or could be processed by distillation to obtain essential oils and other herbal formulations. The plant parts used for these medical preparations were flowers, roots, leaves, seeds, aerial part and whole plants. Leaves were found to be the most frequently used parts of the medicinal plants studied in this chapter, although they belonged to different botanical families (65 families). Leaves are more favoured in preparation of herbal medicines and are the most widely used plant parts accounting for 80 species in a total of 205. Moreover, most of the documented plants were involved in the cure of more than one disease; however, information about their manipulation is necessarily synthetic without specifications about doses or the time required for the cure. Usually, the Tunisian people do not store remedies for a prolonged period of time. When the need arises, they go out and collect the plant and prepare the remedy. Most of the remedies were reported as being prepared from a single species. Moreover, most of the medicinal plants were widely used in the treatment of gastrointestinal disorders, dermatological diseases and respiratory diseases.

With increased demands for the resources available, a number of important plant species have become scarce in areas where they were previously abundant. Unfortunately, if their collection and use is not regulated, some species may be threatened with extinction. The following species represent a high priority for production and conservation: *Allium roseum* L., *Artemisia herba-alba* Asso., *Artemisia campestris* L., *Capparis spinosa* L., *Globularia alypum* L., *Juniperus phoenicea* L., *Laurus nobilis* L., *Matricaria chamomilla*, *Myrtus communis* L., *Periploca laevigata* Ait., *Rosmarinus officinalis* L., *Ruta chalepensis* L. and *Thymus capitatus* (L.) Hoffm. (Fig. 12.2), (Neffeti 1994, Neffeti and Ouled Belgacem. 2006).

In view of the importance of traditional medicine as a valuable source for the development of new drugs providing health services to the population, increased studies concerning natural plant resources were established. It is necessary to document the medicinal utility of lesser known plants available in remote areas of the country. Consequently, many chemical and pharmacological studies were carried out in order to valorise these medicinal plants (Bouraoui et al. 1995, El Bahri et al. 1999, Elbetieha et al. 2000, Hayder et al. 2004, 2008, Saddoud et al. 2007, Bahloul et al. 2008, Bouderbula et al. 2008, Boulila et al. 2008, Chograni et al. 2008, Ghaly et al. 2008, Haouala et al. 2008, Marzouk et al. 2009, Matsuyama et al. 2009, Sassi 2009). As a result, the most studied species from these plants were *Capparis spinosa* L. (Al-Said et al. 1988, Ali-Shtayeh and Abu Ghdeib 1999, Inocencio et al. 2000, Sharaf et al. 2000, Calis et al. 2002, Germano et al. 2002), *Centaurium pulchellum* (Khafagy and Mnajed 1970, Ahmed et al. 1971, Khafagy and Metwally 1971, Kostens and Willuhn 1973, Zirvi et al. 1977), *Ferula communis* L. (Lamnaouer et al. 1987, 1989, 1991a, 1991b, Valle et al. 1987, Appendinno et al. 1988), *Zizyphus lotus* (Ghedira et al. 1995, Renault et al. 1997, Maciuk et al. 2004, Borgi et al. 2007a, 2007b,

Fig. 12.2. List of medicinal plants which are threatened and in high demand in Tunisia.
Color image of this figure appears in the color plate section at the end of the book.

2008, Borgi and Chouchane 2009) and *Coriandrum sativum* L. (Msaada et al. 2007a, 2007b, 2009a, 2009b, 2009c, Neffati and Marzouk 2008).

For the promotion, organization and establishment of a national strategy for medicinal plants, a research development project was implemented by research organizations, plant and food science institutions, pharmacists and other health professionals from Tunisian Universities during the period 2002–2004 in order to preserve biodiversity, promote the use of medicinal plants and to create business opportunities for Tunisians (Neffeti and Ouled Belgacem 2006).

In conclusion, the present study of Tunisian plants reveals a total of 205 species from which a small number was scientifically studied. There is an urgent need not only to protect these species, but also to revive and reinvent such traditional practices of nature conservation. The development on the use of medicinal plants can be only be assured through a multi-disciplinary approach implying different operators since it concerns actors from several departments: Health, agriculture, research, and industry. Therefore, this knowledge is guaranteed in the generation of research focussing on screening programmes dealing with the isolation of bioactive principles and the development of new drugs.

References

Abdelwahed, A., I. Bouhlel, I. Skandrani, K. Valenti, M. Kadri, P. Guiraud, R. Steiman, A.M. Mariotte, K. Ghedira, F. Laporte, M.G Dijoux-Franca, and L. Chekir-Ghedira. 2007. Study of antimutagenic and antioxidant activities of Gallic acid and 1, 2, 3, 4, 6-pentagalloylglucose from *Pistacia lentiscus*: Confirmation by microarray expression profiling. Chem. Biol. Interact. 165: 1–13.

Ahmed, Z.F., F.M. Hammouda, A.M. Rizk, and S.I. Ismail. 1971. Phytochemical studies of certain *Centaurea* species Lipid. Planta Med. 19: 264–9.

Al-Easa, H.S., A.M. Rizk, and H.A. Hussiney. 1995. Phytochemical Investigation of *Rumex vesicarius*. Int. J. Chem. 6: 21–25.

Ali-Shtayeh, M.S. and S.I. Abu Ghdeib 1999. Antifungal activity of plant extracts against dermatophytes. Mycoses 42: 665–672.

Al-Said, MS., E.A. Abdelsattar, S.I. Khalifa, and F.S. El-Feraly. 1988. Isolation and identification of an anti-inflammatory principle from *Capparis spinosa*. Pharmazie 43: 640–641.

Amer, M.A., A.M. Dawidar, and M.B. Fayez. 1974. Constituents of local plants. The triterpenoid constituents of *Cornulaca monacantha*. Planta Med. XVII. 289.

Appendinno, G., S. Tagliapietra, G.M. Nano, and V. Picci. 1988. Ferprenin a prenylated coumain from *Ferula communis*. Phytochemistry 27: 944–946.

Bahloul, N., N. Boudhrioua, and N. Kechaou. 2008. Moisture desorption-adsorption isotherms and isosteric heats of sorption of Tunisian olive leaves (*Olea europaea* L.). Ind. Crop Prod. 28: 162–176.

Bahri-Sahloul, R., S. Ammar, R.B. Fredj, S. Saguem, S. Grec, F. Trotin, and FH. Skhiri. 2009. Polyphenol contents and antioxidant activities of extracts from flowers of two *Crataegus azarolus* L. varieties. Pak. J. Biol. Sci. 1: 660–668.

Baraket, G., O. Saddoud, K. Chatti, M. Mars, M. Marrakchi, M. Trifi, and A. Salhi-Hannachi. 2009. Sequence analysis of the internal transcribed spacers (ITSs) region of the nuclear ribosomal DNA (nrDNA) in fig cultivars (*Ficus carica* L.). Sci. Hortic. 120: 34–40.

Barrero, A.F., J.F. Sanchez, J.E. Oltra, J. Altarejos, N. Ferrol, and A. Barragan. 1991. Oxygenated sesquiterpenes from the wood of *Juniperus oxycedrus*. Phytochemistry 30: 1551–1554.

Barrero, A.F., J.E. Oltra, J. Altarejos, A. Barragan, A. Lara, and R. Laurent. 1993. Minor components in the essential oil of *Juniperus oxycedrus* L. wood. Flavour. Frag. J. 8: 185–189.

Barrero, A.F., J. Molina, J.E. Oltra, J. Altarejos, A. Barragan, A. Lara, and M. Segura. 1995. Stereochemistry of 14-hydroxy-beta-caryophyllene and related compounds. Tetrahedron 51: 3813–3822.

Begum, S., S. Razika, and S.B. Siddiqui. 1998. Triterpenoides from the leaves of *Nerium oleander*. Phytochemistry 44: 329–332.

Bellakhdar, J. 1997. La pharmacopée marocaine traditionnelle. Médecine arabe ancienne et savoirs populaires. IBIS Press.

Ben Salah, H, R. Jarraya, M.T. Martin, N.C.Veitch, R.J. Grayer, M.S. Simmonds , and M. Damak. 2002. Flavonol triglycosides from the leaves of Hammada scparia (POMEL) ILJIN. Chem. Pharm. Bull. (Tokyo). 50: 1268–1270.

Bezanger-Beauquesne L. and M. Pinkas. 2000. Plantes médicinales des régions tempérées. Ed. Maloine.

Bhouri, W., S. Derbel, I. Skandrani, J. Boubaker, I. Bouhlel, M.B. Sghaier, S. Kilani, A.M. Mariotte, M.G. Dijoux-Franca, K. Ghedira, and L. Chekir-Ghedira. 2009. Study of genotoxic, antigenotoxic and antioxidant activities of the digallic acid isolated from *Pistacia lentiscus* fruits. Toxicol *in Vitro*. In Press, Uncorrected Proof, Available online.

Bonnier, G. 1990. La grande flore en couleurs: France, Belgique, Suisse et pays voisins. Ed.Belin, Paris.

Borgi, W. and N. Chouchane. 2009. Antispasmodic effects of *Zizyphus lotus* (L.) Desf. extracts on isolated rat duodenum. J. Ethnopharmacol. 126: 571–573

Borgi, W., A. Bouraoui, and N. Chouchane. 2007a. Antiulcerogenic activity of *Zizyphus lotus* (L.) extracts. J Ethnopharmacol. 112: 228–231.

Borgi, W., K. Ghedira and N. Chouchane. 2007b. Antiinflammatory and analgesic activities of *Zizyphus lotus* root barks. Fitoterapia. 78: 16–19.

Borgi, W., M.C. Recio, J.L. Ríos and N. Chouchane. 2008. Anti-inflammatory and analgesic activities of flavonoid and saponin fractions from *Zizyphus lotus* (L.) Lam. S Afr J. Bot. 74: 320–324.

Bouderbala, S., M. Lamri-Senhadji, J. Prost, M.A. Lacaille-Dubois, and M. Bouchenak. 2008. Changes in antioxidant defence status in hypercholesterolemic rats treated with *Ajuga iva*. Phytomedicine 15: 453–461.

Boukef, K. 1986. Les plantes dans la médecine traditionnelle tunisienne. Agence de Coopération Culturelle et Technique. ISBN: 92-9028-085-9.

Boulila, A., A. Béjaoui, C. Messaoud, and M. Boussaid. 2008. Variation of volatiles in Tunisian populations of *Teucrium polium* L. (Lamiaceae). Chem. Biodivers. 5: 1389–1400.

Bouraoui, A., J.L. Brazier, H. Zouaghi, and M. Rousseau. 1995. Theophylline pharmacokinetics and metabolism in rabbits following single and repeated administration of Capsicum fruit.. Eur. J. Drug. Metab. Pharmacokinet 20: 173–178.

Brooker, R.M., J.N. Eble, and N.A. Starkovsky. 1967. Calepensin, chalepin and Calepin acetate, three novel furocoumarins from *Ruta chalepensis*. Loyda 30: 73–77.

Bruneton J. 1993. Pharmacognosie : phytochimie, Plantes médicinales. Tech. Doc. 2e éd. Paris, France.

Bruneton J. 1999. Pharmacognosie : Phytochimie, Plantes Médicinales. Tech. Doc. 3ème éd. Paris, France.

Buchbauer, G., L. Jirovetz, and W. Jager. 1991. Aromatherapy: Evidence for sedative effects of the essential oil of lavender after inhalation. Z. Naturforsch 46: 1067–1072.

Calis I, A. Kuruuzum-Uz, P.A. Lorenzetto, and P. Ruedi. 2002. (6S)-Hydroxy-3-oxo-alpha-ionol glucosides from *Capparis spinosa* fruits. Phytochemistry 59: 451–7.

Chaieb, M. and Boukhris, M. 1998. Flore succincte et illustrée des zones arides et sahariennes de Tunisie l'Or du Temps, Tunis.

Cham, B.E. and H. M. Meares. 1987. Glycoalcaloides from *Solanum sodomeum* are effective in the treatement of skin cancers in man. Cancer. Lett. 36: 111–118.

Chograni, H., C. Messaoud, and M. Boussaid. 2008. Genetic diversity and population structure in Tunisian *Lavandula stoechas* L. and *Lavandula multifida* L. (Lamiaceae). Biochem. Syst. Ecol. 36: 349–359.

Cuenod, A. 1954. Flore analytique et synoptique de la Tunisie (Cryptogames vasculaires-Gymnospermes et Monocotylédones). Office de l'expérimentation et de la vulgarisation agricole en Tunisie. Imprimerie Sefan, Tunis.

Dawidar, A.M., M.A. Metwally, M. Abou-Elzahab, and M. Abdel-Mogib. 1990. Sesquiterpene lactones from *Echinops spinosissimus*. Pharmazie 45: 70–74.

El Akrem, H., M. Abedrabba, M. Bouix and M. Hamdi. 2007. The effects of solvents and extraction method on the phenolic contents and biological activities *in vitro* of Tunisian *Quercus coccifera* L. and *Juniperus phoenicea* L. fruit extracts. Food Chem. 105: 1126–1134.

El Bahri, L., M. Djegham, and H. Bellil. 1999. *Retama raetam* W: a poisonous plant of North Africa. Vet. Hum. Toxicol. 41: 33–35.

Elbetieha, A., S.A. Oran, A. Alkofahi, H. Darmani, and A.M. Raies. 2000. Fetotoxic potentials of *Globularia arabica* and *Globularia alypum* (Globulariaceae) in rats. J. Ethnopharmacol. 72: 215–219.

El Hassany, B., F. El Hanbali, M. Akssira, F. Mellouki, A. Haidour, and AF. Barrero. 2004. Germacranolides from *Anvillea radiata*. Fitoterapia. 75: 573–576.

El-Mousallamy, A.M.D., H.H. Barakat, A.M.A. Souleman, and S. Awadallah. 1991. Polyphenols of *Acacia raddiana*. Phytochemistry 30: 11.

El Sayed, K., M.S. Al-Said, F.S. El-Feraly, and S.A. Ross. 2000. New quinoline alkaloids from *Ruta chalepensis*. J. Nat. Prod. 63: 995–997.

El-Shazly, A., A. Ateya, L. Witte, and M. Wink. 1996. Quinolizidine alkaloid profiles of *Retama raetam*, *R. sphaerocarpa* and *R. monosperma*. Z. Naturforsch 51: 301–308.

Fahey J.W., A.T. Zacman, and P. Talay. 2001. The chemical diversity and distribution of glucosinolates and isothiocynates among plants. Phytochemistry 56: 5–51.

Gadgoli, C, and S.H. Mishra. 1999. Antihepatotoxic activity of p-methoxy benzoic acid from *Capparis spinosa*. J. Ethnopharmacol. 66: 187–92.

Germano, M.P., R. De Pasquale, V. D'Angelo, S. Catania, V. Silvari, and C. Costa. 2002. Evaluation of extracts and isolated fraction from *Capparis spinosa* L. buds as an antioxidant source. J. Agric. Food Chem. 50: 1168–71.

Ghaly, S.I., A. Said, and M.A. Abdel-Wahhab. 2008. *Zizyphus jujuba* and *Origanum majorana* extracts protect against hydroquinone-induced clastogenicity. Environ Toxicol Pharmacol. 25:10–19.

Ghedira, K., R. Chemli, C. Caron, J.M. Nuzilard, M. Zeches, and L. Le Men-Olivier. 1995. Four cyclopeptide alkaloids from *Zizyphus lotus*. Phytochemistry 38: 767–772.

Goren, A., G. Topcu, G. Bilsel, M. Bilsel, Z. Aydogmus, and J.M. Pezzuto. 2002. The chemical constituents and biological activity of essential oil of *Lavandula stoechas* ssp. stoechas. Z Naturforsch 57: 797–800.

Ham, A., A. Hole and D. Willett. 2004. Lonely Planet. 3rd ed. Tunisia.

Hamada, H., B. Mohamed, G. Massiot, C. Long, and C. Lavaud. 2004. Alkylated isocoumarins from *Pituranthos scoparius*. Nat. Prod. Res. 18: 409–413.

Hamrouni-Sellami, I., H.B. Salah, M.E. Kchouk, and B. Marzouk. 2007. Variations in phytosterol composition during the ripening of Tunisian safflower (*Carthamus tinctorius* L.) seeds. Pak. J. Biol. Sci. 10: 3829–34.

Hamrouni Sellami, I., E. Maamouri, T. Chahed, W. Aidi Wannes, M.E. Kchouk, and B. Marzouk. 2009. Effect of growth stage on the content and composition of the essential oil and phenolic fraction of sweet marjoram (*Origanum majorana* L.). Ind Crop Prod, In Press, Corrected Proof, Available online.

Haouala, S., A. Elayeb, R. Khanfier, and N. Boughanmi. 2008. Aqueous and organic extracts of *Trigonella foenum-graecum* L. inhibit the mycelia growth of fungi. J. Environ. Sci. 20: 1453–1457.

Harborne, J.B. and C.A. Williams. 2001. The identification of orcinol in higher plants in the family ericaceae. Phytochemistry 8: 2223–2222.

Harborne, J.B., S. Norio and C.H. Detoni. 2003. Anthocyanins of Cephais, Cynomorium, Euterpe, Lavatera and Pinanga. Biochem. Syst. Ecol. 22: 835–836.

Hayder, N., A. Abdelwahed, S. Kilani, R.B. Ammar, A. Mahmoud, K. Ghedira, and L. Chekir-Ghedira L. 2004. Anti-genotoxic and free-radical scavenging activities of extracts from (Tunisian) *Myrtus communis*. Mutat. Res. 564: 89–95.

Hayder, N., I. Skandrani, S. Kilani, I. Bouhlel, A. Abdelwahed, R. Ben Ammar, A. Mahmoud, K. Ghedira, and L. Chekir-Ghedira. 2008. Antimutagenic activity of *Myrtus communis* L. using the *Salmonella* microsome assay. S. Afr. J. Bot. 74: 121–125.

Inocencio, C., D. Rivera, F. Alcaraz, and F.A. Tomás-Barberán. 2000. Flavonoid content of commercial capers (*Capparis spinosa*, *C. sicula* and *C. orientalis*) produced in mediterranean countries. Eur. Food. Res. Technol. 212. 70 74

Ismaili, H., S. Sosa, D. Brkic, S. Fkih-Tetouani, A. Il Idrissi, D. Touati, R.P. Aquino, and A. Tubaro. 2002. Topical anti-inflammatory activity of extracts and compounds from *Thymus broussonetii*. J. Pharm. Pharmacol. 54: 1137–1140.

Issaoui, A., A. Kallala, M. Neffati, and N. Akrimi. 1996. Plantes naturelles du Sud Tunisien. Institut des régions arides. Imprimerie Officielle de la République.

Kamel, M.S., K.Othani, H.A. Hassanean, A.A Khalifa, R Kasai, and K.K. Yamaskaki. 2000. Triterpenoidal saponins from *Cornulaca monacantha*. Pharmazie 55: 460–462.

Kandil, F.E. and H.A. Husseiny. 1998. A new flavonoid from *Cornulaca monacantha*. Oriental J. Chem. 14: 215–227.

Kandil, F.E. and M.H. Grace. 2001. Polyphenols from *Cornulaca monacantha*. Phytochemistry 58: 611–613.

Karray-Bouraoui, N., M. Rabhi, M. Neffati, B. Baldan, A. Ranieri, B. Marzouk, M. Lachaâl, and A. Smaoui. 2009. Salt effect on yield and composition of shoot essential oil and trichome morphology and density on leaves of *Mentha pulegium*. Ind. Crops Prod. 30: 338–343.

Khafagy, S.M. and H. Mnajed. 1970. Phytochemical investigation of *Centaurium pulchellum* (Sw.) Druce. Acta Pharm Suec. 7: 667–672.

Khafagy, S.M. and A.M. Metwally. 1971. Phytochemical investigation of *Xanthium occidenta*. Planta Med. 19: 234–240.

Kopp, B., L. Krenn, M. Draxler, A. Hoyer, R. Terkola, P. Vollasterand, and W. Robien. 1996. Bufadiernolides from *Urginea maritime*. J. Nat. Prod. 59: 612–613.

Kostens, J. and G. Willuhn. 1973. Sterol glycosides and fatty acid esters of sterol glycosides in the leaves of *Solanum dulcamara*. Planta Med. 24: 278–85.

Krenn, L., R. Ferth, W. Robien, and B. Kopp. 1991. Bufadienolides from bulbs of *Urginea maritima* sensu strictu. Planta Med. 57: 560–565.

Lamnaouer, D., B. Bodo, M.T. Martin, and D. Molho. 1987. Ferulenol and omega-hydroxyferulenol, toxic coumarins from *Ferula communis* var.genuina. Phytochemistry 26: 1613–1615.

Lamnaouer, D., M.T. Martin, D. Molho, and B. Bodo. 1989. Isolation of daucane esters from *Ferula communis* var. *brevifolia*. Phytochemistry 28: 2711–2716.

Lamnaouer, D., O. Fraigui, M.T. Martin, and B. Bodo. 1991a. Structure of ferulenol derivatives from *Ferula communis* var. *genuina*. Phytochemistry 30: 2383–2386.

Lamnaouer, D., O. Fragui, M.T. Martin, B. Bodo, and D. Molho. 1991b. Structure of isoferprenin, a 4- hydroxycoumarin derivative from *Ferula communis* var. *genuina*. J. Nat. Prod. 54: 576–578.

Lapidus, I.M. 2002. A History of Islamic Societies. 2nd ed. Cambridge University Press, Cambridge (UK).

Laribi, B., I. Bettaieb, K. Kouki, A. Sahli, A. Mougou, and B. Marzouk. 2009. Water deficit effects on caraway (*Carum carvi* L.) growth, essential oil and fatty acid composition. Ind Crop Prod. In Press, Corrected Proof, Available online.

Larsen, E., A. Kharazami, LP. Christensen, and SB. Christensen. 2003. An antiinflammatory galac tolipid from rose hip (*Rosa canina*) that inhibits chemotaxis of human peripherbal blood neutrophils *in vitro*. J. Nat. Prod. 66: 994–995.

Le Floc'h, E. 1983. Contribution à une étude ethnobotanique de la flore tunisienne. Publications Scientifiques Tunisiennes, Programme "Flore et Végétation Tunisiennes". Imprimerie Officielle de la République Tunisienne.

Le Houerou, H. N. 1995. Bioclimatologie et biogéographie des steppes arides du Nord de l'Afrique. Diversité biologique, développement durable et désertification. Options méditerranéennes. Série B: Etudes et recherches. Numéro 10.

Lemordant, D., K. Boukef, and M. Ben Salem. 1977. Plantes utiles et toxiques de Tunisie. Fitoterapia 5: 191–214.

Leporatti, M.L. and Ghedira, K. 2009. Comparative analysis of medicinal plants used in traditional medicine in Italy and Tunisia. J. Ethnobiol. Ethnomed. 5: 31.

Maciuk, A., C. Lavaud, P. Thépenier, M.J. Jacquier, K. Ghédira, and M. Zèches-Hanrot. 2004. Four new dammarane saponins from *Zizyphus lotus*. J. Nat. Prod. 67: 1639–1643.

Maire, R. 1980. Flore de l'Afrique du Nord. Vol. XV. Dicotylédones, Rosales: Saxifragaceae, Pittosporaceae, Plantanaceae, Rosaceae. Edition Lechevalier SARL. Paris.

Marzouk, B., Z. Marzouk, R. Décor, H. Edziri, E. Haloui, N. Fenina, and M. Aouni. 2009. Antibacterial and anticandidal screening of Tunisian *Citrullus colocynthis* Schrad. from Medenine. J. Ethnopharmacol. 125: 344–349.

Matsuyama, K., M.O. Villareal, A. El Omri, J. Han, M.E. Kchouk, and H. Isoda. 2009. Effect of Tunisian *Capparis spinosa* L. extract on melanogenesis in B16 murine melanoma cells. Nat. Med. (Tokyo). 63: 468–72.

Morazzoni, P. and E. Bombardelli. 1994. *Silybum marianum* (*Carduus marianus*), Fitotrapia LXVI: 33–42. Msaada, K., K. Hosni, M.B. Taarit, T. Chahed, and B. Marzouk. 2007a. Variations in the essential oil composition from different parts of *Coriandrum sativum* L. cultivated in Tunisia. Ital. J. Biochem. 56: 47–52.

Msaada, K., K.Hosni, M. Ben Taarit, T. Chahed, M.E. Kchouk, and B. Marzouk. 2007 b. Changes on essential oil composition of coriander (*Coriandrum sativum* L.) fruits during three stages of maturity. Food Chem. 102: 1131–1134.

Msaada, K., K. Hosni, M. Ben Taarit, M. Hammami, and B. Marzouk. 2009a. Effects of growing region and maturity stages on oil yield and fatty acid composition of coriander (*Coriandrum sativum* L.) fruit. Sci. Hortic. 120: 525–531.

Msaada, K., K. Hosni, M. Ben Taarit, T. Chahed, M. Hammami, and B. Marzouk. 2009b. Changes in fatty acid composition of coriander (*Coriandrum sativum* L.) fruit during maturation. Ind Crop Prod. 9: 269–274.

Msaada, K., M. Ben Taarit, K. Hosni, M. Hammami, and B. Marzouk. 2009c. Regional and maturational effects on essential oils yields and composition of coriander (*Coriandrum sativum* L.) fruits. Sci. Hortic. 122: 116–124.

Nagendra, Kr. S. 2000. International encyclopaedia of Islamic dynasties, vol. 4: A Continuing Series. 4: A Continuing Series. Anmol Publications PVT. LTD. ISBN 8126104031.

Najjaa, H., M. Neffati, S. Zouari, and E. Ammar. 2007. Essential oil composition and antibacterial activity of different extracts of *Allium roseum* L., a North African endemic species. C.R. Chim. 10: 820–826.

Neffati, M. 1994. Caractérisation morpho-biologique de certaines espèces végétales nord africaines. Implications pour l'amélioration pastorale. University of Ghent, dissertation.

Neffeti, M. and A. Ouled Belgacem. 2006. A multidisciplinary study of herbal, medicinal and aromatic plants in Southern Tunisia: a new approach. Regional Consultation on Linking Producers to Markets: Lessons Learned and Successful Practices. Cairo, Egypt.

Neffati, M. and B. Marzouk. 2008. Changes in essential oil and fatty acid composition in coriander (*Coriandrum sativum* L.) leaves under saline conditions. Ind. Crop Prod. 28: 137–142.

Pottier-Alapetite, G. 1979. Flore de la Tunisie. Angiospermes, dicotylédones, Apétales, dialypétales. Imprimerie Officielle de la République.

Pottier-Alapetite, G. 1981. Flore de la Tunisie. Angiospermes, dicotylédones, gamétopétales. Imprimerie Officielle de la République.

Renault, J.H., K. Ghedira, P. Thepenier, C. Lavaud, M. Zeches-Hanrot, and L. Le Men-Olivier. 1997. Dammarane saponins from *Zizyphus lotus*. Phytochemistry 44: 1321–1327.

Saddoud, O., K. Chatti, A. Salhi-Hannachi, M. Mars, A. Rhouma, M. Marrakchi, and M. Trifi. 2007. Genetic diversity of Tunisian figs (*Ficus carica* L.) as revealed by nuclear microsatellites. Hereditas. 144: 149–157.

Salido, S., J. Altarejos, M. Nogueras, A. Sanchez, C. Pannecouque, M. Witvrouw, and E. De Clercq. 2002. Chemical studies of essential oils of *Juniperus oxycedrus* ssp. badia. J. Ethnopharmacol. 81: 129–134.

Sassi, A., H. Khaled, M. Wafa, B. Sofiane, D. Mohamed, M. Jean-Claude, and F. Abdelfattah. 2009. Oral administration of *Eucalyptus globulus* extract redues the alloxan-induced oxidative stress in rats. Chem. Biol. Interact. 181: 71–76.

Sayed, M.D., S.I. Balbaa, and M.S. Afifi. 1973. The lipid content of the seeds of *Citrullus colocynthis*. Planta Med. 24: 41–45.

Sayed, M.D., S.I. Balbaa, and M.S. Afifi. 1974. The glycosidal content of the different organs of *Citrullus colocynthis*, Planta medica, 26: 293–298.

Sharaf, M., M.A. El-Ansari, and N.A. Saleh. 2000. Quercetin triglycoside from *Capparis spinosa*. Fitoterapia 71: 46–9.

Soliman, M.A., A.A. El Sawy, H.M. Fadel, F. Osman, and A.M. Gad. 1985. Volatile components of roasted *Citrullus colocynthis* var. *colocynthoides*. Agr. Biol. Chem. Tokyo 49: 269–275.

Stearns, P.N. and W. Leonard Langer. 2001. The Encyclopedia of World History: Ancient, Medieval, and Modern, Chronologically Arranged. 6th ed. Houghton Mifflin Harcourt, Boston, USA.

Tabazik-Wlotzka, C., J.L. Imbert, and P. Pistre. 1967. *Pistacia lentiscus, Pistacia terebintus*, étude comparative des composants de l'extrait éthéro-pétrolique. C.R. Acad. Sc. Paris. 708–710.

Tessier, A. 1994. Phytothérapie Analytique. Phytochimie et Pharmacologie. Ed. Marc Aurèle.

Topcu, G., M.N. Ayral, A. Aydin, A.C. Goren, H.B. Chai, and J.M. Pezzuto. 2001. Triterpenoids of the roots of *Lavandula stoechas* ssp. stoechas. Pharmazie 56: 892–895.

Ulubelen, A. and Y. Olcay. 1989. Triterpenoids from *Lavandula stoechas* subsp. *stoechas*. Fitoterapia, 60: 475–476

Ulubelen, A., N. Goren, and Y. Olcay. 1988. Longipinene derivatives from *Lavandula stoechas* subsp. stoechas. Phytochemistry 27: 3966–3967.

Ulubelen, A., H. Guner, and M. Cetindag. 1988. Alkaloids and coumarins from the roots of *Ruta chalepensis* var. *latifolia*. Planta. Med. 54: 551–2.

Ulubelin, A., L. Ertugrul, H. Birma, R. Yigit, G. Erseven, and V. Olgac. 1994. Antifertility effects of some coumarins isolated from *Ruta chalepensis* and *Ruta chalepensis* var. latifolia in rodents. Phyto. Res. 8: 233–236.

Valle, M.G., G. Appendino, G.M. Nano, and V. Picci. 1987. Prenylated coumarins and sesquiterpenoids from *Ferula communis*. Phytochemistry 1: 253–256.

Valnet, J. 2001. Phytothérapie. Se soigner par les plantes. Maloine S. A.

Van Hellemont J. 1986. Compendium de phytothérapie.Ed. ABP.

Verite, P., A. Nacer, Z. Kabouche, and E. Seguin. 2004. Composition of seeds and stems essential oil of *Pituranthos scoparius* (Coss.&Dur.) Schinz, Flavour. Frag. J. 19: 562–564.

Wannes, W.A., B. Mhamdi, and B. Marzouk. 2009. Variations in essential oil and fatty acid composition during *Myrtus communis* var. *italica* fruit maturation. Food Chem. 112: 621–626.

Wessner, M., B. Champion, J.P. Girault, N. Kaouadji, B. Saidi, and R. Lafont. 1992. Ecdysteroids from *Ajuga iva*. Phytochemistry-Oxford 31: 3785–3788.

Yaniv, Z., Y. Elber, M. Zur, and D. Schafferman. 1991. Differences in fatty acids composition of oils of wild Cruciferae seeds. Phytochemistry 30: 841–843.

Zaouali, Y. and M. Boussaid. 2008. Isozyme markers and volatiles in Tunisian *Rosmarinus officinalis* L. (Lamiaceae): A comparative analysis of population structure. Biochem. Syst. Ecol. 36: 11–21.

Zirvi, K.A., K. Jewers, and M.J. Nagler. 1977. Phytochemical investigation of *Prosopis glandulosa* stem. Planta Med. 32: 244–246.

Chapter 13

Medicinal Plants used in Folk Medicine for Digestive Diseases in Central Spain

M.E. Carretero Accame, M.P. Gómez-Serranillos Cuadrado, M.T. Ortega Hernández-Agero and *O.M. Palomino Ruiz-Poveda*

Introduction

Ethnopharmacology and ethnobotany have been used as tools for the discovery of new active principles and as a way to assure conservation, and also against a threat to traditional cultures (Al-Qura'n 2009). The development of modern medicine involves the healing of several diseases, but at the same time also the disappearance of the traditional knowledge of medicinal plants that were usually orally transmitted (Heinrich et al. 2006). The Convention for Safeguarding Intangible Cultural Heritage held in 2003 (UNESCO 2005) stated that knowledge and practices concerning nature and the universe are part of our cultural heritage: ethnobotany, ethnobiology, ethnoecology, folk medical and pharmaceutical knowledge, among others, are recognized as inextricable components of culture and therefore should be protected and sustained (Pardo de Santayana et al. 2005, Guarrera 2005, Pieroni et al. 2006, Guarrera 2006).

The Iberian Peninsula shows a widely varied vegetation and species biodiversity and so does the centre of Spain. More than 2,000 species of vascular plants have been described in this area (Morales 2003), many of

Department of Pharmacology, School of Pharmacy, Universidad Complutense de Madrid. Pza Ramón y Cajal s/n, 28040 Madrid, Spain.
E-mail: *meca@farm.ucm.es*

them with a traditional use for the relief of several diseases. The studied area corresponds to a circular area within a 110 km radius within the city of Madrid as the geographic centre (Fig. 13.1). The geographic coordinates are comprised between 39° and 42° northern latitude and include the following autonomous communities: Madrid, North of Castilla-La Mancha (Toledo, Cuenca and Guadalajara provinces) and South of Castilla-León (Avila and Segovia provinces).

The region is not uniform from a topographical, geological or climatological point of view. The main relief units at the South and Southwest are the Toledo Mountains, a mountain range that separates Tajo and Guadiana rivers' basins. Western and northern limits are the mountain ranges of the Central System, the foothill of Gredos, Guadarrama and Somosierra mountains. The rest of the studied area corresponds to the central plateau, northern sub-plateau (700–800 m altitude) and southern sub-plateau (600–700 m altitude).

Lithologic characteristics are also variable: granite and gneiss in the western mountain range, slate and quartz in Somosierra, limestone, gypsum and margues along the Tajo river depression; sands, gypsum margues, gravel and clay in the countryside and finally, sand, gravel and slime along the rivers' fertile lowland.

The diversity of geographic characteristics, from high mountains to low depressions (Tajo river depression) determines a wide range of ecosystems

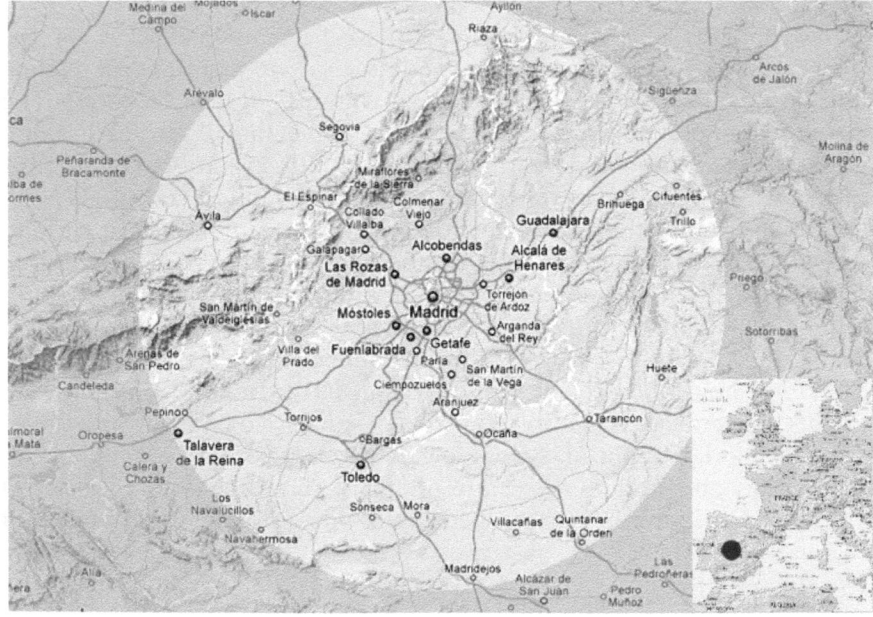

Fig. 13.1. Spanish area of the research.

and vegetation derived from the different climatic parameters. This area belongs to the Mediterranean biogeographic region, with a characteristic drought period coinciding with the higher temperatures in summer. Strong variations exist between areas in terms of rainfall and temperature: the medium year' rainfall value varies from 400 to 2000 mm whereas temperature range is between −8°C (minimum Winter value) and + 44 °C (maximum Summer value). These different environmental conditions have led to a very rich and varied flora in central Spain that is dominated by evergreen forests, especially holm oak (*Quercus ilex* subsp. *ballota* (Desf.) Samp.). However, central Spain has been inhabited since ancient times and the landscape has been greatly transformed for agricultural and stock raising activities (Tardío et al. 2005). Nowadays, it is a highly industrialized area where the original landscape is substituted by industries, housing and garden areas in which native species have been replaced by ornamental species from around the world. Nevertheless, several areas remain intact as natural reserves in the Guadarrama mountain range. The popular use of plants in this region results from the floral diversity and the way its inhabitants have exploited the available natural resources, the most frequently reported are those used for treating gastrointestinal ailments.

In this chapter, an overview of the medicinal plants that are used for digestive diseases in the centre of Spain is presented, in order to contribute to the preservation of traditional knowledge and uses of these species.

Methodology

The investigation was carried out by three different methods: firstly, books from ethnopharmacology, botany, phytochemistry and phytotherapy were reviewed (Leclerc 1983, Fauron and Moratti 1984, Van Hellemont 1986, Sagrera 1987, Bézanger-Beauquesne et al. 1989, Newall et al 1996, Mills and Bone 2000, Vanaclocha and Cañigueral 2003), secondly, several publications from ethnobotany and plant uses in the different autonomous communities were collected (Vázquez et al. 1997, Blanco Castro 1998, Fajardo et al. 2000, Rivera el al. 2005, Fajardo et al. 2007, Vallejo et al. 2009); finally, direct interviews were held with farmers, shepherds and housewives in the countryside. The informants were requested to collect specimens of the plants they knew by their vernacular name, or to show the plant species on site.

The identification and nomenclature of the listed plants were based on those published in Flora Iberica (Real Jardín Botánico CSIC). The following data are reported for each species in the botanical list: family, scientific name, vernacular name(s) (Spanish/ English), parts used, chemical composition, traditional digestive use and other information (toxicity).

Results and conclusions

The plants traditionally used in central Spain for digestive purposes are presented in alphabetical order of their family and botanical names, with the relevant information. Table 13.1 presents those species for which traditional use was obtained by the cited literature and, in some cases, by interviews with farmers, shepherds and housewives in the countryside. Table 13.2 includes those species that contain digestive activity although we have not got any information about their traditional use in central Spain; nevertheless, many of them are used in different areas from the Iberian Peninsula or even in the Mediterranean region for many gastrointestinal diseases.

As a result of this study, 78 genera have been recorded that are used in 14 digestive medicinal uses. The most common families are Lamiaceae and Asteraceae with 30 and 22 species, respectively. Among them, *Mentha* sp., *Thymus* sp., *Jasonia* sp. and *Matricaria* are greatly valued by consumers as being 'beneficial for all gastrointestinal diseases'.

Infusion is the most widespread form to obtain the medicinal preparation and the most common use is to improve the digestion (digestive).

People from the countryside collected different parts of the plant species to obtain the medicinal preparation, the aerial parts, leaves, fruits and seeds being the most frequently used. Roots and bulbs were also used in many of the remedies. Sometimes the vegetal was mixed with other ingredients, such as sugar, honey, alcohol or flour.

There exists a rich traditional knowledge of medicinal plants in Central Spain, especially for those species used for digestive diseases. More than a 100 species are known and used. Many of them were initially used for medicinal purposes, but now are used mainly as food, additives or beverages.

This chapter shows that in most cases, the traditional use is supported by the active principles present in the drug which pharmacological activity has been proved by *in vitro* and *in vivo* assays (Sánchez de Rojas et al. 1995, Gadhi et al. 2001, Duru et al. 2003, Ustün et al. 2006). Although small in number, the clinical trials carried out proved the therapeutic effectiveness of several species in digestive diseases; this occurs with tannins-rich plants in the treatment of diarrhoea: they bind to proteins and induce their precipitation with an astringent effect. Flavonoids-rich plants proved their antioxidant, anti inflammatory and anti ulcerogenic activities (Mota et al. 2009).

The spasmolytic and carminative effect of anetol isolated from the essential oils from *Pimpinella anisum* or *Foeniculum vulgare* has also been proved by scientific research; the results justify the use of these plants and other aromatic species from the Apiaceae or Lamiaceae families to treat colic pain associated with dyspepsia. Nevertheless, some cases were

Table 13.1. Wild medicinal plants traditionally used for digestive diseases in central Spain

FAMILY / Species	Local Spanish/English name	Part used	Chemical composition	Popular use	Other information
AQUIFOLIACEAE					
Ilex aquifolium L.	Acebo/ Holly	Leaves	Flavonoids, bitter principles, triterpenes, cyanogenic glycosides	*Infusion:* laxative, digestive	Toxicity
ASTERACEAE					
Achillea millefolium L.	Milenrama/ Yarrow	Flowered aerial parts, leaves	Essential oil, flavonoids, phenolic acids	*Infusion:* digestive, gastrointestinal disorders, liver failure	
Anthemis sp.	Manzanilla amarga/ Chamomile	Flower heads	Essential oil, flavonoids, phenolic acids	*Infusion:* digestive, carminative	
Artemisia absinthium L.	Ajenjo, asensio/ Wormwood	Flowered aerial parts, leaves	Essential oil, flavonoids, sesquiterpene lactones	*Infusion:* Digestive, tonic, carminative	Toxic at high doses
A. alba L.	Manzanilla, manzanilla de la sierra	Flowered aerial parts, leaves	Essential oil, flavonoids	*Infusion:* digestive, stomach ache	
Bidens aurea (Aiton) Sherff	Té, té mateo/Arizona beggarticks	Leaves	Essential oil, flavonoids	*Infusion:* digestive, stomach ache	
Centaurea ornata Willd.	Arzolla	Roots	Bitter principles	*Infusion:* cholagogue	
Chamaemelum nobile (L.) All.	Manzanilla/ Chamomile	Flower heads	Essential oil, flavonoids	*Infusion:* digestive, anti inflammatory	
Cichorium intybus L.	Achicoria/ Chicory	Roots, leaves	Sesquiterpene lactones, inulin, cumarines, phenolic acids	*Infusion:* digestive, aperitif, liver failure, soft laxative	

Table 13.1. contd...

Table 13.1. contd...

FAMILY / Species	Local Spanish/English name	Part used	Chemical composition	Popular use	Other information
Helichrysum stoechas (L.) Moench	Siempreviva, perpetua, manzanilla/ Everlasting	Flower heads, flowered aerial parts	Essential oil, flavonoids	*Infusion:* digestive	
Inula sp.	Arnica, tabaquera	Roots	Essential oil, inuline	*Decoction:* digestive depurative	
I. helenium L.	Énula, helenio/ Elecampane, elfdock	Roots, rhizomes	Essential oil, bitter principles, inuline	*Decoction:* tonic, digestive	
I. salicina L.	Té de prado/ Willowleaf, yellowhead	Not found	Not found	*Infusion:* digestive, antidiarrhoeic	
Jasonia glutinosa (L.) DC.	Hierba de té, té de Aragón, té de roca, té de piedra, té de peña, té de risco	Flowered aerial parts	Essential oil, flavonoids, phenolic acids, sesquiterpene lactones	*Infusion:* digestive, antidiarrhoeic, carminative	
J. tuberosa (L.) DC.	Té de burro, té de tierra, té de roca	Flowered aerial parts	Essential oil, flavonoids	*Infusion:* digestive, carminative	
Leuzea conifera (L.) DC.	Cuchara de pastor	Flower heads	Not found	*Infusion:* digestive. *Decoction:* gastric disturbance	
Matricaria recutita L.	Manzanilla dulce/ Chamomile	Flower heads	Essential oil, flavonoids, coumarins	*Infusion:* digestive, stomach ache, gastrointestinal disorders (colics, antiseptic)	
Santolina chamaecyparissus L.	Abrótano hembra, cagamilras, manzanilla amarga/ Lavender cotton	Flower heads	Essential oil, flavonoids	*Infusion:* stomach ache, digestive, coleretic. *Decoction:* laxative	

FAMILY / Species	Local Spanish/English name	Part used	Chemical composition	Popular use	Other information
Scolymus hispanicus L.	Cardillo/ Golden thistle, Spanish oyster thistle	Leaves, roots	Flavonoids, hydroxycinnamic derivatives, phenolic acids, triterpenes, essential oil	*Decoction:* digestive, coleretic, antidiarrhoeic	
Silybum marianum (L.) Gaertn.	Cardo María, cardo lechal/ Milk thistle	Fruits	Flavanolignanes	*Infusion:* aperitif, liver protective	
Solidago virgaurea L.	Vara de oro, té de Gredos/ Goldenrod	Aerial parts	Inulin, saponines	*Infusion:* digestive, antidiarrhoeic	
Tanacetum parthenium (L.) Schultz Bip.	Matricaria, manzanilla de huerta/ Feverfew	Flower heads	Essential oil, flavonoids, sesquiterpene lactones	*Infusion:* digestive, stomach ache	
Taraxacum obovatum (Wild) DC.	Pitones/ Dandelion	Leaves	Sesquiterpene lactones	Liver protective	
BORAGINACEAE					
Lithospermum officinale L.	Mijo del sol, té de acequia/ Grom well, European stoneseed	Fruits	Flavonoids, alkaloids	*Infusion:* digestive	Toxic when chronic use
CANNABACEAE					
Humulus lupulus L.	Lúpulo, espárrago de zarza/ Hop, common hop	Cones (inflorescences)	Oleoresin, flavonoids, amines	*Infusion:* aperitif	
CAPRIFOLIACEAE					
Sambucus nigra L.	Saúco/ Black elderberry	Flowers	Phenolic acids, flavonoids	*Infusion:* soft laxative	
CARYOPHYLLACEAE					
Paronychia argentea Lam.	Nevadilla, té de campo/ Algerian tea	Aerial parts	Saponins, flavonoids	*Infusion:* stomach ache, anti ulcerogenic, antidiarrhoeic	

Table 13.1. contd...

Table 13.1. contd...

FAMILY / Species	Local Spanish/English name	Part used	Chemical composition	Popular use	Other information
CHENOPODIACEAE					
Chenopodium ambrosioides L.	Pazote, té de España, de nueva España/ Wormseed	Aerial parts	Essential oil	*Infusion:* digestive, laxative	
CISTACEAE					
Cistus laurifolius L.	Jara, jara de hoja, jara de laurel	Leaves, flowers	Flavonoids, lignans, essential oil	*Infusion, decoction:* stomach ache, anti ulcerogenic	
CUPRESSACEAE					
Juniperus communis L.	Enebro/ Common juniper	Fruits	Terpenoids, tannins, oleoresin	*Whole fruit:* digestive	
DIOSCOREACEAE					
Tamus communis L.	Brionia negra, nueza negra	Roots	Mucilages, starch	*Decoction:* purgative	Highly toxic
GLOBULARIACEAE					
Globularia vulgaris L.	Cabezuelas/ Globe daisy	Leaves	Not found	*Decoction, infusion:* bleeding ulcer	
GRAMINEAE (POACEAE)					
Hordeum vulgare L.	Cebada/ Common barley, barley	Seeds	Starch, enzimes, minerals, alkaloids	*Decoction:* digestive, antidiarrhoeic. *Grinded seeds:* laxative.	
GROSSULARTIACEAE					
Ribes uva-crispa L.	Uva espina, Grosellero espinoso/ Gooseberry	Fruits	Organic acids, vitamins, glucides, minerals	*Whole fruit:* aperitif	

FAMILY / Species	Local Spanish/English name	Part used	Chemical composition	Popular use	Other information
IRIDACEAE					
Iris germanica L.	Lirio azul, lirio cárdeno, lirio común, lirio morado, lirio pascual, lirio barbado /Orris root, German iris	Rhizomes	Flavonoids (isoflavones), triterpenes, glucides	*Infusion, decoction:* anti ulcerogenic	
LAMIACEAE					
Ajuga reptans L.	Búgula/ Common bugle, blue bugle	Flowered aerial parts	Tannins, antocyanins	*Infusion:* antidiarrhoeic	
Hyssopus officinalis L.	Hisopo, guisopo/ Hyssop	Flowered aerial parts	Flavonoids, essential oil, triterpenes	*Infusion:* digestive depurative	
Lavandula latifolia Medik.	Espliego/ Broadleaved lavender	Aerial parts (inflorescence)	Essential oil, flavonoids	*Infusion:* gastrointestinal spasm, meteorism	
Melissa officinalis L.	Toronjil, citronela, hoja de limón/ Lemon balm	Leaves	Essential oil, triterpenes, phenolic acids, flavonoids	*Infusion:* digestive, antispasmodic	
Mentha aquatica L.	Hierbabuena, té de río/ Water mint	Flowered aerial parts	Essential oil, phenolic acids, flavonoids	*Infusion:* digestive	
M. arvensis L.	Menta, japonesa /Wild mint	Flowered aerial parts	Essential oil, flavonoids	*Infusion:* digestive anstiseptic	
M. cervina L.	Poleo, poleo amarillo, poleo cervino	Flowered aerial parts	Essential oil	*Infusion:* digestive	
M. latifolia L.	Té silvestre	Flowered aerial parts	Essential oil, flavonoids	*Infusion:* digestive	
M. longifolia (L.) Huds.	Hierbabuena silvestre/ Spearmint	Flowered aerial parts	Essential oil, flavonoids	*Infusion:* spasmolytic	

Table 13.1. contd...

Table 13.1. contd...

FAMILY / Species	Local Spanish/English name	Part used	Chemical composition	Popular use	Other information
M. pulegium L.	Poleo, té de poleo/ Pennyroyal	Flowered aerial parts	Essential oil, flavonoids	*Infusion:* digestive	
M. spicata L.	Hierbabuena/ Spearmint	Flowered aerial parts	Essential oil, flavonoids	*Infusion:* digestive	
M. suaveolens Ehrh.	Mastranto/ Applemint	Flowered aerial parts	Essential oil	*Infusion:* digestive	
Micromeria fruticosa (L.) Druce	Poleo, té de poleo, ajedrea blanca, albaca silvestre	Whole plant	Essential oil, flavonoids	*Infusion:* digestive	
Nepeta cataria L.	Hierba gatera/ Catnip	Flowered aerial parts	Essential oil, flavonoids	*Infusion:* digestive, abdominal ache	
N. nepetella L.	Hierba gatera/ Catnip	Flowered aerial parts	Essential oil, flavonoids	*Infusion:* digestive	
Origanum virens L.	Orégano, orégano blanco	Flowered aerial parts	Essential oil, flavonoids, phenolic acids	*Infusion:* digestive, carminative	
O. vulgare L.	Orégano, mejorana silvestre, orenga/ Oregano	Flowered aerial parts	Essential oil, flavonoids	*Infusion:* digestive, carminative	
Phlomis lychnitis L.	Hierba de la candelaria, oreja de liebre/ Lampwickplant	Aerial parts	Polyphenols	*Infusion:* stomach depurative, colics	
Rosmarinus officinalis L.	Romero/ Rosemary	Whole plant, flowered aerial parts	Phenolic acids, flavonoids, essential oil, diterpenes, triterpenes	*Infusion:* digestive, carminative, abdominal ache, gall bladder-liver diseases	
Salvia lavandulifolia Vahl.	Salvia de Castilla, jalvia, salvia, blanquilla/ Spanish sage	Flowered aerial parts	Essential oil, flavonoids, phenolic acids	*Infusion:* digestive, soft laxative, abdominal ache	

FAMILY / Species	Local Spanish/English name	Part used	Chemical composition	Popular use	Other information
S. verbenaca L.	Cresta de gallo/ Wild clary	Aerial parts, seeds, leaves	Flavonoids, essential oil, terpenes	*Infusion:* digestive, soft laxative, abdominal ache	
Sideritis hirsuta L.	Rabogato	Aerial parts	Flavonoids, terpenes, essential oil	*Infusion:* stomach and duodene ulcers, anti inflammatory, digestive	
Teucrium chamaedrys L.	Beltrónica, carrasquilla/ Wall germander	Flowered aerial parts	Flavonoids, essential oil, triterpenes, lactonic diterpenes	*Infusion:* stomach ache, colics, antidiarrheic. *Decoction:* emetic	Highly toxic
Thymus bracteatus Lange ex C	Tomillo, tamarilla/ Thyme	Flowered aerial parts	Essential oil, phenolic acids, flavonoids	*Infusion:* digestive	
T. lacaitae Pau	Tomillo lagartijero, tomillo de Aranjuez	Flowered aerial parts	Essential oil, phenolic acids, flavonoids	*Infusion:* digestive	
T. mastichina (L.) L.	Mejorana, tomillo blanco, tomillo salsero/ Mastic thyme, Spanish marjoram	Flowered aerial parts	Essential oil	*Infusion:* stomach and belly ache	
T. pulegioides L.	Serpol, té fino, té morado/ Lemon thyme	Flowered aerial parts	Essential oil	*Infusion:* digestive	
T. serpyllum (Mill.) Benth	Tomillo ratero, té de ratero/ Thyme, creeping thyme	Flowered aerial parts	Essential oil, flavonoids, terpenic acids	*Infusion:* digestive	
T. vulgaris L.	Tomillo/ Thyme, garden thyme	Flowered aerial parts	Essential oil, flavonoids, triterpenes, saponosides	*Infusion:* digestive	
T. zygis Löfl. ex L.	Tomillo rojo/ Thyme, Spanish thyme	Flowered aerial parts	Essential oil, flavonoids	*Infusion:* digestive	

Table 13.1. contd...

Table 13.1. contd...

FAMILY / Species	Local Spanish/English name	Part used	Chemical composition	Popular use	Other information
LEGUMINOSAE (FABACEAE)					
Cassia obovata Colladon	Sen/ Senna	Leaves	Anthracenosides	*Infusion:* laxative	
Glycyrrhiza glabra L.	Regaliz/ Licorice	Roots	Saponins, flavonoids	*Infusion:* stomach ulcers, heartburn	
LILIACEAE					
Allium porrum L.	Puerro/ Leek	Aerial parts	Fibre, vitamins, minerals,	*Whole plant:* digestive, soft laxative	
Smilax aspera L.	Zarzaparrilla/ Sarsaparilla	Roots	Saponins, essential oil, tannins	*Infusion:* laxative	
Linum bienne Mill.	Lino/ Pale flax	Not found	Mucilages	Laxative	
LORANTACEAE					
Viscum album L.	Almuérdago, muérdago/ Mistletoe	Flowered aerial parts	Triterpenes, sterols, amines, phenolic acids, flavonoids, proteins, lectins	*Decoction:* gum rinse, bleeding gum	Toxic at high doses
MORACEAE					
Ficus carica L.	Higo, brevas/ Fig, edible fig	Fruits, leaves, barks	Minerals, vitamins, furanocumarins, enzymes	Laxative	
OLEACEAE					
Olea europaea L.	Olivo/ Olive	Fruits, oil	Iridoids, fatty acids	Laxative	
PAPAVERACEAE					
Papaver rhoeas L.	Amapola, ababol/ Corn poppy	Leaves, stems	Alkaloids, mucilages	*Decoction:* antispasmodic	

FAMILY / Species	Local Spanish/English name	Part used	Chemical composition	Popular use	Other information
PLANTAGINACEAE					
Plantago major L.	Llantén mayor/ Plantain	Aerial parts, seeds, leaves	Mucilages, iridoids, flavonoids, phenolic acids	*Decoction, infusion:* gastric disturbance, peptic ulcer	
P. sempervirens Crantz.	Zaragatona, sargatona	Aerial parts	Not found	*Decoction:* gum disturbances and sores. *Infusion:* bleeding ulcers	
POLYGONACEAE					
Rumex acetosa L.	Acedera menor, acederilla/ Garden sorrel	Whole plant	Tannins, organic acids (oxalic acid), flavonoids, anthraquinones	*Infusion, decoction:* digestive, antispasmodic, laxative, antidiarrhoeic	Oxalate toxicity
R. conglomeratus L.	Romanzas, berzoletas/ Clustered dock	Dried stems	Not found	*Decoction:* antidiarrhoeic	
R. crispus L.	Romanzas, berzoletas, acedera, lengua de vaca/ Curly dock, yellow dock/ Fiddle dock	Stems, leaves, inflorescences	Oxalates, oxalic acid, anthraquinones, tannins, flavonoids, essential oil	*Infusion, decoction:* aperitif, laxative, stomach ache, antidiarrhoeic	
R. pulcher L.	Romanzas, berzoletas	Stems, leaves, inflorescences	Not found	*Infusion, decoction:* stomach ache, antidiarrhoeic	
PORTULACEAE					
Portulaca oleracea L.	Verdolaga/ Little hogweed	Whole plant	Mucilagues, saponins, minerals	*Infusion:* gastroprotective, anti-ulcerogenic	
RHAMNACEAE					
Frangula alnus Mil	Arraclán, avellano bravío/ Buckthorn, frangula	Fruits, barks	Anthraquinones	*Infusion, decoction:* laxative, purgative	

Table 13.1. contd...

Table 13.1. contd...

FAMILY / Species	Local Spanish/English name	Part used	Chemical composition	Popular use	Other information
Rhamnus catharticus L.	Espino cerval	Fruits, barks	Anthraquinones	*Infusion, decoction:* laxative, purgative	
ROSACEAE					
Crataegus monogyna Jacq.	Espino blanco, majuelo/ Hawthorn	Flowers	Flavonoids, phenolic acids	*Decoction:* stomach ache	
Cydonia oblonga Mill.	Membrillo/ Quince	Leaves, seeds, fruits	Organic acids, flavonoids, phenolic acids, tannins, essential oil, pectine	*Decoction:* aperitif, antidiarrhoeic (seeds)	
Fragaria vesca L.	Fresa/ Strawberry	Leaves	Tannins, flavonoids	*Infusion:* aperitif, antidiarrhoeic, depurative, diuretic	
Malus sylvestris L.	Manzana amarga/ Apple, crab apple	Fruits	Pectins	*Fresh fruit:* antidiarrhoeic	
Prunus domestica L.	Ciruela, claudia, regañada/ Plum	Fruits	Vitamins, galactomanans	*Fresh fruit:* laxative	
P. spinosa L.	Endrino/ Blackthorn	Flowers, fruits	Tannins, organic acids, vitamins	*Infusion:* laxative (flowers), astringent (fruits)	
Rosa canina L.	Escaramujo, rosa silvestre/ Rose, dog rose	Leaves, flowers, fruits	Tannins, antocyanosides, essential oil, flavonoids, organic acids	*Decoction:* antidiarrhoeic, gallbladder diseases	
Rubus ulmifolius Schott	Zarzamora/ Elmleaf blackberry	Whole plant, young stems, fruits, leaves	Tannins, flavonoids	*Infusion, direct ingestion:* gum bleeding (young stems), antidiarrhoeic (mature fruits)	
Sorbus domestica L.	Serbal comun, jerbo, azarolla	Leaves, barks, fruit	Tannins, organic acids	*Decoction, fresh fruit:* antidiarrhoeic, stomach ache, ulcer, gallbladder ailments, cholesterol lowering	

FAMILY / Species	Local Spanish/English name	Part used	Chemical composition	Popular use	Other information
RUBIACEAE					
Asperula cynanchica L.	Hierba del cólico, hierba del pinillo, hierba de la piedra	Aerial parts	Not found	*Infusion:* stomach depurative, colics	
RUTACEAE					
Ruta angustifolia Pers.	Ruda/ Rue	Leaves	Flavonoids, essential oil, alkaloids, coumarins	*Infusion:* digestive, stomach ache, colics	Toxic at high doses; abortion
R. montana L.	Ruda montesina/ Rue	Flowered aerial parts	Flavonoids, essential oil, alkaloids, coumarins	*Infusion:* digestive, stomach ache. *Decoction:* emetic	Toxic at high doses; abortion
TILIACEAE					
Tilia tomentosa Moench	Tila/ Silver linden	Inflorescences, barks	Polyphenols, flavonoids, mucilages	*Infusion:* antispasmodic	
UMBELLIFERAE					
Anethum graveolens L.	Eneldo/ Dill	Fruits	Essential oil, flavonoids, coumarins, phenolic acids	*Infusion:* stimulant, carminative, diuretic	
Apium graveolens L.	Apio/ Celery	Aerial parts	Essential oil, coumarins	Stimulant, carminative, digestive	
Coriandrum sativum L.	Coriandro, cilantro/ Coriander	Fruits	Essential oil, lipids	*Infusion, chewed with honey:* carminative, stomaquic	
Foeniculum vulgare Mill.	Hinojo/ Fennel	Fruits	Essential oil, flavonoids, lipids	*Infusion:* carminative, digestive	
Petroselinum crispum (Mill.) A.W. Hill.	Perejil/ Parsley	Aerial parts, roots	Essential oil, flavonoids, minerals	*Infusion, decoction:* stomach disorders, anti anemic, tonic, aperitif	
Pimpinella anisum L.	Anís/ Anise	Fruits	Essential oil, flavonoids, phenolic acids, coumarins	*Decoction:* carminative, colics	

Table 13.1. contd...

Table 13.1. contd....

FAMILY / Species	Local Spanish/English name	Part used	Chemical composition	Popular use	Other information
Smyrnium olusatrum L.	Apio caballar/ Alexanders	Roots	Essential oil, sesquiterpene lactones	*Decoction:* carminative, aperitif	
URTICACEAE					
Urtica dioica L.	Ortiga mayor, ortiga verde/ Stinging nettle	Aerial parts, roots, leaves	Flavonoids, mucilages, minerals, tannins	*Fresh, decoction:* stomach ache, baldness, stomach disorders, anti anemic, laxative	
VERBENACEAE					
Aloysia triphylla L'Herit	Hierba Luisa, Hierbaluisa, Maria Luisa, verbena olorosa, Reina Luisa, té de verbena/ Lemon beebrush	Aerial parts	Essential oil, flavonoids	*Infusion:* intestinal pain, antiseptic, colics	

Table 13.2. Wild medicinal plants in central Spain with digestive activity but without traditional use in this area

FAMILY / Species	Local Spanish/English name	Part used	Chemical composition	Pharmacological use	Other information
ANACARDIACEAE					
Pistacia terebinthus L.	Terebinto, charneca, cornicabra/ Cyprus turpentine	Barks, gills	Oleoresin, tannins	Antidiarrhoeic	
Rhus coriaria L.	Zumaque/ Sicilian sumac	Leaves, stems	Tannins	Antidiarrhoeic	
ARACEAE					
Arum maculatum L.	Aro, Alcatraz/ Cuckoo pint	Roots, fruits	Alkaloids, saponins, essential oil, starch	*Tincture, juice:* gastritic disorders, purgative	Caution with berries toxicity
ARISTOLOCHIACEAE					
Aristolochia paucinervis Pomel	Aristoloquia, aristoloquia macho	Rhizomes	Aristolochic acids, fatty acids	Abdominal pain	High toxicity
ASTERACEAE					
Bellis annua L.	Manzanilla	Flower heads	Essential oil, flavonoids, saponins	*Infusion:* laxative	
Carlina acaulis L.	Cardo ajonjero, cardiguera/ Carline thistle	Roots	Essential oil, flavonoids	*Infusion:* gastro intestinal disorders, cholagogue	
Cnicus benedictus L.	Cardo santo, cardo bendito/ Blessed thistle	Flowered aerial parts	Sesquiterpene lactones, flavonoids	*Infusion:* aperitif, digestive, emetic (high doses)	
Ditrichia viscosa (L.) W. Greuter	Olivarda/ False yellowhead	Aerial parts	Essential oil	*Infusion:* gastroprotective	

Table 13.2. contd...

Table 13.2. contd...

FAMILY / Species	Local Spanish/English name	Part used	Chemical composition	Pharmacological use	Other information
Eupatorium cannabinum L.	Eupatorio, eupatorio de los árabes/ Hemp agrimony	Aerial parts, roots	Essential oil, bitter principles, saponins, tannins, alkaloids	Cholagogue, purgative (high doses, root). Aperitif (aerial parts)	Possible toxicity
Hieracium pilosella L.	Oreja de ratón, vellosilla/ Mouseear hawkweed	Flowered aerial parts	Essential oil, flavonoids, coumarins, phenolic acids, tannins	*Infusion, decoction:* antidiarrhoeic	
BERBERIDACEAE					
Berberis vulgaris L.	Agracejo/ Barberry, common barberry	Roots, stems barks, fruits	Alkaloids, tannins	Liver and glad bladder regulator, soft laxative	
BRASSICACEAE					
Iberis amara L.	Carraspique, ibéride/ Annual candytuft	Seeds, stems, roots, leaves	Glycosinolates, flavonoids	Digestive, laxative	
Sinapis alba L.	Mostaza blanca, jenabe, ajenabe/ White mustard	Seeds	Essential oil, lipids, glucosinolates, mucilages	Digestive	
CHENOPODIACEAE					
Beta vulgaris L.	Acelgas de campo/ Beet	Leaves	Betain, saponins	Digestive, laxative	
Chenopodium album L.	Cenizos/ Lambsquarters	Leaves	Flavonoids	Laxative	
CYPERACEAE					

FAMILY / Species	Local Spanish/English name	Part used	Chemical composition	Pharmacological use	Other information
Cyperus esculentus L.	Chufa/ Yellow nutsedge	Tubers	Starch, fatty acids, sucrose, fibre	*Whole tuber:* Intestinal regulator	
GERANIACEAE					
Geranium sp.	Geranio/ Geranium	Peduncles, roots	Essential oil, tannins	Antidiarrhoeic	
GRAMINEAE (POACEAE)					
Secale cereale L.	Centeno/ Rye	Seeds	Mucilages, starch	*Decoction:* Antidiarrhoeic	
Triticum aestivum L.	Trigo/ Wheat	Seeds	Proteins, minerals, aminoacids, fibre, starch	*Whole grain:* antidiarrhoeic, laxative, antidiverticules	
LABIATEAE (LAMIACEAE)					
Lavandula pedunculata (Mill.) Cav.	Azalla, cantueso, lavanda	Flowered aerial parts	Essential oil, antocyanosides	*Infusion:* digestive, antispasmodic	
Nepeta coerulea Aiton	Nepeta	Leaves, flowers	Essential oil, flavonoids	*Infusion:* antidiarrhoeic, antispasmodic	
N. tuberosa L.	Hierba de los gatos/ Catnip	Leaves, flowers	Essential oil, flavonoids	*Infusion:* antidiarrhoeic, antispasmodic	

Table 13.2. contd...

Table 13.2. contd...

FAMILY / Species	Local Spanish/English name	Part used	Chemical composition	Pharmacological use	Other information
Salvia aethiopis L.	Oropesa, salvia etíope/ Mediterranean sage	Leaves	Essential oil, flavonoids, phenolic acids	*Decoction:* digestive	
S. argentea L.	Oropesa, salvia blanca, salvia Silvestre/ Silver sage	Aerial parts	Essential oil	*Infusion:* digestive	
S. sclarea L.	Esclarea, salvia romana/ European sage	Aerial parts	Essential oil, flavonoids, diterpenes	*Infusion:* digestive	
LEGUMINOSAE (FABACEAE)					
Trigonella foenum-graecum L.	Alholva/ Fenugreek	Seeds	Lipids, mucilages, sapogenins, flavonoids, alkaloids	*Decoction:* aperitif	
LILIACEAE					
Asphodelus aestivus Brot.	Asfodelo, gamón común	Roots, rhizomes	Not found	*Decoction:* antispasmodic	
Polygonatum odoratum (Mill.) Druce	Poligonato/ Aromatic Solomon's seal	Rhizomes	Flavonoids, saponins	Antidiarrhoeic	
LINACEAE					
Linum sp.	Lino/ Flax	Seeds	Mucilages, cyanogenic glycosides	Laxative, gastrointestinal disorders	
ONAGRACEAE					

FAMILY / Species	Local Spanish/English name	Part used	Chemical composition	Pharmacological use	Other information
Epilobium sp.	Hierba de San Antonio, adelfilla de hoja estrecha, laurel de San Antonio/ Willowherb	Aerial parts	Flavonoids, phytosterols, tannins	Antidiarrhoeic	
PAPAVERACEAE					
Fumaria officinalis L.	Fumaria, palomilla/ Fumitory	Flowered stems, whole plant	Alkaloids, mucilages	Cholagogue, choleretic, liver protector, antispasmodic, laxative	
PLANTAGINACEAE					
Plantago sp.	Plantago/ Plantain	Aerial parts, seeds	Mucilages	Intestinal regulator; gastritis, peptic ulcers	
P. afra L.	Zaragatona/ Psyllium	Seeds	Mucilages	Intestinal regulator; gastritis, anti ulcerogenic	
P. lanceolata L.	Llantén menor/ English plantain	Leaves, whole part	Mucilages, flavonoids, tannins, iridoids	*Decoction*: stomach ache, anti ulcerogenic	
POLYGONACEAE					
Polygonum aviculare L.	Centinodia, lengua de pájaro/ Knotweed	Roots, flowers	Tannins, silice salts, mucilages, flavonoids	*Decoction*: gingivitis, antidiarrhoeic	
P. bistorta L.	Bistorta/ Meadow bistort	Roots	Tannins	*Decoction*: antidiarrhoeic, anti inflamatory	

Table 13.2. contd...

Table 13.2. contd...

FAMILY / Species	Local Spanish/English name	Part used	Chemical composition	Pharmacological use	Other information
RAFFLESIACEAE					
Cytinus hypocistis L.	Chupamieles, doncella de Europa, colmenilla	Flowers	Tannins	*Juice:* antidiarrhoeic, stomach ache	
RANUNCULACEAE					
Aquilegia vulgaris L.	Aguileña común/ European columbine	Leaves, stems	Flavonoids, phenolic acids, cyanogenic glycosides	*Decoction:* aperitif, liver and bile duct disorders (jaundice)	
ROSACEAE					
Agrimonia eupatoria L.	Hierba de San Guillermo/ Cocklebur; Churchsteeples	Leaves, flowering stems	Tannins, phenolics acids, flavonoids, terpenoids, essential oil	*Infusion, decoction:* antidiarrhoeic, liver diseases	
Alchemilla alpina L.	Alchemilla, alquimila alpina, estellaria/ Alpine lady's mantle	Aerial parts	Tannins, flavonoids	*Decoction:* antidiarrhoeic	
A. xanthochlora Rothm.	Alquimila, manto de Nuestra Señora/ Hairy lady's mantle	Leaves, flowering stems	Tannins, flavonoids	*Infusion:* antidiarrhoeic	
Amelanchier ovalis Medik.	Guillomo, durillo, agrio	Roots, barks	Tannins	*Infusion:* antidiarrhoeic, choleretic	
Geum rivale L.	Hierba de San Benito, cariofilada acuatica/ Purple avens	Aerial parts, rhizomes, roots	Tannins, phenolic acids, essential oil	*Infusion, decoction:* antidiarrhoeic, digestive tonic	
Potentilla argentea L.	Plateada/ Silver cinquefoil	Roots, aerial parts	Tannins, essential oil, saponins	*Infusion:* antidiarrhoeic, anti ulcerogenic, spasmolytic	

FAMILY / Species	Local Spanish/English name	Part used	Chemical composition	Pharmacological use	Other information
Potentilla erecta (L.) Raeusch.	Tormentilla, sietenrama/ Tormentil, erect cinquefoil	Roots	Tannins, phenolic acids	*Decoction*: antidiarrhoeic	
Potentilla reptans L.	Cincoenrama, pie de Cristo/ Creeping cinquefoil	Aerial parts, leaves	Tannins	*Decoction*: stomache ache, anti ulcerogenic	
Prunus padus L.	Cerisuela, Cerezo aliso, cerezo de racimo/ Bird cherry	Barks, flowers	Organic acids, antocyanosides, cyanogenic glycosides	Antispasmodic, digestive	
Rosa gallica L.	Rosa común/ French rose, rose	Flowers	Tannins, antocyanosides, essential oil, flavonoids	*Decoction*: antidiarrhoeic	
Sanguisorba min or Scop	Pimpinela, hierba del cuchillo/ Small burnet	Aerial parts, roots	Tannins, essential oil, flavonoids	*Decoction*: antidiarrhoeic, abdominal ache, anti-ulcerogenic	
Sorbus aucuparia L.	Serbal de los cazadores, serbal silvestre/ Mountain ash	Leaves, fruits	Organic acids, ascorbic acid, tannins, flavonoids, sorbitol	*Decoction*: antidiarrhoeic	
RUBIACEAE					
Galium verum L.	Cuajaleches, hierba cuajadera/ Yellow spring, bedstraw	Aerial parts	Anthraquinones, flavonoids, iridoids	Spasmolytic	

Table 13.2. contd...

Table 13.2. contd...

FAMILY / Species	Local Spanish/English name	Part used	Chemical composition	Pharmacological use	Other information
TAMARICACEAE					
Tamarix gallica L.	Taray, taraje/ French tamarisk	Leaves, flowers	Tannins	*Infusion*: antidiarrhoeic, liver diseases	
UMBELLIFERAE					
Angelica sp.	Angélica/ Angelica	Roots, fruits	Coumarins, acetylenic compounds	*Decoction*: Liver diseases, digestive, carminative	
Heracleum sphondylium L.	Espondilio, lampaza, pie de oso, belleraca/ Eltrot	Fruits, roots	Essential oil	Digestive	
Pimpinella saxifraga L.	Pimpinela blanca/ Pimpinella	Roots, aerial parts	Essential oil	Tonic, aperitif	
Seseli tortuosum L.	Comino rústico	Fruits	Essential oil, coumarins	Tonic, stomach diseases	

observed where the chemical composition did not justify the observed effectiveness as it happens with the use of *Ilex aquifolium* and *Santolina chamaecyparissus* as soft laxatives.

The information given in this chapter could contribute in preserving the traditional knowledge in areas under industrial development that could lead to its disappearance in a short time.

On the other hand, increasing interest in ethnopharmacology as a source of new active principles with therapeutic activity have led to studies on traditionally used medicinal plants; research on several species has recently taken place, for example with essential oil from *Carum carvi* and *Foeniculum vulgare* for the treatment of gastrointestinal disorders such as intestinal dysbiosis (Hawrelak et al. 2009). Some of the species growing in the central Spain did not seem to have been used traditionally, while it occurs in other areas as is the case with the genus *Potentilla* (Rosaceae) which is mainly distributed in temperate, arctic and Alpine zones of the northern hemisphere. This genus has been known since ancient times for its curative properties. Extracts of the aerial and/or underground parts have been applied in traditional medicine for the treatment of inflammations, wounds, certain forms of cancer, infections due to bacteria, fungi and viruses, diarrhoea, diabetes mellitus and other ailments. Most of the pharmacological effects can be explained by the high amount of tannins and to a lesser extent by triterpenes, present in all plant parts (Tomczyk and Latté 2009).

Finally, the use of historical methods in ethnopharmacology and related areas of research could lead to a better understanding of the dynamics and evolution of human plant use.

Acknowledgements

We wish to thank Dr. Ramón Morales from the Real Jardin Botanico de Madrid for his technical support.

References

Al-Qura'n, S. 2009. Ethnopharmacological survey of wild medicinal plants in Showbak, Jordan. J. Ethnopharmacol. 123: 45–50.

Bézanger-Beauquesne, L., M. Pinkas, and M. Torck. 1989. Les Plantes dans la Thérapeutique Moderne. 10th ed. Maloine, Paris. France.

Blanco Castro, E. 1998. Diccionario de Etnobotánica Segoviana. Pervivencia del Conocimiento sobre las Plantas. Colección Hombre y Naturaleza, Segovia. Spain.

Duru, M.E., A. Cakir, S. Kordali, H. Zengin, M. Harmandar, S. Izumi, and T. Hirata. 2003. Chemical composition and antifungal properties of essential oils of three *Pistacia* species. Fitoterapia 74: 170–176.

Fajardo, J., A. Verde, D. Rivera, and C. Obón. 2000. Las Plantas en la Cultura Popular de la Provincia de Albacete. Instituto de Estudio Albacetenses "Don Juan Manuel" de la Excma. Diputación de Albacete, Albacete. Spain.

Fajardo, J., A. Verde, D. Rivera, and C. Obón. 2007. Etnobotánica en la Serranía de Cuenca. Las Plantas y el Hombre. Diputación provincial de Cuenca, Cuenca. Spain.

Fauron, R. and R. Moatti. 1984. Guide Practique de Phytothérapie. Encyclopédie Médicale de Prescription Phytothérapique. Maloine, Paris. France.

Flora Ibérica. Plantus vasculares de la Península Ibérica e Islas Baleares. 2010. Real Jardín Botánico. CSIC, Madrid. Spain.

Gadhi, C.A., A. Benharref, M. Jana, and A. Lozniewski. 2001. Anti-*Helicobacter pylori* activity of *Aristolochia paucinervis* Pomel extracts. J. Ethnopharmacol. 75: 203–205.

Guarrera, P.M. 2005. Traditional phytotherapy in Central Italy (Marche, Abruzzo, and Latium) Fitoterapia. 76: 1–25.

Guarrera, P.M. 2006. Household dyeing plants and traditional uses in some areas of Italy. J. Ethnobiol. Ethnomed. 2: 9–15.

Hawrelak, J.A., T. Cattley, and S.P. Myers. 2009. Essential oils in the treatment of intestinal dysbiosis: A preliminary *in vitro* study. Altern. Med. Rev. 14: 380–384.

Heinrich, M., J. Kufer, M. Leonti, and M. Pardo de Santayana. 2006. Ethnobotany and ethnopharmacology—Interdisciplinary links with the historical sciences 107: 157–160.

Leclerc, H. 1983. Précis de Phytothérapie. Thérapeutique par les Plantes Françaises. Masson, Paris. France.

Mills, S. and K. Bone. 2000. Principles and Practice of Phytotherapy. Modern Herbal Medicine. Churchill Livingstone, London. UK.

Morales, R. 2003. Catalogue of the vascular plants from Madrid Community (Spain). Botanica Complutensis 27: 31–70.

Mota, K.S., G.E. Dias, M.E. Pinto, A. Luiz-Ferreira, A.R. Souza-Brito, C.A. Hiruma-Lima, J.M. Barbosa-Filho, and L.M. Batista. 2009. Flavonoids with gastroprotective activity. Molecules 3; 14: 979–1012.

Newall, C.A., L.A. Anderson, and J.D. Phillipson. 1996. Herbal Medicines. A guide for health-care professionals. The Pharmaceutical Press, London. UK.

Pardo de Santayana, M., E. Blanco, and R. Morales. 2005. Plants known as té in Spain: an ethno-pharmaco-botanical review. J. Ethnopharmacol. 98: 1–19.

Pieroni, A., M.E. Giusti, C. de Pasquale, C. Lenzarini, E. Censorii, M.R. González-Tejero, C.P. Sánchez-Rojas, J.M. Ramiro-Gutiérrez, M. Skoula, C. Johnson, A. Sarpaki, A. Della , D. Paraskeva-Hadijchambi, A. Hadjichambis, M. Hmamouchi, S. El-Jorhi, M. El-Demerdash, M. El-Zayat, O. Al-Shahaby, Z. Houmani, and M. Scherazed. 2006. Circum-Mediterranean cultural heritage and medicinal plant uses in traditional animal healthcare: a field survey in eight selected areas within the RUBIA Project. J. Ethnobiol. Ethnomed. 2: 16–28.

Rivera, D., C. Obón, C. Inocencio, M. Heinrich, A. Verde, J. Fajardo, and R. Llorach. 2005. The ethnobotanical study of local Mediterranean food plants as medicinal resources in Southern Spain. J. Physiol. Pharmacol. 56: 97–114.

Sagrera, J. 1987. Plantas Medicinales. Grupo Líder, Spain.

Sánchez de Rojas, V.R., T. Ortega, and A. Villar. 1995. Inhibitory effects of *Cistus populifolius* on contractile responses in the isolated rat duodenum. J. Ethnopharmacol. 46: 59–62.

Tardío, J., H. Pascual, and R. Morales. 2005. Wild food plants traditionally used in the province of Madrid, Central Spain. Economic Botany 59: 122–136.

Tomczyk, M. and K.P. Latté. 2009. Potentilla: a review of its phytochemical and pharmacological profile. J Ethnopharmacol. 122: 184–204.

UNESCO. Safeguarding of the Intangible Cultural Heritage. Paris 2005. 26th September 2005.

Ustün, O., B. Ozçelik, Y. Akÿon, U. Abbasoglu, and E. Yesilada. 2006. Flavonoids with anti-*Helicobacter pylori* activity from *Cistus laurifolius* leaves. J. Ethnopharmacol. 108: 457–461.

Vallejo, J.R., D. Peral, P. Gemio, M.C. Carrasco, M. Heinrich, and M. Pardo de Santayana. 2009. *Atractylis gummifera* and *Centaurea ornata* in the province of Badajoz (Extremadura,

Spain)—Ethnopharmacological importance and toxicological risk. J. Ethnopharmacol. 126: 366–370.

Van Hellemont, J. 1986. Compendium de Phytotherapie. Association Pharmaceutique Belge, Service scientifique. Belgium.

Vanaclocha, B., and S. Cañigueral. 2003. Fitoterapia. Vademécum de prescripción. 4th ed. Masson, Barcelona. Spain.

Vázquez, F.M., M.A. Suarez, and A. Pérez. 1997. Medicinal plants used in the Barros Area, Badajoz Province (Spain). J. Ethnopharmacol. 55: 81–85.

Chapter 14

Traditional Medicinal Products and their Interaction with Estrogens Receptors—Implications for Human Health

Dieudonné Njamen

Introduction

Plants have provided man with all his needs in terms of shelter, clothing, food, flavors and fragrances and not the least, medicines. Plants have thus formed the basis of sophisticated traditional medicine systems such as Ayurvedic, Unani, Chinese, and African amongst others (Gurib-Fakim 2006). These systems of medicine have led to some important drugs still in use today. Among medicinal plants some are those with nutritional properties which constitute food in certain parts of the world, while others are just for medicinal use. Some of these plants are used for reproductive purposes by women as well as men since they are endowed with estrogenic properties. This estrogenicity is due to the presence of non-steroid compounds which are structurally and functionally similar to endogenous estrogens found in these plants (Burton and Wells 2002). These estrogenic plant compounds, known as 'phytoestrogens', have a mode of action depending on the presence or not of endogenous estrogens. Therefore, they can act either as antiestrogens or as weak estrogens, hence they are named adaptogens (Andersen 2000). This estrogenic and/or antiestrogenic

Department of Animal Biology and Physiology, Faculty of Science, University of Yaounde 1, P.O.Box 812 Yaounde, Cameroon.
E-mail: *dnjamen@Gmail.com*

activity of phytoestrogens may have an impact on normal biological processes which can be beneficial or adverse for human health.

In this chapter, we have tried to enumerate some of the major effects of phytoestrogens at different levels of the body and to show the mechanism of action by which they affect the target organs. Given that phytoestrogens are plant-derived compounds, we also enumerated some of plant compounds and plant preparations endowed with estrogenic activity and, finally, the effects of certain traditional medicinal plants on reproductive biology and physiology have also been mentioned.

Implication of phytoestrogens in human health

Interest in the physiologic and pharmacological role of bioactive compounds present in plants has increased dramatically over the last few decades. The impact of dietary phytoestrogens on normal biological processes was first recognized in sheep. Observations of sheep grazing on fields rich in clover and cheetahs fed high soy diets in zoos suggested that flavonoids and related phytochemicals can affect mammalian health (Usui 2006). Endogenous estrogens have an important role not only in the hypothalamic-pituitary-gonadal axis, but also in various non-gonadal systems such as the cardiovascular system, bone, central nervous system and lipid metabolism. The structural similarity of phytoestrogens with endogenous estrogens accounts for their ability to bind to estrogen receptors (ERs) and exert various estrogenic or antiestrogenic effects. The classical phytoestrogens, such as genistein and daidzein have a higher affinity to ERβ than ERα. ERβ is strongly expressed in the ovary, uterus, brain, bladder, testis, prostate and lungs. Expression of ERβ appears to occur at different sites in the brain than ERα (Kuiper et al. 1997). Moreover, ERβ is also expressed in both bone and the cardiovascular system (Makela et al. 1999). Agonistic actions on ERβ and weak (or little) activation of ERα is beneficial for osteoporosis, cardiovascular protection, lipid metabolism and breast cancer (Usui 2006).

Phytoestrogens and the central nervous system (CNS): the GnRH pulse generator or hot flush generator

The hypothalamus, a fundamental part of the CNS, regulates pituitary hormone release and vegetative functions such as body temperature and heart activity. Many neurons in the hypothalamus are estrogen receptive (Flügge et al. 1986, Pfaff et al. 2000). Neurons producing hypothalamic gonadotropin releasing hormone (GnRH) are controlled by a variety of neurotransmitters (e.g. GABA, glutamate, catecholamines) in such a way that they release GnRH in a synchronous pattern. That is, GnRH is released in a bolus as a prerequisite for normal luteinizing hormone

(LH) and follicle-stimulating hormone (FSH) release by the pituitary gland (Knobil 1980, 1990). The neurons and their neurotransmitters causing the synchronous and phasic activation of GnRH neurons are collectively known as the 'GnRH pulse generator'. Estrogen deprivation provokes overactivation of the GnRH pulse generator. Altered amounts of the hypothalamic neurotransmitters or their temporal release patterns, among others norepinephrine, serotonin, dopamine or GABA (Jarry et al. 1991, 1999, Freedman 2005, Bachmann 2005, Albertazzi 2006, Carroll 2006, Nelson et al. 2006), activate neighboring neurons involved in the regulation of body temperature and the cardiovascular system, respectively. In ovariectomized animal models and postmenopausal women, this leads to hot flushes and the associated tachycardic response. Only small amounts of estradiol or other estrogenic substances are needed to reduce the GnRH pulse generator to levels of activity, at which hot flushes and tachycardic attacks are reduced or abolished (Wuttke et al. 2007). In animal experiments, this can be easily achieved by hormone therapy or estrogen therapy (HT/ET) (Pan et al. 2001). Beside this steroid replacement, the incidence of hot flushes can also be reduced by estrogenic substances extracted from plants. Investigations have demonstrated that large amounts of isoflavones and genistein in particular are required to relax the GnRH pulse generator. In their unpublished studies, Wuttke et al. (2007) have shown that, an intravenous injection of 50 mg of genistein per rat given acutely (i.e. 150 mg/kg) achieved significant LH inhibition, whereas 5, 10, or 35 mg/rat proved to be ineffective. The reduction of LH in animal experiments is assessed as an indication of a negative feedback effect of an estrogenic substance, i.e. of genistein on the hypothalamus. A number of studies demonstrated that phytoestrogens had the same effect in postmenopausal women. In fact, in Asia, only 10% to 20% of postmenopausal women experience hot flushes compared with 70% to 80% of women in western countries (Lock 1991, 1994, Tang 1994 Boulet, et al. 1994). A popular hypothesis to explain this difference is that isoflavones found in soy, a staple in the traditional Asian diet, influences the body's response to the changing hormonal levels of menopause (Adlercreutz 1990). Dietary supplements containing isoflavones from soy are widely marketed for menopausal symptoms and are increasingly being used by women as an alternative to estrogen. In clinical studies, Tice et al. (2003) reported that the isoflavones from clover extract (Promensil) had little important effect on hot flushes or other symptoms of menopause. However, they also reported some evidence for the biological effect of Promensil (it reduced hot flushes more rapidly than placebo). Thus, thanks to their ability to moderate or to suppress the activity of the GnRH pulse generator which results in the reduction of serum LH levels, phytoestrogens may be a registered drug for hot flush complaints.

Phytoestrogens and the mammary gland

Animal experimental studies

Estrogens used in the hormone therapy/estrogen therapy cause the proliferation of epithelial cells in the mammary glands of all investigated mammals including monkeys (Foth and Cline 1998). This results in a higher number of alveoli and milk ducts and can be visualized with proliferation markers, such as BrdU incorporation or immunochemistry of nuclear proteins produced in dividing cells (e.g. PCNA or KI-67) (Wuttke et al. 2007). Since a high degree of mitotic activity is generally considered as a reason for the higher incidence of certain cancers, long-term exposure of rat mammary glands to hormone therapy estrogens may lead to an increase in mammary cancer incidence (Platet et al. 2004). The possible effects of isoflavones on mammary tumor proliferation be they stimulatory or inhibitory, are often attributed to their estrogen receptor binding capacity. There are two studies on the effects of soy on the mammary glands of monkeys, which confirm a proliferating effect of estradiol on mammary tissue (Jones et al. 2002, Wood et al. 2004), raising the possibility that this effect may be antagonized by an isoflavones rich soy protein diet. Soy administration alone had no effect on mammary tissue. In fact, a study published by Wood et al. (2004) showed that soy protein containing 129 mg/d of isoflavones may have a cancer preventive effect. These animals were compared with animals on isoflavones free soy protein or soy protein containing 0.625 mg/d of conjugated estrogens. The isoflavones/soy protein group showed no change in the density of mammary gland tissue, cell proliferation, or progesterone receptor gene expression, while the conjugated estrogen group showed an increase in markers for estrogenic effects on the mammary gland. Thus, these findings suggest no adverse effect of isoflavones on the mammary gland.

A number of investigations into the effects of soy/isoflavones on malignant mammary tumors were done on immune deficient nu/nu or SCID mice and in rats. The mouse model studied most intensively is the MCF-7 cell transplanted nu/nu mouse, which develops solid, fast growing tumors within weeks. Neither tumor development nor their progression could be inhibited by soy/isoflavones in this model. Most studies even described enhanced proliferation (Martin et al. 1978, Hsieh et al. 1998, Allred et al. 2004a and b). Among various soy preparations, pure genistein treatment was the most effective at stimulating growth of these MCF-7 cell tumors (Allred et al. 2004a,b). The same group also demonstrated that the inhibition of such MCF-7 cell mammary tumors by tamoxifen is negated by genistein and the authors therefore conclude that caution is warranted for postmenopausal women consuming dietary genistein while on tamoxifen therapy for estradiol-response breast cancer (Ju et al. 2002).

A study shown that, in the c-erbB, 2/neu transgenic mouse model, the mammary tumor suppressing effects of tamoxifen were nullified by low but not by high doses of isoflavones-enriched diets (Liu et al. 2005). If applicable to humans, not only mammary cancer patients but also high risk patients trying to prevent mammary cancer by undergoing Tamoxifen treatment should not be exposed to isoflavones.

Another animal model often used in mammary tumor studies is the Sprague-Dawley rat, which develops multiple tumors in the mammary region after 7,12-dimethylbenz(*a*)anthracene (DMBA) or *N*-methyl-*N*-nitrosourea (MNU) administration. Some experiments have shown that soy/isoflavones have no or even stimulatory influence on the development and growth of these tumors (Constantinou et al. 2001, Ueda et al. 2003, Pei et al. 2003, Allred et al. 2004a,b, Kijkuokool et al. 2005). However, postpubertal soy treatment before the induction of the tumor had a slightly preventive effect (Upadhyaya and El-Bayoumy 1998, Hakkak et al. 2000). Carcinogen-induced mammary cancers predominantly express ERα, which leads to the assumption that the proliferating action is mediated by ERα. There is also some indication that substances activating mainly ERβ have an antiproliferative effect and can thereby inhibit cancer development in animal experiments (Weihua et al. 2003, Shaaban et al. 2003). Phytoestrogens, in particular isoflavones, which are claimed to bind preferentially ERβ may thus be an efficient drug for mammary cancer patients.

Different mechanism by which isoflavones/genistein supposedly inhibit tumor development and growth have been proposed:

- Antiproliferative activities (Shao et al. 1998, Choi et al. 1998);
- Anti-angiogenic activities (Fotsis et al. 1993, Sasamura et al. 2004);
- Inhibition of metalloproteinase and thereby invasive growth and metastasis (Menon et al. 1998, Kousidou et al. 2005);
- Induction of apoptosis (Li et al. 1999);
- Induction of tumor cell differentiation (Constantinou and Huberman 1995).

A highly significant tumor preventive effect was observed in several studies when soy/isoflavones were given prepubertally, thus prior to any differentiation of the mammary tissue (Pei et al. 2003, Brown and Lamartiniere 1995, Barnes 1997, Gallo et al. 2001, Lamartiniere et al. 2002).

In 1999, Pasqualini and co-workers established the concept of selective estrogen enzyme modulators. This concept was applied to the mammary gland and mammary cancer in particular (Chetrite et al. 2000, Pasqualini and Chetrite 2005). The authors administrated higher concentrations of estradiol in mammary cancer than in the normal tissue. They attributed this finding to elevated aromatase activity and hence more estrogen production in this tissue, and to a greater activity of sulfatases, the enzymes which convert estradiol sulfate to free estradiol, in mammary cancer tissue. This signifies that the tumor can provide its own estrogen for the most part, even if concentrations in the blood stream are low. On the basis of these findings, the observations of Harris et al. (2004) are alarming. They revealed that the inhibition of sulfotransferases by genistein and daidzein is 10 times greater than that of sulfatases. This implies that the degree of inactivation of estradiol through sulfotransferases is inhibited more profoundly than the degree of estradiol production through sulfatases. The authors concluded that the greater inhibition of sulfotransferases through isoflavones and particularly through daidzein may lead to an increase in free estrogen levels in the tumor tissue, which in turn may stimulate tumor growth. Isoflavones therapy may thus be fatal for women with a mammary tumor.

Finding in humans

Immunocytochemical and molecular processing of normal human mammary tissue demonstrates that ERα is the predominant receptor type in this tissue (Warner et al. 2000, Clarke 2000). There is great speculation about an inhibition of proliferation through ERβ or various splice variants of ERα and ERβ (Platet et al. 2004). An increase in mammary carcinoma malignancy appears to be associated with a drop in ERβ expression (Shaaban et al. 2003). This leads to the theory that a high level of ERβ is associated with a reduction in tumor progression, while a low level would increase the progression of invasive tumors (Warner et al. 2000).

The clinical studies on soy/isoflavones and breast cancer are correlative. They correlate the uptake, excretion or blood concentrations of isoflavones with the frequency of mammary cancer incidences. Similar to the animal experimental findings, the outcomes of clinical studies can be divided into three groups: inhibition, no effect, or stimulation of mammary cancers.

Decreased risk of mammary cancers: most studies conducted in Japan, China, and the USA revealed a significantly decreased risk of breast cancer in Far East Asian women, who meet their protein needs mainly through soy products, when compared to American women of Caucasian descent or European women. It has been suggested that these differences are not of a genetic nature but are due to the difference in nutrition, because breast cancer risk in Japanese women born and raised in the USA is more or less equivalent to that of Caucasian American women. Several authors concluded therefore that the isoflavones contained in soy, which bind to both estrogen receptors (ERα and ERβ), are responsible for this phenomenon (Adlercreutz 1998, 2002).

There is a series of correlative studies indicating an inverse correlation between the amount of ingested or excreted isoflavones and the incidence of mammary cancer (Dai et al. 2001, Yamamoto et al. 2003, Wu et al. 2004). These are findings mainly from studies in Far East Asian populations, in which the subjects consumed soy/isoflavones from the beginning of their childhood. This is important, keeping in mind that in animal experimental investigations mentioned earlier, prepubertal genistein exposure had a cancer preventive effect (Brown and Lamartiniere 1995, Lamartiniere et al. 2002). Meanwhile, migration studies have revealed that Japanese women migrating to western countries after puberty bear the same mammary cancer risk as Japanese age cohorts (Ziegler et al. 1993, Shu et al. 2001), indicating that the Japanese lifestyle prior to and/or during puberty has some preventive effect on the development of mammary cancer.

No changes in cancer risk: a correlative study between soy/isoflavones intake or excretion and the incidence of mammary cancer did not show any significant correlation. The authors of this study concluded that there is neither enhancing nor inhibiting effect of soy/isoflavones on mammary cancer (den Tonkelaar et al. 2001). Similarly, in a large prospective study performed in Japan and involving more than 3,500 women, no correlation was found between 50 g intake of soy protein and breast cancer incidence (Key et al. 1999). It is a clinically established observation that a high breast tissue density after menopause, as determined by mammography, represents a risk factor in the development of mammary cancer (Warren 2004, Simick et al. 2004). One year of isoflavones intervention did not stimulate mammary gland density (Maskarinec et al. 2003).

Increased risks: An increased risk of breast cancer is concluded on the basis of the following studies.

An intervention study in which soy/isoflavones were applied over a short period of time leads to an increased density of the mammary gland tissue (Maskarinec and Meng 2001). However, the same authors demonstrated no increase in the density of mammary gland tissue if administered over a year, while preparations containing estrogen increased the density of breast tissue (Maskarinec et al. 2003).

Frankenfeld et al. (2004) related the ratio of equol and o-desmethylangolensin conversion from soy/isoflavones to the density of mammary tissue. They reported a significant decrease in mammary tissue density in women with augmented equol excretion in the urine, while women metabolizing soy/isoflavones to o-desmethylangolensin display an increase in mammary gland tissue density in comparison with patients in whom the density remains unaltered. The latter women would qualify for a group at higher risk of developing mammary cancer.

Cannulation of lactiferous ducts allows ductal lavage through which cytological atypia and receptors can be analyzed (Maskarinec and Meng 2001, Wrensch et al. 2001, Maskarinec et al. 2003). In an intervention study, ductal lavage in postmenopausal women was analyzed after 1-yr administration of 38 g of soy protein containing 38 mg of genistein. While no differences in serum LH levels, lipids and triglycerides were evident, the ductal lavage of about 30% of the women contained cells displaying epithelial hyperplasia which is considered as a sign of cancer endangerment (Petrakis et al. 1996). In this study, the authors also revealed an increase in cytological atypia and elevated expression of progesterone receptors in cells harvested from ductal lavages.

Finally, data on the beneficial or adverse effects of isoflavones on mammary cancer remain contradictory and therefore inconclusive. However, phytoestrogens/isoflavones may be more efficient to prevent than to treat mammary cancer if they are given prepubertally i.e. prior to any differentiation of the mammary tissue. Moreover, thanks to their higher affinity to ERβ, phytoestrogens can inhibit cancer development by stimulating antiproliferative activities. Thus, the quest for substances activating mainly ERβ might be beneficial for patients developing mammary tumor.

Phytoestrogens and the urogenital tract

Vagina

All the estrogens implemented in hormone therapy/estrogen therapy cause desquamation and cornification of the vaginal epithelium of rodents, resulting in the typical estrous vaginal smears. This smear is used in pharmacology for the study of estrogenic substances. Similar effects occur in the vagina of postmenopausal women in which the number of superficial cells is stimulated by estrogen therapy (Wuttke et al. 2003). All the estrogens used in hormone therapy/estrogen therapy bring about an acidic milieu in the vagina. This prevents ascending infections caused by the increase of superficial cells, the glycogen content of which serves lactobacillae as substrate for the production of lactate. Further, estrogen therapy estrogens promote lubrification upon sexual arousal to allow painless intercourse (Wuttke et al. 2007).

Most animal experimental and intervention studies in humans did not expose any influence of soy/isoflavones on the vagina epithelium as measured by vaginal smears. Histological or molecular analysis of estrogen-regulated genes; however, revealed weak estrogenic effects (Wuttke et al. 2007). In a recent study, Lien et al. (2009) showed that soy aglycons of isoflavones protected vaginal epithelium from degeneration due to ovariectomy-caused shortage of estrogens. This was in agreement with Malaivijitnond et al. (2006) who reported that supplementation of genistein (0.25-2.5 mg/kg) induces vaginal cornification. Kim et al. (2004) also reported that vaginal blood flow of rats decrease by ovariectomy but increase by estradiol supplementation. In placebo-controlled study in asymptomatic postmenopausal women, a soy rich diet stimulated the maturation index of vaginal cells (Chiechi et al. 2003).

Urinary bladder

Estrogen deficiency after menopause causes atrophic changes within the urogenital tract and is associated with urinary symptoms, such as frequency, urgency, nocturia, incontinence, and recurrent infection. The urinary bladder or isolated bladder strips of ovariectomized rabbits or rats respond to expansion with contractions, which are reduced by estrogen administration (Seidlova-Wuttke et al. 2004). Yang et al. (2009) demonstrated that ERα and ERβ were both expressed in normal bladder detrusor muscle with various expression levels, and ERβ was the predominant ER subtype. With estrogen deficiency ERα was upregulated and ERβ downregulated. The effect was reversed after estradiol augmentation. ERβ appears therefore to have an important role in mediating estrogen function in the bladder. It was also demonstrated on isolated rabbit detrusor strips that genistein has a calming effect similar to estradiol on the overactive bladder (Ratz et al. 1999). Since phytoestrogens are claimed to bind preferentially ERβ, they could thus be an efficient drug for urinary incontinence.

Uterus

In all mammalian species, the administration of estrogen alone leads to both endo- and myometrial proliferation. This results in an increase in uterine mass (in rats, this happens within hours), first caused by liquid imbibitions of the tissue, and subsequently through proliferation of endo- and myometrial cells (Seidlova-Wuttke et al. 2003a, Acher 2004). This phenomenon has been adopted by the Organisation for Economic Co-operation and Development (OECD 2001) as a screening test for the estrogenicity of chemicals. In women, estrogens are known to lead to hyperplasia of the endometrium, which results in a 10-fold increase in the cancer risk in this organ (reviewed in Archer 2004). In combination with progestins, however, the cancer risk is reduced rather than increased (Davis et al. 2005). The effects of estrogens used in hormone therapy bring up the safety question of soy/isoflavones with regard to the endometrium. Literature on animal experiments and older clinical studies did not indicate that isoflavones may endanger the endometrium. Since most dietary supplements containing isoflavones have 50–100 mg of the substances, the equivalent for the rat dose would be about 5–10 mg. These doses have been shown to have no effect on the uterus in any of the published studies (Wade et al. 2003, Seidlova-Wuttke et al. 2003b). Certain authors even reported that, at the dietary doses, isoflavones and lignans are associated with a reduced risk of endometrial cancer (Horn-Ross et al. 2003). Other studies; however, have revealed increased uterus growth in rats only with 3–6 fold higher doses (Diel et al. 2004a,b, Kim et al. 2005, Erlandsson et al. 2005). The safety margin is thus not very broad in this case. In one study on rats, the authors even reported inhibitory effects of genistein on estrogen stimulated uterine parameters (Erlandsson et al. 2005). On the basis of these data, it is therefore highly likely that correspondingly high doses of isoflavones in postmenopausal women stimulate uterine parameters. Indeed, a large clinical trial with alarming results was published in July 2004 (Unfer et al. 2004). In this study, 179 postmenopausal women with intact uteri were treated with 150 mg of isoflavones. A further 197 participants received a placebo preparation over a period of 5 yr. After endometrium biopsy, the postmenopausal women were compared with the 197 women treated with placebo. Of those patients treated with isoflavones, 3% developed endometrial hyperplasia as compared to 0% in the placebo treated women. The authors concluded: "Long-term treatment (up to 5 yr) with soy phytoestrogens was associated with an increased occurrence of endometrial hyperplasia. These findings call into question the long-term safety of phytoestrogens with regard to the endometrium". Endometrial hyperplasia is considered as an indication of cancer endangerment. This study has been criticized, because the occurrence of not a single case of simple endometrial hyperplasia in a placebo group is an unlikely outcome. In many other studies, no effect on the endometrium was also reported. However, these studies were only of a short duration (3-12 mon) (Penotti et al. 2003, Kaari et al. 2005, Nikander et al. 2005).

Any anti-estrogenic effect on the endometrium occurring in the rat uterus (Erlandsson et al. 2005) was never reported in the human endometrium (Murray et al. 2003).

Prostate

Androgens are implicated in the development of prostate cancer (Cheng et al. 1993). The conversion of testosterone to the more potent metabolite dihydrotestosterone by prostate-specific steroid 5α-reductase is a key mechanism in the action of androgens in the prostate and is important in the promotion and progression of prostate diseases (Pollard et al. 1989). Clinical prostate cancers often respond to androgen deprivation therapy. Thus, a reduction in androgen levels should affect carcinogenesis processes. The administration of 5α-reductase inhibitors results in a substantial decrease in prostatic sections of the normal gland and a substantial increase in cell death in normal and transformed prostatic cells (Lamb et al. 1992). It has been suggested that for men, phytoestrogens may confer some level of protection against prostate cancer (Barnes et al. 1995, Adlercreutz and Mazur 1997). Epidemiological data have consistently reported a relatively low incidence of prostate cancer in Asian populations whose diet is rich in phytoestrogens, especially the isoflavonoids in soy and other legumes (Donn and Muir 1985, Adlercreutz et al. 1991). The highest incidence has been reported in North American black males (Sondik 1988) in whom the age-adjusted incidence is 125 times greater than that for men in Shanghai, China (Miller 1988). This geographic variation is a major feature of prostate cancer, with Asian men generally being much less susceptible to this disease than Europeans and North Americans. However, Japanese men who migrate to America adopt the prostate cancer incidence of the indigenous population within one or two generations (Kolonel et al. 1986). These epidemiological data support the concept that diet may inhibit the promotion and progression of prostatic cancer in Asian men. It has also been reported that lower levels of 5α-reductase activity have been found in Japanese men (Ross et al. 1992). These men also have a higher urinary excretion and higher plasma levels of phytoestrogens than their western counterparts (Adlercreutz et al. 1991, Adlercreutz et al. 1993). In one particular clinical case, a 66-yr-old man took 160 mg phytoestrogens daily for 1 wk before a radical prostatectomy. The prostatectomy specimens revealed mild patchy microvacuolations and prominent apoptosis, whereas no changes were seen in normal prostate cells (Stephens 1997). These degenerative changes in the prostatectomy specimen, especially the apoptosis, were indicative of androgen deprivation and typical of a response to estrogen therapy (Hellstrom et al. 1993). In one animal study, rats maintained on a soy-free diet for 11 wk developed severe inflammation of the lateral prostate, whereas rats maintained on a soy-containing diet or commercial rat chow did not develop any signs of prostatitis. This finding suggests that soy of a dietary source may play a protective role against the pathogenesis of prostatitis. One possible explanation could be that as soybeans contain a number of phytoestrogens that are weak estrogens, the soy-free diet might disturb the androgen-estrogen ratio (Sharma et al. 1992). Several investigators have also reported that phytoestrogens inhibit the growth of cultured prostate cancer cells (Peterson and Barnes 1993, Rokhlin and Cohen 1995). Thus, a diet rich in phytoestrogens may prevent prostate cancer by a variety of mechanisms, including reducing circulating androgen levels, increasing concentrations of sex hormone-binding globulin (SHBG), competitive binding to cellular hormone receptors, and apparent reduction in the production of dihydrotestosterone.

Phytoestrogens and thyroid

A high dose of estradiol affects the serum tyroxine (T4) and triiodothyronine (T3) levels only marginally (Boado et al. 1983, Thomas et al. 1986).

The results of several animal experiments dealing with the adverse effects of isoflavones on the thyroid caused concern. These adverse effects are probably not executed by estrogen receptors and are not due to the tyrosine kinase inhibiting properties of genistein. Thyroid peroxidase is inhibited by genistein in rats (Chang and Doerge 2000, Son et al. 2001, Doerge and Sheehan 2002). Thyroid peroxidase is an enzyme necessary for the iodination of tyrosine in the thyroglobulin molecule. The inhibition of thyroid peroxidase in regions of iodine deficiency may lead to hypothyroidism, which may eventually lead to goiter development. Nevertheless, any goitrogenic effect of physiological doses of isoflavones has not been illustrated in either animal experiments or clinical observations (Chang and Doerge 2000, Son et al. 2001).

Clinically an epidemiological study has shown that the daily intake of isoflavones was correlated to thyroid cancer. Women with a higher isoflavone intake had a higher risk of developing thyroid cancer than those with a lower consumption (Haselkorn et al. 2003). In an intervention study, postmenopausal women were provided with 56 or 90 mg of isoflavones over a period of 6 mon. The higher dose led to a significant increase in biologically active T3 levels, while T4 values were elevated under both doses, though this was only statistically significant under the lower dose. Interestingly, the TSH levels were significantly elevated as well (as opposed to the expected decrease due to the negative feedback mechanism of T3) (Persky et al. 2002). This observation needs to be interpreted as a central nervous (hypothalamic) effect of isoflavones. These findings are diametrically opposed to the findings in rats, in which genistein had a thyroid peroxidase inhibiting effect, thus reducing T3 and T4 production (Chang and Doerge 2000, Doerge and Chang 2002).

Phytoestrogens and the skeletal system

Bone

Several animal studies have provided convincing data on the significant improvement of bone mass or other points after soy protein or isolated isoflavones-enriched soy extract supplementation (Anderson and Garner 1998, Ishimi et al. 1999, Picherit et al. 2000, Arjmandi and Smith 2002). Several observational epidemiologic studies have also examined the link between dietary intake of phytoestrogens and bone mass in humans and reported that soy protein and soy phytoestrogen intake are beneficial for maintaining or modestly improving bone mass in postmenopausal women

(Horiuchi et al. 2000, Mei et al. 2001, Greendale et al. 2002). Some studies revealed that isoflavones-rich soy protein had a modest effect in retarding bone loss in perimenopausal (Alekel et al. 2000) and postmenopausal (Potter et al. 1998, Clifton-Bligh et al. 2001) women, but such effects were not observed in other studies (Hsu et al. 2001). Studies have also reported inconsistent effects of phytoestrogens (or soy protein) on bone markers in postmenopausal women (Arjmandi and Smith 2002). Arjmandi et al. (2003) reported that in postmenopausal women, the soy protein group had significantly reduced urinary deoxypyridinoline excretion (a specific biomarker of bone resorption), and calcium excretion did not change compared to a milk-based protein group. There was also an enhancing effect of soy isoflavones on insulin-like growth factor 1 (IGF-I) synthesis, and the IGF-I concentration is positively related to bone mass in women. Chen et al. reported that in the Chinese population, soy isoflavones have a mild but significant effect on the maintenance of hip bone mineral concentration in postmenopausal women with low initial bone mass (Chen et al. 2003). In postmenopausal monkeys, however, soy phytoestrogens are poor substitutes for mammalian estrogens in protecting against bone loss resulting from estrogen deficiency (Register et al. 2003).

Joints

There is some epidemiologic evidence that estrogens used in hormone therapy may have some joint-protective effect, as there are reports claiming fewer cases of autoimmune arthritis in women undergoing hormone therapy /estrogen therapy treatment. There is a good indication that autoimmune polyarthritis may be positively influenced by soy/isoflavones: severity in ovariectomized cynomolgus monkeys was reduced by estrogen replacement (Ham et al. 2002). In this monkey model, osteoarthritis markers were improved by estradiol but remained unchanged by soy phytoestrogens (Ham et al. 2004). Furthermore, soy protein alleviated symptoms of osteoarthritis in women and men (Arjmandi et al. 2004).

Phytoestrogens and the cardiovascular system

Arteries

Animal experiments in rabbits and hamsters revealed an atheroprotective effect of soy by means of LDL reduction (Balmir et al. 1996, Alexandersen et al. 2001, Kondo et al. 2002, Arjmandi et al. 2003). It was also discovered that soy can prevent the formation of lesion-induced atherosclerotic plaques. Furthermore, atherosclerotic changes induced by a cholesterol rich diet may be prevented by soy/isoflavones (Register et al. 2005, Adams et al. 2005). In fact, soy preparations containing isoflavones are claimed to increase serum HDL and lowered LDL levels, which often leads to the conclusion that soy preparations have an anti-atherosclerotic effect. In one study, the elasticity of the arteries improved under isoflavones treatment; in another trial; however, the arterial parameters remained unchanged. Isoflavones have anti-oxidant effects, which is also suggestive of the beneficial effects on arteries (Miquel et al. 2006).

Veins

It has been known for decades that estradiol-17β and HT/ET estrogens increase coagulation in animals and humans. This leads to an increase in venous thrombo-embolic diseases. Both arms of the WHI study confirmed this perception convincingly. Soy isoflavones on the other hand did not activate the hemostatic system in healthy postmenopausal women. Some studies do not support the existence of any biologically significant estrogenic effect of soy/phytoestrogens on coagulation, fibrinolysis, or endothelial function (Dent et al. 2001, Teede et al. 2005).

Phytoestrogens and the gastrointestinal tract

In various colon cancer cell lines, ERβ has been determined as the predominant receptor type, while both receptor subtypes may be found in healthy colon tissue. In many of these cell lines, both stimulatory as well as inhibitory effects of HT/ET estrogens have been described (Burkman 2003). In 1981, Hoff et al. demonstrated the inhibition of proliferation through estradiol *in vitro*. On the other hand, the growth of a colon carcinoma cell line is stimulated by estradiol both *in vitro* and when transplanted in immune deficient nu/nu mice (Narayan et al. 1992). Data on any colon protective effect of isoflavones are conflicting: soy/isoflavones appear to have a similar protective effect as HT/ET on the development and/or progression in colon carcinoma. Vitamin D balance obviously plays an essential role in this. Vitamin D hormone is the compound 1,25-dihydroxy vitamin D3 which is essential for Ca^{2+} absorption through intestinal epithelial cells (Ramasamy 2006). The substance 25-D3 is present in the blood at high concentrations. It has been shown that genistein increases the hydroxylation of 25-D3 at position 1 in normal colorectal cells. Vitamin D hormone from this source has a cancer preventive effect (Kallay et al. 2002, Cross et al. 2004). A study by Hakkak et al. (2001) also showed that a lifelong diet of soy could protect rats from carcinogen-induced colon cancer.

The significant reduction in colon cancer incidences in women undergoing HT in the WHI study made the hypothesis of tumorigenesis inhibition and/or tumor progression very likely. However, this phenomenon was not apparent in the estrogen only group, and the effect seems to be attributed to the combination of estrogen/progestin (Anderson et al.2004).

In an overview, Messina and Bennink (1998) illustrated that there is little indication of soy/isoflavones having any colon cancer preventive effect. One epidemiological study was able to show an association between fewer colon cancer cases in women with a high soy intake (Messina and Bennink 1998). Nevertheless, there are also reports negating any protective effect of isoflavones (Tajima and Tominaga 1985, Hoshiyama et al. 1993). Thus isoflavones may have few colon cancer preventive effects, which may involve the genistein-stimulated production of vitamin D.

Phytoestrogens and glucose metabolism and obesity

Many studies in animals as well as in humans suggest that soy has beneficial effects on diabetes mellitus, and several studies in obese humans and animals suggest that soy as a source of dietary protein has significant antiobesity effects (Kawano-Takahashi et al. 1986, Hermansen et al. 2001). Mahalko et al. (1984) fed different sources of fiber to type 2 diabetic subjects for 2 to 4 wk and observed beneficial effects of soy hulls on glucose intolerance, lipid indexes, and glycated hemoglobin. Tsai et al. (1987) observed that in obese subjects with type 2 diabetes, soy polysaccharides significantly reduced increases in postprandial serum glucose and triacylglycerol concentrations. One study indicated that genistein may augment glucose stimulated insulin secretion (Liu et al. 2006). In other studies, experimentally induced diabetes was improved by genistein (Lee 2006). Interesting findings reported by Dang et al. demonstrated that a high concentration of genistein acts as an agonist to peroxisome proliferators activated receptor (PPAR) gamma in an *in vitro* reporter gene assay system (Dang et al. 2003). They demonstrated that at low concentrations, genistein acts as estrogen, stimulating osteogenesis and inhibiting adipogenesis, but at high concentrations, genistein acts as a ligand for PPAR gamma, leading to the upregulation of adipogenesis and the downregulation of osteogenesis. This finding is consistent with the phenomenon of a soy diet improving insulin resistance. Heim et al. reported that genistein enhances osteogenesis and represses adipogenic differentiation of human primary bone marrow stromal cells *in vitro* (Heim et al. 2004). The results of these two papers might explain, at least in part, the effects of phytoestrogen (such as genistein) on obesity and insulin sensitivity in humans.

Phytoestrogens and immune system

Immunosuppressive effects are clinically ambiguous. On the one hand, immunosuppression can have a positive impact on autoimmune diseases. On the other hand, a reduction in immune response affects the organism unfavorably in its defence against infections. With reference to estrogenic effects on the immune system, there is a vast amount of published data resulting from cell biology and animal experiments. Immunomodulatory findings and animal experimental models of human diseases such as autoimmune rheumatoid arthritis, autoimmune glomerulonephritis and even malignant diseases such as leukemia can obviously be directly influenced by estrogens. HT/ET estrogens have been shown to influence autoimmune disease in humans as well. Some examples of such diseases are rheumatoid arthritis, systemic lupus erythematosus, scleroderma, and Sjögren's Syndrome (Lahita et al. 1982, Holmdahl et al. 1989, Cutolo et al. 2004, Yoneda et al. 2004, Ramsey-Goldman 2005, Gompel and Piette 2007), which point to differential mechanisms of action of hydroxylated estradiol derivatives, acting in either a pro- or anti-inflammatory way in humans (Rachon et al. 2006).

The positive influences of genistein have been described in animal experiments following induced autoimmune arthritis. As with the other possible pathophysiologically relevant effects of isoflavones, the problem with the immunosuppressive nature of isoflavones has not been sufficiently investigated on an experimental or clinical level.

Mode of action of phytoestrogens

Phytoestrogens are adaptogens. This is their main mechanism of action. They can be beneficial when estrogen levels are either increased or decreased. When they are metabolized, they bind on the same cellular sites as do estrogens. They bind the human estrogen receptor but with an affinity 10–1000 fold lower than estradiol (Morito et al. 2001). That is why they are considered to be weak estrogenic compounds. They compete with estradiol for the receptor site and therefore, can act as agonists or antagonists (Verdeal et al. 1980). In fact, when a phytoestrogen has attached itself to an estrogen receptor, this prevents estrogen from exerting its effects. On an average, phytoestrogens have about 2% of the strength of estrogens. Therefore, when estrogen levels are high, substituting a phytoestrogen for an estrogen means that there will be much less estrogenic activity at a given binding site. Conversely, if estrogen levels are low and estrogen-binding sites are empty, filling them with phytoestrogens that contain 2% estrogen activity will result in a total increase in systemic estrogenic effect (Andersen 2000). Thus, the mode of action of phytoestrogens depends on the presence or not of endogenous estrogens. They act either as antiestrogens or as weak estrogens in relation to endogenous levels of estrogen.

A second mode of action for phytoestrogens may be their ability to affect the endogenous production of estrogen. The pituitary gland releases gonadotrophins that stimulate estrogen synthesis in the ovaries. Phytoestrogens appear to lower gonadotrophin levels, which will lengthen

level of activity as compared to isoflavonoids (Miksicek 1993, Breinholt and Larsen 1998, Collins-Burow et al. 2000). Moreover, some flavonoids like kaempferol and quercetin exhibit antiestrogenic activities (Collins-Burow et al. 2000). Naringenin and hesperetin are flavonoids found in higher amounts in citrus fruits (Breinholt et al. 2004). 8-prenyl naringenin is the most potent phytoestrogen discovered so far. It is found mostly in hops (Schaefer et al. 2003).

Lignans

Lignans are more prevalent in the plant kingdom but their estrogenic properties are lower compared to isoflavones. The estrogenic active lignans enterodiol and enterolactone are produced by bacteria in the colon from secoisolariciresinol and matairesinol respectively (Adlercreutz 1990).

Coumestans

Coumestrol is the most important coumestan consumed by humans. It is found at high concentrations in clover and alfalfa sprouts, but also at lower concentrations in sunflower seeds, lima bean seeds, pinto bean seeds, and round split peas (Franke et al. 1995).

Plant compounds with estrogenic activity

Estrogenic activity of plant preparations and plant compounds can be determined by using uterotrophic assays. In fact, uterotrophic assays are often used as a standard assay for the determination of estrogenic activity (Bachman et al. 1998, OECD 2001). Moreover, estrogenic activity is generally characterized by uterus morphological, histological and biochemical modifications (Diel et al. 2002). Phytoestrogens are gaining popularity for the development of functional food products targeted at women suffering from menopausal symptoms and to prevent osteoporosis. They may also reduce the risk of hormone-dependant cancers, such as breast cancer in women and prostate cancer in men. Because of the structural similarity to estradiol and its binding ability to the human estrogen receptor, they are suspected to act as natural SERMs (Selective Estrogenic Receptor Modulators) (Gurib-Fakim 2006). That is the case of griffonianone C, griffonianone E,7-O-geranylformononetin, 4'-O-geranylisoquiritigenin, 4'-methoxy-7-O-[(E)-3-methyl-7-hydroxymethyl-2,6-octadienyl]isoflavone (7-O-DHF) and 3',4'-dihydroxy-7-O-[(E)-3,7-dimethyl-2,6 octadienyl]isoflavones (7-O-GISO), all derived from *Millettia griffoniana* Baill (Ketcha Wanda et al. 2006).

   Millettia griffoniana is a Fabaceae whose root and stem bark are used in Cameroon folk medicine to treat boils, insect bites, sterility, amenorrhea as well as menopausal disorders, and illnesses with an inflammatory component like pneumonia and asthma. These *Millettia griffoniana* derivatives were tested by Ketcha Wanda et al. (2006) for their potential estrogenic activities in three different estrogen receptor alpha (ERα)-dependant assays. In a yeast-based ERα assay, all tested substances and 17β-estradiol as endogenous agonist, showed a significant induction of β-galactosidase activity. The test compounds at the concentration of 5 x 10^{-6} M could achieve 59–121% of the β-galactosidase induction obtained with 10^{-8} M 17β-estradiol (100%). In the reporter gene assay based on stably transfected MCF-7 cells, the estrogen responsive induction of luciferase was also stimulated by the *Millettia griffoniana* isoflavones. In Ishikawa cells (endometrial adenocarcinoma cells), all substances exhibited estrogenic activity revealed by the induction of alkaline phosphatase (AlkP) activity.

Estrogenic activity of griffonianone C was also evaluated *in vivo* in ovariectomized rats (Ketcha Wanda et al. 2007): results revealed that griffonianone C did not significantly increase the uterine wet weight, although there was a trend towards an increase. Results also revealed that griffonianone C affected the expression of estrogen-responsive genes in the uterus and liver. At the dose of 20 mg/kg/d, it caused a 50-fold upregulation of complement C3 mRNA compared to the control. A significant increase in calcium binding protein 9-kDa mRNA expression was observed in the uterus of ovariectomized rats treated with estradiol (41-fold versus control) or 20 mg/kd/d of griffonianone C (25-fold versus control). In contrast, the expression of clusterin and progesterone receptor in the uterus was strongly decreased by both estradiol and griffonianone C at the highest dose.

Plant Preparations with Estrogenic Activity

Several plant preparations tested in the Animal Physiology laboratory (Faculty of Science, University of Yaounde 1, Cameroon) showed an estrogenic activity. They are, *Brenania brieyi*, *Erythrina lysistemon*, and many others such as *Millettia conraui*, *Millettia drastica*, *Bridelia ferruginea*, *Pseudarthria hookeri* and *Nauclea latifolia*, all belonging to the family of Fabaceae.

The stem bark of *Erythrina lysistemon*, one of the traditionally used 'women remedies', has been assessed for its estrogenic activity. The ethyl-acetate extract of the stem bark of *E. lysistemon* showed estrogenic activities *in vitro* either in a yeast-based estrogen receptor assay or on the estrogen-dependant stimulation of alkaline phosphatase activity in human endometrial carcinoma cell line Ishikawa. The estrogenic activity was investigated *in vivo* in proliferative status of target sex organs such as the uterus and vagina. The results obtained showed that oral administration of 200 mg/kg/d of *E. lysistemon* extract in comparison to untreated ovariectomized rats significantly increased the vaginal epithelial height by 47% and induced a weak increase of uterine epithelial height by 7%. Both were not as pronounced as those elicited in the positive control of 100 μg/kg/d of ethinylestradiol given orally. Overall results suggested that the extract of *E. lysistemon* contains secondary metabolites endowed with estrogenic activity (Tanee et al. 2007).

Fruit of the Cameroonian medicinal tree *Brenania brieyi* (Fabaceae) is used in the treatment of endocrine disorders, including menopausal complaints. To assess the potential estrogenicity, 50 mg/kg/d of a methanol extract from *Brenania brieyi* fruit was administered intravaginally twice a day to ovariectomized female Wistar rats for 3 and 7 d. The uterine weight doubled within 7 d, and the vaginal epithelial height increased both after 3 and 7 d of treatment. The results suggested that fruit of *Brenania brieyi* contains estrogenic secondary metabolites (Magne Ndé et al. 2007).

Five extracts namely the ethyl acetate extract of the stem bark of *Millettia conraui*, the ethyl acetate extract of the stem bark of *Millettia drastica*, the methanol extract of the leaves of *Bridelia ferruginea*, the methanol extract of the roots of *Pseudarthria hookeri* and the methanol extract of the roots of *Nauclea latifolia* showed interesting estrogenic properties, and were therefore further investigated on alkaline phosphase induction in Ishikawa cells. The extracts of *Millettia conraui*, *Millettia drastica*, *Bridelia ferruginea*, *Pseudarthria hookeri* and *Nauclea latifolia* showed significant stimulatory effects at 10 and 100 mg/ml doses. The extract of *Bridelia ferruginea* was not further evaluated because of its toxicity on Ishikawa cells. *In vivo* experiments showed that *per os* administration of 200 mg/kg of the extracts of *Millettia conraui* and *Bridelia ferruginea* significantly increased uterine epithelial height by 18% and 28% respectively compared with the uteri of ovariectomized controls after 7 d of treatment. Uterine epithelial height of animals treated with 100 μg/kg/d of ethinylestradiol increased by 242% in the same experiment. Extracts of *Nauclea latifolia* and *Millettia drastica* had no effect on the uterine epithelial height of ovariectomized rats. The extracts of *Nauclea latifolia*, *Millettia drastica*, *Bridelia ferruginea*, *Millettia conraui* in doses of 200 mg/kg/d given orally, significantly increased vaginal epithelial height by 15%, 24%, 51% and 58% following the same treatment regiment compared to untreated controls. In line with these data was the finding that the vaginal epithelial height and vaginal cornification in the presence of each of these extracts was more advanced than in ovariectomized controls although not as prominent as in response to ethinylestradiol treatment (Njamen et al. 2008).

Pharmacological effects of traditional plant preparations on reproductive biology and physiology nutritional plant preparations

Diet is an essential factor influencing biological processes such as growth, reproduction, survival and responses to xenobiotics (Clarke et al. 1977 Gaillard et al. 1977, Garland et al. 1989, National Research Council 1995, Rao 1996). Many food plants are thus used in traditional medicine for male and female reproductive purposes.

Red ginseng *Panax ginseng* (Araliaceae): roots are taken orally as adaptogens, aphrodisiacs, nourishing stimulants and in the treatment of type 2 diabetes as well as sexual dysfunction in men. De Andrade et al. (2007) reported that the Korean red ginseng can be an effective alternative for treating male erectile dysfunction. A study in impotent men with erectile dysfunction has shown that, patients who received 1,000 mg/kg BW of ginseng 3 times daily reported improved erection, rigidity, penetration and maintenance (de Andrade et al. 2007).

Lepidium meyenii (Brassicaceae): the plant commonly known as 'maca' is cultivated in Perou. The roots of this plant are a source of food and medicine. Maca is claimed to increase sexual and reproductive capacities. For that reason, it is also called 'ginseng andin'. The nutritional value of the dried roots of maca is important and similar to that of rice and wheat. It is made up of 60% of glucose, 10% of proteins, 8.5% of nutritional fibers, and 2.2% of fat substances. Maca contains a chemical substance called 'isothiocyanate of P-methoxybenzyl'which is claimed to be an aphrodisiac (increases manliness and libido). Maca also contains vitamins A, B_1, B_2, B_2, B_{12}, C, D and E and trace elements such as magnesium, potassium, manganese, copper, zinc and many others substances. Arginine, lysine, vitamins E and C, and zinc contribute to the spermatogenesis equilibrium and help for hormonal disequilibrium such as infertility, irregular menstruations, dysmenorrheal, sexual weakness in men and hot flushes. Thus in women, maca increases the libido and stimulates sexual activity, increases fertility, decreases menopausal symptoms, relieves menstrual pains and reduces menstrual cycle disturbances. In men, maca fights against subfertility and infertility by increasing the production of spermatozoids, increases libido and stimulates sexual activity, increases serum testosterone levels, fights against erectile dysfunctions and manliness (Humala-Tasso and Combelles 2003).

Cleome gynandra (Capparaceae) and *Vernonia amygdalina* Del. (Asteraceae), two medicinal plants used traditionally in western Uganda to induce childbirth and hence hasten the labor process. *V. amygdalina* is mainly growing as a wild plant and is eaten as a vegetable in Cameroon. *C. gynandra* is home grown and is widely used as an edible food and vegetable in most parts of Uganda and East Africa. The roots and leaves decoction of *Vernonia amygdalina* are traditionally used in western Uganda to treat various ailments such as treatment of painful uterus, inducing uterine contractions, management of retained placenta and post partum bleeding, malaria, induced abortion, antimicrobes (bacterial and fungal infections), infertility, colic pains and treatment of irregular and painful menstruation. The roots of *Cleome gynandra* are chewed to induce uterine contractions and removal of retained placenta and control post partum bleeding in the childbirth process. The roots, leaves and flowers of *Cleome gynandra* are used in the prevention of miscarriages and treatment of colic pains when boiled or cooked as food. The leaves, roots and flowers of *Cleome gynandra* are chewed, cooked or are sun-dried and drunk in tea to treat sexual impotence or erectile dysfunctions in men (Kamatenesi-Mugisha and Oryem-Origa 2005).

Medicinal plant preparations

Plants used for female reproductive purposes

Many herbal remedies are traditionally used as contraceptives (to prevent ovulation or fertilization), abortifacients (to prevent implantation or to push out unwanted conceptus), emmenagogues (to stimulate uterine flow) or oxytocics (to stimulate uterine contractions, particularly to promote labor) (Ritchie 2001). This is the case with *Mimosa pudica*, *Ageratum conyzoides*, *Cola nitida*, *Coleus barbatus*, and *Hibiscus rosa-sinensis* amongst others.

Ageratum conyzoides (Asteraceae) is used for venereal diseases in El Salvador de Bahia (Brasil). *Ageratum conyzoides* plant extract inhibited uterine contractions induced by 5-hydroxytryptamine suggesting that the extract exhibited specific antiserotonergic activity on isolated uterus plant extract but had no effect on uterine contractions induced by acetylcholine. The results support the popular use of the plant as a spasmolytic (Achola and Munenge 1998, Silva et al. 2000).

Cola nitida (Sterculiaceae) is used for infertility. The oestrous cycles of rats treated with hydroalcohol extracts of *Cola nitida*, were blocked at the dioestrous II stage. Only 50% of the cycles of rats treated with *Cola nitida* were disrupted. The extract contains weak antioestrogen-like activity that provokes a blockage of female rat ovulation and oestrous cycle by acting on the hypophysis and/or hypothalamus secretion. This effect was mediated by oestrogen receptors (Benie et al. 2003).

Coleus barbatus (Lamiaceae) is used to interrupt pregnancy in Brazil and used as an emmenagogue in other countries. *Coleus barbatus* showed an anti implantation effect in the pre implantation period in rats, but after embryo implantation the extract had little effect (Almeida and Lemonica 2000).

Hibiscus rosa-sinensis (Malvaceae) flower decoctions are used in India and Vanuatu as aphrodisiacs, for menorrhagia, uterine hemorrhage and for fertility control (Lans 2006). Flower extracts produced an irregular estrous cycle in mice with prolonged oestrus and metoestrus and other indications of antiovulatory effects, androgenicity and estrogenic activity. *Hibiscus rosa-sinensis* possesses anti-complementary, anti-diarrhetic and anti-phologistic activity. *Hibiscus rosa-sinensis* flower showed anti-spermatogenic, androgenic, anti-tumor and anti-convulsant activities (Murthy et al. 1997, Reddy et al. 1997).

Plants used for male reproductive purposes

Many plants are used to improve erectile function, libido, sexual performance and sexual behavior. This is the case with *Alpinia calcarata, Camelia sinensis*, and *Curculigo orchoides* amongst others.

Alpinia calcarata (Zingiberaceae): *Alpinia calcarata* rhibozomes are claimed to possess a strong and safe oral aphrodisiac activity. In an experimental study, 3 h after an oral administration of hot water extract of *A. calcarata* to male rats at the doses of 150, 250, and 500 mg/kg, their sexual behavior was monitored (for 15 min) 3 h later using receptive females. Fertility was determined in a separate group (with the highest dose) using a non competitive copulation test. In the sexual behavior study, the hot water extract impaired the number of rats ejaculating and markedly prolonged the latency for ejaculation. Further, the number of rats mounting and intromitting, and the latencies for mounting and intromission were inhibited. Collectively, these observations indicate a strong aphrodisiac action. The other parameters remained unchanged indicating non-impairment in libido, sexual arousability, sexual vigor and sexual performance or penile erectile ability. However, a slight impairment was evident in sexual motivation (with the highest dose) in a partner preference test. In the fertility test, hot water extract induced profound oligozoospermia but fertility was uninhibited. The highest dose of hot water extract also elevated the serum testosterone level and the number of spontaneous penile erections rapidly and markedly (Ratnasooriya and Jayakody 2006).

Camelia sinensis (Theaceae): In Sri Lankan traditional medicine black tea brew of this plant is claimed to have male sexual stimulant activity. Experiments in male rats showed that, 3 h after an oral administration of 84, 167, and 501 mg/ml of black tea brew, their sexual behavior were monitored (for 15 min) using receptive females. The overall results showed that black tea brew possesses marked aphrodisiac activity (in terms of prolongation of latency of ejaculation shortening of mount- and intromission latencies and elevation of serum testosterone level). The aphrodisiac action had a rapid onset and appears to be mediated via inhibition of anxiety and elevation of serum testosterone level. Further, this aphrodisiac action was not associated with impairment of other sexual parameters like libido, sexual motivation, sexual arousal, sexual vigor or penile erection (Ratnasooriya and Fernando 2008).

Curculigo orchoides: The rhizomes of *Curculigo orchioides* have been traditionally used as an aphrodisiac. An experimental study showed that an administration of 100 mg/kg of extract to male rats significantly change the sexual behavior assessed by determining parameters such as penile erection, mating performance, mount frequency and mount latency. Moreover a pronounced anabolic and spermatogenic effect was evidenced by weight gains of reproductive organs. The treatment also markedly affected sexual behavior of animals as reflected in reduction of mount latency, an increase in mount frequency and enhanced attractability towards the female. Penile erection index was also incremented in the treated group (Chauhan et al. 2007).

Conclusion

Certain medicinal plants or nutritional plant preparations are widely used for reproductive purposes for women and men. Women traditionally use many plants as contraceptives, abortifacients, emmenagogues or as oxytocics. Plants used by men for reproductive purposes are mainly used for erectile dysfunction, infertility, manliness and for prostate problems.

As regard to phytoestrogens, many plant compounds endowed with estrogenic activity due to their structural and functional similarity with endogenous estrogens allow them to bind estrogen receptors and to exert estrogenic activity. They act either as antiestrogen or as weak estrogen in relation with endogenous estrogen levels. Phytoestrogen can thus influence normal biological processes and therefore, have beneficial or adverse implications to human health. They may relieve hot flushes in postmenopausal women, protect against certain cancers (breast cancer, uterus cancer, prostate cancer) and prevent cardiovascular diseases and osteoporosis. However, an intake of phytoestrogens after menopause or in advanced age, even at higher doses, cannot emulate apparent benefits of a lifelong phytoestrogen-rich diet.

References

Achola, K.J. and R.W. Munenge. 1998. Bronchodilating and uterine activities of *Ageratum conyzoides* extract. Pharm. Biol. 36: 93–96.
Adams, M.R. and D.L. Golden, J.K. Williams, A.A. Franke, T.C. Register, and J.R. Kaplan. 2005. Soy protein containing isoflavones reduces the size of atherosclerosis plaques without affecting coronary artery reactivity in adult male monkeys. J. Nutr. 135: 2852–2856.

Adlercreutz, H. 1990. Western diet and Western diseases: some hormonal and biochemical mechanisms and associations. Scand. J. Clin. Lab. Invest. Suppl. 201: 3–23.

Adlercreutz, H. 1998. Epidemiology of phytoestrogens. Baillieres Clin. Endocrinol. Metab. 12: 605–623.

Adlercreutz, H. 2002. Phytoestrogens and breast cancer. J. Steroid. Biochem. Mol. Boil. 83: 113–118.

Adlercreutz, H. and W. Mazur. 1997. Phyto-oestrogens and western diseases. Ann. Med. 29: 95–120.

Adlercreutz, H., H. Honjo, and A. Higashi. 1991. Urinary excretion of lignans and isoflavonoid phytoestrogens in Japanese men and women consuming a traditional Japanese diet. Am. J. Clin. Nutr. 54: 1093–1100.

Adlercreutz, H., H. Markkanen, and S. Watanabe. 1993. Plasma concentrations of phyto-oestrogens in Japanese men. Lancet 342: 1209–1210.

Akiyama, T., J. Ishida, and S. Nakagawa. 1987. Genistein, a specific inhibitor of tyrosine-specific protein kinases. J. Biol. Chem. 262: 5592–5595.

Albertazzi, P. 2006. Noradrenergic and serotonergic modulation to treat vasomotor symptoms. J. Br. Menopause Soc. 12: 7–11.

Alekel, D.L., A.S. Germain, C.T. Peterson, K.B. Hanson, J.W. Stewart, and T. Toda. 2000. Isoflavones-rich soy protein isolate attenuates bone loss in the lumbar spine of perimenopausal women. Am. J. Clin. Nutr. 72: 844–852.

Alexandersen, P., J. Haarbo, V. Breinholt, and C. Christiansen. 2001. Dietary phytoestrogens and estrogen inhibit experimental atherosclerosis. Climacteric 4: 151–159.

Allred, CD., K.F. Allred, Y.H. Ju, L.M. Clausen, D.R. Doerge, S.L. Schantz, D.L. Korol, M.A. Walling, and W.G. Helferich. 2004a. Dietary genistein results in larger MNU-induced, estrogen-dependent mammary tumors following ovariectomy of Sprague-Dawley rats. Carcinogenesis 25: 211–218.

Allred, CD.,K.F. Allred, Y.H. Ju, T.S. Goeppinger, D.R. Doerge, and W.G. Helferich. 2004b. Soy processing influences growth of estrogen-dependent breast cancer tumors. Carcinogenesis 25: 1649–1657.

Almeida, F.C.G. and I.P. Lemonica. 2000. The toxic effects of *Coleus barbatus* B. on the different periods of pregnancy in rats. J. Ethnopharmacol. 73: 53–60.

Andersen, G.D. 2000. Phytoestrogens: what they are and how they work. Dynamic Chiropractic 18: 21.

Anderson, G.L., M. Limacher, A.R. Assaf, T. Bassford, S.A. Beresford, Black H., D. Bonds, R. Brunner, R. Brzyski, B. Caan, R. Chlebowski, D. Curb, M. Gass, J. Hays, G. Heiss, S. Hendrix, B.V. Howard, J. Hsia, A. Hubbell, R. Jackson, K.C. Johnson, H. Judd, J.M. Kotchen, L. Kuller, A.Z. LaCroix, D. Lane, R.D. Langer, N. Lasser, C.E. Lewis, J. Manson, K. Margolis, J. Ockene, M.J. O'Sullivan, L. Phillips, R.L. Prentice, C. Ritenbaugh, J. Robbins, J.E. Rossouw, G. Sarto, M.L. Stefanick, L. Van Horn, J. Wactawski-Wende, R. Wallace, and S. Wassetheil-Smoller. 2004. Effects of conjugated equine estrogen in postmenopausal women with hysterectomy: the Women's Health Initiative randomized controlled trial. JAMA 291: 1701–1712.

Anderson, J.J. and S.C. Garner. 1998. Phytoestrogens and bone. Baillieres Clin. Endocrinol. Metab. 12: 543–557.

Anderson, J.W., B.M. Johnstone, and M.E. Cook-Newell. 1995. Meta-analysis of the effects of soy protein intake on serum lipids. N. Engl. J. Med. 333 (5): 276–82.

Archer, D.F. 2004. Neoplasia of the female reproductive tract: effects of hormone therapy. Endocrine 24: 259–263.

Arjmandi, B.H. and B.J. Smith. 2002. Soy isoflavones' osteo-protective role in postmenopausal women: mechanism of action. J. Nutr. Biochem. 13: 130–137.

Arjmandi, B.H., L. Alekel, and B.W. Hollis. 1996. Dietary soybean protein prevents bone loss in an ovariectomised rat model of osteoporosis. J. Nutr. 126: 161–167.

Arjmandi, B.H., D.A. Khalil, B.J. Smith, E.A. Lucas, S. Juma, M.E. Payton, and R.A. Wild. 2003. Soy protein has a greater effect on bone in postmenopausal women not on hormone replacement therapy, as evidenced by reducing bone resorption and urinary calcium excretion. J. Clin. Endocrinol. Metab. 88: 1048–1054.

Arjmandi, B.H. and D.A. Khalil, E.A. Lucas, B.J. Smith, N. Sinichi, S.B. Hodges, S. Juma, M.E. Munson, M.E. Payton, R.D. Tivis, and A. Svanborg. 2004. Soy protein may alleviate osteoarthritis symptoms. Phytomedicine 11: 567–575.

Axelson, M., J. Sjovall, B.E. Gustafsson, and K.D. Setchell. 1984. Soya-a dietary source of the non-steroidal estrogen equol in man and animals. J. Endocrinol. 102: 49–56.

Bachman, S., J. Hellwig, R. Jackh, and M.S. Christian. 1998. Uterotrophic assay of two concentrations of migrates of 23 polystyrenes administered orally (by gavage) to immature female wistar rats. Drug Chem. Toxicol. Suppl. 21: 1–30.

Bachmann, G.A. 2005. Menopausal vasomotor symptoms: a review of causes, effects and evidence-based treatment opinions. J. Reprod. Med. 50: 155–165.

Balmir, F., R. Staack, E. Jeffrey, M.D. Jimennez, L. Wang, and S.M. Potter. 1996. An extract of soy flour influences serum cholesterol and thyroid hormones in rats and hamsters. J. Nutr. 126: 3046–3053.

Barnes, S. 1997. The chemopreventive properties of soy isoflavonoids in animal models of breast cancer. Breast Cancer Res. Treat. 46: 169–179.

Barnes, S., T.G. Peterson, and L. Coward. 1995. Rationale for the use of genistein containing soy matrices in chemoprevention trials for breast and prostate cancer. J. Cell. Biochem. 22: 181–187.

Bathena, S.J. and M.T. Velasquez. 2002. Beneficial role of dietary phytoestrogens in obesity and diabetes. Am. J. Clin. Nutr. 76: 1191–1201.

Benie, T., J. Duval, and M.L.Thieulant. 2003. Effects of some traditional plant extracts on rat oestrous cycle compared with Clomid. Phytotherapy Research 17: 748–55.

Boado, R., E. Ulloa, and A.A. Zaninovich. 1983. Effects of oestradiol benzoate on the pituitary-thyroid axis of male and female rats. Acta. Enddocrinol. (Copenh) 102: 386–391.

Boulet, M.J., B.J. Oddens, P. Lehert, H.M. Vemer, and A. Visser. 1994. Climacteric and menopause in seven South-east Asian countries. Maturitas 19: 157–176.

Breinholt, V. and J.C. Larsen. 1998. Detection of weak oestrogenic flavonoids using a recombinant yeast strain and a modified MCF7 cell proliferation assay. Chem. Res. Toxicol. 11: 622–629.

Breinholt, V.M., G.W. Svendsen, L.O. Dragsted, and A. Hossaini. 2004. The citrus-derived flavonoids naringenin exerts uterotrophic effects in female mice at human relevant doses. Basic Clin. Pharmacol. Toxicol. 94: 30–36.

Brown, N.M. and C.A. Lamartiniere. 1995. Xenostrogens alter mammary gland differentiation and cell proliferation in the rat. Environ. Health Perspect 103: 708–713.

Brownson, D.M., N.G. Azios, B.K. Fuqua, S.F. Dharmawardhane, and T.J. Mabry. 2002. Flavonoids effects relevant to cancer. J. Nutr. 132: 3482–3489.

Burkman 2003. Hormone replacement therapy. Current controversies. Minerva Ginecol 55: 107–116.

Burton, J.L. and M. Wells. 2002. The effects of phytoestrogens on the female genital tract. Rev. J. Clin. Path. 55: 401–407.

Carroll, D.G. 2006. Nonhormonal therapies for hot flashes in menopause. Am. Fam. Physician 73: 457–464.

Chang, H.C. and D.R. Doerge. 2000. Dietary genistein inactivates rat thyroid peroxidase in vivo without an apparent hypothyroid effect. Toxicol. Appl. Pharmacol. 168: 244–252.

Chauhan, N.S., C.V. Rao, and V.K. Dixit. 2007. Effect of Curculigo orchioides rhizomes on sexual behaviour of male rats. Fitoterapia.

Chen, Y.M., and S.C. Ho, S.S. Lam, S.S. Ho, and J.L. Woo. 2003. Soy isoflavones have a favorable effect on bone loss in Chinese postmenopausal women with lower bone mass: a double-blind, randomized, controlled trial. J. Clin. Endocrinol. Metab. 88: 4740–4747.

Cheng, E., C. Lee, and J. Grayhack. 1993. Endocrinology of the prostate. In: H. Lepor and R.K. Lauson (eds.). Prostate diseases. Saunders. Philadelphia. pp. 57–71.

Chetrite, G.S., J. Cortes-Prieto, J.C. Philippe, F. Wright, and J.R. Pasqualini. 2000. Comparison of estrogen concentrations, estrone sulfatases and aromatase activities in normal, and in cancerous, human breast tissues. J. Steroid. Biochem. Mol. Boil. 72: 23–27.

Chiechi, L.M., G. Putignano, V. Guerra, M.P. Schiavelli, A.M. Cisternino, and C. Carriero. 2003. The effect of a soy rich diet on the vaginal epithelium in postmenopause: a randomized double blind trial, Maturitas 45: 241–246.

Choi, Y.H., L. Zhang, W.H. Lee, and K.Y. Park. 1998. Genistein-induced G2/M arrest is associated with the inhibition of cyclin B1 and induction of p21 in human breast carcinoma cells. Int. J. Oncol. 13: 391–396.

Clarke, H.E., M.E. Coates, J.K. Eva, D.J. Ford, C.K. Milner, P.N. O'Donoghue, P.P. Scott, and R.J. Ward. 1977. Dietary standards for laboratory animals: report of the Laboratory Animals Centre Diets Advisory Committee. Lab. Anim. 11: 1–28.

Clarke, R. 2000. Introduction and overview: sex steroids in the mammary gland. J. mammary Gland Biol. Neoplasia 5: 245–250.

Clarke, R., L. Hilakivi-Clarke, E. Cho, M.R. James, and F. Leonessa. 1996. Oestrogens, phytoestrogens and breast cancer. Adv. Exp. Med. Biol. 401: 63–85.

Clifton-Bligh, P.B., R.J. Baber, G.R. Fulcher, M.L. Nery, and T. Moreto. 2001. The effect of isoflavones extracted from red clover (Rimostil) on lipid and bone metabolism. Menopause 8: 259–265.

Collins-Burow, B.M., M.E. Burow, B.N. Duong, and J.A. McLachlan. 2000. Oestrogenic and antiestrogenic activities of flavonoids phytochemicals through oestrogen receptor binding-dependent and -independent mechanisms. Nutr. Cancer 38: 229–244.

Constantinou, A. and E. Huberman. 1995. Genistein as an inducer of tumor cell differentiation: possible mechanisms of action. Proc. Soc. Exp. Boil. Med. 208: 109–115.

Constantinou, A.I., D. Lantvit, M. Hawthorne, X. Xu, R.B. van Breemen, and J.M. Pezzuto. 2001. Chemopreventive effects of soy protein and purified soy isoflavones on DMBA-induced mammary tumors in female Sprague-Dawley rats. Nutr. Cancer 14: 75–81.

Cross, H.S., E. Kallay, D. Lecher, W. Gerdenitsch, H. Adlercreutz, and H.J. Armbrecht. 2004. Phytoestrogens and vitamin D metabolism: a new concept for the prevention and therapy of colorectal, prostate, and mammary carcinomas. J. Nutr. 134: 1207–1212.

Cutolo, M., A. Sulli, S. Capellino, B. Villaggio, P. Montagna, B. Seriolo, and R.H. Straub. 2004. Sex hormones influence on the immune system: basic and clinical aspects in autoimmunity. Lupus 13: 635–638.

Dai, Q., X.O. Shu, F. Jin, J.D. Potter, L.H. Kushi, J. Teas, Y.T. Gao, and W. Zheng. 2001. Population-based case-control study of soy food intake and breast cancer risk in Shanghai. Br. J. Cancer 85: 372–378.

Dang, Z.C., V. Audinot, S.E. Papapoulos, J.A. Boutin, and C.W. Lowick. 2003. Peroxisome proliferator-activated receptor gamma (PPARgamma) as a molecular target for the soy phytoestrogen genistein. J. Biol. Chem. 278: 962–967.

Davis, S.R., I. Dinatale, L. Rivera-woll, and S. Davison. 2005. Postmenopausal hormone therapy: from monkey glands to transdermal patches. J. Endocrinol. 185: 207–222.

De Andrade, E., A.A. de Mesquita, J. de Almeida Claro, P.M. de Andrade, V. Ortiz, M. Paranhos, M. Srougi, and G. Kiter. 2007. Study of the efficacy of Korean Red Ginseng in the treatment of erectile dysfunction. Asian J. Androl. 9: 241–244.

den Tonkelaar, I., L. Keinan-Boker, P.V. Veer, C.J. Arts, H. Adlercreutz, J.H. Thijssen, and P.H. Peeters. 2001. Urinary phytoestrogens and postmenopausal breast cancer risk. Cancer Epidemiol. Biomarkers Prev. 10: 223–228.

Dent, S.B., C.T. Peterson, L.D. Brace, J.H. Swain, M.B. Reddy, K.B. Hanson, J.G. Robinson, and D.L. Alekel. 2001. Soy protein intake by perimenopausal women does not affect circulating lipids and lipoproteins or coagulation and fibrinolytic factors. J. Nutr. 131: 2280–2287.

Diel, P., S. Schmidt, and G.Vollmer. 2002. *In vivo* test systems for the quantitative and qualitative analysis of the biological activity of phytoestrogens. J. Chromatogr. B 777: 191–202.

Diel, P., R.B. Geis, A. Caldarelli, S. Schmidt, U.I. Leschowky, A. Voss, and G. Vollmer. 2004a. The differential ability of the phytoestrogen genistein and of estradiol to induce uterine weight and proliferation in the rat is associated with a substance specific modulation of uterine gene expression. Mol. Cell. Endocrinol. 221: 21–32.

Diel, P., S. Schmidt, G. Vollmer, P. Janning, A. Upmeier, H. Michna, H.M. Bolt, and G.H. Degen. 2004b. Comparative responses of three rats strains (DA/Han Sprague-Dawley and Wistar) to treatment with environmental estrogens. Arch. Toxicol. 78: 183–193.

Doerge, D.R. and H.C. Chang. 2002. Inactivation of thyroid peroxidase by soy isoflavones, *in vitro* and *in vivo*. J. Chromatorgr. B Anal. Technol. Biochem. Life Sci. 777: 269–279.

Doerge, D.R. and D.M. Sheehan. 2002. Goitrogenic and estrogenic activity of soy isoflavones. Environ. Health. Perspect 110: 349–353.

Donn, A.S. and C.S. Muir. 1985. Prostate cancer-some epidemiological features. Bull. Cancer 72: 381–390.

Erlandsson, M.C., U. Islander, S. Moverare, C. Ohlsson, and H. Carlsten. 2005. Estrogenic agonism and antagonism of the soy isoflavones genistein in uterus, bone and lymphopoiesis in mice. APMIS 113: 317–323.

Evans, B.A., K. Griffiths, and M.S. Morton. 1995. Inhibition of 5-alpha reductase in genital skin fibroblasts and prostate tissue by dietary lignans and isoflavonoids. J. Endocrinol. 147: 295–302.

Flügge, G, W. Oertel, and W. Wuttke. 1986. Evidence for estrogen-receptive GABAergic neurons in the preoptic/anterior hypothalamic area of the brain. Neuroendocrinology 43: 1–5.

Foth, D. and J.M. Cline. 1998. Effects of mammalian and plant estrogens on mammary glands and uteri of macaques. Am. J. clin. Nutr. 68: 1413–1417.

Fotsis, T., M. Pepper, H. Adlercreutz, G. Fleischmann, T. Hase, R. Montesano, and L. Schweigerer. 1993. Genistein, a dietary-derived inhibitor of in vitro angiogenesis. Proc. Natl. Acad. Sci. USA 90: 2690–2694.

Franke, A.A, L.J. Custer, C.M. Cerna, and K. Narala. 1995. Rapid HPLC analysis of dietary phytoestrogens from legumes and from human urine. Proc. Soc. Exp. Biol. Med. 208: 18–26.

Franke, T.F., S.I. Yang, and T.O. Chan. 1995. The protein kinase encoded by the Akt proto-oncogene is a target of the PDGF-activated phosphatidylinositol 3-kinase. Cell 81: 727–736.

Frankenfeld, F.L., A. McTiernam, E.J. Aiello, W.K. Thomas, K. LaCroix, J. Schramm, S.M. Schwartz, V.L. Holt, and J.W. Lampe. 2004. Mammographic density in relation to daidzein-metabolizing phenotypes in overweight, postmenopausal women. Cancer Epidemiol. Biomarkers Prev. 13: 1156–1162.

Freedman, R.R. 2005. Pathophysiology and treatment of menopausal hot flashes. Semin. Reprod. Med. 23: 117–125.

Gaillard, D.,G. Chamoiseau, and R. Derache. 1977. Dietary effects on inhibition of rat hepatic microsomal drug-metabolizing enzymes by a pesticide (Morestan). Toxicology 8: 23–32.

Gallo, D., S. Giacomelli, F. Cantelmo, G.F. Zannoni, G. Ferrandina, E. Fruscella, A. Riva, P. Morazzoni, E. Bombardelli, S. Mancuso, and G. Scambia. 2001. Chemoprevention of DMBA-induced mammary cancer in rats by dietary soy. Breast Cancer Res. Treat. 69: 153–164.

Garland, E.M., T. Sakata, M.J. Fisher, T. Masui, and S.M. Cohen. 1989. Influences of diet and strain on the proliferative effect on the rat urinary bladder induced by sodium saccharin. Cancer Res. 49: 3783–3794.

Gompel, A. and J.C. Piette. 2007. Systemic lupus erythematosus and hormone replacement therapy. Menopause Int. 13: 65–70.

Greendale, G.A., G. FitzGerald, M.H. Huang, B. Sternfeld, E. Gold, T. Seeman, S. Sherman, and M. Sowers. 2002. Dietary soy isoflavones and bone mineral density: results from the study of women's health across the nation. Am. J. Epidemiol. 155: 746–754.

Gurib-Fakim, A. 2006. Medicinal plants: Traditions of yesterday and drugs of tomorrow. Mol. Asp. Med. 7: 1–93.

Hakkak, R., S. Korourian, S.R. Shelnutt, S. Lensing, M.J. Ronis, and T.M. Badger. 2000. Diets containing whey proteins or soy protein isolate protect against 7,12-dimethylbenz(a) anthracene-induced mammary tumors in female rats. Cancer Epidemiol. Biomarkers Prev. 9: 113–117.

Hakkak, R. and S. Korourian, M.J. Ronis, J.M. Johnston, and T.M. Badger. 2001. Soy protein isolate consumption protects against azoxymethane-induced colon tumors in male rats. Cancer Lett. 166: 27–32.

Ham, K.D., R.F. Loeser, B.R. Lindgren, and C.S. Carlson. 2002. Effects of long-term estrogen replacement therapy on osteoarthritis severity in cynomolgus monkeys. Arthritis Rheum. 46: 1956–1964.

Ham, K.D., T.R. Oegema, R.F. Loeser, and C.S. Carlson. 2004. Effects of long-term estrogen replacement therapy on articular cartilage IGFBP-2, IGFBP-3, collagen and proteoglycan levels in ovariectomized cynomolgus monkeys. Osteoarthritis Cartilage 12: 160–168.

Harris, R.M., D.M. Wood, L. Bottomley, S. Blagg, K. Owen, P.J. Hughes, R.H. Waring, and C.J. Kirk. 2004. Phytoestrogens are potent inhibitors of estrogen sulfation: implications for breast cancer risk and treatment. J. Clin. Endocrinol. Metab. 89: 1779–1787.

Haselkorn, T., S.L. Stewart, and P.L. Horn-Ross. 2003. Why are thyroid cancer rates so high in Southeast Asian women living in the United States? The bay area thyroid cancer study. Cancer Epidemiol. Biomarkers Prev. 12: 144–150.

Heim, M., O. Frank, G. Kampmann, N. Sochocky, T. Pennimpede, P. Fuchs, W. Hunzinker, P. Weber, I. Martin, and I. Bendik. 2004. The phytoestrogen genistein enhances osteogenesis and represses adipogenic differentiation of human primary bone marrow stromal cells. Endocrinology 145: 848–859.

Hellstrom, M., M. Haggman, and S. Branstedt. 1993. Histopathological changes in androgen deprived localised prostatic cancer. A study of total prostatectomy specimens. Eur. Urol. 24: 461–465.

Hermansen, K., M. Sondergaard, L. Hoie, M. Cartensen, and B. Brock. 2001. Beneficial effects of a soy-based dietary supplement on lipid levels and cardiovascular risk markers in type 2 diabetic subjects. Diabetes care 24: 228–233.

Holmdahl, R., H. Carlsten, L. Jansson, and P. Larsson. 1989. Oestrogen is a potent immunomodulator of murine experimental rheumatoid disease. Br. J. Rheumatol. 28: 54–58.

Horiuchi, T., T. Onouchi, M. Takahashi, H. Ito, and H. Orimo. 2000. Effect of soy protein on bone metabolism in postmenopausal Japanese women. Osteoporos. Int. 11: 721–724.

Horn-Ross, P.L., E.M. John, A.J. Canchola, S.L. Stewart, and M.M. Lee. 2003. Phytoestrogen intake and endometrial cancer risk. J. Natl. Cancer Inst. 95: 1158–1161.

Hoshiyama, Y., T. Sekine, and T. Sasaba. 1993. A case-control study of colorectal cancer and its relation to diet, cigarettes, and alcohol consumption in Saitama Prefecture, Japan. Tohoku J. Exp. Med. 171: 153–165.

Hsieh, C.Y., R.C. Santell, S.Z. Haslam, and W.G. Helferich. 1998. Estrogenic effects of genistein on the growth of estrogen receptor-positive human breast cancer (MCF-7) cells *in vitro* and *in vivo*. Cancer Res. 58: 3833–3838.

Hsu, C.S., W.W. Shen, Y.M. Hsuch, and S.L. Yeh. 2001. Sot isoflavones supplementation in postmenopausal women. Effects on plasma lipids, antioxidant enzyme activities and bone density. J. Reprod. Med. 46: 221–226.

Humala-Tasso, K.K. and P.O. Combelles. 2003. La maca, une culture millénaire d'altitude. Article publié dans "Pour la Science" n°311, pp. 25–39.

Ishimi, Y., C. Miyaura, M. Ohmura, Y. Onoe, T. Sato, Y. Uchiyama, M. Ito, X.Wang, T. Suda, and S. Ikegami. 1999. Selective effects of genistein, a soybean isoflavones, on B-lymphopoiesis and bone loss caused by estrogen deficiency. Endocrinology 140: 1893–1900.

Jarry, H., S. Leonhardt, and W. Wuttke. 1991. Gamma-amminobutyric acid neurons in the preoptic/anterior hypothalamic area synchronize the phasic activity of the gonadotropin-releasing hormone pulse generator in ovariectomized rats. Neuroendocrinology 53: 261–267.

Jarry, H., P.M. Wise, S. Leonhardt, and W. Wuttke. 1999. Effects of age on GABA turnover rates in specific hypothalamic areas in female rats. Exp. Clin. Endocrinol. Diabetes 107: 59–62.

Jones, J.L., B.J. Daley, B.L. Enderson, J.R. Zhou, and M.D. Karlstad. 2002. Genistein inhibits Tamoxifen effects on cell proliferation and cell cycle arrest in T47D breast cancer cells. Am. Surg. 68: 575–577.

Ju, Y.H., D.R. Doerge, K.F. Allred, C.D. Allred, and W.G. Helferich. 2002. Dietary genistein negates the inhibitory effect of tamoxifen on growth of estrogen-dependent human breast cancer (MCF-7) cells implanted in athymic mice. Cancer Res. 62: 2474–2477.

Kaari, C., M.A. Haidar, J.M. Junior, M.G. Nunes, L.G. Quadros, C. Kemp, J.N. Stavale, and E.C. Baracat. 2005. Randomized clinical trial comparing conjugated equine estrogens and isoflavones in postmenopausal women: a pilot study. Maturitas 53: 49–58.

Kallay, E., H. Adlercreutz, H. Farhan, D. Lechner, E. Bajna, W. Gerdenitsch, M. Campbell, and H.S. Cross. 2002. Phytoestrogens regulate vitamin D metabolism in the mouse colon: relevance for colon tumor prevention and therapy. J. Nutr. 132: 3490–3493.

Kamatenesi-Mugisha, M. and H. Oryem-Origa. 2005. Traditional herbal remedies used in the management of sexual impotence and erectile dysfunction in western Uganda. African Health Sciences 5: 40–49.

Kavanagh, K.T., L.J. Hafer, and D.W. Kim. 2001. Green tea extracts decrease carcinogen-induced mammary tumor burden in rats and rate of breast cancer cell proliferation in culture. J. Cell. Biochem. 82: 387–398.

Kawano-Takahashi, Y., H. Ohminami, H. Okada, I. Kitagawa, M. Yoshikawa, S. Arichi, and T. Hayashi. 1986. Effect of soya saponins on gold thioglucose (GTG)-induced obesity in mice. Int. J. Obes. 10: 293–302.

Ketcha Wanda, G.J.M., D. Njamen, E. Yankep, F.S. Tagatsing, Z. Tanee Fomum, J. Wober, S. Starcke, O. Zierau, and G. Vollmer. 2006. Estrogenic properties of isoflavones derived from *Millettia griffoniana*. Phytomedicine 13: 139–145.

Ketcha Wanda, G.J.M., S. Starcke, O. Zierau, D. Njamen, T. Richter, and G. Vollmer. 2007. Estrogenic activity of griffonianone C, an isoflavones from the root bark of *Millettia griffoniana*: Regulation of the expression of estrogen responsive genes in uterus and liver of ovariectomized rats. Planta Medica 73: 512–518.

Key, T.J., G.B. Sharp, P.N. Appleby, V. Beral, M.T. Goodman, M. Soda, and K. Mabuchi. 1999. Soya foods and breast cancer risk: a prospective study in Hiroshima and Nagasaki, Japan. Br. J. Cancer 81: 1248–1256.

Kijkuokool, P., I.S. Parhar, and S. Malaivijitnond. 2005. Genistein enhances N-nitrosomethylurea-induced rat mammary tumorigenesis. Cancer Lett. 242: 53–59.

Kim, H.S., L.H. Kang, T.S. Kim, H.J. Moon, I.Y. Kim, H. Ki, K.I. Park, B.M. Lee, S.D. Yoo, and S.Y. Han. 2005. Validation study of OECD rodent uterotrophic assay for the assessment of estrogenic activity in Sprague-Dawley immature female rats. J. Toxicol. Environ. Health A 68: 2249–2262.

Kim, S.W., N.N. Kim, S.J. Jeong, R. Munarriz, I. Goldstein, and A.M. Traish. 2004. Modulation of rat vaginal blood flow and estrogen receptor by estradiol. J. Urol. 172: 1538–1543.

Knight, D.C. and J.A. Eden. 1995. Phytoestrogens-a short review. Maturitas 22: 167–175.

Knight, D.C., J.A. Eden, and G.E. Kelly. 1996. The phytoestrogen content of infant formulas. Med. J. Aust. 164: 575.

Knobil, E. 1980. The neuroendocrine control of the menstrual cycle. Recent Prog. Horm. Res. 36: 53–88.

Knobil, E. 1990. The GnRH pulse generator. Am. J. Obset. Gynecol. 163: 1721–1727.

Kolonel, L.W., J.H. Hankin, and A.M.Y. Nomura. 1986. Multiethnic studies of diet, nutrition, and cancer in Hawaii. In: Y. Hayashi , M. Nagao, T. Sugimura, S. Takayama, L. Tomatis, L.W. Wattenberg, and G.N. Wogan. (eds.). Nutrition and cancer. Japanese Science Society Press, Tokyo. pp. 29–40.

Kondo, K., Y. Suzuki, Y. Ikeda, and K.Umemura. 2002. Genistein, an isoflavones included in soy, inhibits thrombotic vessel occlusion in the mouse femoral artery and *in vitro* platelet aggregation. Eur. J. Pharmacol. 455: 53–57.

Kousidou, O.C., T.N. Mitropoulou, A.E. Roussidis, D. Kletsas, A.D. Theocharis, and N.K. Karamanos. 2005. Genistein suppresses the invasive potential of human breast cancer cells through transcriptional regulation of metalloproteinases and their tissue inhibitors. Int. J. Oncol. 26: 1101–1109.

Kuiper, G.G., B. Carlsson, and K. Gradien. 1987. Comparison of the ligand binding specificity and transcript tissue distribution of oestrogen receptors alpha and beta. Endocr. 138: 863–870.

Kuiper, G.G., B. Carlsson, K. Grandien, E. Enmark, J. Haggblad, S. Nilsson, and J.A. Gustafsson. 1997. Comparison of the ligand binding specificity and transcript tissue distribution of oestrogen receptors alpha and beta. Endocr 138: 863–870.

Lahita, R.G., L. Bradlow, J. Fishman, and H.G. Kunkel. 1982. Estrogen metabolism in systemic lupus erythematosus: patients and family members. Arthritis Rheum 25: 843–846.

Lamartiniere, C.A., M.S. Cotroneo, W.A. Fritz, J. Wang, R.Mento-Marcel, and A. Elgavish. 2002. Genistein chemoprevention: timing and mechanisms of action in murine mammary and prostate. J. Nutr. 132: 552–558.

Lamb, J.C., M.A. Levy, R.K. Johnson, and J.T. Isaacs. 1992. Response of rat and human prostatic cancers to the novel 5a-reductase inhibitors, SK&F 105657. Prostate 21: 15–34.

Lans, C. 2006. Creole remedies of Trinidad and Tobago.*Lulucom*http://www.lulu.com/content/302210].

Le Bail, J.C., F. Varnat, J.C. Nicolas, and G. Habrioux. 1998. Oestrogenic and antiproliferative activities on MCF-7 human breast cancer cells by flavonoids. Cancer Lett. 130: 209–216.

Lee, J.S. 2006. Effects of soy protein and genistein on blood glucose, antioxidant enzyme activities, and lipid profile in streptozotocin-induced diabetic rats. Life Sci. 79: 1578–1584.

Li, Y., S. Upadhyay, M. Bhuiyan, and F.H. Sarkar. 1999. Induction of apoptosis in breast cancer cells MDA-MB-231 by genistein. Oncogene 18: 3166–3172.

Lien, T.F., Y.L. Hsu, D.Y.Lo, and R.Y.Y. Chiou. 2009. Supplementary health benefits of soy aglycons of isoflavone by improvement of serum biochemical attributes, enhancement of liver antioxidative capacities and protection of vaginal epithelium of ovariectomized rats. Nutr. Metab. 6:15.

Liu, D., W. Zhen, Z. Yang, J.D. Carter, H. Si, and K.A. Reynolds. 2005. Genistein acutely stimulates insulin secretion in pancreatic beta-cells through a cAMP-dependent protein kinase pathway. Diabetes 55: 1043–1050.

Lock, M. 1991. Contested meanings of the menopause. Lancet 337: 1270–1272.

Lock, M. 1994. Menopause in cultural contest. Exp. Gerontol. 29: 307–317.

Magne, Ndé C.B., D. Njamen, J.C. Mbanya, Z.T.Fomum, O.Zierau, and G.Vollmer. 2007. Estrogenic effects of the methanol extract of the fruits of *Brenania brieyi* (Rubiaceae). J. Nat. Medicine 61: 86–89.

Mahalko, J.R. H.H. Sandstead, L.K. Johnson, L.F. Inman, D.B. Milne, R.C. Warner, and E.A. Haunz. 1984. Effect of consuming fiber from corn bran, soy hulls, or apple powder on glucose tolerance and plasma lipids in type II diabetes. Am. J. Clin. Nutr. 39: 25–34.

Makela, S., H. Savolainen, E. Aavik, M. Myllarniemi, L. Strauss, E. Taskinen, J.A. Gustafsson, and P. Hayry. 1999. Differentiation between vasculoprotective and uterotrophic effects of ligands with different binding affinities to estrogen receptors alpha and beta. Proc. Natl. Acad. Sci. USA 7077–7082.

Malaivijitnond, S., K. Chansri, P. Kijkuokul, N. Urasopon, and W. Cherdshewasart. 2006. Using vaginal cytology to assess the estrogenic activity of phytoestrogen-rich herb. J. Ethnopharmaol. 107: 354–360.

Martin, P.M., K.B. Horwitz, D.S. Ryan, and W.L. McGuire. 1978. Phytoestrogen interaction with estrogen receptors in human breast cancer cells. Endocrinology 103: 1860–1867.

Maskarinec, G. and L. Meng. 2001. An investigation of soy intake and mammographic characteristics in Hawaii. Breast Cancer Res. 3: 134–141.

Maskarinec, G., A.E. Williams, and L. Carlin. 2003. Mammographic densities in a one-year isoflavones intervention. Eur. J. Cancer Prev. 12: 165–169.

Mazur, W.M., Wahalak, S. Rasku, A. Salakka, T. Hase, and H. Adlercreutz. 1998. Lignan and isoflavonoid concentration in tea and coffee. Br. J. Nutr. 127: 1260–1268.

Mei, J., S.S. Yeung, and A.W. Kung. 2001. High dietary phytoestrogen intake is associated with higher bone mineral density in postmenopausal but not premenopausal women. J. Clin. Endocrinol. Metab. 86: 5217–5221.

Menon, L.G., R. Kuttan, M.G. Nair, Y.C. Chang, and G. Kuttan. 1998. Effects of isoflavones genistein and daidzein in the inhibition of lung metastasis in mice induced by B16F-10 melanoma cells. Nutr. Cancer 30: 74–77.

Messina, M. and M. Bennink. 1998. Soyfoods, isoflavones and risk of colonic cancer: a review of the in vitro and in vivo data. Baillieres Clin. Endocrinol. Metab. 12: 707–728.

Messina, M.J. and C.L. Loprinzi. 2001. Soy for breast cancer survivors: a critical review of the literature. J. Nutr. 131: 3095–3108.

Miksicek, R.J. 1993. Commonly occurring plant flavonoids have oestrogenic activity. Mol. Pharmacol. 44: 37–43.

Miksicek, R.J. 1995. Oestrogenic flavonoids: structural requirements for biological activity. Proc.Soc. Exp. Biol. Med. 208: 44–50.

Miller, SG. 1988. Diagnosis of stage A prostatic cancer in People's Republic of China. In: D.S.Coffey, M.I. Resnick, F.A.Dorr,and J.P. Karr (eds.). Multidisciplinary analysis of controversies in the management of prostatic cancer. Plenum Press, New York. pp. 17–24.

Miquel, J.,A. Ramirez-Bosca, J.V. Ramirez-Bosca, and J.D. Alperi. 2006. Menopause: a review on the role of oxygen stress and favorable effects of dietary antioxidants. Arch. Gerontol. Geriatr. 42: 289–306.

Morito, K., T. Hirose, and J. Kinjo. 2001. Interaction of phytoestrogens with estrogen receptors alpha and beta. Biol. Pharm. Bull. 24: 351–356.

Murray, M.J., W.R. Meyer, B.A. Lessey, R.H. Oi, R.E. DeWire, and M.A. Fritz. 2003. Soy protein isolate with isoflavones does not prevent estradiol-induced endometrialhyperplasia in postmenopausal women: a pilot trial. Menopause 10: 456–464.

Murthy, D.R., C.M. Reddy, and S.B. Patil. 1997. Effect of benzene extract of *Hibiscus rosa sinensis* on the estrous cycle and ovarian activity in albino mice. Biol. Pharm. Bull. 20: 756–758.

Narayan, S., G. Rajakumar, H. Prouix, and P. Singh. 1992. Estradiol is trophic for colon cancer in mice: effect on ornithine decarboxylase and c-myc messenger RNA. Gastroenterology 103: 1823–1832.

National Research Council, Committee on Animal Nutrition. 1995. Nutrient requirements of the Laboratory Animals. National Academy Press, Washington DC. pp.11–79.

Nelson, H.D., K.K. Vesco, E. Haney, R. Fu, A. Nedrow, J. Miller, C. Nicolaidis, M. Walker, and L. Humphrey. 2006. Nonhormonal therapies for menopausal hot flashes: systematic review and meta-analysis. JAMA 295: 2057–2071.

Nikander, E., E.M. Rutanen, P. Nieminen, T. Wahlstrom, O.Ylikorkala, and A. Titinen. 2005. Lack of effect of isoflavonoids on the vagina and endometrium in postmenopausal women. Fertile Steril 83: 137–142.

Njamen, D., C.B. Magne Ndé, Z.Tanee Fomum, and G. Vollmer. 2008. Effects of the extracts of some tropical medicinal plants on estrogen inducible yeast and Ishikawa screens, and on ovariectomized rats. Pharmazie 63: 164–168.

OECD. 2001. Third meeting of the validation management group for the screening and testing of endocrine disrupters (mammalian effects). Joint meeting of the chemicals committee and the working party on chemicals, pesticides and biotechnology. http://www.oecd.org.

Pan, Y., M.S. Anthony, M. Binns, and T.B. Clarkson. 2001. A comparison of oral micronized estradiol with soy phytoestrogen effects on tail skin temperatures of ovariectomized rats. Menopause 8: 171–174.

Pasqualini, J.R. and G.S. Chetrite. 2005. Recent insight on the control of enzymes involved in estrogen formation and transformation in human breast cancer. J. Steroid. Biochem. Mol. Boil. 93: 221–236.

Pei, R.J., M. Sato, T. Yuri, N. Danbara, Y. Nikaido, and A. Tsubura. 2003. Effect of prenatal and prepubertal genistein exposure on N-methyl-N-nitrosourea-induced mammary tumorigenesis in female Sprague-Dawley rats. *In Vivo* 17: 349–357.

Penotti, M., E. Fabio, A.B. Modena, M. Rinaldi, U. Omodei, and P.Vigano. 2003. Effect of soy-derived isoflavones on hot flushes, endometrial thickness, and the pulsatility index of the uterine and cerebral arteries. Fertile Steril 79: 1112–1117.

Persky, V.W., M.E. Turyk, L. Wang, S. Freels, Jr. R. Chatterton, S. Barnes, Jr. J. Erdman, D.W. Sepkovic, H.L. Bradlow, and S. Potter. 2002. Effect of soy protein on endogenous hormones in postmenopausal women. Am. J. Clin. Nutr. 75: 145–153.

Peterson, G. and S. Barnes. 1993. Genistein and biochanin A inhibit the growth of human prostate cancer cells but not epidermal growth factor receptor tyrosine autophosphorylation. Prostate 22: 335–345.

Petrakis, N.L., S. Barnes, E.B. King, J. Lowenstein, J. Wiencke, M.M. Lee, R. Miike, M. Kirk, and L. Coward. 1996. Stimulatory influence of soy protein isolate on breast secretion in pre- and postmenopausal women. Cancer Epidemiol. Biomarkers Prev. 5: 785–794.

Pfaff, D.W., N. Vasudevan, H.K. Kia, Y.S. Zhu, J. Chan, J. Garey, M. Morgan, and S. Ogawa. 2000. Estrogens, brain and behavior: studies in fundamental neurobiology and observations related to women's health. J. Steroid. Biochem. Mol. Biol. 74: 365–373.

Picherit, C., V. Coxam, C. Bennetau-Pelissero, S. Kati-Coulibaly, M.J. Davicco, P. Lebecque, and J.P. Barlet. 2000. Daidzein is more efficient than genistein in preventing ovariectomy-induced bone loss in rats. J. Nutr. 130: 1675–1681.

Platet, N., A.M. Cathiard, and M. Gleizes. 2004. Estrogens and their receptors in breast cancer progression: a dual role in cancer proliferation and invasion. Crit. Rev. oncol. Hematol. 51: 55–67.

Pollard, M., P.H. Luckert, and D.L. Snyder. 1989. The promotional effect of testosterone on induction of prostate cancer in MNU-sensitive L-W rats. Cancer Lett. 45: 209 –212.

Potter, S.M., J.A. Baum, H. Teng, R.J. Stillman, N.F. Shay, and J.W. .Erdman Jr. 1998. Soy protein and isoflavones: their effects on blood lipids and bone density in postmenopausal women. Am. J. Clin. Nutr. 68: 1375–1379.

Rachon, D., G. Rimoldi, and W. Wuttke. 2006. In vitro effects of genistein and resveratrol on the production of interferon-gamma (IFNgamma) and interleukin-10 (IL-10) by stimulated murine splenocytes. Phytomedicine 13: 419–424.

Ramasamy, I. 2006. Recent advances in physiological calcium homeostasis. Clin. Chem. Lab. Med. 44: 237–273.

Ramsey-Goldman, R. 2005. Does hormone replacement therapy affect activity in patients with systemic erythematosus? Nat. Clin. Pract. Rheumatol. 1: 72–73.

Rao, G.N. 1996. New diet (NTP-2000) for rats in the National Toxicology Program toxicity and carcinogenicity studies. Fundam. Appl. Toxicol. 32: 102–108.

Ratnasooriya, W.D. and J.R.A.C. Jayakody. 2006. Effects of aqueous extract of Alpinia calcarata rhizomes on reproductive competence of male rats. Acta. Biol. Hung. 57: 23–35.

Ratnasooriya, W.D. and T.S.P. Fernando. 2008. Effect of black tea brew of Camellia sinensis on sexual competence of male rats. J. Ethnopharmacol. 118: 373–377.

Ratz, P.H., K.A. McCammon, D. Altstatt, P.F. Blackmore, O.Z. Shenfeld, and S.M. Schlossberg. 1999. Differential effects of sex hormones and phytoestrogens on peak and steady contractions in isolated rabbit detrusor. J. Urol. 162: 1821–1828.

Reddy, C.M., D.R. Murthy, and S.B. Patil. 1997. Antispermatogenic and androgenic activities of various extracts of *Hibiscus rosa sinesis* in albino mice. Indian J. Exp. Biol. 35: 1170–4.

Register, T.C., M.J. Jayo, and M.S. Anthony. 2003. Soy phytoestrogens do not prevent bone loss in postmenopausal monkeys. J. Clin. Endocrinol. Metab. 88: 4362–4370.

Register, T.C. and J.A. Cann, J.R. Kaplan, J.K. Williams, M.R. Adams, T.M. Morgan, M.S. Anthony, R.M. Blair, J.D. Wagner, and T.B. Clarkson. 2005. Effects of soy isoflavones and conjugated equine estrogens on inflammatory markers in atherosclerotic, ovariectomized monkeys. J. Clin. Endocrinol. Metab. 90: 1734–1740.

Ritchie, H.E. 2001. The safety of herbal medicine use during pregnancy. Frontiers in Fetal Health 3: 259–266.

Rokhlin, O.W. and M.B. Cohen. 1995. Differential sensitivity of human prostatic cancer cell lines to the effects of protein kinase and phosphatase inhibitors. Cancer Lett. 98: 103–110.

Ross, R.K., L. Bernstein, and R.A. Lobo. 1992. 5-a-Reductase activity and risk of prostate cancer among Japanese and US white and black males. Lancet 339: 887– 889.

Sasamura, H., A. Takahashi, J. Yuan, H. Kitamura, N. Masumori, N. Miyao, N. Itoh, and T. Tsukamoto. 2004. Antiproliferative and antiangiogenic activities of genistein in human renal cell carcinoma. Urology 64: 389–393.

Schaefer, O., M. Humpel, K.H. Fritzemeier, R. Bohlmann, and W.D. Schleuning. 2003. 8-Prenyl naringenin is a potent ERalpha selective phytoestrogen present in hops and beer. J. Steroid Biochem. Mol. Biol. 84: 359–360.

Seidlova-Wuttke, D., T. Becker, V. Christoffel, H. Jarry, and W. Wuttke. 2003a. Silymarin is a selective estrogen receptor beta (ERbeta) agonist and has estrogenic effects in the metaphysic of the femur but no or antiestrogenic effects in the uterus of ovariectomized (ovx) rats. J. Steroid Biochem. Mol. Biol. 86: 179–188.

Seidlova-Wuttke, D., H. Jarry, T. Becker, J. Christoffel, and W.Wuttke. 2003b. Pharmacology of Cimicifuga racemosa extract BNO 1055 in rats: bone, fat and uterus. Maturitas 44: 39–50.

Seidlova-Wuttke, D., A. Schultens, H. Jarry, and W. Wuttke. 2004. Urodynamic effects of estradiol (E2) in ovariectomized (ovx) rats. Endocrine 23: 25–32.

Setchell, K.D. 1998. Phytoestrogens: the biochemistry, physiology, and implication for human health of soy isoflavones. Am. J. Clin. Nutr. 68: 1333–1346.

Shaaban, A.M., P.A. O'Neill, M.P. Davies, R. Sibson, C.R. West, P.H. Smith, and C.S. Foster. 2003. Declining estrogen receptor-beta expression defines malignant progression of human breast neoplasia. Am. J. Surg. Pathol. 27: 1502–1512.

Shao, Z.M., J. Wu, Z.Z. Shen, and S.H. Barsky. 1998. Genistein inhibits both constitutive and EGF-stimulated invasion in ER-negative human breast carcinoma cell lines. Anticancer Res. 18: 1435–1439.

Sharma, O.P.,H. Adlercrueutz, J.D. Strandberg, B.R. Zirkin, D.S. Coffey, and L.L. Ewing. 1992. Soy of dietary source plays a preventive role against the pathogenesis of prostatitis in rats. J. Steroid Biochem. Mol. Biol. 43: 557–564.

Shu, X.O., F. Jin, Q. Dai, W. Wen, J.D. Potter, L.H. Kushi, Z. Ruan, Y.T. Gao, and W. Zheng. 2001. Soy food intake during adolescence and subsequent risk of breast cancer among Chinese women. Cancer Epidemiol. Biomarkers Prev. 10: 483–488.

Silva, M.J., F.R.Capaz, and M.R. Valc. 2000. Effects of the water soluble fraction from leaves of *Ageratum conyzoides* on smooth muscle. Phytotherapy Research 14: 130–132.

Simick, M.K., R. Jong, B. Wilson, and L. Lilge. 2004. Non-ionizing near-infrared radiation transillumination spectroscopy for breast tissue density and assessment of breast cancer risk. J. Biomed. Opt. 9: 794–803.

Son, H.Y., A. Nishikawa, T. Ikeda, T. Imazawa, S. Kimura, and M. Hirose. 2001. Lack of effect of soy isoflavones on thyroid hyperplasia in rats receiving an iodine-deficient diet. Jpn. J. Cancer. Res. 92: 103–108.

Sondik, E. 1988. Incidence, survival and mortality trends in the United States. In: D.S.Coffey,M.I. Resnick,, F.A. Dorr, and J.P. Karr (eds.). A multidisciplinary analysis of controversies in the management of prostate cancer. Plenum Press, New York. pp. 9–16.

Sonnenschein, C. and A.M. Soto. 1998. An update review of environmental estrogen and androgen mimics and antagonists. J. Steroid Biochem. Mol. Biol. 65: 143–150.

Stephens, F.O. 1997. Phytoestrogens and prostate cancer: possible preventive role. Med. J. Australia 167: 138–140.

Tajima, K. and S.Tominaga. 1985. Dietary habits and gastro-intestinal cancers: a comparative case-control study of stomach and large intestinal cancer in Nagoya, Japan. Jpn. J. Cancer Res. 76: 705–716.

Tanee, F.S.F., D. Njamen, C. Magne Ndé, J. Wanji, O. Zierau, Z.T. Fomum, and G. Vollmer. 2007. Estrogenic effects of the ethyl-acetate extract of the stem bark of *Erythrina lysistemon* Hutch (Fabaceae). Phytomedicine 14: 222–226.

Tang, G.W. 1994. The climacteric of Chinese factory workers. Maturitas 19: 177–182.

Teede, H.J., F.S. Dalais, D. Kotsopoulos, B.P. McGrath, E. Malan, T.E. Gan, and R.E. Peverill. 2005. Dietary soy containing phytoestrogens does not activate the hemostatic system in postmenopausal women. J. Clin. Endocrinol. Metab. 90: 1936–1941.

Thomas, D.K., L.H. Storlien, W.P. Bellingham, and K. Gillette. 1986. Ovarian hormone effects on activity, glucoregulation and thyroid hormones in the rat. Physiol. Behav. 36: 567–573.

Tice, J.A., B. Ettinger, K. Ensrud, R. Wallace, T. Blackwell, and S.R. Cummings. 2003. Phytoestrogen supplements for the treatment of hot flashes: the Isoflavones Clover Extract (ICE) Study: a randomized controlled trial. JAMA 290: 207–214.

Tsai, A.C., A.I. Vinik, A. Lasichak, and G.S. Lo. 1987. Effects of soy polysaccharide on postprandial plasma glucose, insulin, glucagon, pancreatic polypeptide, somatostatin, and triglyceride in obese diabetic patients. Am. J. Clin. Nutr. 45: 596–601.

Ueda, M., N. Niho, T. Imai, M. Shibutani, K. Mitsumori, T. Matsui, and M. Hirose. 2003. Lack of significant effects of genistein on the progression of 7,12-dimethylbenz(a) anthracene-induced mammary tumors in ovariectomized Sprague Dawley rats. Nutr. Cancer 47: 141–147.

Umland, E.M., J.S. Cauffield, J.K. Kirk, and T.E. Thomason. 2000. Phytoestrogens as therapeutic alternatives to traditional hormone replacement in postmenopausal women. Pharmacotherapy 20: 981–990.

Unfer, V., M.L. Casini, L. Costabile, M.Mignosa, S.Gerli, and G.C. Di Renzo. 2004. Endometrial effects of long-term treatment with phytoestrogens: a randomized, double-blind, placebo-controlled study. Fertile Steril 82: 145–148.

Upadhyaya, P. and K. El-Bayoumi. 1998. Effect of dietary soy protein isolate, genistein, and 1,4-phenylenebis(methylene)selenocyanate on DNAbinding of 7,12-dimethylbenz(a) anthracene in mammary glands of CD rats. Oncol. Rep. 5: 1541–1545.

Usui ,T. 2006. Pharmaceutical prospects of phytoestrogens. End. J. 53: 7–20.

Verdeal, K., R.R. Brown, T. Richardson, and D.S. Ryan. 1980. Affinity of phytoestrogens for oestradiol-binding proteins andeffect of coumestrol on growth of 7,12-dimethylbenz[a] anthracene-induced rat mammary tumors. J. Natl. Cancer. Inst. 64: 285–290.

Wade, M.G., A. Lee, A. McMahon, G. Cooke, and I. Curran. 2003. The influence of dietary isoflavones on the uterotrophic response in juvenile rats. Food Chem. Toxicol. 41: 1517–1525.

Warner, M., S. Saji, and J.A. Gustafsson. 2000. The normal and maglinant mammary gland: a fresh look with ER beta onboard. J. Mammary Gland Biol. Neoplasia 5: 289–294.

Warren, M.P. 2004. A comparative review of the risks and benefits of hormone replacement therapy regimens. Am. J. Obstet. Gynecol. 190: 1141–1167.

Weihua, Z., S. Anderson, G. Cheng, E.R. Simpson, M. Warner, and J.A. Gustafsson. 2003. Update on estrogen signaling. FEBS Lett. 546: 17–24.

Whitten, P.L., C. Lewis, and E. Russell. 1995. Potential adverse effects of phytoestrogens. J. Nutr. 125: 771–776.

Wood, C.E., T.C. Register, M.S. Anthony, N.D. Kock, and J.M. Cline. 2004. Breast and uterine effects of soy isoflavones and conjugated equine estrogens in postmenopausal female monkeys. J. Clin. Endocrinol. Metab. 89: 3462–3468.

Wrensch, M.R., N.L. Petrakis, R. Miike, E.B. King, K. Chew, J. Neuhaus, M.M. Lee, and M. Rhys. 2001. Breast cancer risk in women with abnormal cytology in nipple aspirates of breast fluid. J. Natl. Cancer Inst. 93: 1791–1798.

Wu, A.H., M.C. Yu, C.C. Tseng, N.C. Twaddle, and D.R. Doerge. 2004. Plasma isoflavones levels versus self-reported soy isoflavones levels in Asian- American women in Los Angeles County. Carcinogenesis 25: 77–81.

Wuttke, W., H. Jarry, T. Becker, A. Schultens, V. Christoffel, C. Gorkow, and D. Seidlova-Wuttke. 2003. Phytoestrogens: endocrine disrupters or replacement for hormone replacement therapy? Maturitas 44: 9–20.

Wuttke, W., H. Jarry, and D.Seidlová-Wuttke. 2007. Isoflavones-Safe food additives or dangerous drugs? Ageing Research Reviews 6: 156.

Yamamoto, S., T. Sobue, M. Kobayashi, S. Sasaki, and S. Tsugane. 2003. Soy, isoflavones, and breast cancer risk in Japan. J. Natl. Cancer Inst. 95: 906–913.

Yang, X., Y.Z. Li, Z. Mao, P. Gu, and M. Shang. 2009. Effects of estrogen and tibolone on bladder istology and estrogen receptors in rats. Chinese Med. J. 122: 381–385.

Yoneda, T., N. Ishimaru, R. Arakaki, M. Kobayashi, T. Izawa, K. Moriyama, and Y. Hayashi. 2004. Estrogen deficiency accelerates murine autoimmune arthritis associated with receptor activator of nuclear factor-kappa B ligand-mediated osteoclastogenesis. Endocrinology 145: 2384–2391.

Ziegler, R.G., R.N. Hoover, M.C. Pike, A. Hildesheim, A.M. Nomura, D.W. West, A.H. Wu-Williams, L.N. Kolonel, P.L. Horn-Ross, J.F. Rosenthal, and M.B. Hyer. 1993. Migration patterns and breast cancer risk in Asian-American women. J. Natl. Cancer Inst. 85: 1819–1827.

Chapter 15

Applications of Microarray Technology in Ethnomedicinal Plant Research

Mahmoud Youns,[1] Jörg D. Hoheisel,[2] and *Thomas Efferth[1]**

Introduction

Nowadays, a variety of DNA array and DNA chip devices and systems are available. Application of such devices allows DNA and/or RNA hybridization analysis to be carried out in highly parallel formats. Applications of DNA microarray hybridization are mostly directed at gene expression analysis or screening samples for single nucleotide polymorphisms (SNPs). In addition to the molecular biologically related analyses and genomic research applications, such microarray systems are also being applied for infectious and genetic diseases, cancer diagnostics and forensic and genetic identification purposes. Array technology continues to improve in performance aspects regarding sensitivity and selectivity and in becoming a more economical research tool. DNA microarrays will continue to revolutionize genetic analysis and many important diagnostic areas. Additionally, microarray technology is now being applied to new areas of proteomic and cellular analysis (Heller 2002).

[1]Department of Pharmaceutical Biology, Institute of Pharmacy, University of Mainz, Staudinger Weg 5; 55099 Mainz, Germany.
[2]Functional Genome Analysis, German Cancer Research Center (DKFZ), Im Neuenheimer Feld 580, 69120 Heidelberg, Germany.
Corresponding address: Professor Thomas Efferth, Department of Pharmaceutical Biology, Institute of Pharmacy, University of Mainz, Staudinger Weg 5; 55128 Mainz; Germany.
E-mail: *efferth@uni-mainz.de*

The continuous gain of information on the sequence of entire genomes has increasingly motivated researchers to identify the functions of these genes and their interaction pathways in health and disease (Golub et al. 1999). Genes of the human genome project constitute approximately the whole possible and promising therapeutic targets for medicine (Lander et al. 2001, Venter et al. 2001). The emergence of new tools and technologies enables investigators to address previously intractable problems and to reveal novel potential targets for therapies. Methods to measure gene expression include serial analysis of gene expression (SAGE) differential display, northern blotting, and dot-blot analysis. All previously mentioned techniques are inappropriate for the high throughput analysis of the expression of multiple genes at once. Nowadays, microarray technology can be used to test tens of thousands of genes at the same time and to monitor the expression of those genes. Microscopic arrays of large sets of DNA sequences immobilized on solid substrates are becoming a standard technology applied all over the world. Arrays are ordered samples of DNA sequences with each sample representing a particular gene. These arrays can then be assayed for changes in the gene expression of the representative genes after various treatments, different conditions or tissue origins, thus providing a functional aspect to sequence information in a given sample (Eisen and Brown 1999). DNA microarrays are precious tools in the identification or quantification of many specific DNA sequences in complex nucleic acid samples (Masino et al. 2000). Therefore, they have been used to identify cardinal aspects of growth and development, as well as to explore the underlying genetic causes of many human diseases (Debouck and Goodfellow 1999). Since its first application (Schena et al. 1995), microarray technologies have been productively functional to almost each and every aspect of biomedical research (Cole et al. 1999, Debouck and Goodfellow 1999, Gerhold et al. 2002, Jayapal and Melendez 2006, Perou et al. 2000).

Microarray based studies have enormous potential in the exploration of disease processes such as cancer (Cole et al. 1999) and in drug response, design and development (Gerhold et al. 2002). In addition, the technology is applied to a considerable extent to investigate many pathological conditions, such as pulmonary fibrosis (Kaminski et al. 2000), inflammation (Heller et al. 1997), breast cancer (Perou et al. 2000) and colon cancer (Alon et al. 1999). Microarrays generate gene expression 'profiles'. Such profiles are comprehensive patterns that are characteristic of the responses of cells or tissues to drug treatment, environmental changes, differentiation into specialized tissues, or to dedifferentiation into tumor cells. Thus, microarrays documents detailed responses of cells and tissues to both disease and the intended and unintended effects of drug treatments and hence improving medical research (Golub et al. 1999,

Pomeroy et al. 2002, Gerhold et al. 2002). Moreover, such microarray trials expand the size of existing gene families, discover new patterns of coordinated gene expression across gene families and disclose entirely new classes of genes.

Microarray design

DNA microarrays are generally fabricated on glass, plastic, or silicon substrates. Such microarrays comprise from a hundred to many thousands of test sites that can range in size from 10 to 500 microns. High density microarrays may have up to 106 test sites in a 1–2 cm² area. DNA probes are selectively spotted to individual test sites by a variety of techniques. Probes can include synthetic oligonucleotides, larger DNA/RNA fragments, or amplicons. The DNA probes can be either covalently or noncovalently attached to support material. Depending on the array format, probes can be the target DNA or RNA sequences to which other 'reporter probes' would subsequently be hybridized (Heller 2002). The actual construction of microarrays involves immobilization or in situ synthesis of DNA probes onto the specific test sites of the substrate material or the solid support. High-density DNA arrays can be fabricated using physical delivery techniques (e.g. microjet deposition technology) that allow the dispensing and spotting of nano/picoliter volumes onto the specific test site locations on the microarray. In some cases, the probes on the microarrays are synthesized in situ using a photolithographic process (Heller 2002).

cDNA and oligonucleotide microarrays

Most arrays used for gene expression profiling and molecular targets analysis in the biological sciences today can be divided into two groups: oligonucleotide microarrays and complementary DNA (cDNA) microarrays (Schulze and Downward 2001). This division refers to characteristics of the array probes, the individual pieces of gene-specific DNA that are spotted on the array surface. Synthetic oligonucleotides have a maximum length of around 80 nucleotides, thus conferring greater specificity among members of gene families. On the other hand, cDNA probes are usually products of the polymerase chain reaction (PCR) generated from genomic DNA or cDNA libraries, and are typically more than 150 nucleotides in length (Kuo et al. 2002, Lipshutz et al. 1999).

Array fabrication involves either spotting of presynthesized probes using highly precise robotic spotters, or in situ synthesis on glass slides (Schulze and Downward 2001). High-density spotted cDNA microarrays may contain up to 40 000 probes on a single microscope slide. In contrast, oligonucleotide arrays, consisting of gene-specific oligonucleotides, are synthesized directly onto a solid surface by either photolithography or inkjet technology (Schena et al. 1998). Probes can be designed to represent the most unique part of a given transcript, allowing the detection of closely related genes (Schulze and Downward 2001). A major advantage of oligonucleotide arrays over cDNA arrays is that there is no handling and tracking of cDNA resources (Schena et al. 1998). Furthermore, the use of synthetic reagents in the manufacturing of oligonucleotide arrays minimizes variation among arrays, thus ensuring a high degree of reproducibility between microarray experiments.

Sample preparation is similar for cDNA and oligonucleotide microarrays. In both cases, mRNA is extracted, purified, reverse transcribed to cDNA, then labeled and hybridized to probes on the surface of the array slide (Schulze and Downward 2001). Two different fluorescent dyes allow cDNA from two different treatment populations to be labeled with different colors (Fig. 15.1). When mixed and hybridized to the same array, the differentially labeled cDNA results in competitive binding of the target to the probes on the array. After hybridization, the slide is imaged using a confocal laser scanner and fluorescence measurements are made separately for each dye at each spot on the array (Fig. 15.1). This dual labeling enables the ratio of transcript levels for each gene on the array to be determined (Schulze and Downward 2001, Wullschleger and Difazio 2003). Specialized software and data management tools are used for data extraction, normalization, filtering and analysis (Youns et al. 2009a).

Importance of microarray technology

An enormous number of publications have now appeared on the uses of microarrays for gene expression, SNPs, point mutations, and for pharmacogenomic and diagnostic applications. Examples of which are: general applications of microarrays for gene expression analysis (Seo et al. 2000, Ziauddin and Sabatini 2001); gene expression analysis for cancer (Brem et al. 2001, Graveel et al. 2001); gene expression analysis for drug discovery, metabolism, and toxicity (Madden et al. 2000, Gerhold et al. 2001); gene expression analysis on high-density microarrays (Miki et al. 2001, Zarrinkar et al. 2001); gene expression analysis for neuroscience applications (Geschwind 2000, Zirlinger et al. 2001, Cavallaro et al. 2001); general articles on use of microarrays for cancer (Chen et al. 2001, Kannan et al. 2001); infectious disease diagnostics (Cummings and Relman 2000, Chizhikov et al. 2001); use of microarrays for disease diagnostics (Helmberg, 2001, Johnston-Wilson et al. 2001, Petrik 2001); gene expression analysis for microbiological and infectious disease (Kagnoff and Eckmann 2001, Mehrotra and Bishai 2001, Simmen et al. 2001); use of microarrays for genotyping (Hacia and Collins 1999, Hacia et al. 2000, Mahalingam and Fedoroff 2001); and applications of microarrays for plant biology

(Hertzberg et al. 2001, Hihara et al. 2001). In addition to microarrays for DNA hybridization analysis, efforts are now being directed for proteomic applications (Fung et al. 2001, Kodadek 2001, Moerman et al. 2001) and microarrays for the immobilization of cell and tissue samples (Moch et al. 2001, Rimm et al. 2001).

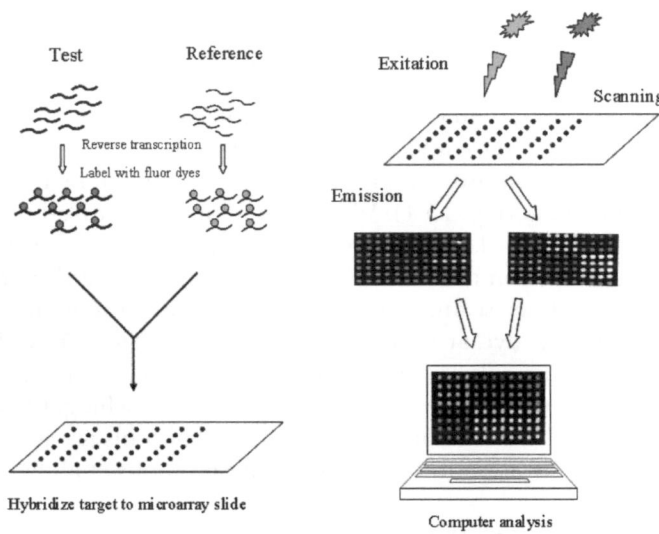

Fig. 15.1. cDNA microarray procedure. Templates for genes of interest are amplified by PCR then printed on coated glass microscope slides. Total RNA from both the test and reference sample is fluorescently labeled with different fluorescent dyes. The fluorescent targets are pooled and allowed to hybridize to the clones on the array slide. Laser excitation of the incorporated targets yields an emission with characteristic spectra, which is measured using a scanning confocal laser microscope. Monochrome images are imported into software in which the images are merged. Data from a single hybridization experiment is viewed as a normalized ratio. In addition, data from multiple experiments can be examined using any number of data mining tools (Youns et al. 2009b).

Color image of this figure appears in the color plate section at the end of the book.

Microarray for the identification of potential therapeutic targets

Recently, the search for discovering new therapeutic options and disclosing therapeutic targets has relied mainly on approaches based on large-scale genomics including the sequencing of expressed-sequence tags (ESTs), serial analysis of gene expression (SAGE) differential display, homology cloning and related approaches (Fryer et al. 2002). These EST sequencing efforts had resulted in availability of databases comprising information on the majority of human genes (Gerhold and Caskey 1996), most of which were of unknown therapeutic significance. Although these bioinformatics approaches are extraordinarily valuable, investigators still need an experimental approach to prioritize the potential therapeutic targets. Microarray technology provides an excellent solution because it facilitates the identification of novel potential therapeutic targets from thousands of genes in a single experiment. Using microarray technology to screen for differentially expressed genes is widely acknowledged as a valuable approach in the target discovery process.

Forward pharmacology is an approach that assists in identifying the mechanisms of action of poorly understood drugs. It involves using microarrays to monitor mRNA changes induced by drugs or other bioactive compounds in order to deduce previously unknown actions of these compounds (Gerhold et al. 2002). For example, Hughes et al. (2000) assessed levels of more than 6,000 transcripts including 279 gene knockout strains under 300 experimental conditions. The global expression profiles of these diverse experimental conditions were then used for monitoring the effects of several drugs on yeast where dyclonine, a topical anesthetic of unknown action, was found to be implicated in perturbation of the pathway of ergosterol metabolism. This hypothesis was further confirmed by similar observations of the effects haloperidol on the same pathway (Hughes et al. 2000).

Application of array technology after therapeutic target validation

Among the most important approaches that can be taken to validate and to prioritize candidate therapeutic targets are gene knockout and knock-in strategies in cells, model organisms and mice. In this context, microarrays can be used to assess the primary and secondary consequences of the genetic manipulation.

Application of microarray technology in Traditional Chinese Medicine (TCM)

TCM has been used for thousands of years in China. Nowadays, TCM is widely practiced in Chinese cancer centers. Nevertheless, it is a brand new area for formal scientific evaluation. TCM represents a holistic approach and lacks high-quality scientific evidence on its effectiveness. Therefore, TCM is frequently regarded with some skepticism by western academic medicine (Efferth et al. 2007a, b). Since DNA microarray technologies have been introduced into medical sciences (DeRisi et al. 1996, Schena et al. 1995), many researchers have applied this technology to pharmacological analyses (Eisen and Brown 1999, Heller et al. 1997). Recent studies tried to cross the bridge between TCM and modern western medicine through applying recently developed technologies to identify molecular mechanisms and novel targets involved in TCM action on different disorders. Some of these studies are presented here to pinpoint the significance of applying high throughput technologies in medicinal plant research.

Studies of PC-SPES using microarray technology

PC-SPES, a preparation of eight Chinese herbs, is used in the treatment of prostate cancer and exhibits a promising antiproliferative and antitumor activity *in vivo* and *in vitro* in diverse cancer types (Schmidt et al. 2009). Using DNA microarray technology, Bonham et al. (2002) investigated the molecular effects of the herbal compound PC-SPES on prostate carcinoma cells, where cDNA microarray analysis was utilized to identify expression profiling changes in LNCaP prostate carcinoma cells after treatment with PC-SPES and estrogenic agents including diethylstilbestrol. Interestingly, PC-SPES altered the expression of 156 genes following 24 h of exposure. Of particular interest, transcripts encoding cell cycle-regulatory proteins, alpha- and beta-tubulins, and the androgen receptor were significantly down-regulated. After comparing the gene expression profiles patterns resulting from these treatments, the authors concluded that PC-SPES exhibits activities distinct from those of diethylstilbestrol and postulated that alterations in specific genes involved in modulating the cell cycle, cell structure , and androgen response could be responsible for PC-SPES-mediated antiproliferative effect (Bonham et al. 2002).

Previous studies of curcumin using microarray technology

Curcumin, diferuloylmethane, is a major chemical component of turmeric (*Curcuma longa*) and is used as a spice to give a specific flavor and yellow color to curry. It has been used as a cosmetic and in some medical preparations (Govindarajan 1980). Curcumin has been shown to display anticancer activities in a mouse model system, as indicated by its ability to inhibit phorbol ester-induced skin cancers (Huang et al. 1988). In addition to its anticarcinogenic effect, it has been shown to exhibit antioxidant, antiproliferative, anti angiogenic, and anti-inflammatory properties (Aggarwal et al. 2003). The previously mentioned effects of curcumin were thought to be mediated by its inhibitory effects on host cell-signaling factors, including NF-κB, protein kinase C, Egr-1, c-Myc, AP-1 transcription factor, epidermal growth factor receptor tyrosine kinase, c-Jun N-terminal kinase, protein tyrosine kinases, protein serine/threonine kinases, and IκB kinase (Hong et al. 1999, Huang et al. 1988, Lin et al. 2000).

Gene expression profiles were used to characterize the anti-invasive mechanisms of curcumin in the highly invasive lung adenocarcinoma cells (CL1-5) (Chen et al. 2004). Using microarray chips containing 9600 PCR-amplified cDNA fragments, 81 genes were found to be significantly down-regulated and 71 genes were significantly up-regulated after curcumin treatment. Interestingly, at lower concentrations of curcumin (10 μM), several invasion-related genes were down-regulated, including matrix metalloproteinase 14 (MMP14; 0.65-fold), neuronal cell adhesion molecule (0.54-fold) and integrins α6 (0.67-fold) and β4 (0.63-fold). In addition, several heat-shock proteins (Hsp) [Hsp70 (3.75-fold), Hsp27 (2.78-fold) and Hsp40-like protein (3.21-fold)] were up-regulated after curcumin treatment (Chen et al. 2004).

Using whole-genome microarray chips, Nones et al (2009) demonstrated the effects of dietary curcumin on colonic inflammation and gene expression in multidrug resistance gene-deficient (*mdr1a*−/−) mice, a model of inflammatory bowel disease (Nones et al. 2009). Microarray and pathway analyses suggested that the effect of dietary curcumin on colon inflammation could be mediated via an induction of xenobiotic metabolism and suppression of pro-inflammatory pathways, probably mediated by pregnane X receptor (Pxr) and peroxisome proliferator-activated receptor α (PPARα) activation of retinoid X receptor (Rxr). Moreover, they documented the potential role of global gene expression and pathway analyses to study and better understand the effect of foods in modulating colonic inflammation (Nones et al. 2009).

Yan et al. (2005) undertook a transcriptional profiling study to identify novel key player genes modulating the effect of curcumin. A cDNA microarray comprised of 12,625 probes was used to compare total RNA extracted from curcumin-treated and untreated MDA-1986 cells for differential gene expression. 202 up-regulated mRNAs and 505 transcripts down-regulated were identified. The proapoptotic activating transcription factor 3 (ATF3) was found to be significantly up-regulated >4-fold. In addition, two negative regulators of growth control [antagonist of myc transcriptional activity (Mad) and p27kip1] were induced 68- and 3-fold, respectively. Furthermore, two dual-activity phosphatases (CL 100 and MKP-5), which inactivate the c-*jun*-NH$_2$-kinases, showed augmented expression, matching the reduced expression of the upstream activators of c-jun-NH$_2$-kinase (MEKK and MKK4). Among the down-regulated genes, the expression of Frizzled-1 (Wnt receptor) was found to be most strongly suppressed (8-fold). Additionally, two genes implicated in growth control (*K-sam*, encoding the keratinocyte growth factor receptor, and *HER3*) as well as the E2F-5 transcription factor, which regulates genes controlling cell proliferation, also showed decreased expression. Moreover, the authors identified ATF3 as a novel contributor to the proapoptotic effect of curcumin (Yan et al. 2005).

Sentrix Human WG-6 BeadChips were recently used to perform a large-scale gene-expression profiling during curcumin-triggered apoptosis (8–36 h) in follicular lymphoma HF4.9 cells (Skommer et al. 2007). The comprehensive microarray response included differential expression of genes encoding transcription and splicing factors, apoptotic signaling proteins, proteins involved in regulation of cell adhesion, tumor and metastasis suppressors, lymphoid development, migration (e.g. CXCR4) or B-cell activation (e.g. CD20), and others (Skommer et al. 2007).

In another study, in order to extend the knowledge on pathways or molecular targets already reported to be affected by curcumin (cell cycle arrest, phase-II genes) and to explore potential new candidate genes and pathways that may play a role in colon cancer prevention, a gene expression analysis in response to curcumin treatment in two human colon cancer cell lines (HT29 and Caco-2) were investigated. Using microarrays containing four thousand human genes, HT29 colon cancer cells were exposed to two different concentrations of curcumin and gene expression changes were monitored after different time points. Changes in gene expression after short-term treatment (3 or 6 h) with curcumin were also investigated in a second cell type, Caco-2 cells. Gene expression changes (>1.5-fold) were observed at all time points. Results showed that HT29 cells were more sensitive to curcumin than Caco-2 cells. Early response genes were involved in signal transduction, cell cycle, gene transcription, DNA repair, xenobiotic metabolism and cell adhesion. A number of cell cycle genes, among them several have a role in transition through the G2/M phase, were modulated in HT29 cells after curcumin treatment. Furthermore, the observed changes in G2/M cell cycle arrest genes were confirmed by flow cytometry. Moreover, the authors showed that some cytochrome P450 genes were down-regulated by curcumin in both cell lines. In addition, curcumin affected expression of tubulin genes, p53, metallothionein genes and other genes involved in colon carcinogenesis (Van Erk et al. 2004).

Ramachandran et al compared the expression profiles of apoptotic genes induced by curcumin in MCF-7 human breast cancer cell line and MCF-10A mammary epithelial cell line (Ramachandran et al. 2005). Microarray hybridization of Clonetech apoptotic arrays consisted of 214 apoptosis-associated genes was performed. Of the 214 apoptosis-associated genes, the expression of 104 genes was significantly altered after curcumin exposure. They reported that the gene expression was altered up to 14 fold levels in MCF-7 as compared to MCF-10A (up to 1.5 fold) and concluded that the effect of curcumin was higher in MCF-7 cells compared to MCF-10A mammary epithelial cell line. In MCF-7 cells, curcumin up-regulated 22 genes and down-regulated 17 genes. The up-regulated genes included CRAF1, GADD45, TRAF6, HIAP1, CASP2, CASP1, CASP3, CASP4, and TRAP3. The down-regulated genes included TNFR, TRIAL, PKB, and TNFRSF5 (Ramachandran et al. 2005).

At microRNA level, array analysis showed that curcumin alters the expression profiles of microRNAs in human pancreatic cancer cells. An oligonucleotide microarray chip was used to profile microRNA (miRNA) expressions in pancreatic cells treated with curcumin. Curcumin altered miRNA expression in human pancreatic cells, up-regulating miRNA-22 and down-regulating miRNA-199a. They suggested that the biological effects of curcumin may be mediated through modulation of miRNA expression (Sun et al. 2008).

Studies of berberine using microarray technology

The natural isoquinoline alkaloid, berberine, has been found in many clinically important plants including *Berberis aquifolium* (Oregon grape), *Berberis vulgaris* (barberry), *Coptis chinensis* (golden thread or Coptis) and *Coscinium fenestratum* (McIsaac et al. 1997, Rojsanga et al. 2006). Berberine exhibits a wide range of biochemical and pharmacological activities (Kuo et al. 2004, Schmeller et al. 1997, Efferth 2005) and has been reported to be used as an anti-arrhythmia, anti-hypertension, anti-diarrhea, and anti-inflammatory agent (Ckless et al. 1995, Kaneda et al. 1991, Taylor and Baird 1995). Additionally, the natural isoquinoline alkaloid product was reported to possess an anti-tumor activity against different tumor types (Choi et al. 2009, Ho et al. 2009, Kim et al. 2008). In order to better understand the physiological actions and the molecular targets of berberine regarding pancreatic cancer, Iizuka et al (2003) utilized oligonucleotide arrays composed of approximately 11,000 genes and undertook a gene expression profiling study to monitor the expression changes associated with sensitivities to berberine and Coptidis rhizoma in 8 human pancreatic cancer lines. From their oligonucleotide array data, 20 and 13 genes with strong correlations (r2 > 0.81) to ID50 values for berberine and C. rhizoma were selected, respectively. Among those 33 genes, the levels of expression of 12 were correlated with the ID_{50} values of both berberine and C. rhizoma. In addition, Iizuka et al concluded that these genes are associated with tumor-killing activity of berberine in C. rhizoma (Iizuka et al. 2003). Moreover, expression of the remaining 21 genes was correlated with the ID50 value of either purified berberine or C. rhizoma. Such an interesting approach allowed common and distinct genes responsible for anti proliferative activities of purified berberine and C. rhizoma to be identified (Iizuka et al. 2003).

Recently, DNA microarray chips were utilized to investigate the gene expression changes of *Yersinia pestis*, a Gram negative coccobacillar bacterium that causes plague, in response to berberine. The analysis was done after exposing *Y. pestis* to berberine. The authors concluded that a total of 360 genes were differentially expressed; 333 genes were up-regulated, and 27 were down-regulated. The up-regulation of genes that encode proteins involved in metabolism was a remarkable change. Genes encoding cellular envelope and transport/binding functions represented the majority of the altered genes in addition to a number of genes of unknown encoding or unassigned functions. Furthermore, number of genes related to iron uptake were also induced (Zhang et al. 2009).

Previous studies of artesunate using microarray technology

Artesunate is a semisynthetic derivative of artemisinin, the active principle of *Artemisia annua* L. artesunate and other artemisinin derivatives are novel drugs in the treatment of malaria (Price, 2000). Artesunate is well tolerated, with insignificant side effects (Hien et al. 1992). In addition to the well known antimalarial activity of artesunate, we have previously identified a profound cytotoxic action of artesunate against cancer cell lines of different tumor types (Efferth et al. 2001).

Previously, we identified mRNA expression profiles associated with the response of tumor cells to artesunate, arteether, and artemether (artemisinin derivatives) (Efferth et al. 2002) and performed a hierarchical cluster analyses of the inhibition concentration 50% (IC_{50}) values and mRNA expression levels of 464 genes deposited in the database of the National Cancer Institute (USA) (Scherf et al. 2000). The mRNA expression of 208 out of 464 genes (45%) correlated significantly with IC_{50} values of at least one artemisinin derivative. These genes belonged to different biological classes (oncogenes, DNA damage and repair genes, apoptosis-regulating genes, drug resistance genes, proliferation-associated genes, tumor suppressor genes and cytokines). Hierarchical cluster analysis identified two different gene clusters. One cluster contained genes significantly correlated to all artemisinin derivatives. The second cluster contained genes differentially associated with the response of artemisinin derivatives to cancer cells (Efferth et al. 2002).

Additionally, in order to stress on the importance of the genome-wide microarray analyses and to confirm that it provides an attractive approach to identify genes involved in the response of cancer cells to natural products, Anfosso et al (2006) investigated artemisinin and six other derivatives and found a significant correlation of the sensitivity of these compounds to genes regulating angiogenesis. Indeed, the inhibition of angiogenesis is among mechanisms governing effects of artemisinins towards tumors as shown by us and others (Chen et al. 2003, Dell'Eva et al. 2004).

Recently, we performed a gene expression profiling study to identify novel molecular targets modulating the effect of artesunate on MiaPaCa-2 and BxPC-3 pancreatic cancer cell lines. cDNA microarray chips containing about 7000 genes representing apoptotic, angiogenic, growth factors, anti-apoptotic, and metastasis-associated genes were used. Results showed that artesunate mediates growth inhibitory effects and induces apoptosis in pancreatic cancer cells through modulation of multiple signaling pathways. Moreover we introduced artesunate as a novel topoisomerase IIα inhibitor. Several molecular targets involved in the intrinsic and extrinsic apoptotic pathways were affected after treatment with artesunate. Among those molecular targets are *CASP2, CASP 3, CASP 4, CASP 5, CASP 6, CASP 8, CASP 9, APAF1, BAX, BAK* and *CASP 10.* It was also shown that the cytotoxic effect of ART on pancreatic cancer cells could be mediated through up-regulation of *DDIT3, NAG-1*and down-regulation of *PCNA* and *RRM2* genes (Youns et al. 2009a).

Previous studies of cantharidin using microarray technology

Cantharidin is a vesicant product of the Chinese blister beetles and Spanish flies (Wang 1989). All body fluids of blister beetles have cantharidin, and the dried bodies have been used as an anticancer agent in traditional Chinese medicine for a long time (Chang et al. 2008). Cantharidin was found to induce apoptosis of human multiple myeloma cells via inhibition of the JAK/STAT pathway (Sagawa et al. 2008). Zhang et al (2004) used cDNA microarrays (12,800 chip; United Gene Holdings, Ltd., PRC) to identify gene expression changes in HL-60 promyeloid leukemia cells after treatment with cantharidin. Cantharidin-treated cells decreased expression of genes coding for proteins involved in DNA repair (e.g. *ERCC, FANCG*), DNA replication (e.g. *DNA polymerase delta*), energy metabolism (e.g. *isocitrate dehydrogenase alpha, ADP/ATP translocase*). Moreover, cantharidin decreased expression of genes coding for proteins that have oncogenic activity (e.g. *GTPase, c-myc,*) or show tumor-specific expression (e.g. *phosphatidylinositol 3-kinase*). In addition, the authors reported that cantharidin could be used as an oncotherapy sensitizer after demonstrating that exposure of HL-60 cells to cantharidin resulted in the decreased expression of MDR (multidrug resistance-associated) protein genes (e.g. *ABCA3, MOAT-B*). Finally, they concluded that the increased expression of genes involved in modulating cytokine production and inflammatory response (e.g. *NFIL-3, N-formylpeptide receptor*), partly explained the stimulating effects on leukocytosis (Zhang et al. 2004).

In our group, in an attempt to identify key molecular determinants that explore sensitivity or resistance of tumor cells to cantharidin, we analyzed the microarray database of the National Cancer Institute in 60 tumor cell lines (Efferth 2005, Efferth et al. 2005, Rauh et al. 2007). Among 9706 genes identified, 21 genes whose mRNA expression correlated with the highest correlation coefficients to IC_{50} values were selected by COMPARE analysis and false discovery rate calculation (Efferth 2005). These genes were further subjected to hierarchical cluster analysis to reveal whether the expression profiles of those genes could be used to predict sensitivity or resistance of cell lines to cantharidin. It is intriguing that many of them are involved in DNA repair, DNA damage response, and/or apoptosis (Efferth 2005).

Gene expression profiling of plant responses to abiotic stress

In addition to the previously mentioned applications, expression profiling has become an important tool to investigate how an organism responds to environmental changes. Plants, being sessile, have the ability to dramatically alter their gene expression patterns in response to environmental changes such as temperature, water availability or the presence of deleterious levels of ions (Hazen et al. 2003). Sometimes these genetic alterations are successful adaptations leading to tolerance while in other instances the plant ultimately fails to adapt to the new environment and is considered sensitive to that condition. Microarray expression profiling can define both tolerant and sensitive responses. These profiles of plant response to environmental extremes (abiotic stresses) are expected to lead to regulators that will be useful in biotechnological approaches to improve stress tolerance as well as to new tools for studying regulatory genetic circuitry (Hazen et al. 2003). Drought, temperature extremes, and saline soils are three of the most common abiotic stresses plants encounter. A previous study showed that gene expression profiles of either drought- or salt-stressed barley plants showed very few similarities (Oztur et al. 2002). However, while not the same genes, it appears as though genes encoding isoforms were differentially regulated in response to different stresses, possibly inducing a similar defense mechanism (Oztur et al. 2002). A very strong correlation between the genes induced by high salinity and those induced by drought stress was found using an Arabidopsis full-length cDNA microarray (Seki et al. 2002).

Inflammatory disorders represent a substantial health problem. Medicinal plants belonging to the Burseraceae family, including Boswellia, are especially known for their anti-inflammatory properties. The gum resin of *Boswellia serrata* contains boswellic acids, which inhibit leukotriene biosynthesis (Boden et al. 2001). Although Boswellia extract has proven to be anti-inflammatory in clinical trials (Gupta et al.1997, 1998, 2001), the underlying mechanisms remain to be characterized. TNFα represents one of the most widely recognized mediators of inflammation. Roy et al. (2005) conducted the first whole genome screen for TNF-inducible genes in human microvascular cells (HMEC). Results showed that TNFα induced 522 genes and down-regulated 141 genes. Of the 522 genes induced by TNFα in HMEC, 113 genes were clearly sensitive to BE (standardized Boswellia extract, 5-Loxin®) treatment. Such genes were directly related to inflammation, cell adhesion, and proteolysis.

Future perspectives

The transcriptional-wide data provided after microarray application always gives potential information that helps to find molecular mechanisms of drug action, causes of disease and the discovery of gene products that are targets for therapy in various diseases. Using microarray approach, a novel molecular targets and new therapeutic options could be well characterized.

Conclusions

In conclusion, DNA microarrays are a powerful and easy-to-use genomic technique. In this chapter, we mentioned some of the recent research studies done using microarray technology taking mainly medicinal plants derived from TCM as our example candidates. Furthermore we showed that expression analysis was effective in identifying novel pathways and molecular targets mediating their effects.

References

Aggarwal, B.B., A. Kumar, and A.C. Bharti. 2003. Anticancer potential of curcumin: preclinical and clinical studies. Anticancer Res. 23: 363–98.

Alon, U., N. Barkai, D.A. Notterman, K. Gish, S. Ybarra, D Mack, and A.J. Levine. 1999. Broad patterns of gene expression revealed by clustering analysis of tumor and normal colon tissues probed by oligonucleotide arrays. Proc. Natl. Acad. Sci. USA 96: 6745–50.

Anfosso, L., T. Efferth, A. Albini, and U. Pfeffer. 2006. Microarray expression profiles of angiogenesis-related genes predict tumor cell response to artemisinins. Pharmacogenomics J. 6: 269–78.

Boden, S.E., S. Schweizer, T. Bertsche, M. Dufer, G. Drews, and H. Safayhi. 2001. Stimulation of leukotriene synthesis in intact polymorphonuclear cells by the 5-lipoxygenase inhibitor 3-oxo-tirucallic acid. Mol. Pharmacol. 60: 267–73.

Bonham, M., H. Arnold, B. Montgomery, and P.S. Nelson. 2002. Molecular effects of the herbal compound PC-SPES: identification of activity pathways in prostate carcinoma. Cancer Res. 62: 3920–4.

Brem, R., U. Certa, M. Neeb, A.P. Nair, and C. Moroni. 2001. Global analysis of differential gene expression after transformation with the v-H-ras oncogene in a murine tumor model. Oncogene 20: 2854–8.

Cavallaro, S., B.G. Schreurs, W. Zhao, V. D'Agata, and D.L. Alkon. 2001. Gene expression profiles during long-term memory consolidation. Eur. J. Neurosci. 13: 1809–15.

Chang, C.C., D.Z. Liu, S.Y. Lin, H.J. Liang, W.C. Hou and W.J. Huang, C.H. Chang, F.M. Ho, and Y.C. Liang. 2008. Liposome encapsulation reduces cantharidin toxicity. Food Chem. Toxicol 46: 3116–21.

Chen, H., J.Liu, B.A. Merrick, and M.P. Waalkes. 2001. Genetic events associated with arsenic-induced malignant transformation: applications of cDNA microarray technology. Mol. Carcinog 30: 79–87.

Chen, H.H., H.J. Zhou, and X. Fang. 2003. Inhibition of human cancer cell line growth and human umbilical vein endothelial cell angiogenesis by artemisinin derivatives in vitro. Pharmacol. Res. 48: 231–6.

Chen, H.W., S.L. Yu, J.J. Chen, H.N. Li, Y.C. Lin, P.L. Yao, H.Y. Chou, C.T. Chien, W.J. Chen, Y.T. Lee, and P.C. Yang. 2004. Anti-invasive gene expression profile of curcumin in lung adenocarcinoma based on a high throughput microarray analysis. Mol. Pharmacol. 65: 99–110.

Chizhikov, V., A. Rasooly, K. Chumakov, and D.D. Levy. 2001. Microarray analysis of microbial virulence factors. Appl. Environ. Microbiol. 67: 3258–63.

Choi M.S., J.H. Oh, S.M. Kim, H.Y. Jung, H.S. Yoo, Y.M. Lee, D.C. Moon, S.B. Han, and J.T. Hong. 2000. Berberine inhibits p53-dependent cell growth through induction of apoptosis of prostate cancer cells. Int. J. Oncol. 34: 1221–30.

Ckless, K., J.L. Schlottfeldt, M. Pasqual, P. Moyna, J.A. Henriques, and M. Wajner. 1995. Inhibition of in-vitro lymphocyte transformation by the isoquinoline alkaloid berberine. J. Pharm. Pharmacol. 47: 1029–31.

Cole, K.A., D.B. Krizman, and M.R. Emmert-Buck. 1999. The genetics of cancer—a 3D model. Nat. Genet. 21: 38–41.

Cummings, C.A. and D.A. Relman. 2000. Using DNA microarrays to study host-microbe interactions. Emerg. Infect Dis. 6: 513–25.

Debouck, C. and P.N. Goodfellow. 1999. DNA microarrays in drug discovery and development. Nat. Genet. 21: 48–50.

Dell'Eva, R., U. Pfeffer, R. Vene, L. Anfosso, A. Forlani, A. Albini, and T.Efferth. 2004. Inhibition of angiogenesis in vivo and growth of Kaposi's sarcoma xenograft tumors by the anti-malarial artesunate. Biochem. Pharmacol. 68: 2359–66.

DeRisi, J., L. Penland, P.O. Brown, M.L. Bittner, P.S. Meltzer, M. Ray, Y. Chen, Y.A. Su, and J.M. Trent. 1996. Use of a cDNA microarray to analyse gene expression patterns in human cancer. Nat. Genet. 14: 457–60.

Efferth, T. 2005. Microarray-based prediction of cytotoxicity of tumor cells to cantharidin. Oncol. Rep. 13: 459–63.

Efferth, T., H. Dunstan, A. Sauerbrey, H. Miyachi, and C.R. Chitambar. 2001. The antimalarial artesunate is also active against cancer. Int. J. Oncol. 18: 767–73.

Efferth, T., A. Olbrich, and R. Bauer. 2002. mRNA expression profiles for the response of human tumor cell lines to the antimalarial drugs artesunate, arteether, and artemether. Biochem. Pharmacol. 64: 617–23.

Efferth, T., Z.Chen, B. Kaina, and G. Wang. 2005. Molecular Determinants of Response of Tumor Cells to Berberine. Cancer Genomics & Proteomics 2: 115–125.

Efferth, T., R. Rauh, S. Kahl, M. Tomicic, H. Bochzelt, M.E. Tome, M.M. Briehl, R. Bauer, and B. Kaina. 2005. Molecular modes of action of cantharidin in tumor cells. Biochem. Pharmacol. 69: 811–8.

Efferth, T., Y.J. Fu, Y.G. Zu, G. Schwarz, V.S. Konkimalla, and M. Wink. 2007a. Molecular target-guided tumor therapy with natural products derived from traditional Chinese medicine. Curr. Med. Chem. 14: 2024–32.

Efferth, T., P.C. Li, V.S. Konkimalla, and B. Kaina. 2007b. From traditional Chinese medicine to rational cancer therapy. Trends Mol. Med. 13: 353–61.

Eisen, M.B. and P.O. Brown. 1999. DNA arrays for analysis of gene expression. Methods Enzymol 303: 179–205.

Fryer, R.M., J. Randall, T. Yoshida, L.L. Hsiao, J. Blumenstock, K.E. Jensen, and S.R. Gullans. 2002. Global analysis of gene expression: methods, interpretation, and pitfalls. Exp. Nephrol. 10: 64–74.

Fung, E.T., V. Thulasiraman, S.R. Weinberger, and E.A. Dalmasso. 2001. Protein biochips for differential profiling. Curr. Opin. Biotechnol. 12: 65–9.

Gerhold, D. and C.T. Caskey. 1996. It's the genes! EST access to human genome content. Bioessays 18: 973–81.

Gerhold, D., M. Lu, J. Xu, C. Austin, C.T. Caskey, and T. Rushmore. 2001. Monitoring expression of genes involved in drug metabolism and toxicology using DNA microarrays. Physiol. Genomics 5: 161–70.

Gerhold, D.L., R.V. Jensen, and S.R. Gullans. 2002. Better therapeutics through microarrays. Nat. Genet. 32 Suppl: 547–51.

Geschwind, D.H. 2000. Mice, microarrays, and the genetic diversity of the brain. Proc. Natl. Acad. Sci. USA 97: 10676–8.

Golub, T.R., D.K. Slonim, P. Tamayo, C. Huard, M. Gaasenbeek, J.P. Mesirov, H. Coller, M.L. Loh, J.R. Downing, M.A. Caligiuri, C.D. Bloomfield, and E.S. Lander. 1999. Molecular classification of cancer: class discovery and class prediction by gene expression monitoring. Science 286: 531–7.

Govindarajan, V.S. 1980. Turmeric—chemistry, technology, and quality. Crit. Rev. Food Sci. Nutr. 12: 199–301.

Graveel, C.R., T. Jatkoe, S. J. Madore, A.L. Holt, and P.J. Farnham. 2001. Expression profiling and identification of novel genes in hepatocellular carcinomas. Oncogene 20: 2704–12.

Gupta, I., A. Parihar, P. Malhotra, G.B. Singh, R. Ludtke, H. Safayhi, and H.P. Ammon. 1997. Effects of Boswellia serrata gum resin in patients with ulcerative colitis. Eur. J. Med. Res. 2: 37–43.

Gupta, I., V. Gupta, A. Parihar, S. Gupta, R. Ludtke, H. Safayhi and H.P. Ammon. 1998. Effects of Boswellia serrata gum resin in patients with bronchial asthma: results of a double-blind, placebo-controlled, 6-week clinical study. Eur. J. Med. Res. 3: 511–4.

Gupta, I., A. Parihar, P. Malhotra, S. Gupta, R. Ludtke, H. Safayhi, and H.P. Ammon. 2001. Effects of gum resin of Boswellia serrata in patients with chronic colitis. Planta Med. 67: 391–5.

Hacia, J.G. and F.S. Collins. 1999. Mutational analysis using oligonucleotide microarrays. J. Med. Genet. 36: 730–6.

Hacia, J.G., K. Edgemon, N. Fang, R.A. Mayer, D. Sudano, N. Hunt, and F.S. Collins. 2000. Oligonucleotide microarray based detection of repetitive sequence changes. Hum. Mutat. 16: 354–63.

Hazen, S.P., Y. Wu, and J.A. Kreps. 2003. Gene expression profiling of plant responses to abiotic stress. Funct. Integr. Genomics 3: 105–11.

Heller, M.J. 2002. DNA microarray technology: devices, systems, and applications. Annu. Rev. Biomed. Eng. 4: 129–53.

Heller, R.A., M. Schena, A. Chai, D. Shalon, T. Bedilion, J. Gilmore, D.E. Woolley, and R.W. Davis. 1997. Discovery and analysis of inflammatory disease-related genes using cDNA microarrays. Proc. Natl. Acad. Sci. USA 94: 2150–5.

Helmberg, A. 2001. DNA-microarrays: novel techniques to study aging and guide gerontologic medicine. Exp. Gerontol. 36: 1189–98.

Hertzberg, M., M. Sievertzon, H. Aspeborg, P. Nilsson, G. Sandberg, and J. Lundeberg. 2001. cDNA microarray analysis of small plant tissue samples using a cDNA tag target amplification protocol. Plant 25: 585–91.

Hien, T.T., N.H. Phu, N.T. Mai, T.T. Chau, T.T. Trang, P.P. Loc, B.M. Cuong, N.T. Dung N, H. Vinh, and D.J. Waller. 1992. An open randomized comparison of intravenous and intramuscular artesunate in severe falciparum malaria. Trans. R. Soc. Trop. Med. Hyg. 86: 584–5.

Hihara, Y., A. Kamei, M. Kanehisa, A. Kaplan, and M. Ikeuchi. 2001. DNA microarray analysis of cyanobacterial gene expression during acclimation to high light. Plant Cell 13: 793–806.

Ho, Y.T., J.S. Yang, T.C. Li, J.J. Lin, J.G. Lin, K.C. Lai, C.Y. Ma, W.G. Wood, and J.G. Chung. 2009. Berberine suppresses in vitro migration and invasion of human SCC-4 tongue squamous cancer cells through the inhibitions of FAK, IKK, NF-kappaB, u-PA and MMP-2 and -9. Cancer Lett. 279: 155–62.

Hong, R.L., W.H. Spohn, and M.C. Hung. 1999. Curcumin inhibits tyrosine kinase activity of p185neu and also depletes p185neu. Clin. Cancer Res. 5: 1884–91.

Huang, M.T., R.C. Smart, C.Q. Wong, and A.H. Conney. 1988. Inhibitory effect of curcumin, chlorogenic acid, caffeic acid, and ferulic acid on tumor promotion in mouse skin by 12-O-tetradecanoylphorbol-13-acetate. Cancer Res. 48: 5941–6.

Hughes, T.R., M.J. Marton, A.R. Jones, C.J. Roberts, R. Stoughton, C.D. Armour, H.A. Bennett, E. Coffey, H. Dai, Y.D. He, M.J. Kidd, A.M. King, M.R. Meyer, D. Slade, P.Y. Lum, S.B. Stepaniants, D.D. Shoemaker, D. Gachotte, K. Chakraburtty, J. Simon, M. Bard, and S.H. Friend. 2000. Functional discovery via a compendium of expression profiles. Cell 102: 109–26.

Iizuka, N., M. Oka, K. Yamamoto, A. Tangoku, K. Miyamoto, T. Miyamoto, S. Uchimura, Y. Hamamoto and K. Okita. 2003. Identification of common or distinct genes related to antitumor activities of a medicinal herb and its major component by oligonucleotide microarray. Int. J. Cancer 107: 666–72.

Jayapal, M. and A.J. Melendez. 2006. DNA microarray technology for target identification and validation. Clin. Exp. Pharmacol. Physiol. 33: 496–503.

Johnston-Wilson, N.L., C.M. Bouton, J. Pevsner, J.J. Breen, E.F. Torrey, and R.H. Yolken. 2001. Emerging technologies for large-scale screening of human tissues and fluids in the study of severe psychiatric disease. Int J Neuropsychopharmacol. 4: 83–92.

Kagnoff, M.F. and L. Eckmann. 2001. Analysis of host responses to microbial infection using gene expression profiling. Curr. Opin. Microbiol. 4: 246–50.

Kaminski, N., J.D. Allard, J.F. Pittet, F. Zuo, M.J. Griffiths, D. Morris, X. Huang, D. Sheppard, and R.A. Heller. 2000. Global analysis of gene expression in pulmonary fibrosis reveals distinct programs regulating lung inflammation and fibrosis. Proc. Natl. Acad. Sci. USA 97: 1778–83.

Kaneda, Y., M. Torii, T. Tanaka, and M. Aikawa. 1991. *In vitro* effects of berberine sulphate on the growth and structure of Entamoeba histolytica, Giardia lamblia and Trichomonas vaginalis. Ann. Trop. Med. Parasitol. 85: 417–25.

Kannan, K., N. Amariglio, G. Rechavi, J. Jakob-Hirsch, I. Kela, N. Kaminski, G Getz, E. Domany, and D. Givol. 2001. DNA microarrays identification of primary and secondary target genes regulated by p53. Oncogene 20: 2225–34.

Kim, S., J.H. Choi, J.B. Kim, S.J. Nam, J.H. Yang, J.H. Kim, and J.E. Lee. 2008. Berberine suppresses TNF-alpha-induced MMP-9 and cell invasion through inhibition of AP-1 activity in MDA-MB-231 human breast cancer cells. Molecules 13: 2975–85.

Kodadek, T. 2001. Protein microarrays: prospects and problems. Chem. Biol. 8: 105–15.

Kuo, C.L., C.W. Chi, and T.Y. Liu. 2004. The anti-inflammatory potential of berberine *in vitro* and *in vivo*. Cancer Lett. 203: 127–37.

Kuo, W.P., T.K. Jenssen, A.J. Butte, L. Ohno-Machado, and I.S. Kohane. 2002. Analysis of matched mRNA measurements from two different microarray technologies. Bioinformatics 18: 405–12.

Lander, E.S., L.M. Linton, B. Birren, C. Nusbaum, M.C. Zody, J. Baldwin, K. Devon, K. Dewar, M. Doyle, W. FitzHugh, R. Funke, D. Gage, K. Harris, A. Heaford, J. Howland, L. Kann, J. Lehoczky, and R. LeVine. 2001. Initial sequencing and analysis of the human genome. Nature 409: 860–921.

Lin, J.K., M.H. Pan, and S.Y. Lin-Shiau. 2000. Recent studies on the biofunctions and biotransformations of curcumin. Biofactors 13: 153–8.

Lipshutz, R.J., S.P. Fodor, T.R. Gingeras, and D.J. Lockhart. 1999. High density synthetic oligonucleotide arrays. Nat. Genet. 21: 20–4.

Madden, S.L., C.J. Wang, and G. Landes. 2000. Serial analysis of gene expression: from gene discovery to target identification. Drug. Discov. Today 5: 415–425.

Mahalingam, R. and N. Fedoroff. 2001. Screening insertion libraries for mutations in many genes simultaneously using DNA microarrays. Proc. Natl. Acad. Sci. USA 98: 7420–5.

Masino, L., S.R. Martin, and P.M. Bayley. 2000. Ligand binding and thermodynamic stability of a multidomain protein, calmodulin. Protein. Sci. 9: 1519–29.

McIsaac, W., V. Goel, and D. Naylor. 1997. Socio-economic status and visits to physicians by adults in Ontario, Canada. J. Health Serv. Res. Policy 2: 94–102.

Mehrotra, J. and W.R. Bishai. 2001. Regulation of virulence genes in Mycobacterium tuberculosis. Int. J. Med. Microbiol. 291: 171–82.

Miki, R., K. Kadota, H. Bono, Y. Mizuno, Y. Tomaru, P. Carninci M. Itoh, K. Shibata, J. Kawai, H. Konno, S. Watanabe, K. Sato, Y. Tokusumi, N. Kikuchi, Y. Ishii, Y. Hamaguchi, I. Nishizuka, H. Goto, H. Nitanda, S. Satomi, A. Yoshiki, M. Kusakabe, J.L. DeRisi, M.B. Eisen, V.R. Iyer, P.O. Brown, M. Muramatsu, H. Shimada, Y. Okazaki, and Y. Hayashizaki. 2001. Delineating developmental and metabolic pathways in vivo by expression profiling using the RIKEN set of 18,816 full-length enriched mouse cDNA arrays. Proc. Natl. Acad. Sci. USA 98: 2199–204.

Moch, H., T. Kononen, O.P. Kallioniemi, and G. Sauter. 2001. Tissue microarrays: what will they bring to molecular and anatomic pathology? Adv. Anat. Pathol. 8: 14–20.

Moerman, R., J. Frank, J.C. Marijnissen, T.G. Schalkhammer, and G.W. van Dedem. 2001. Miniaturized electrospraying as a technique for the production of microarrays of reproducible micrometer-sized protein spots. Anal. Chem. 73: 2183–9.

Nones, K., Y.E. Dommels, S. Martell, C. Butts, W.C. McNabb, Z.A. Park ZA, S. Zhu, D. Hedderley, M.P. Barnett, and N.C. Roy. 2009. The effects of dietary curcumin and rutin on colonic inflammation and gene expression in multidrug resistance gene-deficient (mdr1a-/-) mice, a model of inflammatory bowel diseases. Br. J. Nutr. 101: 169–81.

Oztur, Z.N., V. Talame, M. Deyholos, C.B. Michalowski, D.W. Galbraith, N. Gozukirmizi, R. Tuberosa, and H.J. Bohnert. 2002. Monitoring large-scale changes in transcript abundance in drought- and salt-stressed barley. Plant Mol. Biol. 48: 551–73.

Perou, C.M., T. Sorlie, M.B. Eisen, M. van de Rijn, S.S. Jeffrey, C.A. Rees, J.R. Pollack, D.T. Ross, H. Johnsen, L.A. Akslen, O. Fluge, A. Pergamenschikov, C. Williams, S.X. Zhu, P.E. Lønning, A.L. Børresen-Dale, P.O. Brown, and D. Botstein. 2000. Molecular portraits of human breast tumours. Nature 406: 747–52.

Petrik, J. 2001. Microarray technology: the future of blood testing? Vox Sang 80: 1–11.

Pomeroy, S.L., P. Tamayo, M. Gaasenbeek, L.M. Sturla, M. Angelo, M.E. McLaughlin, J.Y. Kim, L.C. Goumnerova, P.M. Black, C. Lau, J.C. Allen, D. Zagzag, J.M. Olson, T. Curran, C. Wetmore, J.A. Biegel, T. Poggio, S. Mukherjee, R. Rifkin, A. Califano, G. Stolovitzky, D.N. Louis, J.P. Mesirov, E.S. Lander, and T.R. Golub. 2002. Prediction of central nervous system embryonal tumour outcome based on gene expression. Nature 415: 436–42.

Price, R.N. 2000. Artemisinin drugs: novel antimalarial agents. Expert Opin. Investig. Drugs 9: 1815–27.

Ramachandran, C., S. Rodriguez, R. Ramachandran, P.K. Raveendran Nair, H. Fonseca, Z. Khatib, E. Escalon, and S.J. Melnick. 2005. Expression profiles of apoptotic genes induced by curcumin in human breast cancer and mammary epithelial cell lines. Anticancer Res. 25: 3293–302.

Rauh, R., S. Kahl, H. Boechzelt, R. Bauer, B. Kaina, and T. Efferth. 2007. Molecular biology of cantharidin in cancer cells. Chin. Med. 2: 8.

Rimm, D.L., R.L. Camp, L.A. Charette, D.A. Olsen, and E. Provost. 2001. Amplification of tissue by construction of tissue microarrays. Exp. Mol. Pathol. 70: 255–64.

Rojsanga, P., W. Gritsanapan, and L. Suntornsuk. 2006. Determination of berberine content in the stem extracts of Coscinium fenestratum by TLC densitometry. Med. Princ. Pract. 15: 373–8.

Roy, S., S. Khanna, H. Shah, C. Rink, C. Phillips, H. Preuss, G.V. Subbaraju, G. Trimurtulu, A.V.Krishnaraju, M. Bagchi, D. Bagchi, and C.K. Sen. 2005. Human genome screen to identify the genetic basis of the anti-inflammatory effects of Boswellia in microvascular endothelial cells. DNA Cell Biol. 24: 244–55.

Sagawa, M., T. Nakazato, H. Uchida, Y. Ikeda, and M. Kizaki. 2008. Cantharidin induces apoptosis of human multiple myeloma cells via inhibition of the JAK/STAT pathway. Cancer Sci. 99: 1820–6.

Schena, M., D. Shalon, R.W. Davis, and P.O. Brown. 1995. Quantitative monitoring of gene expression patterns with a complementary DNA microarray. Science 270: 467–70.

Schena, M., R.A. Heller, T.P. Theriault, K. Konrad, E. Lachenmeier, and R.W. Davis. 1998. Microarrays: biotechnology's discovery platform for functional genomics. Trends Biotechnol. 16: 301–6.

Scherf, U., D.T. Ross, M. Waltham, L.H. Smith, J.K. Lee, L. Tanabe, K.W. Kohn, W.C. Reinhold, T.G. Myers, D.T. Andrews, D.A. Scudiero, M.B. Eisen, E.A. Sausville, Y. Pommier, D. Botstein, P.O. Brown, and J.N. Weinstein. 2000. A gene expression database for the molecular pharmacology of cancer. Nat. Genet. 24: 236–44.

Schmeller, T., B. Latz-Bruning, and M. Wink. M. 1997. Biochemical activities of berberine, palmatine and sanguinarine mediating chemical defence against microorganisms and herbivores. Phytochemistry 44: 257–66.

Schmidt, M., C. Polednik, P. Gruensfelder, J. Roller, and R. Hagen. 2009. The effects of PC-Spes on chemosensitive and chemoresistant head and neck cancer cells and primary mucosal keratinocytes. Oncol. Rep. 21: 1297–305.

Schulze, A. and J. Downward. 2001. Navigating gene expression using microarrays—a technology review. Nat. Cell Biol. 3: E190–5.

Seki, M., M. Narusaka, J. Ishida, T. Nanjo, M. Fujita, Y. Oono, A. Kamiya, M. Nakajima, A. Enju, T. Sakurai, M. Satou, K. Akiyama, T. Taji, K. Yamaguchi-Shinozaki, P. Carninci, J. Kawai, Y. Hayashizaki, and K. Shinozaki. 2002. Monitoring the expression profiles of 7000 Arabidopsis genes under drought, cold and high-salinity stresses using a full-length cDNA microarray. Plant J. 31: 279–92.

Seo, J., M. Kim, and J. Kim. 2000. Identification of novel genes differentially expressed in PMA-induced HL-60 cells using cDNA microarrays. Mol. Cells 10: 733–9.

Simmen, K.A., J. Singh, B.G. Luukkonen, M. Lopper, A. Bittner, N.E. Miller, M.R. Jackson, T. Compton, and K. Früh. 2001. Global modulation of cellular transcription by human cytomegalovirus is initiated by viral glycoprotein B. Proc. Natl. Acad. Sci. USA 98: 7140–5.

Skommer, J., D. Wlodkowic, and J. Pelkonen. 2007. Gene-expression profiling during curcumin-induced apoptosis reveals downregulation of CXCR4. Exp. Hematol. 35: 84–95.

Sun, M., Z. Estrov, Y. Ji, K.R. Coombes, D.H. Harris, and R. Kurzrock. 2008. Curcumin (diferuloylmethane) alters the expression profiles of microRNAs in human pancreatic cancer cells. Mol. Cancer Ther. 7: 464–73.

Taylor, C.T. and A.W. Baird. 1995. Berberine inhibition of electrogenic ion transport in rat colon. Br. J. Pharmacol. 116: 2667–72.

Van, Erk M.J., E. Teuling, Y.C. Staal, S. Huybers, P.J. Van Bladeren, J.M. Aarts, and B. Van Ommen. 2004. Time- and dose-dependent effects of curcumin on gene expression in human colon cancer cells. J. Carcinog. 3: 8.

Venter, J.C., M.D. Adams, E.W. Myers, P.W. Li, R.J. Mural, G.G. Sutton, H.O. Smith, M. Yandell, C.A. Evans, R.A. Holt, J.D. Gocayne, P. Amanatides, R.M. Ballew, D.H. Huson, and J.R. Wortman. 2001. The sequence of the human genome. Science 291: 1304–51.

Wang, G.S. 1989. Medical uses of mylabris in ancient China and recent studies. J. Ethnopharmacol. 26: 147–62.

Wullschleger, S.D. and S.P. Difazio. 2003. Emerging use of gene expression microarrays in plant physiology. Comp. Funct. Genomics 4: 216–24.

Yan, C., M.S. Jamaluddin, B. Aggarwal, J. Myers, and D.D. Boyd. 2005. Gene expression profiling identifies activating transcription factor 3 as a novel contributor to the proapoptotic effect of curcumin. Mol. Cancer Ther. 4: 233–41.

Youns, M., T. Efferth, J. Reichling, K. Fellenberg, A. Bauer, and J.D. Hoheisel. 2009a. Gene expression profiling identifies novel key players involved in the cytotoxic effect of Artesunate on pancreatic cancer cells. Biochem. Pharmacol. 78: 273–83.

Youns, M., T. Efferth, and J.D. Hoheisel. 2009b. Microarray analysis of gene expression in medicinal plant research. Drug. Discov. Ther. 3(5): 200–207.

Zarrinkar, P.P., J.K. Mainquist, M. Zamora, D. Stern, J.B. Welsh, L.M. Sapinoso, G.M. Hampton, and D.J. Lockhart. 2001. Arrays of arrays for high-throughput gene expression profiling. Genome. Res. 11: 1256–61.

Zhang, J.P., K. Ying, Z.Y. Xiao, B. Zhou, Q.S. Huang, H.M. Wu, M. Yin, Y. Xie, Y.M. Mao, and Y. C. Rui. 2004. Analysis of gene expression profiles in human HL-60 cell exposed to cantharidin using cDNA microarray. Int. J. Cancer 108: 212–8.

Zhang, J., G. Zuo, Q. Bai, Y. Wang, R. Yang, and J. Qiu. 2009. Microarray expression profiling of Yersinia pestis in response to berberine. Planta Med. 75: 396–8.

Ziauddin, J. and D.M. Sabatini 2001. Microarrays of cells expressing defined cDNAs. Nature 411: 107–10.

Zirlinger, M., G. Kreiman, and D.J. Anderson. 2001. Amygdala-enriched genes identified by microarray technology are restricted to specific amygdaloid subnuclei. Proc. Natl. Acad. Sci. USA 98: 5270–5.

Chapter 16

Combining Ethnobotany and Informatics to Discover Knowledge from Data

Jitendra Gaikwad,[1] Karen Wilson,[2] Jim Kohen,[3] Subramanyam Vemulpad,[4] Joanne Jamie[5] and *Shoba Ranganathan[6,*]*

Introduction

Mankind has relied on plants since time immemorial, as a source of food, clothing, fuel, shelter and primary health care. The study of the complex interactions between plants and humans is described as 'ethnobotany', a term coined by Harshberger in 1895 (Harshberger 1896). Ethnobotany has evolved as an interdisciplinary area incorporating data from disciplines such as chemistry, anthropology, ecology, environmental sciences, geography,

[1]Department of Chemistry and Biomolecular Sciences, Macquarie University, Sydney, NSW 2109, Australia.
E-mail: *jitendra.gaikwad@mq.edu.au*
[2]National Herbarium of NSW, Royal Botanic Gardens, Sydney NSW 2000, Australia.
E-mail: *karen.wilson@rbgsyd.nsw.gov.au*
[3]Department of Biological Sciences, Macquarie University, Sydney, NSW 2109, Australia.
E-mail: *jim.kohen@mq.edu.au*
[4]Department of Chiropractic, Macquarie University, Sydney, NSW 2109, Australia.
E-mail: *subramanyam.vemulpad@mq.edu.au*
[5]Department of Chemistry and Biomolecular Sciences, Macquarie University, Sydney, NSW 2109, Australia.
E-mail: *joanne.jamie@mq.edu.au*
[6]Department of Chemistry and Biomolecular Sciences and ARC Centre of Excellence in Bioinformatics, Macquarie University, Sydney, NSW 2109, Australia and Department of Biochemistry, Yong Loo Lin School of Medicine, National University of Singapore, Singapore.
E-mail: *shoba.ranganathan@mq.edu.au*
*Corresponding author

medicine, economics, linguistics, pharmacology, and indigenous law (Fig. 16.1) (Alcorn 1995). Ethnobotanical studies have resulted in the discovery of important bioactive compounds including the antimalarial artemisinin, from *Artemisia annua* (sweet sagewort) (WHO 2008) and the anti-inflammatory curcumin, from *Curcuma longa* (turmeric) (Padma 2005). According to the World Health Organization (WHO), an estimated 80% of the population in developing countries depend on traditionally used medicinal plants for their primary health care (WHO 2007). Much of the world's biodiversity is also concentrated in developing countries, with unique taxa and great species diversity (Peterson et al. 2003). Today, indigenous traditional knowledge is on the verge of extinction due to acculturation, loss of biodiversity, demise of the knowledge custodians and apathy of younger generations towards the centuries-old knowledge (Brouwer et al. 2005). To combat diminishing biodiversity and dwindling traditional knowledge, access to and exchange of information is crucial amongst researchers, scientists, policy makers and indigenous populations (Ramirez 2007). To facilitate the exchange of information between these diverse users, significant information management systems are required (O'Neill et al. 2003). Information technology (IT) has dramatically changed the way scientific research is conducted (Buneman 2005), giving rise to multidisciplinary fields such as biodiversity informatics (Bisby 2000, Berendsohn 2007). Biodiversity informatics is a new discipline, integrating species level information from diverse domains using the scientific name of the organism as the linking thread (Sarkar 2007).

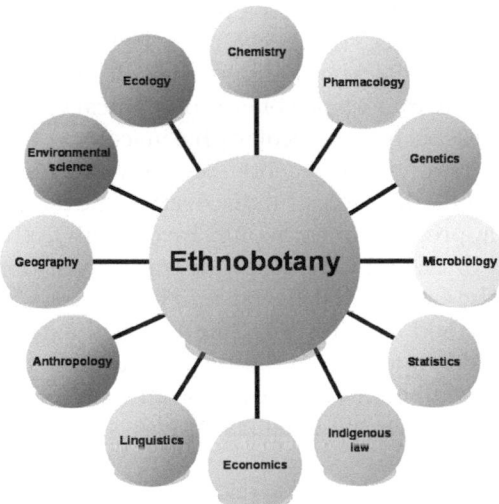

Fig. 16.1. Ethnobotany as a field and some of its associated disciplines.

Color image of this figure appears in the color plate section at the end of the book.

In the recent past, researchers have begun harnessing IT as a tool for documenting, disseminating and analyzing ethnobotanical data. (Thomas 2003). The advent of the internet has enabled the ethnobotanist to access, share and disseminate information more effectively, thus contributing to global knowledge. However, much of the public domain ethnobotanical data remains inaccessible, due to the paucity of best data management practices and interoperability tools.

In the present chapter, we review the current status of ethnobotanical data and the application of information technology for efficient information management and conservation. After a brief overview of data available in ethnobotany and the importance of information technology in managing this data, we discuss data integration strategies for the efficient use of ethnobotanical information.

Ethnobotanical data

The use of plants in different cultures globally precedes written human history. Ancient Ayurveda texts, originating around 6000 BC in India, describe more than 2000 medicinal plant species (Thatte and Dahanukar 1986, Narayana et al. 1998, Mukherjee and Wahile 2006). Known botanical studies documenting the use of plants by humans date back to the work of Theophrastus in 375 BC (Kokwaro 1995). Pedanius Dioscorides, a Greek surgeon, wrote *De Materia Medica* around 77 AD, a compendium containing medicinal information on 600 plant species from the Mediterranean region (Osbaldeston 2000). In the 19th century, intensive ethnobotanical data was gathered and compiled as a result of colonialism and ambitious botanical explorations (Davis 1995, Mauro 1997). However, in the 20th century, when traditional knowledge worldwide was being lost (Turner 1995, Cox 2000), the importance of ethnobotanical studies decreased due to advances in areas such as molecular pharmacology (Cox 2000). Interest in ethnobotanical studies has recently been revived, due to a number of factors which include the successful discovery of novel drugs following an ethnobotanical approach (Fabricant and Farnsworth 2001) as exemplified by the discovery of the antiviral drug prostratin from *Homalanthus nutans* (Cox 2000) and the antimalarial artemisinin from *Artemisia annua* (Klayman 1985), threatened food security (FAO 2009), and climate change (Cavaliere 2009). Additionally, concerns for diminishing traditional knowledge and biodiversity along with increasing drug resistance in infectious agents have rekindled an interest in ethnobotanical studies (Lewis 2003, Sanon et al. 2003, Singh 2007).

A vast amount of ethnobotanical data is scattered in different languages and formats with limited accessibility and application (Alcorn 1981, Ellen 1996, Buenz et al. 2004). It is estimated that there are more than 300,000 higher plants in the world (Govaerts 2001) and approximately 25000 to 30000 plants are used in traditional medicines (Heywood 1995). There are, however, less than 90 plant species used worldwide for obtaining plant-derived drugs (Farnsworth 1988). In addition, globally, traditional knowledge is unevenly distributed, similar to the uneven distribution of biodiversity (Chavan and Krishnan 2003). The world's biodiversity and the indigenous populations that are dependent on it are concentrated in tropical developing countries such as Brazil, India, Mexico, Peru, Vietnam and tropical African countries. Ironically, information related to the traditional use of plants in these countries is available through major data resources located in developed countries such as the United States of America (USA) and United Kingdom (UK) (Table 16.1). Various factors have contributed to this knowledge and content divide. Firstly, in developing countries, a significant amount of information related to biological resources is not easily accessible and is stored in heterogeneous formats (Chavan et al. 2004). Secondly, during the colonization period, most specimens and information were deposited in museums and herbaria in developed countries such as the Royal Botanic Gardens, Kew, UK. Insufficient funds, an orthodox mindset and research agencies in the developed world commercializing biodiversity and related traditional knowledge without sharing the benefits with the custodians, have made the situation more complicated (Drahos 2000, Masood 2004, Gaikwad and Chavan 2006). To ameliorate this situation, it is essential for researchers and knowledge custodians to embrace information technology for digitization, efficient conservation and judicious sharing of knowledge. For complete global coverage of ethnobotanical data, there are innumerable reports, voucher specimens, scientific articles, books, theses and oral histories that are yet to be digitized.

Using information technology for ethnobotanical data—complexities and challenges

From simple cataloguing of traditionally used plants, ethnobotany has evolved to address complex human-plant interactions. Ethnobotanical data are crucial for a variety of scientific studies such as pharmacology, veterinary science, ecology and traditional knowledge conservation. The impediments to accessing ethnobotanical information limit the ability of researchers and policy-makers to address issues in the areas of analysis, conservation, sustainable use and decision making.

The diverse nature of available ethnobotanical data and formats has posed challenges for data access, management and integration. At present, the greatest challenge is to use informatics tools to digitize both unpublished and published ethnobotanical data (historical and contemporary) held in diverse formats for ease of access. A second challenge is to facilitate accession and seamless integration of data from different domains such as taxonomy, ecology, phytochemistry, pharmacology, microbiology and molecular biology. Such an integrated information system can potentially correlate ethnobotanical data with datasets from various other disciplines, thereby facilitating better understanding for taking informed decisions. For example, integrating ethnobotanical information with taxonomic datasets, such as the International Plant Name Index will help to resolve the scientific name and related issues concerning ethnobotanical plant species of significance (Buenz et al. 2004).

In the recent past, there has been an increase in the availability of ethnobotanical data over the web (WWW). Examples of ethnobotanical databases accessible on the internet are listed in Table 16.1. Despite an increase in the number of initiatives providing data in digital formats such as relational databases and HTML web pages, the total data available online is fragmented and only a small fraction of what is known. Moreover, these efforts do not commensurate with the amount of ethnobotanical data collected over the centuries and the rate at which it is generated in electronic formats. Further, due to the lack of informatics infrastructure facilitating easy access and limitations in data management software, enormous amounts of data cannot be integrated, leading to redundancy and inefficient use. For example, a database such as NAPRALERT, with worldwide coverage, provides ethnobotanical information on 20,000 medicinal plants (Table 16.1) including some Indian species. In comparison, the Encyclopedia of Indian Medicinal Plants provides ethnobotanical information on 6,198 medicinal Indian plant species (Table 16.1), but not all of these are included in NAPRALERT. Since these datasets are not linked there is no easy way of checking the overlap between them.

Thus, as with portals, such as the Global Biodiversity Information Facility (GBIF) data portal, Species 2000 and ITIS Catalogue of Life, the Consortium for the Barcode of Life (CBOL), Australia's Virtual Herbarium (AVH) and the Universal Biological Indexer and Organizer (uBio), a unified or federated data portal is required for accessing ethnobotanical information globally.

Ethnobotanical data standards

The heterogeneous nature of traditional plant knowledge and its associated data, held in a range of digital and non-digital formats across the globe, has made its utilization cumbersome. Comprehensive data standards are required to overcome these impediments and to simplify acquisition, management and sharing. To date, there have been a few efforts in this direction. An early attempt led to the Economic Botany Data

Table 16.1. Examples of ethnobotanical databases available on the Web

Data Source and URL	From	No. of species	Coverage
Prelude Medicinal Plants Database http://www.metafro.be/prelude	Africa	251	Africa
Customary Medicinal Knowledgebase http://biolinfo.org/cmkb	Australia	500	Australia
Samoan Medicinal Plant Database http://www.dittmar.dusnet.de/english/esamoa.html	Germany	371	Samoa
FRLHT Encyclopedia of Indian Medicinal Plants http://www.frlht.org/newenvis/indian-medicinal-plants-database.php	India	6198	India
TKDL - Traditional Knowledge Digital Library	India	2147	India
Yuthog Foundation for Tibetan Medicine http://www.yuthog.org/?Home	India	200	Tibet
TradiMed http://www.tradimed.com/	Korea	502	China and Korea
Ethnobotany of the Peruvian Amazon http://www.biopark.org/peru/plants-amazon.html	Peru	50	Peru
SEPASAL www.rbgkew.org.uk/ceb/sepasal/	UK	6200	Tropical and subtropical drylands
Plants For A Future http://www.pfaf.org/database/index.php	UK	7000	Worldwide
NAPRALERT http://www.napralert.org/Default.aspx	USA	20,000	Worldwide
EthnoMedicinals http://www.ethnomedicinals.com/index.htm	USA	5400	Worldwide
Dr. Duke's Phytochemical and Ethnobotanical Databases http://www.ars-grin.gov/duke/	USA	4000	Worldwide
A Modern Herbal http://www.botanical.com/botanical/mgmh/mgmh.html	USA	800	Worldwide
Raintree http://www.rain-tree.com/index.html	USA	94	Amazon rainforest
TCM Herb Database http://www.rmhiherbal.org/herblib/a-index.html	USA	290	China
Hawaiian Ethnobotany online database http://www2.bishopmuseum.org/ethnobotanydb/index.asp	USA	145	Hawaii
Native American Ethnobotany http://herb.umd.umich.edu/	USA	4029	USA

Collections Standard (EBDCS) (Cook 1995), which did not sufficiently address ethnobotanical and traditional knowledge issues, and this in turn led to the current Databases and Registries of TK and Biological/Genetic Resources Standard.

Economic Botany Data Collections Standard (EBDCS)

Between 1989 and 1992, the Taxonomic Database Working Group (TDWG) developed the Economic Botany Data Collections Standard (EBDCS). TDWG, which is now known as Biodiversity Information Standards, is an international scientific body involved in the development of data standards for the exchange of biodiversity and associated data. EBDCS was developed to standardize descriptors and terms describing the economic, social, and cultural values of plant species (Cook 1995). However, this standard has been adopted by only a handful of organizations such as the New York Botanical Garden, the Food and Agricultural Organization, the Royal Botanic Gardens at Kew and Edinburgh, the National Commission for Knowledge and Use of Biodiversity (CONABIO) and the Mediterranean Agronomic Institute of Chania in Crete (Daphne 2002). Although EBDCS covers some aspects of ethnobotany, this standard has not been widely accepted due to problems in its implementation in a relational database structure, its being non-intuitive for beginners and its being unavailable in readily useable electronic format (Cook 1999, Daphne 2002). These difficulties have also been documented as far back as 1999 as a TDWG Economic Botany Subgroup report by Cook (1999). Another drawback of this standard is its limited scope, which does not cover traditional aspects of plant use such as cultural significance, ownership and medicine preparation methods. In 2006, TDWG adopted a new process for data standards development, and all older standards are being revised. Thus, the biodiversity informatics researcher and user community recognizes the need to revisit and revise EBDCS to make it more comprehensive and user friendly.

Databases and Registries of Traditional Knowledge and Biological/ Genetic Resources Standard (DRTKBGRS)

To protect against misappropriation of traditional knowledge (TK) in the form of patents on non-original innovations, the Traditional Knowledge Digital Library (TKDL) was initiated in India in 2001. For developing TK databases, such as TKDL and registries, an international data standard addressing objectives such as positive and defensive knowledge protection, complete stakeholder involvement, decentralized functioning of local and national databases, conservation and preservation is essential. To address these objectives, DRTKBGRS was developed by experts dealing with traditional knowledge from the Asia-Pacific region. In 2003, the Intergovernmental Committee of the World Intellectual Property Organization adopted DRTKBGRS as a technical standard for the documentation of traditional knowledge and its associated biological/genetic resources. The standard comprehensively covers various aspects of TK documentation such as data integration, support for multiple languages, culturally suitable documentation, issues related to vocabularies, and capacity building (Intergovernmental Committee 2002). Additionally, the Traditional Knowledge Resource Classification System was developed based on the structure of International Patent Classification for Indian systems of medicine such as Ayurveda, Unani, Yoga and Siddha. Currently, the European Patent Office accesses TKDL for evaluating patent applications. Mongolia, Malaysia, Thailand and some African countries are in the process of replicating TKDL for protecting traditional knowledge from misappropriation.

Important components of the data integration strategies

The key components for integrating data (primary and secondary), while protecting sensitive data are content building, knowledge organization and knowledge dissemination, each of which requires input from indigenous communities as well as researchers and scientists. These components need to be inter-linked, as shown in Fig. 16.2, to facilitate efficient interchange between ethnobotanical data generators, managers and users. During the implementation of the data integration strategies, to protect intellectual property rights, it is vital that indigenous communities determine the nature of data collection, management and presentation.

Content building

Ethnobotanical information exists in both digital and non-digital formats spread across a variety of institutions and indigenous communities globally. For content building, efficient and globally agreed informatics tools such as TKDL, which can be used as a stand-alone or web-based application, as required, for locating, digitizing, integrating and analyzing traditional information. With due recognition and respect, indigenous communities and stakeholders must be encouraged to participate in all aspects of the content building (Fig. 16.2), as the content will be used in making informed decisions such as for land and resource management as well as for research and conservation activities. As the knowledge base grows, information gaps can be identified, so that specific solutions can be explored.

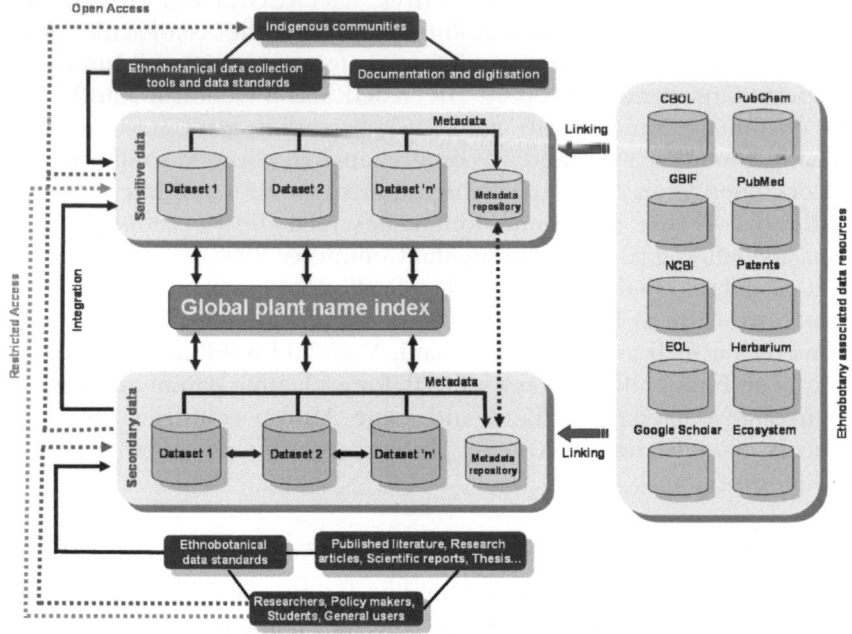

Fig. 16.2. A schematic representation of the key components and the ethnobotanical data integration platform using information technology.

Color image of this figure appears in the color plate section at the end of the book.

Knowledge Organization

Efficient utilization of the knowledge generated and its enrichment is difficult without proper organization of the information. To facilitate discovery and retrieval of relevant information, schemas, data standards, metadata repositories, ontologies and controlled vocabularies needs to be developed and adopted.

It is necessary to have simple, multilingual and easy to understand data standards for implementation by domain experts with different skill levels and linguistic background. Comprehensive suites of ethno-biodiversity data standards and schemas would facilitate interoperability and integration of ethnobotanical databases and portals with non-ethnobotanical data portals such as the Global Biodiversity Information Facility (GBIF) data portal, Encyclopedia of Life (EOL) and NCBI PubChem. Similarly, use of common standards and integration tools will also enable ethnobotanical datasets to be linked and their data made available in a united form via a single federated portal for easy access by all users. Note that this does not mean that a single global database is needed. It will be impractical and would not work as well as setting up a web portal for easy user access that links many databases around the world. This approach ensures that each dataset is kept by its originating group of custodians, thus facilitating maintenance and updating of the records by these knowledgeable custodians as needed over time. The development of such standards and schema as well as a unified portal is thus a matter of priority.

Extracting pertinent information from semantically heterogeneous resources can be facilitated by developing ethnobotanical ontologies and controlled vocabularies. As an integral part of the knowledge organization, a metadata repository should be developed for primary (first-hand) and secondary ethnobotanical datasets. The metadata will provide basic information to users about ownership, methods and standards adopted for data generation, documentation and limitations of the stored information.

Along with data interoperability and retrieval it is necessary to maintain the quality of the ethnobotanical data for multiple uses. Feedback and validation mechanisms are required among the data generators, disseminators, users and domain experts, for maintaining the data quality of ethnobotanical datasets.

There is a need for a global ethnobotanical plant name index, along the lines of the Global Names Index (GNI), capturing scientific names from primary and secondary ethnobotanical datasets (Fig. 16.2). This index will serve as a catalogue for determining the plant species having ethnobotanical value while assisting in the validation of the scientific names and in biodiversity conservation (Sarkar 2007).

Knowledge Dissemination

It is important to provide information to a wide audience in the language that they understand, for conducting research, policy making and conservation. Ironically, much of the available ethnobotanical information is only in English, thus restricting its usability and leading to content and knowledge divide. The native language of many countries rich in biodiversity and traditional knowledge is not English. For the sustenance and conservation of this knowledge, it is crucial to digitize and disseminate information in the respective native languages as well. For example, information from TKDL is available in English, French, Japanese, German, and Spanish as well as local native languages. Further, to protect against misappropriation, access to data shared by indigenous communities should be accessible only to authorized users. Conversely, the secondary data collated from public domain resources such as reports, publications and theses can be made publicly available. Simultaneously, similar to federated data resources such as NCBI, GBIF and AVH, there is a need for federated ethnobotanical data portals for accessing distributed and multi-disciplinary datasets.

Conclusion

Immense ethnobotanical data having cultural significance and socio-economic value is distributed among different indigenous communities and research organizations. Quick access and dissemination of this data, while respecting intellectual property rights is crucial for addressing global problems such as diminishing biodiversity, food security, dwindling traditional knowledge and the need for new medicines to combat increasing drug resistance. To protect and acknowledge the intellectual property rights of the knowledge custodians, a mutually agreeable code of conduct needs to be established, especially related to benefit sharing. The code of conduct will need to be agreed collaboratively in light of respective national and international laws. Moreover, active participation and consent of indigenous populations need to be the key aspects in moving from data acquisition to data integration and from documentation to dissemination. For efficient utilization of information, coordinated efforts should be formulated for developing better standards, digitization tools and integrative ethnobotanical information management systems with multi-lingual support. Due to the emergence and rapid development in the field of information technologies, it has now become possible to digitize, manage and make ethnobotanical data available to a wider audience. By embracing the power of information technology, ethnobotanists and policy makers can help to bridge the content gap between the developing and the developed worlds. The key to achieving these goals would be the initiation of collaborative partnerships and awareness of the significance of IT among ethnobotanists, indigenous communities and policy makers from developing countries. It is important to ensure conservation and sustainable use of this information for human well-being and for future generations.

Acknowledgements

JG is grateful to Macquarie University for the award of an MQRES PhD scholarship.

Reference

Australia's Virtual Herbarium (AVH). Available from: http://www.anbg.gov.au/avh/
Biodiversity Information Standards (TDWG). Available from: http://www.tdwg.org/about-tdwg/
Consortium for the Barcode of Life (CBOL). Available from: http://www.barcoding.si.edu/
Economic Botany Data Standard. Available from: http://www.kew.org/tdwguses/index.htm
Global Biodiversity Information Facility (GBIF) Data Portal. Available from: http://data.gbif.org/welcome.htm
Global Names Index. Available from: http://www.globalnames.org/
International Plant Name Index. Available from: http://www.ipni.org/

NCBI PubChem. Available from: http://pubchem.ncbi.nlm.nih.gov/

Species2000. Available from: http://www.sp2000.org/

Traditional Knowledge Digital Library. Available from: http://www.tkdl.res.in/tkdl/ LangDefault/Common/Home.asp

Universal Biological Indexer and Organizer (uBio). Available from: http://www.ubio.org/

Alcorn, J.B. 1981. Factors influencing botanical resource perception among the Huastec: suggestions for future ethnobotanical inquiry. J. Ethnobiol. 1: 221–230.

Alcorn, J. B.1995. The scope and aims of ethnobotany in a developing world. In: R. E. Schultes and S. von Reis. (eds.). Ethnobotany: Evolution of a Discipline. Chapman and Hall, London. pp. 23–39.

Berendsohn, W. 2007. "Biodiversity Informatics", The Term. Available from: http://www. bgbm.org/BioDivInf/TheTerm.htm

Bisby, F.A. 2000. The Quiet Revolution: Biodiversity Informatics and the Internet. Science. 289: 2309–2312.

Brouwer, N., Q. Liu, D. Harrington, J. Kohen, S. Vemulpad, J. Jamie, M. Randall, and D. Randall. 2005. An Ethnopharmacological Study of Medicinal Plants in New South Wales. Molecules. 10: 1252–1262.

Buenz, E.J., D.J. Schnepple, B.A. Bauer, P.L. Elkin, J.M. Riddle, and T.J. Motley. 2004. Techniques: Bioprospecting historical herbal texts by hunting for new leads in old tomes. Trends Pharmacol. Sci. 25: 494–498.

Buneman, P.2005. What the Web Has Done for Scientific Data—and What It Hasn't. In: W. Fan, Z. Wu and J. Yang. (eds.). Advances in Web-Age Information Management: 6th International Conference, WAIM 2005, Hangzhou, China. Springer, Berlin/Heidelberg. pp. 1–7.

Cavaliere, C. 2009. The Effects of Climate Change on Medicinal and Aromatic Plants. HerbalGram: 44–57.

Chavan, V. and S. Krishnan. 2003. Biodiversity Information in India: Challenges and Potentials. In: J. Shimura. (ed.). Joint international forum on biodiversity information, building capacity in Asia and Oceania. National Institute for Environmental Studies. Tsukuba, Japan. pp. 114–120.. Available from: http://circa.gbif.net/Public/irc/gbif/ocb/n ewsgroups?n=representative&a=da&art=13&att=BuildingCapacity_2003_P114–120. pdf

Chavan, V., A.V. Watve, M.S. Londhe, N.S. Rane, A.T. Pandit, and S. Krishnan. 2004. Cataloguing Indian biota: The Electronic catalogue of known Indian fauna Curr. Sci. 87: 749–763

Cook, F.E.M. 1999. TDWG Economic Botany Subgroup Report, October 1999. Available from: http://www.nhm.ac.uk/hosted_sites/tdwg/1999rep3.html

Cook, F.E.M. 1995. Economic Botany Data Collection Standard. Prepared for the International Working Group on Taxonomic Databases for Plant Sciences (TDWG). Kew: Royal Botanic Gardens, Kew, United Kingdom.

Cox, P. A. 2000. Will Tribal Knowledge Survive the Millennium? Science. 287: 44–45.

Daphne, C., 2002. Economic Botany Subgroup Report. Taxonomic Databases Working Group (TDWG) Annual Meeting (Indaiatuba, Brazil). Available from: www.cria.org.br/ eventos/tdbi/tdwg/daphne.doc

Davis, E.W. 1995. Ethnobotany: An old practice, a new discipline. In: R.E. Schultes and S. von Reis. (eds.). Ethnobotany: Evolution of a Discipline. Chapman and Hall, London. pp. 40–51.

Drahos, P. 2000. Indigenous knowledge, intellectual property and biopiracy: is a global bio-collecting society the answer. Eur. Intellect. Prop. Rev. 22. 245–250.

Ellen, R. 1996.Putting plants in their place: Anthropological approaches to understanding the ethnobotanical knowledge of rainforest populations. In: D. S. Edwards, W. E. Booth and S.C. Choy (eds.). Tropical Rainforest Research-Current Issues. Kluwer Academic Publishers, Dordrecht; Boston. pp. 457–465.

Fabricant, D.S. and N.R. Farnsworth. 2001. The value of plants used in traditional medicine for drug discovery. Environ. Health Perspect. 109: 69–75.

FAO, 2009. FAO and Traditional Knowledge: the linkages with sustainability, food security and climate change impacts. Food and Agriculture Organization of the United Nations. Rome, Italy. pp. 1–16. Available from: ftp://ftp.fao.org/docrep/fao/011/i0841e/i0841e00. pdf

Farnsworth, N.R. 1988. Screening plants for new medicines. In. E.O. Wilson (ed.), Biodiversity. National Academy Press, Washington D.C.

Gaikwad, J. and V. Chavan. 2006. Open access and biodiversity conservation: Challenges and potentials for the developing world. Data Sci. J. 5: 1–17.

Govaerts, R. 2001. How Many Species of Seed Plants Are There? Taxon. 50: 1085–1090.

Harshberger, J.W. 1896. The purpose of ethnobotany. Botanical Gazette. 21: 146–154.

Heywood, V.H. 1995. Global Biodiversity Assessment. Cambridge University Press, Great Britain.

Intergovernmental Committee. 2002. Technical Proposals on Databases and Registries of Traditional Knowledge and Biological/Genetic Resources. World Intellectual Property Organization Geneva. Available from: http://www.wipo.int/edocs/mdocs/tk/en/wipo_grtkf_ic_4/wipo_grtkf_ic_4_14.pdf

Klayman, D.L. 1985. Qinghaosu (artemisinin): an antimalarial drug from China. Science. 228: 1049–1055.

Kokwaro, J.O. 1995. Ethnobotany in Africa. In: R.E. Schultes and S. von Reis. (eds.). Ethnobotany: Evolution of a Discipline. Chapman and Hall, London, UK pp. 216–225.

Lewis, W.H. 2003. Pharmaceutical discoveries based on ethnomedicinal plants: 1985 to 2000 and beyond. Econ. Bot. 57: 126–134.

Masood, E. 2004. Developing world slow to share biodiversity data. SciDev.Net News. Accessible at: http://www.scidev.net/en/news/developing-world-slow-to-share-biodiversity-data.html

Mauro, A. 1997. The Wild and the Sown: Botany and Agriculture in Western Europe. Cambridge University Press, Cambridge, UK.

Mukherjee, P.K. and A. Wahile. 2006. Integrated approaches towards drug development from Ayurveda and other Indian system of medicines. J. Ethnopharmacol. 103: 25–35.

Narayana, D.B.A., C.K. Katayar, and N.B. Brindavanam. 1998. Original system: search, research or re-search. Indian Drug Manufacturer's Association (IDMA) Bulletin 29: 413–416.

O'Neill, J., B. Bauldock, and B. Lawlor. 2003. Global access for biodiversity through integrated systems. MetaDiversity: III Global Biodiversity Information Facility, Denmark. pp. 1–43. Available from: http://www.nbii.gov/images/uploaded/154294_1118011048355_Metadiversity3.pdf

Osbaldeston, T.A. 2000. Dioscorides. De materia medica. IBIDIS Press, Johannesburg, South Africa.

Padma, T.V. 2005. Ayurveda. Nature 436: 486–486.

Peterson, A.T., D.A. Vieglais, A.G.N. Sigüenza, and M. Silva. 2003. A global distributed biodiversity information network: building the world museum. Bull. B.O.C. 123A: 186–196.

Ramirez, C.R. 2007. Ethnobotany and the Loss of Traditional Knowledge in the 21st Century. Ethnobot. Res. App. 5: 245–247.

Sanon, S., E. Ollivier, N. Azas, V. Mahiou, M. Gasquet, C.T. Ouattara, I. Nebie, A. S. Traore, F. Esposito, G. Balansard, P. Timon-David, and F. Fumoux. 2003. Ethnobotanical survey and in vitro antiplasmodial activity of plants used in traditional medicine in Burkina Faso. J. Ethnopharmacol. 86: 143–147.

Sarkar, I.N. 2007. Biodiversity informatics: organizing and linking information across the spectrum of life. Brief Bioinform. 8: 347–357.

Singh, A. 2007. A Short Note on Designing Curriculum for Medicinal Phytochemistry. Ethnobotanical Leaflets 2007.

Thatte, U.M. and S.A. Dahanukar. 1986. Ayurveda and Contemporary Scientific Thought. Trends Pharmacol. Sci. 7: 247–251.

Thomas, M.B. 2003. Emerging Synergies Between Information Technology and Applied Ethnobotanical Research. Ethnobot. Res. App. 1: 65–74.

Turner, N.J. 1995. Ethnobotany today in northwestern North America. In: R.E. Schultes and S. von Reis. (eds.). 1995. Ethnobotany: Evolution of a Discipline. Chapman and Hall, London. pp. 264–283.

WHO. 2007. Health of indigenous peoples. Factsheets No 326 World Health Organisation. Geneva, Switzerland. Available from: http://www.who.int/mediacentre/factsheets/fs326/en/print.html

WHO. 2008. Traditional medicine. Factsheets No 134 World Health Organisation. Geneva, Switzerland. Available from: http://www.who.int/mediacentre/factsheets/fs134/en/index.html

Chapter 17

Conservation Strategies for Ethnomedicinal Plants

Arun Rijal

Introduction

Medicinal plants are very important as according to the World Health Organization (WHO), over 80% of the world's population (4.3 billion people) relies on traditional plant-based medicines as their primary form of health care (Bannerman et al. 1983). Traditional/indigenous communities have depended on plants to meet their basic needs over centuries and people around the world use between 50,000 to 80,000 flowering plants as medicinal herbs (Marinelli 2005, IUCN Species Survival Commission, 2007, Robertson 2008). Even today, traditional forms of native medicines are used extensively and medicinal plants have an important role in the health care of millions of people from developing countries (Hamilton 2008). In China, traditional medicine accounts for around 40% of total health care services. Similarly, traditional medical systems support many people including a large number of the poor population in Africa, South Asia, South America and Latin America for their health cure. But due to various threats medicinal plants and knowledge based on them are endangered. The revitalization of traditional medical heritage is crucial for improving the health security of millions of people and to maintain natural interdependence of people and plants.

Medicinal plants are considered a source of income or plants of wealth by collectors, traders and pharmaceutical industries while allopathic or modern medicine considers these plants as a remedy for many ills. In developing countries there is widespread use of traditional medicines

Director, Plant People Protection, P.O.Box 4326, Bauddha, Mahankal-6, Kathmandu, Nepal.
E-mail: *arunrijal@yahoo.com*

and this is mainly due to the diverse and rich medicinal plant resource base, traditional practices, flexibility, easy accessibility, affordability, lack of alternatives or being the only available source of health care and faith on traditional practices or herbal treatments. Rural people especially the traditional healers and the elder member of the community are rich with traditional knowledge related to medicinal plants for human health, veterinary and plant health care (bio-pesticide and bio-fertilizer) (Listorti and Doumani 2001, Rijal 2007). Medicinal plants are distributed across various bio-geographic regions, vegetation types and landscape elements from the high Himalayas down to the coast and therefore need to be conserved in their diverse habitats. Countries like China (Tibet), India, Nepal, Sri Lanka, Thailand and Malaysia have the richest medicinal plant related, health cultures both a codified and verbal (folk) tradition (Hamilton 2008). The verbal culture which is rooted by tradition in ethnic communities particularly indigenous /tribal groups is the most active carrier of the indigenous knowledge of medicinal plants. The written traditions are documented in the medical manuscripts like Ayurveda, Siddha, Unani, Yajurbeda, Rgyud-bzhi, Sowa Rigpa and Nighantas. Both of these traditions or knowledge are very important as they have been helping millions of people to cure illnesses and have contributed to the self-reliance of millions of people as herbal cure is cheap and available to all regardless of their economic situation. Despite such importance, no country has developed and implemented any strategy to conserve medicinal plants and knowledge related to medicinal plants use and conservation.

Although the effort to conserve medicinal plants in most of the developing countries is very poor, traditional practitioners or indigenous communities in some countries have started to conserve medicinal plants by cultivating home garden as a response to scarcity (Rijal 2007, Lulekal et al. 2008). Traditional beliefs and practices have also played an effective role in conservation and sustainable utilization of medicinal plants, however to maintain traditional knowledge, conservation of medicinal plants is very important because if the medicinal plant becomes extinct then knowledge related to it could be lost (Saul 1992, Brockman et al. 1997). Therefore, to conserve medicinal plants, first their habitat and related knowledge need to be protected and the dangers need to be identified then based on these informations conservation strategy could be developed to address the problems.

Importance of conservation of medicinal plants

Human beings have always been very dependent on plants. People and plants share the same ecosystem and have delicate interactions or inter-dependence. Many people around the world use herbal medicine for everyday healthcare, with as many as 80% relying on it in some countries (Abcbe 2001, Hamilton 2008). Most of the plants are harvested by poor communities in developing countries whose livelihoods would suffer if the plants die out. Traditional knowledge that builds upon the long experiences of people was adopted in social, economic, environmental, spiritual and political practices to maintain this delicate balance. Therefore, people's perception and use of their natural environment play an important role in the conservation of resources and the consequent sustainable development (Benz et al. 1996, Gerritsen 1998), due to this preservation of traditional knowledge has become an important aspect of biodiversity conservation (Alcorn 1993). However, throughout the world the loss of traditional knowledge has in recent years become a matter of concern (Brockman et al. 1997, Benz et al. 2000, Ladio and Lozada 2000, Byg and Balsev 2001, Ladio and Lozada 2003). The Chiang Mai Declaration 'Saving lives by saving plants'(WHO 1993) focusing on primary health care, also expressed concern over the loss of medicinal plant diversity and the need for international cooperation and conservation concepts.

Knowledgeable about the use of plants among indigenous people indicates their close affinity with plants of the area and with nature. Forest products are very important to rural communities and are harvested in significant quantities by a large number of households across literally all forest types in developing countries (Scoones et al. 1992, Perez and Arnold 1996, Neumann and Hirsch 2000, Cunningham 2001). Moreover, income from medicinal plants contributes significantly to rural household cash resources. Most people worldwide rely on herbal medicines collected mainly from the wild (Xiao 1991, Srivastava et al. 1996, Lange 1997) and this dependency is up to 80% in some countries (Hamilton 2008, Yineger et al. 2008). In rural areas lacking alternatives of living, forest products remain the only easily accessible resources available to poor people and have a 'safety net' function.

Approximately 250,000 flowering plants exist on the earth and of them, only a few (less than 1% of all tropical plants) have been screened so far for possible pharmaceutical use while the habitat of these plants are being destroyed faster than scientists can research the plants (Sheldon 2009). This indicates that if the current rate of extinction continues, we could lose major potential plants before their pharmaceutical properties are discovered to cure diseases like cancer, HIV and many other diseases.

Threats to medicinal plants

Due to the anthropogenic causes of different natures and magnitude, medicinal plants are threatened. Such threats differ for various species in different countries and within the country in different locations. Over exploitation of important medicinal plants that have been used to make traditional remedies including drugs to cure cancer, HIV, malaria and many other diseases has threatened the health of millions of people dependent on them. Some 15,000 of 50,000 medicinal species are under threat of extinction and shortages have already been experienced in China, India, Kenya, Nepal, Tanzania and Uganda (Hamilton 2008).

The various factors affecting medicinal plants are deforestation, habitat destruction due to agricultural expansion, fire, overgrazing, unsustainable harvest practices, climate change (drought), unmanaged market, invasive alien species, unplanned tourism, lack of awareness, information gap, lack of incentives to conserve, urban and industrial development, lack of management and loss of traditional knowledge. Of these, commercial over-harvesting, pollution, competition from invasive species and habitat destruction does the most harm while others also contribute. Studies indicated that in developing countries, poverty is the root cause of threat to medicinal plants. Due to limited cultivable land and increased demand for food owing to population growth, there is rapid encroachment of forests and pastures. Besides, large scale encroachment has taken place due to clearing of forests for cash crops in countries from Asia, Africa, Australia and South America (Choppra 2009). Such conversion of natural land cover to agricultural land and livestock grazing areas is leading to degradation of natural flora including medicinal plants.

Despite the immense importance of plant resources to social, ecological and economic aspects of rural livelihood, little attention has been given to these dimensions in land use planning and other rural development activities, and the knowledge of such plant-people interrelation is limited. Due to threats, a large number of medicinal species are endangered and the 2007 IUCN Red data list reveals that the number of threatened plant species is increasing gradually. This indicates the need for developing a conservation strategy based on analysis of threats to conserve ethno-medicinal plants and to revitalize knowledge related to them. The major threats to medicinal plants are as follows:

Habitat destruction

Habitat destruction is the major threat to the plant communities. This is the main cause of threat to 83% endangered plant species (BGCI 2009). Commercial harvest and agricultural expansion are the main causes of habitat destruction (Lulekal et al. 2008). Other human induced habitat destruction includes degradation of habitats, fragmentation, pollution, urban sprawl, infrastructure development, and changes to the characteristics of land, forest and other vegetation fire, desertification and deforestation. Brazil is the world's greatest rainforest destroyer and it is estimated that burning or felling destroy more than 2.7 million acres each year (Raintree Nutrition Inc. 2009).

Studies indicated that unsustainable agriculture and unstable governments (politics) also accelerate habitat destruction (Primack 2006). It is noticed that commercial felling of forests has destroyed most of the contained medicinal plants in the tropics and subtropics while agricultural practices destroy every plant species. A study indicated that areas with high agricultural productivity are facing more habitat destruction problem (Wikipedia 2009). From the previous approximately 16 million square kilometers of tropical rainforest, less than 9 million square kilometers remain today (Primack 2006). The current rate of deforestations (160,000 square kilometers per year) is very high and the loss is approximately 1% of original forest habitat per year (Laurance 1999). Habitat destruction is a very serious global issue as it reduces the carrying capacity and thereby leads to extinction of species (Barbault and Sastrapradja 1995).

Not only tropical rainforests but also other forests and different ecosystems have suffered from destruction of a different nature. Anthropogenic activities like agricultural practices and commercial logging have seriously affected about 94% of temperate broadleaf forests which includes many old growth stands that lost more than 98% of their area (Primack 2006) while 10–20% of the world's drylands including temperate grasslands, savannas, shrublands, scrub and deciduous forests are somewhat degraded (Kauffman and Pyke 2001) and of these about 10–20% of land (9 million square kilometers) of seasonally drylands that are converted to deserts by humans through the process of desertification (Primack 2006). Wetlands and marine areas have also experienced high levels of habitat destruction. More than 50% of wetlands in the U.S. have been destroyed in just the last 200 years (Stein et al. 2000) and between 60% and 70% of European wetlands have been completely destroyed (Ravenga et al. 2000). About one-fifth (20%) of marine coastal areas have been highly modified by humans (Burke et al. 2000). One-fifth of coral reefs have also been destroyed, and another fifth have been severely degraded by overfishing, pollution, and invasive species. Of these, a serious case is from the Philippines where 90% of the coral reefs have been destroyed (MEA 2005). Similarly, over 35% mangrove ecosystems worldwide have been destroyed (MEA 2005).

Geist and Lambin (2002) from 152 case studies of net losses of tropical forest cover found the share of agricultural expansion 96%, infrastructure expansion 72%, and wood extraction 67%. Specific insight into the precise causes of tropical deforestation reveal that contribution of transport extension 64%, commercial wood extraction 52%, permanent cultivation 48%, cattle ranching 46%, shifting (slash and burn) cultivation

41%, subsistence agriculture 40%, and fuel wood extraction for domestic use 28%.

Excessive grazing and reclamation of grassland

Free grazing of livestock is also a threat to medicinal plants in developing countries. Uncontrolled grazing has lead to degradation of pastures containing numerous medicinal herbs and shrubs (FAO 2003). Grazing also affects regeneration of trees and shrubs. Excessive herding in alpine meadows, arid steppe and desert regions, has destroyed the primary vegetation and because of this many wild medicinal plants have been reduced and some of them have even became extinct (ibid).

In traditional living, the number of livestock used to be limited while forest and pasture land areas used to be big. Shrinkage of natural habitats due to encroachment by increased population and industrial development intensified and concentrated grazing pressure leaving an adverse effect on plants and plant communities.

Depending on the grazing pattern, intensity and frequency, the effect of grazing varies from individual plants to plant communities. Selective grazing decreases species diversity and also affects the soil temperature, nutrient cycle and energy flow (Kala et al. 2002).

Invasive alien species (IAS)

Invasive alien species (IAS) are considered to be one of the main direct drivers of biodiversity loss at the global level. IAS have social, economical and ecological effects. Travel, trade, and tourism associated with globalization and the expansion of the human population intentionally or unintentionally and by natural causes like floods, storms, animals or birds have spread invasive alien species beyond its biogeographical niches. Threat from IAS is very serious and is considered the second most serious threat to biodiversity after habitat destruction (Nature.ca 2009). IAS can be predators, competitors, parasites, hybridizers, or diseases of native and domestic plants. They threaten many species with extinction while some interfere with the ecosystem changing its function. Moreover, domination by invasive species will decrease local plant diversity and distinctiveness and change community structure and species composition of native ecosystems. In some cases they may also affect plants or the plant community by changing the nutrient cycle. Through effects like competitive exclusion, niche displacement hybridization, predation and extinction invasive species could also alter the evolutionary pathway of native species (CBD 2009). Invasive species have several effects on the environment and on human beings. IAS by affecting important species on which people are dependent for food, cloth, shelter, medicine and clean

464 *Ethnomedicinal Plants*

water or some times by transmitting diseases which harm human beings. Their effect is more severe among the indigenous people who are highly dependent on forest products for their livelihoods.

Over exploitation

Over extraction of medicinal plants has threatened the future of several species. Over extraction is mainly due two reasons, one is increased market demand and the other is increased subsistence use owing to population increase. This not only affects people's access to traditional medicines, but also threatens commercially valuable wild species.

Most of the medicinal plants from developing countries are harvested from the wild. Some estimates indicate that 15,000 of the 50,000 – 80,000 plant species used for medicinal purposes and mostly the ones collected from the wild are threatened due to unsustainable collection practices (Farnsworth et al. 1991, Laloo et al. 2006). The threat is more serious when plant parts like the root, tuber, bulb and whole plants are harvested. In traditional harvest practice some tuber, roots, bulbs or plants used to be left to maintain those species but due to market pressure and/or migration traditional control over forests or other natural habitats is broken making open access to the plant resources and reckless or non-existent collection is threatening the survival of many of the plant species that are used in both traditional and modern medicines. Over exploitation of some of the species for its latex, has resulted in reduction of bees and other insects populations which are the main pollinating agents. The reduction of pollinators also affects medicinal and other plants in the formation of seeds and fruits which will affect regeneration.

Besides, several medicinal plants have multiple uses and are also used as food, fodder, timber, firewood, fiber, resin, spices and for religious and cultural uses etc. Similarly, many medicinal plants are used to cure more than one illness (Kala et al. 2004, Kala 2005, Rijal 2008). Increase in demand of these daily use materials owing to a larger population has magnified the threats to medicinal plants.

Effect of climate change

Climate change increases the heat stress on the plant which will reduce growth and production mainly in the tropical and subtropical areas. In some plants it may effect reproduction as increased heat causes sterility of pollen (Moneo and Iglesias 2006). Effects of climate change like changes in seasonal patterns, weather events, temperature ranges and various consequences related to these have affected living organisms including medicinal plants. Effects on the life cycle and distribution of plants has been experienced and medicinal plants endemic to specific geographic regions or ecosystems are more vulnerable to climate change (IPCC 2007). Reports from Artic and the Himalayas have already indicated problems of climate change on medicinal plants (Kirakosyan et al. 2003, Barnett et al. 2005, Zobayed et al. 2005, Hamilton 2008). Shifting phenologies and ranges may not seem serious at the moment but it could threaten the existence of species because these factors could ultimately endanger wild medicinal plants by disrupting synchronized phenologies of interdependent species, exposing some early-blooming medicinal plants to the dangers of late cold spells, allowing invasives to enter into their habitats and compete for resources, and initiating migratory challenges which are among the other threats (Walther et al. 2002, Gore 2006, Nickens 2007).

In the Arctic and Alpine areas medicinal plants are facing challenges due to the rapid change in environments, and some researchers have raised concerns regarding the possible losses of local plant populations and genetic diversity in those areas. Studies also proved that temperature stress can affect the secondary metabolites and other compounds that plants produce (Kirakosyan et al. 2003, Zobayed et al. 2005) which are usually the basis for their medicinal activity. Arctic plants produce phenolic compounds, anthocyanins and flavonoids that help to protect them from photoinhibition and UV damages. Decrease in production of such compounds owing to increase in temperature will reduce the plant's ability to serve as antioxidants for human health (Zobayed 2005, Gore 2006). Similarly, decrease in rainfall has affected annual herbs and several of them are now difficult to find (Grabherr 2009, Salick et al. 2009).

Millions of poor people from developing countries depend on medicinal plants for their primary healthcare and also for income. The potential loss of medicinal plants from the effects of climate change is likely to have major ramifications on the livelihoods of large numbers of vulnerable populations across the world (Hawkins et al. 2008). Further, the problems associated with climate change are likely to be much more difficult to combat than other threats as it cannot be resolved in so short a time.

Market threat

The WHO has estimated the present demand for medicinal plants is approximately US $14 billion per year and the demand is growing at the rate of 15 to 25% annually and according to WHO's estimate projection this demand might exceed US$5 trillion in 2050 (FRLHT, 1996, Sharma 2004). In North America and Europe alone, the demand for herbal remedies has been increasing by about 10% a year for the last decade (Hamilton 2008).

Plants are harvested in significant quantities by a large number of households across virtually all forest types in developing countries (Scoones et al. 1992; Perez and Arnold 1996; Neumann and Hirsch 2000; Cunningham 2001) and forest-derived income, especially from non-timber forest products (NTFPs), is important in helping households meet current consumption needs and covering costs (Scherr et al. 2004; Sunderlin et al. 2005). Recent studies show that medicinal plants have been harvested and traded in the Himalayas for millennia (Olsen 1998) and that they remain important to rural harvesters even today. For instance, the annual trade in medicinal plants in and from Nepal has been estimated at 7000–27,000 tonnes with an export value of USD 7–30 million, involving around 323,000 harvesting households (Olsen 2005) while in the Tibetan autonomous region of China up to 80% income comes from NTFPs (Salik et al. 2006).

Due to the increasing demand owing to population growth and commercialization, the future of wild plants that had helped human beings from the unknown past and still important to millions of people is becoming uncertain. Commercial collectors generally harvest medicinal plants with little care for sustainability and this happens mainly because such collections are competitive and not organized. The rapid increase in the market for herbal medicine is threatening to wipe out up to a fifth of the plant species on which it depends, destructing their natural habitats and jeopardizing the health of millions of people in developing countries (Edwards 2004). But two-thirds of the medicinal plants in use are still harvested from the wild, and research suggests that between 4000 and 10,000 of them may now be endangered (Hamilton 2003). This is serious in the case of species that involves destructive collection such as whole plants, bark, stem, heart wood, tuber, root, bulb or resin. Besides, household demands for medicinal herbs has also increased due to growth in population.

Information of the market for medicinal plants is very limited and not monitored. An unplanned market is inequitable and opaque and oligopoly is reported from many developing countries (Olsen and Helles 1997, Rijal 2007). Difficulties in price prediction in immature markets lead to unfair distribution of profits. Rural people and indigenous communities that are mainly involved in collection (harvest) do not get a good price for their products or get a very low price. In such a situation they find ways to enhance the income from commercial NTFPs to meet their household needs by enlarging the number and quantity of products which increase pressure on the resource base (Rijal 2007). Some households in developing countries were found to take money in advance due to their poor economic state,which made them unable to bargain while paying back with herbs (Olsen 1998, Pandit et al. 2004, Pethiya 2005). Besides, lack of information, unawareness of market skills, lack of storage facilities in the village and lack of financial assets to trade by themselves makes collectors dependent on the local trader who continues to monopolize local trade (Rijal 2007). It is unfortunate that herbal industries or traders that are benefiting from medicinal plants are not doing any thing for the conservation of medicinal plants.

Lack of incentives

Poverty, limited cultivable land and lack of alternatives coupled with the need for longer term investment and difficult marketing discourages the locals from medicinal plant cultivation. Moreover, due to difficult access and absence of primary processing units the transportation cost of medicinal plants is high, leaving little profit margin for the growers (Rasool 1998). Promotion of cultivation, processing and utilization of medicinal plants needs financial and other incentives. Business communities or individuals interested in cultivation, processing and trading of medicinal plants could be encouraged by providing credit facilities, soft loans, grants and technical support. The government could help to establish infrastructure and cooperatives in the areas of tribal or rural communities. Similarly, providing training on resource assessment, cultivation, sustainable harvest techniques, processing, storage, manufacture and marketing will be useful. This will contribute in conservation of medicinal plants by making utilization sustainable. It will also help to avoid overexploitation, collection, processing, storage or transportation wastage and marketing knowledge and improvement of product quality will help to make the best use of available resources.

Governments either initiate themselves or encourage private parties to establish herbaria and medicinal plant gardens to protect critically threatened medicinal plant species. Such species not only need to be protected for medicinal use but also for educational purposes.

Another serious reason that discourages traditional herbal practices is the decrease in interest among the younger generation (Kala 2000, 2002, Rijal 2008) and the low economic return from these practices. This decrease in interest among the younger generation has decreased the number of visitors to traditional herbalists. In a personal communication (2001) I was told by traditional herbalists that their earnings had decreased so much that it was now less than the labor wage while the price of livelihood materials had increased several folds. Similar conditions were observed in India and Thailand (Bhatiasevi 1998, Kala et al. 2006).

Lack of information and awareness

Lack of information about the status of medicinal plants in the wild, impact of pressure on the status of wild medicinal plants, the trend of transmission of traditional knowledge, information on the herbal market and climate change effects are obstructing conservation of medicinal plants. For sustainable harvest information on resource base is very important (Peters 1994). Information on yield rate or growth rate of most of the medicinal plants is not available. Moreover, in many developing countries harvest permits are issued without information on the resource base while there is a lack of field monitoring during harvest seasons. Due to such practices several medicinal plants have disappeared locally from many places (Rijal 2007) and continuation of such practices could lead to extinction of several important medicinal plants including many endemic species. Lack of such information has also affected conservation management plan development in many developing countries (ibid.).

It seems that industries are also not aware about the threat that they are imposing on medicinal plants and probably because of that they have not taken any initiative for conservation. Similarly, consumers of herbal medicines are also not aware about the threatened situation of different medicinal plants otherwise they would have put pressure to impose certification of sustainable harvest products. Similarly, awareness of the environmental problems seems limited with government officials, politicians, policy makers and traders.

Loss of knowledge

Traditional knowledge is transmitted vertically from the elder to younger members of the family and also horizontally, through oral communication, imitation and participation in communal activities (Gispert and Gomez Campos 1986). But with the introduction of modern allopathic medicine and the activities of missionaries in the indigenous territory, the traditional/indigenous knowledge about medicinal plants is being lost rapidly (Sindiga 1995, Hill 2003, Rijal 2007). Missionaries encouraged the indigenous people to use modern medicine and change their life style, due to this the younger generation stopped practicing spiritual and cultural activities including herbal treatment (Gurung 1995, Sindiga 1995, Hill 2003, Kiringe 2005, Rijal 2007). Similarly, studies also indicated that migration to the indigenous territories has also affected indigenous medicinal plant knowledge. Quantitative comparison of medicinal plant use knowledge between villages with homogenous indigenous communities and heterogeneous communities showed that people from homogenous communities were more knowledgeable than those from heterogeneous communities (Rijal 2008). Traditional livelihood supported sharing of knowledge but social and economic influences of migrated people influenced the livelihood of indigenous communities that affected transfer of traditional knowledge (ibid.). Changes in life style and socio-economic status of people are reflected in a declining use of wild plants (Uniyal et al. 2003) among some indigenous communities which ultimately affects the knowledge transfer (Ladio and Lozada 2000, 2001, 2004). When knowledge regarding use of medicinal plants is lost, it will make medicinal plants out of context for the people. This will also erode knowledge of conservation of medicinal plants and also the value for its conservation because indigenous peoples and local communities have developed a traditional perception and use of their natural environment that plays an important role in the conservation of resources (Benz et al. 1996; Gerritsen 1998).

The loss of knowledge may affect biodiversity conservation as well as the future of millions of people including indigenous/rural communities. Knowledge related to several important medicinal plants that are environmentally also important is vulnerable because it is limited to traditional healers and elders. Traditional harvest used to be sustainable and attention was paid to avoid damage to important plants or the resource base. But loss of traditional knowledge/practices regarding the harvest method, distribution practice and conservation of resource base resulted in unsustainable harvest which has not only affected the resource base but also pushed some species to the verge of extinction (Rijal 2007).

The importance of documenting such knowledge is recognized , e.g. in the Convention on Biological Diversity (CBD 1992, article 8j) and the subsequent Global Strategy for Plant Conservation and Economic Development (Twang and Kapoor 2004). Together with the recognition of the importance of traditional knowledge of plant use, serious concern about the loss of such knowledge has recently been expressed (Johannes 1989, Brockman et al. 1997, Benz et al. 2000, Ruddle 2000, Ladio and Lozada 2000, Byg and Balslev 2001, Matavele and Habib 2002, Ladio and Lozada 2003, Twang and Kapoor 2004).

Tourism

Several biologically important places from different countries have opened up for tourism, due to this local medicinal plants in such places have suffered damage in various degrees. They are either threatened due to collection by tourists, grazed by animals from the tourism market (horses, oxen, elephants, mules, camels), used as firewood for campfire, cooking or heating for the tourists etc. Besides, safari trips from these animals also destroy medicinal plants by trampling them. Safari animals and jungle walks or treks by tourists will also leave a trampling effect and soil compaction and clearance of vegetation and accelerated soil erosion due to land exposure from trekking trail development which also affect medicinal plants. Such threats are more severe in mountaineous areas, where due to the cold, large amounts of plants are destroyed for firewood which lead to extinction of some species. Species such as the Rhododendron is especially threatened as even its green stems are used for firewood among trekkers in Nepal and India (Chhetri et al. 2002, Stevens 2003). Similarly, collection of the orchid *Gymnadenia conospsea* in Wuling Mountain State Nature Reserve, Hebei Province has led to the extinction of this species (GTZ ADB 2007).

Urban and industrial development

Urbanization and industrial development have also affected forests and other habitats of medicinal plants. In many developing countries due to economic development, industrialization and urbanization several new cities have been developed. Similarly, population growth has also contributed in the development of cities and more houses in villages, which has resulted in accelerated encroachment of different habitats reclaiming vast wild areas and has led to the destruction of the resource base of wild medicinal plants. For example, Jianqiao in Hangzhou, Zhejiang Province, used to be the cultivation base for medicinal plants such as *Rehmannia glutinosa* var. *lutea*, and *Ophiopogon japonicus* and these were damaged by industrialization resulting in disappearance of the germplasm resources (Anonymous 2009).

Development of roads and electric transmission lines, ropeways (cable car) and establishment of industry has threatened medicinal plants by damaging their habitats and also by encroachment of such habitats. Moreover, construction of reservoirs for high dam hydropower projects inundated large areas of forests and other vegetation types with massive destruction of medicinal plant habitats.

Pollution

Pollution effects such as acid rain and eutrophication are affecting medicinal plants in many countries. Air pollution affects plants directly while other pollutants affects basic substances of survivability like water, habitat, nutrients etc. Combination of acidic air pollutants with water droplets makes acid rain and this either damages leaves or the entire plants. When it is absorbed by the soil or mixed in aquatic habitats, these habitats becomes harmful for all living organisms (Gardiner 2006). Similarly, damage of the ozone layer by chlorofluorocarbons (CFCs) exposed plants and other living organisms to the ultraviolet (UV) rays from the sun which otherwise used to be shielded by ozone layers (ibid).

Emissions from transport have detrimental effects on plants. This kind of pollution results in changes in growth and phenology, with a consistent trend for accelerated senescence and delayed flowering. This also affects the leaf surface characteristics like changes in surface wax structure (Honour et al. 2009).

It was also noticed that lack of pollinators either due to use of pesticides or pollution affected pollination which resulted in failure of seed (or fruit) formation for regeneration of several species (Gonbo 2006).

Lack of proper management

Lack of proper management systems or arrangement is also responsible for uncontrolled extraction of medicinal plants. In the past, indigenous people practiced traditional/indigenous management which was efficient to manage plant habitats and meet medicinal plant needs of each household. Generally, forests and other habitats are either managed by declaring them protected areas or handing over management responsibility to community institutions of different types or to intergovernmental/non-governmental organizations (I/NGO) or managed by the government itself. Due to uncontrolled collection of medicinal plants in the absence of proper management arrangements, the locals usually collect these plants prematurely using improper techniques like uprooting of plants, collection of wrong parts of the plant etc. that result in poor quality of plant material (Rasool 1998). In some countries either the Forest Department or the community institution made attempts to control through permits and contracts the extraction and transportation of medicinal plants, but lack of information on resource base has had a negative impact on plant species. Lack/limitation of technical manpower to assess resource base commercial exploitation of medicinal plants has affected the resource base and also threatened several medicinal plants.

Policy weakness

Some medicinal plant species that are not listed in any threat category of International Union for Conservation of Nature (IUCN) are threatened in a few countries due to accelerated harvest owing to the market force in unplanned situations. Such species though need to be listed as nationally threatened categories and their trade banned to protect them from overexploitation. But due to lack of initiatives from the governments to list these as 'protected species' in some countries and weakness in the implementation of trade ban in others have put at risk the future of several medicinal plants. Political instability is also a reason in several Asian, African and South American countries.

In traditional management practices, harvest, management and resource distribution was very sustainable as it used be based on actual needs of households. It also paid attention to ecological aspects of plants and such knowledge could be used in species as well as ecosystem conservation. But forest policies of many countries do not recognize merits of traditional practices and even community management policies are unable to address the rights of traditional people. Lack of policy to protect such precious knowledge resulted in erosion of traditional knowledge as transmission to the younger generation is affected by migrated culture, preaching against traditional medical practices by Christian missionaries and the influence of modern allopathic medicine (Hill 2003, Rijal 2008). China and India have recognized traditional knowledge and established institutions to study, document and promote such knowledge but in other developing countries no initiation could be seen. In Nepal the practice of Ayurveda is recognized and Ayurvedic colleges, hospitals and traditional medicine factories (Baidhyakhana) have been established, but no initiation could be found to document, promote and utilize other traditional practices which are limited only among indigenous communities from different geographic locations.

Several developing countries still lack policies to protect medicinal plants and practices to monitor the market which is either weak or lacking in many countries to support poor people who are dependent on wild medicinal plants (Kala et al. 2006, Aguirre-Bastos 2008). Moreover, there is no policy to assist the rural poor with either providing financial and technical assistance or soft loan facilities to develop storage resources and to establish processing plants which help to add value to the medicinal plants or its products. Similarly, lack of pollution control policy or weak implementation also affect medicinal plants in several ways.

Conservation Strategy

Conservation of medicinal plants is very important because it is connected to survivability of a large number of people. Loss of medicinal plants will not only affect healthcare and household economy but also destroy the hope for discovering new medicines for diseases like HIV-AIDS, cancer and other serious diseases. To conserve medicinal plants, all of the threats identified above need to be addressed in a correct manner. Some activities are suggested here to address the problem and safeguard the future of the medicinal plants and the people dependent on them:

Genetic conservation of medicinal plants

To rescue plants from extinction, the establishment of gene banks of medicinal plants is necessary. Besides developed countries, some developing countries (China and India) already have such arrangements while many developing countries do not have any such facilities. These gene banks collect and conserve important medicinal plants that are rare/endangered/

threatened/vulnerable or which are commercially threatened. The gene bank maintains important medicinal plant species as live material in field gene banks, in the form of seed, *in vitro* material and DNA. For long-term conservation, the accessions are stored under cryogenic conditions.

Afforestation/farming/ex-situ conservation

Demand for medicinal plants is increasing every year due to the rapid growth of the herbal industry, while more than 95% of the supply is still from wild collections. This indicates that a large number of wild medicinal plants are under pressure from overexploitation due to trade demand. To protect threatened medicinal plants species from extinction and to also maintain a supply for human needs, several measures for conservation and proliferation of medicinal plants are necessary. Various international organizations are also involved in such activities with an objective of promoting conservation of medicinal plants and traditional knowledge for enhancing health and livelihood security and revitalizing local healthcare traditions.

To address the increased threat to medicinal wild plants governments need to develop a strategy to meet household and commercial demands of medicinal plants. Programs to encourage people in establishing home gardens of medicinal plants would help to address the household demand of medicinal plants while to meet the commercial demand, farming of medicinal plant species would be appropriate. Afforestation or farming helps to establish a habitat of species and conserve threatened tree and shrub species. Similarly, cultivation of species that are threatened in natural habitats helps to divert pressure from the natural population of such species. Studies indicated that agro-forestry offers cultivation as well as conservation through integration of shade tolerant medicinal plants at a lower strata (Rao et al. 2004). Farming of medicinal plants ensures authentication, reliability (standardized quality and quantity) and continuity, while lack of quality planting materials, poor development and extension support in the cultivation, processing and an unorganized market discourage farming of medicinal plants (Gonbo 2006). Similarly, lack of knowledge of the market, unfair marketing strategies and the lengthy cultivation cycle of some mountain species of medicinal plants, have made farmers reluctant to cultivate medicinal plants. To encourage them to cultivate medicinal plants several arrangements like training on farming practices, market information, buy back arrangement, harvest training, training on drying/processing, storage, soft financial loans etc. are needed. To produce large scale of planting materials (seedlings), tissue culture would be a viable option. Besides, the support of irrigation, pest control etc. are also needed for the farmers.

Studies carried out in six different countries indicated there have been some initiatives from the community level to cultivate medicinal plants (Hamilton 2008). But comparing these with the magnitude and geographical distribution of threats indicates that it is not sufficient to address the problem. Several community level conservation initiatives could be learnt from developing countries in Asia and Africa, but they are not able to address medicinal plant issues sufficiently. For example, Nepal is very successful in community forestry but its forestry policies are not able to address medicinal plant issues. Hence medicinal plant conservation needs to be integrated in such community forestry initiatives. This capacity of the community needs to be enhanced in terms of technical knowledge of planting, harvesting, processing, storing, marketing etc.

Besides, botanical gardens could also contribute in conservation of medicinal plants especially the highly endangered ones. There are more than 2500 botanic gardens in 150 countries that conserve 6 million accessions of living plants representing around 80,000 species i.e. only about 28%. This clearly indicates that many species still await conservation support.

In-situ conservation

Cultivation ensures immediate supply of herbal materials but will not confirm their long term conservation. Therefore, besides cultivation to meet commercial and subsistence demands it is also necessary to think of long term conservation. The best option for long term conservation of medicinal plants is *in-situ* conservation. *In-situ* conservation involves conservation of medicinal plants in its natural habitats by arranging protection through declaration of protected areas of different types (conservation areas, national parks, reserve forests/shrublands/grasslands or medicinal plant hotspots). *In-situ* conservation helps to conserve millions of species through the protection of natural areas and is the primary means for the maintenance of these resources in the absence of other reliable options. Due to economic reasons and also weak technical manpower, several developing countries are not able to bring all important natural habitats of important medicinal plants into the protected area system. The protected area system in the world covers about 2,279,192,012 ha area (both terrestrial and marine) including different habitats (IUCN 2008).

While declaring new protected areas or developing management plans for the existing protected areas, provision of sustainable utilization of forest products by traditionally dependent indigenous groups or rural communities needs to be addressed. If these traditional rights are not addressed in the conservation plans then this could raise acerbity among these people which could risk the future of protected areas. This could also lead to the loss of traditional plant use knowledge as such knowledge disappears when plants becomes out of context. Besides, collection of medicinal or other useful plants wherein the collection is non-destructive, for example fruits, flowers, leaves, seeds, twigs etc. should be allowed on a sustainable quantity from the wild because this will provide economic incentives to rural communities and attract them to conservation generating local stewardship for conservation of endangered biological resources of the area. A manual to make such a sustainable wild collection of medicinal plants has been developed with the support of various organizations (MPSG 2007).

Religious and cultural institutions (temples, monasteries, sacred grooves, Guthi) from different parts of the world have also been contributing in conservation of forests and different medicinal and aromatic plants (Israel et al. 1997, Jeeva and Anusuya 2005, Jeeva et al. 2005, Bhakat and Sen 2008). Recognizing the rights of such institutions will encourage them to help in conservation of habitats and important plant diversity within them. One example of such recognition is the Forest Policy of Nepal (Forest Act 1993) which recognized Religious Forest (HMGN 1995) and it has provision of handing over management authorities of such religious forests to religious institutions.

Protection of indigenous knowledge

Traditional people are well aware about the ecosystem of their area and familiar with conservation methods of different medicinal plants in their use. It is also found that they have been conserving important plants and ecosystems from the unknown past and are the best conservationists. Hence protection of their knowledge and encouraging traditional people and rural communities will make conservation sustainable.

Traditional practices contributed in conservation of medicinal and other plants very effectively. Such practices include domestication, prohibition on grazing in specific areas, belief that holy spirits live in forests and plants, sacredness of trees, beliefs on sacred or religious forests, protection of plants at burial sites, near religious monuments, selective harvesting, collection of only dead wood for firewood, not harvesting in summer which helps to distinguish live and dead trees, protecting seedlings from fire and using energy saving stoves. Such knowledge could be used to deal with threats related to plants and also to address issues related to the daily lives of the people. Effectiveness of indigenous knowledge and the ability of indigenous communities to conserve biodiversity is also accepted by the Convention of Biological Diversity (CBD) and it encouraged signatory countries to protect indigenous knowledge related to biodiversity conservation. The ecological management of indigenous/traditional communities was developed over several years of observations and experiments and to make it work the concept of conservation was inbuilt and interwoven in the traditional and religious beliefs of the indigenous communities.

Since indigenous knowledge is very useful for conservation, it should be studied and documented so that it can be used for conservation and economic development. To protect knowledge and provide benefit, it is important that the state should acknowledge folklore and legitimize its role. Forest-related traditional knowledge should also be included in the national policy and institutional setups be made for access to resources and benefit sharing.

Need of research

Ethno-botanical studies help to generate knowledge regarding use of different plant species, the ecological role of different plants and knowledge related to conservation. Such information could be useful to pharmaceutical research, to make supply of herbs sustainable, to conserve medicinal plants and their habitat and to protect traditional knowledge.

Most of the countries that supply herbs to industries issue permits to collectors, but without having resource base information. The continuation of such practices could lead species to extinction. Hence, governments of all countries which supply medicinal herbs need to carry out research on resource base and yield rate or growth rates and need to practice issuing harvest (collection) permits only based on such information. Similarly, regular study of the medicinal plant market is also needed to assess pressure on medicinal plants. Such information on resource base or information on status of species and market pressure will be useful to develop a conservation strategy for threatened species.

Cultivation of medicinal plants is one option to meet the demand of the market, but information on the propagation of medicinal plants is available for less than 10% and agro-technology is available only for 1% of the total known plants globally (Lozoya 1994, Khan and Khanum 2000). This indicates that there is need to conduct research on agro-technology.

When the status of any species becomes very vulnerable and needs strict protection then as an alternative the threatened herb could be substituted. But information on the biochemistry of most plants is unknown. Therefore, to support conservation and also meet the market demand by supplying substitutes, information on biochemistry will be needed. Therefore, research on biochemistry of close relatives of all important medicinal plants need to be carried out.

Many countries do not have information on distribution of different medicinal plants. Information on distribution of plants is important to declare any area as protected or medicinal plant 'hot spots' for promoting *in-situ* conservation. Therefore, a study on distribution of medicinal plants is needed which could be used to design protected areas or arrange other protection arrangement.

Incentives

Rural communities and indigenous (tribal) people are very much linked to the plants and ecology of their areas. Moreover, they also have sustainable harvest knowledge and conservation methods. Therefore, providing incentives to these people to conserve medicinal plants of their areas will be effective and would also interest them as these plants are very important to them. Ten grass-roots projects studied plant life in India, Pakistan, China, Nepal, Uganda and Kenya showed that this approach can succeed. In Uganda, the project has ensured the sustainable supply of low-cost malaria treatment, and in China a community-run medicinal plant reserve has been created for the first time (Hamilton 2008).

The younger generation of herbal practitioners are not keen to adopt the tradition as a profession. To address this problem available herbal formulations need to be standardized for their efficacy and to establish a social capital trust for herbal practitioners in order to promote the tradition.

Regarding the problem related to erosion of traditional knowledge or knowledge transmission to the younger generation in the uses of medical plants, incentives should be given to the traditional herbal healers for preparation of herbal formulations, and attempts should be made to organize them. Traditional knowledge related to many lesser known medicinal plants has been declining rapidly and to address this it is necessary to document these lesser known medicinal plants. It has been experienced that to motivate people—benefits play an important role. If the link between health benefits and economic benefits with cultural practices can convince the younger generation of traditional people or rural communities then this will help to conserve traditional knowledge or the practice of traditional herbal culture. Moreover, securing patent rights and paying royalties for such knowledge will encourage people to maintain traditional practices and also conserve medicinal plants and their habitats.

Market Improvement

Increasing commercialization of medical plants provides opportunities to increase rural economy and thereby enhance their livelihoods. But harvesters are not able to get a good price for their products due to the lack of market information available to the farmers/collectors and existence of oligopoly of the few middlemen (Rijal 2007). The market chain of medicinal plants is very large and due to lack of financial assets, information on markets at different levels and lack of skills/tricks, producers (harvesters, collectors, farmers) have to rely on the middlemen. This problem could be resolved by arranging to sell the products directly by the producers to the Industry. Bio-partnership and buy back arrangements between farmers and companies/industries would be useful in this regard.

It is observed that price of medicinal herbs are generally cheaper during the harvest season which increases later. By delaying the sale of products for a few months could provide more benefits to the producers, however lack of storage facilities unables them to hold on to their products a for long time. Moreover, value addition to the products could also increase benefits to producers but lack of knowledge on processing and different options of value addition prevents them from increasing profits of their products. If the government or I/NGOs help to establish storage facilities, processing plants, accessible roads etc. or provide soft loans to the villagers' cooperatives to develop such facilities, then this could help to increase benefits to producers (harvester, collector, farmers) and improve rural economy.

The high price for herbal products from the wild and the low price for the same material produced from cultivation is the main reason that discourages farmers from cultivation of several medicinal plants. To address this problem policy arrangement should be made to levy extra charge (conservation tax) on the wild herbal products. This will make herbal products collected from the wild more expensive as compared to the products from the farm and will create a market environment for the cultivated products. The revenue generated from the extra levy (conservation tax) could be used in medicinal plant conservation programs or for programs to assist farmers.

Similarly, it is important to discourage illegal harvesting from the wild. By controlling such unauthorized harvesting will help to make the harvest sustainable or at least decrease pressure on the wild medicinal plants. Certification of herbal plants from different sources (wild or cultivated) will help to discourage harvest that does not meet the standards of sustainability. There are various types of certification in practice for the forest products which focus on different areas of the supply chain like production, processing, trade, manufacturing, marketing etc. Certification schemes like forest management certification (e.g. Forest Stewardship Council FSC), social certification (e.g. Fair Trade Federation FTF), organic certification (e.g. International Federation of Organic Agriculture IFOAM), and product quality certification are relevant for MAP products (Walter 2002).

Prevention of IAS

It is very difficult to control IAS once it spreads and it is also expensive. The best option is to check entry of such pests by strengthening quarantine. But if the species has already entered as in the case of islands, one option would be to eradicate them entirely from the island (Tye 2008). This could be done only if the invasive species is detected in its early stage of arrival and before it spreads widely. If the pest has already spread widely, beyond the eradication level then efforts should be made to keep their population under limits through various measures like regular uprooting and burning, identifying the predator of that species, etc. In the case of protected areas or other natural areas, invasive species can be prevented from spreading into such areas by regular monitoring activities of people, goods carried into the park and regular field observation to detect such species and to destroy them.

Biological control is also one of the measures used to control invasive species. This includes the introduction of a natural enemy of the pest to control its population. But studies need to be carried out to identify the appropriate natural enemy so that it only attacks the target species and no other damage is done.

Bio-partnership

Establishment of the relationship between consumers and suppliers will also be useful to safeguard the future of medicinal plants. Bio-partnership helps to satisfy both the short-term and long-term goals of the parties involved in the management of the resources. Traders/industrialists will be interested in such partnerships as this ensures the regularity of supply and the quality and quantity of herbal materials. However the traditional/indigenous people should not be excluded from such partnerships, rather the agreement should be with them and the rural communities and their knowledge could be useful in production and utilization of herbal plants. Bio-partnership will also encourage industrialists and traders to invest in conservation and cultivation of medicinal plants.

Policy solution

Conserving and protecting medicinal plants is being carried out through the enforcement of the Acts and Regulations. Some countries have policies that favor conservation of threatened medicinal plants and also provide economic benefits through sustainable harvest, however poor implementation still remains a problem. In some African countries and many Asian countries there are regulations to control collection, transport and trade, but have not been implemented (Hamilton 2008).

In many countries there is a need to establish institutions to protect medicinal plants and knowledge related to it with mandates clarified by policy. Religious institutions such as, monasteries have been contributing towards the conservation of traditional Tibetan medicinal plant knowledge. Institutions of most religions have been contributing in the conservation of plants (Salick et al. 2007, Hamilton 2008) and they could be utilized in conservation of medicinal plants. Such institutions should be encouraged by acknowledging their role in policy documents.

In most countries, the patent right on knowledge of medicinal plants and its properties is not secured. Securing such rights and providing royalties from medical industries or other sectors for using such knowledge will help to improve the economy of indigenous/rural communities and also the national economy of developing countries which will have a positive impact on medicinal plants and knowledge related to them. Besides, to protect the knowledge of traditional medicinal plants, a policy should be formulated to introduce such topics in the school curriculum of rural areas. This will not only increase the interest of children in traditional knowledge of plants, but also generate pride among those from indigenous communities.They would be willing to follow such practices and transmission of traditional knowledge related to the use and conservation of medicinal plants and ecosystem management would continue.

To regulate the herbal market, provisions for certification and introduction of a specific mark to identify sustainably harvested products is needed. This will help to control reckless collection and supply to the market. In the case of critically threatened species, a ban on trade can also be created as an immediate and short term solution which could be relaxed later after an improvement in the situation. Such a ban could be either complete or only on export of unprocessed products based on the status of the species and the nature of threat. Bans prohibit over-exploitation of wild endangered species.

Pollution problems are affecting human as well as plant life, either because of a lack of a proper policy or weakness in implementation. Therefore, a policy to check pollutions from industries, transports and other pollutants needs to be developed and effectively endorsed. Besides, regulations need to be developed and implemented to check pollution related to the agricultural sector, that is pesticides and herbicides.

Awareness

People should be made aware of the ecological, economical and social importance of medicinal plants. They should also be alerted of the different types of threats to medicinal plants. People should also be cautioned of existing harvest related problems and be trained in sustainable harvest methods. Students from schools and colleges can be taught the value of medicinal plants and also be familiarized with traditional herbal practices and threats. Field trips could be made to observe different valuable medicinal plants. Interaction programs could also be arranged for students and others with indigenous communities who have knowledge of herbal uses and conservation. Similarly, the electronic media could be used to generate awareness through airing documentaries on herbal practices and traditional knowledge, medicinal plants, their ecology and conservation status and other awareness audio-visual clips.

Conclusion

Wild plant species used for medicinal purposes continue to support indigenous and local communities that have relied on them for centuries for their traditional medicines. Indigenous knowledge related to medicinal plants has a significant role in making rural communities self-reliant in primary health care, in supporting livelihoods, creation of knowledge resources and generating rural employment. Loss in knowledge will make future generations ignorant of them and will not protect from grazing or other exploitation or make any special conservation arrangements. Similarly, loss of species will lead to loss of knowledge of its use. Both loss of species or knowledge will affect indigenous people and many others who are dependent on them for treating illnesses.

Several human induced and some natural threats have threatened the future of several medicinal plants that are very important to human beings. The extinction or threat to medicinal plants has also threatened loss of traditional knowledge related to them and their ecology. To address these threats initiation to implement long term strategies at the global, national and local levels for conservation of medicinal plant resources and using their rich associated traditional knowledge, for social, cultural and economic benefits. For this, commitment is needed from all sectors of the society including the government, I/NGO, industries, political bodies/ leaders, scientists, civil societies and rural communities. There should be an integrated approach for holistic conservation management so that the harvest will be sustainable, collection regularized by certifying products, to formulate policies to conserve habitats, supply by encouraging cultivation and decrease wild collections and protect traditional knowledge. But to develop such holistic conservation efforts it is necessary to generate information on the status of different medicinal plants and traditional knowledge, dependency of indigenous and rural communities on wild plants in different countries.

Reference

Abebe, D. 2001. The role of medicinal plants in healthcare coverage of Ethiopia, the possible benefits of integration. In: Proceedings of the National Workshop. Conservation and sustainable use of medicinal plants in Ethiopia. 28 April–01 May 1998. M. Zewdu, A. Addis, and A. Demissie (eds.). Institute of Biodiversity Conservation and Research Ethiopia. pp. 6–21.

Aguirre-Bastos, C. 2008. A Roadmap for the Commercial Development of Medicinal Plants of the Andean region of South America. The European Foresight Monitoring Network (EFMN), Foresight Brief no. 155.   http://docs.google.com/gview?a=vand q=cache:091p1zAgGj4J:www.foresight-network.eu/index.php%3Foption%3Dcom_ docman%26task%3Ddoc_view%26gid%3D376+lack+policy+to+monitor+market+of +medicinal+plantsandhl=en

Alcorn, J.B. 1993. Indigenous economic institutions and ecological knowledge: A Ghanaian case study. The Environmentalist 19: 217–227.

Anonymous. 2009. Medicinal Plants. http://english.biodiv.gov.cn/images_biodiv/resources/ medicinal-en.htm

Bannerman, R.H., B. John, and W.C. Ch'en 1983. Traditional medicine and health care coverage: a reader for health administrators and practitioners. WHO. Geneva.

Barnett, T.P., J.C. Adam, and D.P. Lettenmaier. 2005. Potential impacts of a warming climate on water availability in snow-dominated regions. Nature 438: 303–308.

Barbault, R. and S. D. Sastrapradja. 1995. Generation, maintenance and loss of biodiversity. Global Biodiversity Assessment, Cambridge University Press, Cambridge. pp. 193–274.

Benz, B., J.Cevallos, E. Munoz, and F. Santana . 1996. Ethnobotany serving society: a case study from the Sierra de Manantlan Biosphere Reserve. Sida 17:1–16.

Benz B, J.Cevallos, F. Santana, and S. Graf . 2000. Losing knowledge about plant use in the Sierra de Manantlan Biosphere Reserve, Mexico. Economic Botany 54: 183–191.

BGCI. 2009. The Threat Posed by Habitat Loss and Degradation. http://www.bgci.org/ ourwork/habitat_loss/

Bhakat, R.K. and U.K. Sen. 2008. Ethnomedicinal Plant Conservation through Sacred Groves. In: K.Bose. (ed.). Health and Nutritional Problems of Indigenous populations. Kamla—Raj Enterprises, Delhi, India, Special Volume No. 2: 55–58.

Bhatiasevi, A. 1998. NGOs aim to revive traditional therapy: Interest in herbal medicine increasing. The Bangkok Post. http://www.geocities.com/RainForest/7813/0216_med. htm

Brockman A, B. Masuzumi, and S. Augustine. 1997. When all people have the same story, humans will cease to exist, protecting and conserving traditional knowledge. A Report for the Biodiversity Conservation Office Dene Cultural Institute, Canada.

Burke, L., Y. Kura, K. Kassem, C. Ravenga, M. Spalding, and D. McAllister. 2000. Pilot Assessment of Global Ecosystems: Coastal Ecosystems. World Resources Institute, Washington, D.C.

Byg, A. and H. Balslev. 2001. Traditional knowledge of *Dypsis fibrosa* (Arecaceae) in Eastern Madagascar. Economic Botany 55: 263–275.

CBD. 1992. Convention on Biological Diversity (CBD): Traditional Knowledge and the Convention on Biological Diversity. (http://www.biodiv.org./programmes/socio-eco/ traditional/)

CBD 2009. Invasive Alien Species: What's the Problem? http://www.cbd.int/invasive/problem. shtml

Chhetri, N., E. Sharma, D.C. Deb, and R.C. Sundriyal. 2002. Impact of Firewood Extraction on Tree Structure, Regeneration and Woody Biomass Productivity in a Trekking Corridor of the Sikkim Himalaya. Mountain Research and Development 22(2): 150–158.

Choppra V. 2009. Vanishing Tropical Forests. Boloji.com, Boloji Media Inc. http://www.boloji. com/environment/59.htm

Cunningham, A.B. 2001. Applied ethnobotany: people, wild plant use and conservation. People and Plants Conservation Series, Earthscan, London.

Edwards, R. 2004. Herbal medicine boom threatens plants. New Scientist. http://www. newscientist.com/article/dn4538-herbal-medicine-boom-threatens-plants.html

FAO. 2003. Transhuman grazing systems in temperate Asia, by J.M. Suttie. and. S.G. Reynolds. FAO Plant Production and Protection Series No. 31. Rome. http://www.fao. org/documents/show_cdr.asp?url_ile=/DOCREP/006/Y4856E/y4856e0m.htm

Farnsworth, N.R, O. Akerele. V. Heywood, and D.D. Soejarto. 1991. Global Importance of Medicinal Plants. In: O.Akerele, V. Heywood, and H.Synge. (eds.). Conservation of medicinal plants. Cambridge University Press, Cambridge, UK. pp. 25–51.

Foundation for Revitalization of Local Health Tradition (FLRHT). 1996. Encyclopedia of Indian Medicinal Plants. FLRHT, Bangalore. www.frlht-india.org

Gardiner, L. 2006. Air Pollution Affects Plants, Animals, and Environments. University Corporation for Atmospheric Research (UCAR). The Regents of the University of Michigan. http://www.windows.ucar.edu/tour/link=/milagro/effects/wildlife_forests.html

Geist, H.J. and E.E. Lambin. 2002. Proximate causes and underlying driving forces of tropical deforestation. Bio. Science 52(2): 143–150.

Gerritsen, P.R.W. 1998. Community development, natural resource management and biodiversity conservation in the Sierra de Manantlan biosphere reserve, Mexico. Community Development Journal 33: 314–324.

Gispert, M. and A. Gomez Campos. 1986. Plantas medicinales silvestres: el proceso de adquisicion, transmision y colectivizacion del conocimiento vegetal. Biotica. 11: 113–125.

Gonbo, T. 2006. Report of Medicinal Plants Conservation Campaign, Kanji Village, 30th September to 2nd October 2006.

Gore, A. 2006. An Inconvenient Truth. Rodale Inc. New York.

Grabherr, G. 2009. Biodiversity in the high ranges of the Alps: Ethnobotanical and climate change perspectives. ScienceDirect 19(2): 167–172.

GTZ A.D.B. 2007. Disregarding Traditional Knowledge and Culture. Sustainable Agrobiodiversity Management in the Mountain Areas of Southern China. http://www. agrobiodiversity.cn/index.php?id=208

Gurung G.M. 1995. Report from a Chepang Village (Society, Culture and Ecology). Kathmandu, Nepal.

Hamilton, A. 2003. Medicinal Plants and Conservation: Issues and Approaches. WWF Report, UK.

Hamilton, A.C. (ed.). 2008. Medicinal plants in conservation and development: case studies and lessons learnt. Plantlife International, Salisbury, UK.

Hawkins, B., S. Sharrock and K. Havens. 2008. Plants and climate change: which future? Botanic Gardens Conservation International. Richmond, UK.

Hill D.M. 2003. Traditional medicine in contemporary contexts, protecting and respecting indigenous knowledge and medicine. National Aboriginal Health organisation (NAHO) (http://www.naho.ca/english/pdf/research_tradition.pdf)

HMGN. 1995. Forest Act. 1993. Ministry of Forests and Soil Conservation, His Majesty's Government of Nepal, Kathmandu, Nepal.

Honour, Sarah L., J.N.B., T.A. Ashenden, J.N. Cape, and S.A. Power. 2009. Responses of herbaceous plants to urban air pollution: Effects on growth, phenology and leaf surface characteristics. Environmental Pollution 157(4): 1279–1286.

Intergovernmental Panel on Climate Change (IPCC) 2007. Climate Change 2000: Synthesis Report. http://www.ipcc.ch/ pdf/assessment-report/ar4/syr/ar4_syr.pdf

Israel, E., V. Chitra, and D. Narasimhan. 1997. Sacred Groves: Traditional Ecological Heritage. International Journal of Ecology and Environmental Sciences 23: 463–470.

IUCN Species Survival Commission Medicinal Plant Specialist Group. 2007. "Why Conserve and Manage Medicinal Plants?" www.iucn.org/themes/ssc/sgs/mpsg/main/Why.html.

IUCN. 2008. World Heritage and Protected Areas. 2008. edition. IUCN Programme on Protected Areas, Gland, Switzerland.

Jeeva, S. and R. Anusuya. 2005. Ancient ecological heritage of Meghalaya. Magnolia 3: 20–22.

Jeeva, S., B.P. Mishra, N. Venugopal, and R.C. Laloo. 2005. Sacred forests: Traditional ecological heritage in Meghalaya. Journal of Scott Research Forum 1: 93–97.

Johannes, R.E. 1989. Fishing and traditional knowledge conservation. In: R.E. Johannes (ed.). Traditional Eological Knowledge: a collection of essays. IUCN Publications Services, Gland, Switzerland. pp. 39–43.

Kala, C.P. 2000. Status and conservation of rare and endangered medicinal plant in the Indian trans-Himalaya. Biological Conservation. 93: 371–379.

Kala, C.P. 2002. Medicinal Plants of Indian Trans-Himalaya. Bishen Singh Mahendra Pal Singh. Dehradun, India.

Kala, C.P. 2005. Current status of medicinal plants used by traditional Vaidyas in Uttaranchal state of India. Ethnobotany Research and Applications 3: 267–278.

Kala, C.P., S.K. Singh, and G.S. Rawat. 2002. Effect of sheep and goat grazing on the species diversity in the alpine meadows of Western Himalaya. The Environmentalist 22: 183–189.

Kala, C.P., N.A. Farooquee, and U. Dhar. 2004. Prioritization of medicinal plants on the basis of available knowledge, existing practices and use value status in Uttaranchal, India. Biodiversity and Conservation 13: 453–469.

.Kala, C.P., P.P. Dhyani, and B.S. Sajwan. 2006. Developing the medicinal plants sector in northern India: challenges and opportunities. Journal of Ethnobiology and Ethnomedicine 2: 32–46.

Kauffman, J.B. and D.A. Pyke. 2001. Range ecology, global livestock influences. In: S.A. Levin (ed.). Encyclopedia of Biodiversity. Academic Press, San Diego, CA. 5: 33–52

Khan, I.A. and A. Khanum. 2000. Role of Biotechnology in Medicinal and Aromatic Plants Ukaaz Publications. Hyderabad.

Kirakosyan, A., E. Seymour, P.B. Kaufman, S. Warber, S. Bolling, and S.C. Chang. 2003. Antioxidant capacity of polyphenolic extracts from leaves of *Crataegus laevigata* and *Crataegus monogyna* (hawthorn) subjected to drought and cold stress. Journal of Agricultural and Food Chemistry 51: 3973–3976.

Kiringe, J.W. 2005. Ecological and Anthropological Threats to Ethno-Medicinal Plant Resources and their Utilization in Maasai Communal Ranches in the Amboseli Region of Kenya. Ethnobotany Research and Application 3: 231–241.

Ladio A.H. and M. Lozada. 2000. Edible wild plant use in a Mapuche community of north-western Patagonia. Human Ecology 28, 53–71.

Ladio A.H. and M. Lozada. 2001. Non-timber forest product use in two human populations from NW Patagonia: a quantitative approach. Human Ecology 29: 367–380.

Ladio, A.H. and M. Lozada. 2003. Comparison of wild edible plant diversity and foraging strategies in two aboriginal communities of north-western Patagonia. Biodiversity and Conservation 12: 937–951.

Ladio A.H. and M. Lozada. 2004. Patterns of use and knowledge of wild edible plants in distinct ecological environments: a case study of a Mapuche community from north-western Patagonia. Biodiversity and Conservation. 2004. 13: 1153–1173.

Laloo, R.C., L. Kharlukhi, S. Jeeva, and B.P. Mishra. 2006. Status of medicinal plants in the disturbed and the undisturbed sacred forests of Meghalaya, northeast India: population structure and regeneration efficacy of important tree species. Current Science 90: 225–232.

Lange, D. 1997. Trade in plant material for medicinal and other purposes. TRAFFIC Bulletin 17: 21–32.

Laurance, W.F. 1999. Reflections on the tropical deforestation crisis. Biological Conservation 91: 109–117.

Listorti, Y.A. and F.M. Doumani. 2001. Environmental Health, Bridging the Gap. World Bank Discussion Paper 422. The International Bank of Reconstruction and Development/ The World Bank, Washington DC.

Lozoya, X. 1994. Ethnobotany and the Search of New Drugs. John Wiley and Sons, London.

Lulekal, E., E. Kelbessa, T. Bekele, and H. Yineger. 2008. An ethnobotanical study of medicinal plants in Mana Angetu District, southeastern Ethiopia. Ethnobiology and Ethnomedicine 4(10): 1746–4269.

Marinelli, J. (ed.). 2005. Plant: The Ultimate Visual Reference to Plants and Flowers of the World. DK Publishing, Inc. New York.

Matavele, J. and M. Habib. 2002. Ethnobotany in Cabo Delgado, Mozambique: use of medicinal plants. Environment Development and Sustainability 2: 227–234.

MEA. 2005. Ecosystems and Human Well-Being. Millennium Ecosystem Assessment. Island Press, Covelo, CA.

Medicinal Plant Specialist Group (MPSG) 2007. Internatinoal standard for sustainable wild collection of medicinal and aromatic plants (ISSC-MAP). Version 1.0. Bundesamt für Naturschutz (BfN), MPSG/SSC/IUCN, WWF Germany, and TRAFFIC, Bonn, Gland, Frankfurt, and Cambridge (BfN-Skripten 195). www.bfn.de/fileadmin/MDB/documents/service/skript195

Moneo, M. and A. Iglesias. 2006. Food and Climate: What are the effects of climate change on plants. Environmental Science Published for Everybody Round the Earth (ESPERE), http://www.atmosphere.mpg.de/enid/266.html

Nature.ca. 2009. Conservation Issues. Issues in native plants conservation. Native Plants crossroads, website of Canadian Museum of Nature http://nature.ca/plnt/ci/ci_e.cfm

Neumann, R.P. and E. Hirsch. 2000. Commercialisation of non-timber forest products: review and analysis of research. Centre for International Forestry Research, Bogor, Indonesia.

Nickens, T.E. 2007. Walden warming. National Wildlife. October/November: 36–41.

Olsen, C.S. 1998. The trade in medicinal and aromatic plants from Central Nepal to Northern India. Economic Botany 52: 279–292.

Olsen, C.S. 2005. Valuation of commercial central Himalayan medicinal plants. Ambio 34: 607–610.

Olsen, C.S. and F. Helles. 1997 Medicinal plants, market and margins in the Nepal Himalaya: trouble in Paradise. Mountain Research and Development 17: 363–374.

Pandit B.H., G.B. Thapa, and M.A. Zoebisch. 2004. Promoting marketing of cinnamon tree products in Palpa District of Nepal. National Workshop on Agroforestry Management in Vietnam's Uplands: Marketing, Research and Development, Organized by Hanoi Agricultural University and World Agroforestry Centre at Hoa Binh Province Vietnam, April 22–24. 2004.

Pérez, M.R. and J.E.M. Arnold. (eds.). 1996. Current issues in non-timber forest products research. Centre for International Forestry Research, Bogor, Indonesia.

Pethiya, B.P. 2005 Strategies for poverty alleviation through dovetailing the potential of microfinance practices with non-timber forest products from dipterocarps: Lessons from India. Indian Institute of Forest Management, Bhopal, India. http://www.apafri.org/8thdip/Session%203/S3_Pethiya.doc

Peters, C.M. 1994. Sustainable harvest of non-timber plant resources in tropical moist forest. An ecological primer. Biodiversity Support Program. Washington DC.

Primack, R.B. 2006. Essentials of Conservation Biology. 4th ed. Habitat destruction. Sinauer Associates, Sunderland, MA. pp. 177–188.

Rao, M.R., M.C. Palada, and B.N. Becker 2004. Medicinal and aromatic plants in agroforestry systems. Agroforestry System, vol. 61–62 (1-3): 107–122.

Rasool, G. 1998. Saving the plants that save us. Medicinal plants of the northern areas of Pakistan. Basdo, Gilgit. Pakistan.

Ravenga, C., J. Brunner, N. Henninger, K. Kassem, and R. Payne. 2000. Pilot Analysis of Global Ecosystems: Wetland Ecosystems. World Resources Institute, Washington, D.C.

Robertson, E. 2008. Medicinal Plants at Risk. Center for Biological Diversity (CBD).

Raintree Nutrition Inc. 2009. Rain Forest Facts. Raintree Nutrition, Inc., Carson City, NV. http://www.rain-tree.com/facts.htm

Rijal, A. 2007. Multifaceted Forest Management for Biodiversity Conservation and Poverty Alleviation: a case study of mid hills of Nepal, Ph.D. Dissertation, University of Copenhagen, Denmark.

Rijal, A. 2008. A Quantitative Assessment of Indigenous Plant uses among two Chepang Communities in the Central Mid-hills of Nepal. Ethnobotanical Research and Application, 2008.

Ruddle, K. 2000. Systems of knowledge: dialogue, relationships and process. Environment, Development and Sustainability 2: 227–304.

Salick, J., A. Byg, A. Amend, B. Gunn, W. Law, and H. Schmidt 2006. Tibetan Medicine Plurality. Economic Botany 60: 227–253. http://www.bioone.org/archive/0013-0001/60/3/pdf/i0013-0001-60-3-227.pdf

Salick, J., A. Amend, D. Anderson, K. Hoffmeister, B. Gunn, and Z.D. Fang. 2007. Tibetan Sacred Sites Conserve Old Growth Trees in the Eastern Himalayas. Biodiversity and Conservation. http://www.springerlink.com/content/b210277k24813t1l/?p=1668110ec0b74b43a54d154a9f3845adandpi=0

Salick, J., F. Zhendong, and A. Byg. 2009. Eastern Himalayan alpine plant ecology, Tibetan ethnobotany, and climate change. ScienceDirect 19(2): 147–155.

Sharma, A.B. 2004. Global Medicinal Plants Demand May Touch $5 Trillion By 2050. Indian Express, March 29, 2004.

Saul, R. 1992. Indigenous Forest Knowledge: Factors Influencing its Social Distribution. In: M. Allen (ed.). Anthropology of Nepal: People, Problems and Processes. Mandala Book Point, Kathmandu, Nepal.

Scherr, S.J., A. White, and D. Kaimowitz. 2004. A new agenda for forest conservation and poverty reduction: making markets work for low income producers. Forest Trends, CIFOR & IUCN. Washington, DC.

Scoones, I., M. Melnyk, and J.N. Pretty. 1992. The hidden harvest: wild foods and agricultural systems. IIED, London.

Sheldon, R.H. 2009. Medicinal Plants in Danger, Thousands of Medicinal Species Worldwide Face Extinction. http://medicinal-plants.suite101.com/article.cfm/medicinal_plants_under_threat

Sindiga, I. 1995. Traditional medicine in Africa: An introduction. In: I. Sindiga, C. Nyaigotti-Chacha, and M.P. Kanunah (eds.) Traditional Medicine in Africa. East African Educational Publishers Ltd., Nairobi. pp. 1–15

Srivastava, J., J. Lambert. and N. Vietmeyer. 1996. Medicinal Plants: an Expanding Role in Development, World Bank, Washington, DC.

Stein, B. A., L. S. Kutner, and J. S. Adams (eds.). 2000. Precious Heritage: The Status of Biodiversity in the United States. Oxford University Press, New York.

Stevens, S. 2003. Tourism and deforestation in the Mt. Everest region of Nepal. Geographical Journal. 169(3): 255–277. http://medicinal-plants.suite101.com/article.cfm/medicinal_plants_under_threat

Sunderlin, W.D., A. Angelsen, B. Belcher, P. Burgers, R. Nasi, L. Santoso, and S. Wunder. 2005. Livelihoods, Forests, and Conservation in Developing Countries: An Overview. World Development 33(9): 1383–1402.

Twang, S. and P. Kapoor. 2004. Protecting and Promoting Traditional Knowledge: Systems, National Experiences and International Dimensions. United Nations Conference on Trade and Development (UNCTAD), United Nations, New York and Geneva.

Tye, A. 2008. Invasive alien species : the biggest threat to Pacific biodiversity. Pacific Regional Environment Programme. http://www.sprep.org/topic/Invasive.htm

Uniyal S.K., A. Awasthi, and G. Rawat. 2003. Developmental Processes, Changing Lifestyle and Traditional Wisdom: Analysis from Western Himalaya. The Environment 23: 307–312.

Walter, S. 2002. Certification and benefit-sharing mechanisms in the field of non-wood forest products. An overview.—Medicinal Plant Conservation. 8: 3–9.

Walther, G-R., E. Post, P. Convey, A. Menzel, C. Parmesan, T.J.C. Beebee, J-M. Fromentin, O. Hoegh-Guidberg, and F. Bairlein. 2002. Ecological responses to recent climate change. Nature. 416: 389–395.

WHO. 1993. Guidelines on the conservation of medicinal plants. World Health Organization, Geneva. [http://www.wwf.org.uk/filelibrary/pdf/guidesonmedplants.pdf]

Wikipedia 2009. Habitat Destruction. http://en.wikipedia.org/wiki/Habitat_destruction#cite_note-Stein-6

Xiao, Pei-Gen, 1991. The Chinese approach to medicinal plants. Their utilization and conservation. In: O. Akerele, V. Heywood, and H. Synge (eds.). Conservation of medicinal plants. University Press, Cambridge, UK. pp. 305–313.

Yineger H, D. Yewhalaw, and D. Teketay. 2008. Ethnomedicinal plant knowledge and practice of the Oromo ethnic group in Southwestern Ethiopia. Journal of Ethnobiology and Ethnomedicine 4: 11.

Zobayed, S.M.A., F. Afreen, and T. Kozai. 2005. Temperature stress can alter the photosynthetic efficiency and secondary metabolite concentrations in St. John's wort. Plant Physiology and Biochemistry 43: 977–984.

Index

Color Plate Section

Chapter 3

Fig. 3.1. *Pterodon emarginatus* Vog. Fruits without the outside layer (5 cm), and with dark outside layer (epicarp–5.5 cm).

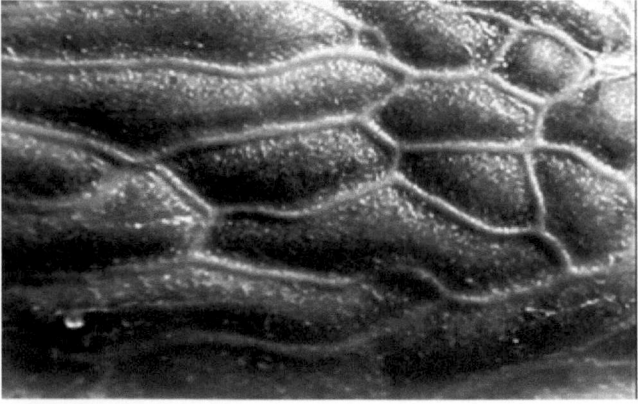

Fig. 3.2. Veins of the fruits of *Pterodon emarginatus* outlining small salient and elongated regions that correspond to the oil producing glands.

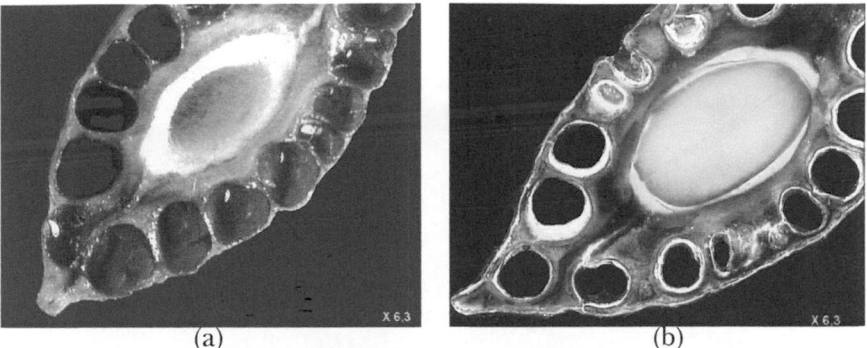

(a)   (b)

Fig. 3.3. *Pterodon emarginatus* Vog. (fruit) (a) Cross section (6.3 x), showing details of the transversal cut of the schizogenous glands. (b) Cross section (6.3 x) of the schizogenous glands, with the seed observed at the center. Coloration of aqueous methyl green solution 1:20.

Chapter 6

Fig. 6.1. *Pai-de-santo* smoking with a *defumador* at Centro Caboclo Ubirajara e Exú Ventania, August 2008.

Chapter 9

Fig. 9.1. Flowering twigs of *Woodfordia fruticosa* (Linn.) Kurz.

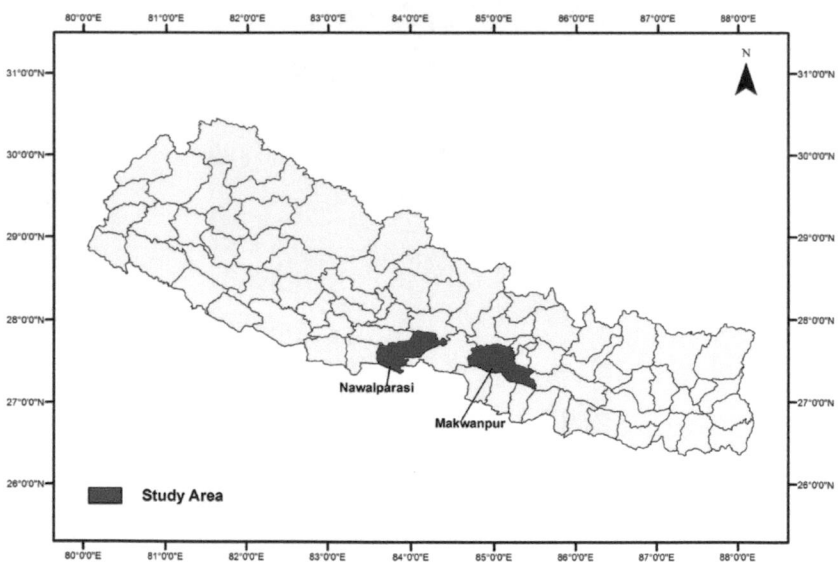

Fig.9.2 Map of Nawalparasi and Makawanpur Districts of Nepal.

Fig. 9.3. Sucking of nectars of flowers of *Woodfordia fruticosa* (Linn.) Kurz.

Fig. 9.4. *Woodfordia fruticosa* flowers in the market (in circle).

Chapter 12

Fig. 12.2. List of medicinal plants which are threatened and in high demand in Tunisia.

Chapter 15

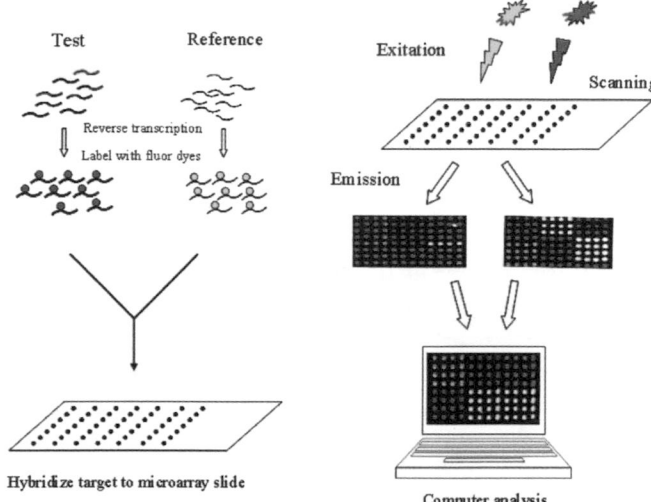

Fig. 15.1. cDNA microarray procedure. Templates for genes of interest are amplified by PCR then printed on coated glass microscope slides. Total RNA from both the test and reference sample is fluorescently labeled with different fluor dyes. The fluorescent targets are pooled and allowed to hybridize to the clones on the array slide. Laser excitation of the incorporated targets yields an emission with a characteristic spectra, which is measured using a scanning confocal laser microscope. Monochrome images are imported into software in which the images are merged. Data from a single hybridization experiment is viewed as a normalized ratio. In addition, data from multiple experiments can be examined using any number of data mining tools (Youns et al. 2009b).

Chapter 16

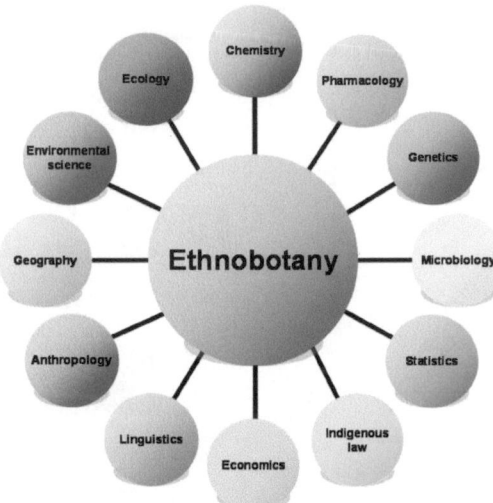

Fig. 16.1. Ethnobotany as a field and some of its associated disciplines.

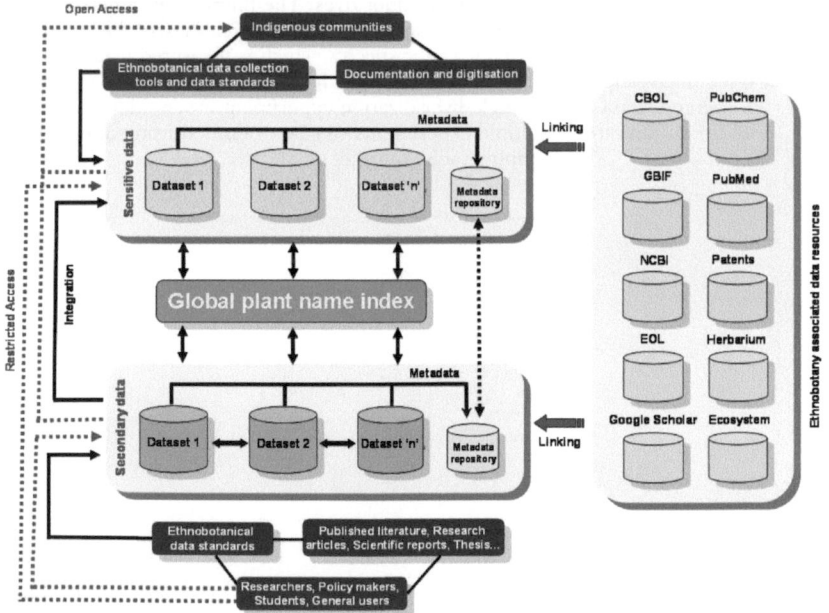

Fig. 16.2. A schematic representation of the key components and the ethnobotanical data integration platform using information technology.

T - #0312 - 071024 - C7 - 234/156/23 - PB - 9780367383077 - Gloss Lamination